Methods in Enzymology

Volume 158

METALLOBIOCHEMISTRY

Part A

METHODS IN ENZYMOLOGY

Methods in Enzymology

Volume 158

Metallobiochemistry

Part A

EDITED BY

James F. Riordan

CENTER FOR BIOCHEMICAL AND BIOPHYSICAL SCIENCES AND MEDICINE
HARVARD MEDICAL SCHOOL
BOSTON, MASSACHUSETTS

Bert L. Vallee

CENTER FOR BIOCHEMICAL AND BIOPHYSICAL SCIENCES AND MEDICINE
HARVARD MEDICAL SCHOOL
BOSTON, MASSACHUSETTS

ACADEMIC PRESS, INC.
Harcourt Brace Jovanovich, Publishers
San Diego New York Berkeley Boston
London Sydney Tokyo Toronto

ACADEMIC PRESS, INC.
1250 Sixth Avenue
San Diego, California 92101

United Kingdom Edition published by
ACADEMIC PRESS INC. (LONDON) LTD.
24-28 Oval Road, London NW1 7DX

LIBRARY OF CONGRESS CATALOG CARD NUMBER: 54-9110

ISBN 0-12-182059-9 (alk. paper)

PRINTED IN THE UNITED STATES OF AMERICA
88 89 90 91 9 8 7 6 5 4 3 2 1

Table of Contents

Section I. Sample Preparation

Section II. Analytical Techniques

Section III. Analysis of Metals

Contributors to Volume 158

Article numbers are in parentheses following the names of contributors.
Affiliations listed are current.

DAVID S. AULD (3, 8, 11), *Department of Pathology and Center for Biochemical and Biophysical Sciences and Medicine, Harvard Medical School, Boston, Massachusetts 02115*

JOHN O. BAKER (6), *Biotechnology Research Branch, Solar Fuels Research Division, Solar Energy Research Institute, Golden, Colorado 80401*

ROGER L. BERTHOLF (21), *Department of Pathology, University of Virginia Medical Center, Charlottesville, Virginia 22908*

GEORGE N. BOWERS, JR. (22, 23), *Clinical Chemistry Laboratory, Hartford Hospital, Hartford, Connecticut 06115*

SUE BROWN (21), *Department of Pathology, University of Virginia Medical Center, Charlottesville, Virginia 22908*

SUBHASH CHANDRA (14), *Department of Chemistry, Cornell University, Ithaca, New York 14853*

N. DENNIS CHASTEEN (32), *Department of Chemistry, University of New Hampshire, Durham, New Hampshire 03824*

M. CRISTINA CRISOSTOMO (30), *Departments of Laboratory Medicine and Pharmacology, University of Connecticut School of Medicine, Farmington, Connecticut 06032*

MERLE A. EVENSON (26), *Departments of Pathology-Laboratory Medicine and Medicine, University of Wisconsin, Madison, Wisconsin 53706*

K. H. FALCHUK (33), *Center for Biochemical and Biophysical Sciences and Medicine, and Department of Medicine, Harvard Medical School, Boston, Massachusetts 02115*

WAYNE W. FISH (27), *Biotechnology Research Division, Research and Development, Phillips Petroleum Company, Bartlesville, Oklahoma 74004*

DONITA L. GARLAND (13), *National Eye Institute, National Institutes of Health, Bethesda, Maryland 20892*

JAMES M. HARNLEY (13), *Beltsville Human Nutrition Research Center, United States Department of Agriculture, Beltsville, Maryland 20705*

K. L. HILT (33), *Department of Biochemistry and Biophysics, University of California, Davis, California 95616*

BARTON HOLMQUIST (2), *Center for Biochemical and Biophysical Sciences and Medicine and Department of Biological Chemistry and Molecular Pharmacology, Harvard Medical School, Boston, Massachusetts 02115*

SIDNEY M. HOPFER (30), *Departments of Laboratory Medicine and Pharmacology, University of Connecticut School of Medicine, Farmington, Connecticut 06032*

JEAN L. JOHNSON (29), *Department of Biochemistry, Duke University Medical Center, Durham, North Carolina 27710*

S. A. LEWIS (31), *Analytical Chemistry Department, Hazelton Laboratories America, Inc., Vienna, Virginia 22180*

WOLFGANG MARET (9), *Center for Biochemical and Biophysical Sciences and Medicine, Harvard Medical School, Brigham and Women's Hospital, Boston, Massachusetts 02115*

DONNA M. MARTIN (32), *Department of Physical Sciences, Rhode Island College, Providence, Rhode Island 02908*

MARK T. MARTIN (4, 25, 28), *Sir William Dunn School of Pathology, University of Oxford, Oxford OX2 6S1, England*

NANCY MENDOZA (21), *Department of Pathology, University of Virginia Medical Center, Charlottesville, Virginia 22908*

ROBERT G. MICHEL (18), *Department of Chemistry, University of Connecticut, Storrs, Connecticut 06268*

GEORGE H. MORRISON (14), *Department of Chemistry, Cornell University, Ithaca, New York 14853*

JOSÉ A. OLIVARES (17), *Environmental Technology Division, Savannah River Laboratory, E. I. DuPont DeNemours and Co., Aiken, South Carolina 29808*

JANET OSTERYOUNG (19), *Department of Chemistry, State University of New York at Buffalo, Buffalo, New York 14214*

THEODORE C. RAINS (22), *Department of Chemistry, National Bureau of Standards, Gaithersburg, Maryland 20899*

JAMES F. RIORDAN (1), *Center for Biochemical and Biophysical Sciences and Medicine, Harvard Medical School, Boston, Massachusetts 02115*

TERENCE H. RISBY (15), *Division of Environmental Chemistry, Department of Environmental Health Sciences, The Johns Hopkins University School of Hygiene and Public Health, Baltimore, Maryland 21205*

JOHN SAVORY (21), *Department of Pathology and Biochemistry, University of Virginia Medical Center, Charlottesville, Virginia 22908*

SALVADOR F. SENA (23), *Department of Laboratory Medicine, Danbury Hospital, Danbury, Connecticut 06810*

ROBERT SHAPIRO (25, 28), *Center for Biochemical and Biophysical Sciences and Medicine, Harvard Medical School, and Brigham and Women's Hospital, Boston, Massachusetts 02115*

WALTER SLAVIN (12), *Perkin-Elmer Corporation, Ridgefield, Connecticut 06877*

F. WILLIAM SUNDERMAN, JR. (30), *Departments of Laboratory Medicine and Pharmacology, University of Connecticut School of Medicine, Farmington, Connecticut 06032*

B. L. VALLEE (1, 33), *Center for Biochemical and Biophysical Sciences and Medicine, Harvard Medical School, Boston, Massachusetts 02115*

HAROLD E. VAN WART (10), *Department of Chemistry and Institute of Molecular Biophysics, Florida State University, Tallahassee, Florida 32306*

CLAUDE VEILLON (7, 24), *Vitamin and Mineral Nutrition Laboratory, Beltsville Human Nutrition Research Center, United States Department of Agriculture, Beltsville, Maryland 20705*

JACQUES VERSIECK (20), *Department of Internal Medicine, Division of Gastroenterology, University Hospital, De Pintelaan 185, B-9000 Ghent, Belgium*

FRED W. WAGNER (5), *Department of Biochemistry, University of Nebraska-Lincoln, Lincoln, Nebraska 68583*

MICHAEL R. WILLS (21), *Department of Pathology and Internal Medicine, University of Virginia Medical Center, Charlottesville, Virginia 22908*

KAREN A. WOLNIK (16), *Elemental Analysis Research Center, U.S. Food and Drug Administration, Cincinnati, Ohio 45202*

MICHAEL ZEPPEZAUER (9), *Fachbereich 15.2 Biochemie, Universität des Saarlandes, 6600 Saarbrücken, Federal Republic of Germany*

Preface

Although it has been suspected since ancient times that metals participate in biological processes, details of the manner in which they might exert a function have until recently remained unknown. It is now appreciated that metals play roles in catalysis, hormone action, gene and other regulatory functions, structural stabilization of macromolecules, muscle contraction, nerve conduction, and transport. The alkali and alkaline earth metals as well as cobalt, copper, iron, manganese, molybdenum, nickel, zinc, and others are known to be essential, many of them in the mechanism of action of specific enzymes.

Most of these elements occur in biological matter in very low concentrations, a fact that long dominated the thinking about the subject and presented not so much intellectual as analytical challenges. The presence or absence of a particular "trace" element in a given biological matrix was the crucial experimental question preoccupying many investigators, and its answer became almost an end in itself; as a consequence, little thought needed to or could be given to how an answer in the affirmative would be pursued. Recent advances in methodology and instrumentation have solved most of the analytical problems that impeded progress. Nevertheless, it is important to realize that the difficulties and frustrations of the past left an imprint on the field that is only slowly giving way to the new realities.

The designation "trace element" has been employed variously to indicate tissue content, the total knowledge of the subject, or—by implication—even its potential importance. Whatever the connotation or viewpoint, this historically conditioned nomenclature has categorized pertinent metals based entirely on once poor detection limits and signal-to-noise ratios of analytical methods which are now so excellent that in this field the very definition of a "trace" has become almost meaningless.

Distinct biological roles for most essential metals are now well recognized, and approaches to the exploration of their functions have become standardized to the point where they have merged with those employed throughout all of biochemistry. As a consequence, what was once the province of the analytical chemist has not only become part of the mainstream of biochemical knowledge and thought but can even be considered a scientific discipline in its own right. "Metallobiochemistry" is a field which has grown and matured rapidly over the past decade to encompass a wide variety of scientific subspecialties but with primary emphasis on the role of a metal or metals in a biochemical system. Metalloenzymes have been the central investigatory targets of this field with most efforts directed toward the participation of the metal in catalysis. Importantly, the very recognition, purification, and characterization of metalloenzymes have depended as much on progress in the physical chemistry of proteins and the methodology for their isolation and characterization as on advances in spectroscopic, electrochemical, isotopic, and other methods for the detection of metals.

The appropriate analytical method depends, among other things, on the specific metal in question, the nature of the biological matrix, the amount of material available, and whether other metals may be present as well. No method of analysis is helpful unless proper precautions have been taken during the enzyme isolation procedure to avoid either loss of intrinsic metal, addition of extraneous metal, or substitution of the intrinsic by an extraneous metal. Hence a healthy paranoia based on long experience has led to a number of practices generally regarded as safe for avoiding such contamination problems. All of these are detailed in this volume of Metallobiochemistry, Part A, which provides both the emerging and the practicing metallobiochemist with the tools to answer what should be the number one and two questions: Is this enzyme a metalloenzyme? What metal is it? Subsequent volumes will consider the multiplicity of techniques that can be applied to the study of metalloenzymes by virtue of the fact that they contain a metal and to specific classes of metallobiochemicals. In particular, the role of metals in nonenzymatic proteins will be emphasized as will their roles in metalloenzymes other than as components of the catalytic mechanism.

We are deeply indebted to all of the authors who contributed to this volume for their patience, cooperation, and enthusiasm. We also appreciate the advice and suggestions from all those who participated in our initial survey. To those at Academic Press who were so diligent and indulgent we express our sincere thanks.

Finally, we would like to acknowledge both our affection and kindred spirit for the founding editors, Nate Kaplan and Sid Colowick, who will long be remembered for their contributions to science, humanity, and the dignity of man.

Bert L. Vallee
James F. Riordan

METHODS IN ENZYMOLOGY

EDITED BY

Sidney P. Colowick and Nathan O. Kaplan

VANDERBILT UNIVERSITY
SCHOOL OF MEDICINE
NASHVILLE, TENNESSEE

DEPARTMENT OF CHEMISTRY
UNIVERSITY OF CALIFORNIA
AT SAN DIEGO
LA JOLLA, CALIFORNIA

I. Preparation and Assay of Enzymes
II. Preparation and Assay of Enzymes
III. Preparation and Assay of Substrates
IV. Special Techniques for the Enzymologist
V. Preparation and Assay of Enzymes
VI. Preparation and Assay of Enzymes (*Continued*)
Preparation and Assay of Substrates
Special Techniques
VII. Cumulative Subject Index

METHODS IN ENZYMOLOGY

EDITORS-IN-CHIEF

Sidney P. Colowick and Nathan O. Kaplan

VOLUME XLVI. Affinity Labeling
Edited by WILLIAM B. JAKOBY AND MEIR WILCHEK

VOLUME XLVII. Enzyme Structure (Part E)
Edited by C. H. W. HIRS AND SERGE N. TIMASHEFF

VOLUME XLVIII. Enzyme Structure (Part F)
Edited by C. H. W. HIRS AND SERGE N. TIMASHEFF

VOLUME XLIX. Enzyme Structure (Part G)
Edited by C. H. W. HIRS AND SERGE N. TIMASHEFF

VOLUME L. Complex Carbohydrates (Part C)
Edited by VICTOR GINSBURG

VOLUME LI. Purine and Pyrimidine Nucleotide Metabolism
Edited by PATRICIA A. HOFFEE AND MARY ELLEN JONES

VOLUME LII. Biomembranes (Part C: Biological Oxidations)
Edited by SIDNEY FLEISCHER AND LESTER PACKER

VOLUME LIII. Biomembranes (Part D: Biological Oxidations)
Edited by SIDNEY FLEISCHER AND LESTER PACKER

VOLUME LIV. Biomembranes (Part E: Biological Oxidations)
Edited by SIDNEY FLEISCHER AND LESTER PACKER

VOLUME LV. Biomembranes (Part F: Bioenergetics)
Edited by SIDNEY FLEISCHER AND LESTER PACKER

VOLUME LVI. Biomembranes (Part G: Bioenergetics)
Edited by SIDNEY FLEISCHER AND LESTER PACKER

VOLUME LVII. Bioluminescence and Chemiluminescence
Edited by MARLENE A. DELUCA

VOLUME LVIII. Cell Culture
Edited by WILLIAM B. JAKOBY AND IRA PASTAN

VOLUME LIX. Nucleic Acids and Protein Synthesis (Part G)
Edited by KIVIE MOLDAVE AND LAWRENCE GROSSMAN

VOLUME LX. Nucleic Acids and Protein Synthesis (Part H)
Edited by KIVIE MOLDAVE AND LAWRENCE GROSSMAN

VOLUME 61. Enzyme Structure (Part H)
Edited by C. H. W. HIRS AND SERGE N. TIMASHEFF

VOLUME 62. Vitamins and Coenzymes (Part D)
Edited by DONALD B. MCCORMICK AND LEMUEL D. WRIGHT

VOLUME 63. Enzyme Kinetics and Mechanism (Part A: Initial Rate and Inhibitor Methods)
Edited by DANIEL L. PURICH

VOLUME 64. Enzyme Kinetics and Mechanism (Part B: Isotopic Probes and Complex Enzyme Systems)
Edited by DANIEL L. PURICH

VOLUME 65. Nucleic Acids (Part I)
Edited by LAWRENCE GROSSMAN AND KIVIE MOLDAVE

VOLUME 66. Vitamins and Coenzymes (Part E)
Edited by DONALD B. MCCORMICK AND LEMUEL D. WRIGHT

VOLUME 67. Vitamins and Coenzymes (Part F)
Edited by DONALD B. MCCORMICK AND LEMUEL D. WRIGHT

VOLUME 68. Recombinant DNA
Edited by RAY WU

VOLUME 69. Photosynthesis and Nitrogen Fixation (Part C)
Edited by ANTHONY SAN PIETRO

VOLUME 70. Immunochemical Techniques (Part A)
Edited by HELEN VAN VUNAKIS AND JOHN J. LANGONE

VOLUME 71. Lipids (Part C)
Edited by JOHN M. LOWENSTEIN

VOLUME 72. Lipids (Part D)
Edited by JOHN M. LOWENSTEIN

VOLUME 73. Immunochemical Techniques (Part B)
Edited by JOHN J. LANGONE AND HELEN VAN VUNAKIS

VOLUME 74. Immunochemical Techniques (Part C)
Edited by JOHN J. LANGONE AND HELEN VAN VUNAKIS

VOLUME 75. Cumulative Subject Index Volumes XXXI, XXXII, XXXIV–LX
Edited by EDWARD A. DENNIS AND MARTHA G. DENNIS

VOLUME 76. Hemoglobins
Edited by ERALDO ANTONINI, LUIGI ROSSI-BERNARDI, AND EMILIA CHIANCONE

VOLUME 77. Detoxication and Drug Metabolism
Edited by WILLIAM B. JAKOBY

VOLUME 78. Interferons (Part A)
Edited by SIDNEY PESTKA

VOLUME 79. Interferons (Part B)
Edited by SIDNEY PESTKA

VOLUME 80. Proteolytic Enzymes (Part C)
Edited by LASZLO LORAND

VOLUME 81. Biomembranes (Part H: Visual Pigments and Purple Membranes, I)
Edited by LESTER PACKER

VOLUME 82. Structural and Contractile Proteins (Part A: Extracellular Matrix)
Edited by LEON W. CUNNINGHAM AND DIXIE W. FREDERIKSEN

VOLUME 83. Complex Carbohydrates (Part D)
Edited by VICTOR GINSBURG

VOLUME 84. Immunochemical Techniques (Part D: Selected Immunoassays)
Edited by JOHN J. LANGONE AND HELEN VAN VUNAKIS

VOLUME 85. Structural and Contractile Proteins (Part B: The Contractile Apparatus and the Cytoskeleton)
Edited by DIXIE W. FREDERIKSEN AND LEON W. CUNNINGHAM

VOLUME 86. Prostaglandins and Arachidonate Metabolites
Edited by WILLIAM E. M. LANDS AND WILLIAM L. SMITH

VOLUME 87. Enzyme Kinetics and Mechanism (Part C: Intermediates, Stereochemistry, and Rate Studies)
Edited by DANIEL L. PURICH

VOLUME 88. Biomembranes (Part I: Visual Pigments and Purple Membranes, II)
Edited by LESTER PACKER

VOLUME 89. Carbohydrate Metabolism (Part D)
Edited by WILLIS A. WOOD

VOLUME 90. Carbohydrate Metabolism (Part E)
Edited by WILLIS A. WOOD

VOLUME 91. Enzyme Structure (Part I)
Edited by C. H. W. HIRS AND SERGE N. TIMASHEFF

VOLUME 92. Immunochemical Techniques (Part E: Monoclonal Antibodies and General Immunoassay Methods)
Edited by JOHN J. LANGONE AND HELEN VAN VUNAKIS

VOLUME 93. Immunochemical Techniques (Part F: Conventional Antibodies, Fc Receptors, and Cytotoxicity)
Edited by JOHN J. LANGONE AND HELEN VAN VUNAKIS

VOLUME 94. Polyamines
Edited by HERBERT TABOR AND CELIA WHITE TABOR

VOLUME 95. Cumulative Subject Index Volumes 61–74, 76–80
Edited by EDWARD A. DENNIS AND MARTHA G. DENNIS

VOLUME 96. Biomembranes [Part J: Membrane Biogenesis: Assembly and Targeting (General Methods; Eukaryotes)]
Edited by SIDNEY FLEISCHER AND BECCA FLEISCHER

VOLUME 97. Biomembranes [Part K: Membrane Biogenesis: Assembly and Targeting (Prokaryotes, Mitochondria, and Chloroplasts)]
Edited by SIDNEY FLEISCHER AND BECCA FLEISCHER

VOLUME 98. Biomembranes (Part L: Membrane Biogenesis: Processing and Recycling)
Edited by SIDNEY FLEISCHER AND BECCA FLEISCHER

VOLUME 99. Hormone Action (Part F: Protein Kinases)
Edited by JACKIE D. CORBIN AND JOEL G. HARDMAN

VOLUME 100. Recombinant DNA (Part B)
Edited by RAY WU, LAWRENCE GROSSMAN, AND KIVIE MOLDAVE

VOLUME 101. Recombinant DNA (Part C)
Edited by RAY WU, LAWRENCE GROSSMAN, AND KIVIE MOLDAVE

VOLUME 102. Hormone Action (Part G: Calmodulin and Calcium-Binding Proteins)
Edited by ANTHONY R. MEANS AND BERT W. O'MALLEY

VOLUME 103. Hormone Action (Part H: Neuroendocrine Peptides)
Edited by P. MICHAEL CONN

VOLUME 104. Enzyme Purification and Related Techniques (Part C)
Edited by WILLIAM B. JAKOBY

VOLUME 105. Oxygen Radicals in Biological Systems
Edited by LESTER PACKER

VOLUME 106. Posttranslational Modifications (Part A)
Edited by FINN WOLD AND KIVIE MOLDAVE

VOLUME 107. Posttranslational Modifications (Part B)
Edited by FINN WOLD AND KIVIE MOLDAVE

VOLUME 108. Immunochemical Techniques (Part G: Separation and Characterization of Lymphoid Cells)
Edited by GIOVANNI DI SABATO, JOHN J. LANGONE, AND HELEN VAN VUNAKIS

VOLUME 121. Immunochemical Techniques (Part I: Hybridoma Technology and Monoclonal Antibodies)
Edited by JOHN J. LANGONE AND HELEN VAN VUNAKIS

VOLUME 122. Vitamins and Coenzymes (Part G)
Edited by FRANK CHYTIL AND DONALD B. MCCORMICK

VOLUME 123. Vitamins and Coenzymes (Part H)
Edited by FRANK CHYTIL AND DONALD B. MCCORMICK

VOLUME 124. Hormone Action (Part J: Neuroendocrine Peptides)
Edited by P. MICHAEL CONN

VOLUME 125. Biomembranes (Part M: Transport in Bacteria, Mitochondria, and Chloroplasts: General Approaches and Transport Systems)
Edited by SIDNEY FLEISCHER AND BECCA FLEISCHER

VOLUME 126. Biomembranes (Part N: Transport in Bacteria, Mitochondria, and Chloroplasts: Protonmotive Force)
Edited by SIDNEY FLEISCHER AND BECCA FLEISCHER

VOLUME 127. Biomembranes (Part O: Protons and Water: Structure and Translocation)
Edited by LESTER PACKER

VOLUME 128. Plasma Lipoproteins (Part A: Preparation, Structure, and Molecular Biology)
Edited by JERE P. SEGREST AND JOHN J. ALBERS

VOLUME 129. Plasma Lipoproteins (Part B: Characterization, Cell Biology, and Metabolism)
Edited by JOHN J. ALBERS AND JERE P. SEGREST

VOLUME 130. Enzyme Structure (Part K)
Edited by C. H. W. HIRS AND SERGE N. TIMASHEFF

VOLUME 131. Enzyme Structure (Part L)
Edited by C. H. W. HIRS AND SERGE N. TIMASHEFF

VOLUME 132. Immunochemical Techniques (Part J: Phagocytosis and Cell-Mediated Cytotoxicity)
Edited by GIOVANNI DI SABATO AND JOHANNES EVERSE

VOLUME 133. Bioluminescence and Chemiluminescence (Part B)
Edited by MARLENE DELUCA AND WILLIAM D. MCELROY

VOLUME 134. Structural and Contractile Proteins (Part C: The Contractile Apparatus and the Cytoskeleton)
Edited by RICHARD B. VALLEE

VOLUME 135. Immobilized Enzymes and Cells (Part B)
Edited by KLAUS MOSBACH

VOLUME 136. Immobilized Enzymes and Cells (Part C)
Edited by KLAUS MOSBACH

VOLUME 137. Immobilized Enzymes and Cells (Part D)
Edited by KLAUS MOSBACH

VOLUME 138. Complex Carbohydrates (Part E)
Edited by VICTOR GINSBURG

VOLUME 139. Cellular Regulators (Part A: Calcium- and Calmodulin-Binding Proteins)
Edited by ANTHONY R. MEANS AND P. MICHAEL CONN

VOLUME 140. Cumulative Subject Index Volumes 102–119, 121–134

VOLUME 141. Cellular Regulators (Part B: Calcium and Lipids)
Edited by P. MICHAEL CONN AND ANTHONY R. MEANS

VOLUME 142. Metabolism of Aromatic Amino Acids and Amines
Edited by SEYMOUR KAUFMAN

VOLUME 143. Sulfur and Sulfur Amino Acids
Edited by WILLIAM B. JAKOBY AND OWEN GRIFFITH

VOLUME 144. Structural and Contractile Proteins (Part D: Extracellular Matrix)
Edited by LEON W. CUNNINGHAM

VOLUME 145. Structural and Contractile Proteins (Part E: Extracellular Matrix)
Edited by LEON W. CUNNINGHAM

Volume 146. Peptide Growth Factors (Part A)
Edited by David Barnes and David A. Sirbasku

Volume 147. Peptide Growth Factors (Part B)
Edited by David Barnes and David A. Sirbasku

Volume 148. Plant Cell Membranes
Edited by Lester Packer and Roland Douce

Volume 149. Drug and Enzyme Targeting (Part B)
Edited by Ralph Green and Kenneth J. Widder

Volume 150. Immunochemical Techniques (Part K: *In Vitro* Models of B and T Cell Functions and Lymphoid Cell Receptors)
Edited by Giovanni Di Sabato

Volume 151. Molecular Genetics of Mammalian Cells
Edited by Michael M. Gottesman

Volume 152. Guide to Molecular Cloning Techniques
Edited by Shelby L. Berger and Alan R. Kimmel

Volume 153. Recombinant DNA (Part D)
Edited by Ray Wu and Lawrence Grossman

Volume 154. Recombinant DNA (Part E)
Edited by Ray Wu and Lawrence Grossman

Volume 155. Recombinant DNA (Part F)
Edited by Ray Wu

Volume 156. Biomembranes (Part P: ATP-Driven Pumps and Related Transport: The Na,K-Pump)
Edited by Sidney Fleischer and Becca Fleischer

Volume 157. Biomembranes (Part Q: ATP-Driven Pumps and Related Transport: Calcium, Proton, and Potassium Pumps)
Edited by Sidney Fleischer and Becca Fleischer

Volume 158. Metalloproteins (Part A)
Edited by James F. Riordan and Bert L. Vallee

VOLUME 159. Initiation and Termination of Cyclic Nucleotide Action (in preparation)
Edited by JACKIE D. CORBIN AND ROGER A. JOHNSON

VOLUME 160. Biomass (Part A: Cellulose and Hemicellulose) (in preparation)
Edited by WILLIS A. WOOD AND SCOTT T. KELLOGG

VOLUME 161. Biomass (Part B: Lignin, Pectin, and Chitin) (in preparation)
Edited by WILLIS A. WOOD AND SCOTT T. KELLOGG

VOLUME 162. Immunochemical Techniques (Part L: Chemotaxis and Inflammation) (in preparation)
Edited by GIOVANNI DI SABATO

VOLUME 163. Immunochemical Techniques (Part M: Chemotaxis and Inflammation) (in preparation)
Edited by GIOVANNI DI SABATO

VOLUME 164. Ribosomes (in preparation)
Edited by HARRY N. NOLLER, JR. AND KIVIE MOLDAVE

VOLUME 165. Microbial Toxins: Tools for Enzymology (in preparation)
Edited by SIDNEY HARSHMAN

Section I

Sample Preparation

[1] Preparation of Metal-Free Water

By James F. Riordan and Bert L. Vallee

It has been and should be emphasized repeatedly that the single most critical factor in determining the trace element content of a biological sample is contamination. The problem is analogous to that encountered in microbiology and is equally significant since failure to maintain chemical sterility is every bit as disastrous to metallobiochemistry as lack of microbial sterility is to bacteriology. Anyone contemplating working in the trace element field has to accept the necessity of purity. In fact it is necessary to acquire absolute fanaticism in this regard and to view every item of equipment, every container, every reagent as a serious threat to experimental success. It is not an impossible task nor is it prohibitively expensive. It merely requires strict attention to detail and a keen sense of compulsion. Since water is not only the most important but also the cheapest and most common laboratory reagent, it usually constitutes the principal source of contamination. It is therefore fitting that the first chapter in this volume addresses the problem of water purification.

Methods for the preparation of high-purity laboratory water have been described in detail in previous volumes in this series[1-3] and elsewhere.[4-6] The reader is referred to these sources. However, there are a number of important considerations that have to be stressed, in large reason because misconceptions continue to persist. Beyond that, many new approaches are becoming dependent on these techniques, and many scientists now interested in metalloenzymes, for example, need not necessarily be specialists in analytical chemistry.

The American Society for Testing Materials (ASTM) has set standards[7] for four grades of reagent water (Table I). The grade recommended for trace element determination is Type I or reagent-grade ultrapure water. Type IV water can be used in situations where trace element considerations do not apply as in most glassware washing, much as Type III water is preferred for this purpose. Type II or analytical grade water is

[1] P. E. Hare, this series, Vol. 47, p. 13.

[2] C. Veillon and B. L. Vallee, this series, Vol. 54, p. 472.

[3] G. C. Ganzi, this series, Vol. 104, p. 391.

[4] R. Thiers, *Methods Biochem. Anal.* **5,** 273 (1957).

[5] M. Zief and J. W. Mitchell, "Contamination Control in Trace Element Analysis," p. 113. Wiley, New York, 1976.

[6] K. Griffiths, *Lab. Dyn.* **May** (1985).

[7] S. A. Fisher and V. C. Smith, *Mater. Res. Stand.* **12,** 27 (1972).

TABLE I
SOME ASTM SPECIFICATIONS FOR REAGENT GRADE WATER

Property	Specifications			
	Type I	Type II	Type III	Type IV
Total matter (mg/liter, max)	0.1	0.1	1.0	2.0
Conductivity (μmho/cm, max 25°)	0.06	1.0	1.0	5.0
Specific resistance [megohm (MΩ)/cm, min 25°]	16.66	1.0	1.0	0.20
pH (25°)	6.8–7.2	6.6–7.2	6.5–7.5	5.0–8.0
$KMnO_4$ reduction time (min)	>60	>60	>10	>10

usually prepared by double distillation and while not recommended for preparation of solutions for trace element work, it is acceptable when freedom from organic material is the main requirement. Other societies such as the National Committee for Clinical Laboratory Standards, the American College of Pathologists, and the American Chemical Society have also issued specifications for reagent grade water. However, within the present context, the only criterion of significance is not whether the water to be used meets certain guidelines but whether in the system under study it contributes no unnecessary problems. Most important, perhaps, is the fact that use of high-purity water for all buffers and solutions during the course of purification of a metalloenzyme will not provide adequate compensation if the reagents, containers, chromatographic media, etc. are themselves a source of contamination. Nevertheless, clean water is essential to ultimate analytical validity.

It used to be that repeated distillation was the method of choice for preparing water of high purity.[4] Distilled water produced in many laboratories can still contain substantial amounts of inorganic impurities depending on the type of still employed and the container used for collecting and storing the product. The effectiveness of distillation and, indeed, of any purification procedure can generally be assessed by measuring conductivity or, more typically, resistivity. The theoretical limit of resistivity for pure water is 18.3 MΩ/cm at 25° and a borosilicate glass still can generate a 1.8 MΩ/cm distillate from tap water. It should be emphasized, however, that the resistivity of the original distillate is much less significant than that of the water drawn from a storage supply immediately before use. A noncontaminating polyolefin storage container of limited capacity is recommended. Large-volume storage containers that can hold more than a few days supply are notoriously prone to contamination.

High-purity water with very low cation concentration can be obtained with a vitreous silica still in which the water is heated to subboiling temperature by a silica-coated infrared radiator mounted just above the surface of the liquid. This arrangement reduces formation of the aerosol spray that occurs at a bubbling surface and is capable of producing up to 1.5 liters/hr. The entire apparatus and storage container are kept in a vertical laminar-flow hood to prevent airborne contamination at the time of use.

A central distillation facility designed to service several laboratories or an entire building often produces water that is useful for many experimental purposes. For trace element determination, however, distilled water is definitely unsatisfactory. Aside from the fact that maintenance of such a facility is outside of one's control and not always of the highest standards, the distribution system becomes a source of many problems which, in most cases, begin with a central holding tank connected to feed lines with many stagnant areas. Both of these components invariably support growth of microorganisms. The quality of the water varies with usage and therefore becomes unreliable. Any repairs or changes within the system almost certainly diminish the subsequent water quality.

The use of reverse osmosis to provide purified water for a central facility has many advantages over distillation, primarily low maintenance and cost. Water is forced under pressure through membranes that exclude impurities on the basis of size and charge. More than 90% of the inorganic components can be removed and even greater efficiency for organic and particulate contents can be achieved. Water can be supplied continuously thus reducing some of the problems of storage.

Centrally purified water can serve an important function as input for a local, dedicated redistillation or demineralization unit. Therefore a stand-alone system is recommended. Installing a connection for such a feed line might be problematic, and the long-term savings may not offset reduced flexibility. Moreover, the system remains dependent on a reliably operating central supply and an appropriate storage container.

For most laboratories that specialize in trace element work, a unit consisting of ion-exchange resins and microporous filters will be the best way to produce high-purity water. Generally the water is equal to, or better than, that produced by distillation and is obtained with greater convenience, directly from tap water and at sufficient rates to eliminate the need for storage.

One system that has been found to produce water of exceptional quality, both in terms of trace metals and trace organic materials, uses a series of multiple filters.[2] Tap water is fed to a Continental water treatment system which consists of a 1.0-μm particulate filter, an organic absorber, and a mixed-bed ion-exchange resin. The output has a resistivity of 1 MΩ/cm, equivalent to triple-distilled water. This is used to supply a Millipore Milli-Q system consisting of a prefilter, an organic absorber, two mixed-bed ion-exchange columns, and a final 0.2-μm filter. Purified water can be obtained essentially on demand at a rate of 1.5 liters/min, with a resistivity of 18 MΩ/cm, and virtually free of particulates, minerals, and organics. This is generally adequate for the majority of purposes. The prepurification of the tap water uses replaceable cartridges which prolong the life of the more expensive furnishing resins.

A similar procedure is employed in the Barnstead NANOpure system[6] which combines mixed-bed deionization, carbon adsorption, membrane microfiltration, and special resins to produce high-purity water having below the limit-of-detection concentrations of 35 major elements as measured by graphite furnace atomic absorption. The key feature of these and other resin cartridge devices is that they purify the water only as needed, obviating the requirement for noncontaminating storage. However, in order to keep bacterial growth within the unit to a minimum they should be used on a more or less continual basis. If they are not used for even a few days it is possible that sterility problems could arise.

[2] Elimination of Adventitious Metals

By BARTON HOLMQUIST

Introduction

Despite the best efforts at contamination control including the cleaning of storage containers and other handling equipment, the control of airborne contamination, reagent purity and purification, and water and its purification, for which several excellent reviews are presented in this volume and elsewhere,[1-3] the ubiquity of metals, especially zinc, iron, and copper, require that as a final step before use solutions must be treated to remove trace metal contaminants. In this chapter we present two methods to minimize adventitious metal ion contamination inherent in solutions

[1] M. Zief and J. W. Mitchell, "Contamination Control in Trace Element Analysis," p. 113. Wiley, New York, 1976.

[2] R. E. Thiers, *Methods Biochem. Anal.* **5,** 273 (1957).

[3] C. Veillon and B. L. Vallee, this series, Vol. 54, p. 446.

used in the laboratory. Although the basic solvent water is easily purified and readily available as distilled or deionized, once the water is "contaminated" by addition of buffers, salts, acids, bases, or other components in the making up of solutions, a highly variable, and sometimes unbelievable, amount of metal contamination results. With few exceptions all such adjuvants add metals that may have deleterious effects on the experiment. Before such solutions can be used they must be cleansed of those metals of potential concern. The two methods reported here, dithizone extraction and ion exchange with a chelating resin, have been used extensively in numerous laboratories and have proven to be necessary in all aspects of work involving the replacement or depletion of the metals in metalloenzymes. Without their use much of the progress in metallobiochemistry could not have been achieved.

Dithizone Extraction

Diphenylthiocarbazone (dithizone) is a complexing agent soluble in organic solvents that reacts with various metals to form organic soluble chelates. Upon metal complexation, the bright green color of its solutions (ε_{618} = 29,500 in carbon tetrachloride) turns red, the color change serving as a sign of metal contamination. The extreme insolubility of dithizone in aqueous media versus its solubility in CCl_4 or $CHCl_3$ provides the basis of the extraction procedure.

```
        NH—NH—φ
       /
  S=C
       \
        N=N—φ
```

Dithizone is sensitive to oxygen and recrystallization is recommended. This is achieved by filtering a saturated chloroform solution (~1 g/50 ml), then evaporating one-half of the solvent with a nitrogen stream. The crystals are collected by filtration, washed with carbon tetrachloride, and dried under vacuum, mp 167–169°.[1] Alternatively, hexane can be added to the chloroform solution, and the product allowed to crystallize.

The reagent is said to be very suitable for the extraction of aqueous solutions of alkali and alkaline earth salt solutions to remove 23 elements (Table I).[1]

With due caution buffer solutions commonly employed in the laboratory, for example, Tris, HEPES, MES, phosphate, are amenable to extraction as are those containing substrates and cofactors. Some caution must be employed, however, especially if chloroform is used. Many organic compounds especially those containing no charge will partition into

TABLE I
ELEMENTS EXTRACTED BY ORGANIC SOLUTIONS OF DITHIZONE[a]

H																
Li	Be											B	C	N	O	F
Na	Mg											Al	Si	P	S	Cl
K	Ca	Sc	Ti	V	Cr	Mn	Fe	Co	Ni	Cu	Zn	Ga	Ge	As	Se	Br
Rb	Sr	Y	Zr	Nb	Mo	Tc	Ru	Rh	Pd	Ag	Cd	In	Sn	Sb	Te	I
Cs	Ba	La	Hf	Ta	W	Re	Os	Ir	Pt	Au	Hg	Ti	Pb	Bi	Po	
	Ra	Ac	Th	Pa	U											

[a] Underlined elements are extracted. From Ref. 1.

the organic phase depleting its concentration of the aqueous solution. If in doubt, a preextraction with pure organic solvent followed by analysis for the component in question in the solvent should be made. In general the extraction can be performed on aqueous solutions in the pH range between 3 and 7.5 but it should be recognized that the efficiency and rate of extraction of all metals are not identical at all pH values, a factor that is best overcome by very vigorous shaking for at least 5 min for each extraction.

Procedure

All manipulations involving carbon tetrachloride or chloroform should be done in a hood and rubber gloves worn to protect against the toxicity of halogenated solvents. A fresh dithizone solution, 0.02% in either carbon tetrachloride (Fisher 99 mol% is preferred) or chloroform, is added to the solution to be extracted contained in a Teflon-stopcock separatory funnel (halogenated solvents dissolve stopcock grease). The amount of organic solution should be approximately one-tenth to one-twentieth of the aqueous phase. The funnel is shaken vigorously for 5 min. If there is extensive metal contamination, the dithizone will change from green through gray to red. The phases are allowed to separate, the more dense halocarbon solvent removed through the stopcock, and the extraction repeated with a second aliquot of dithizone. Generally, this is sufficient for most solutions; the green color of the second aliquot should remain unchanged. If any color change is detected, additional extractions are necessary. The phases are separated, one-twentieth of the volume of neat solvent is added, and the funnel shaken again. The pale green extract is removed as before and the process of extraction with the neat solvent is repeated until no detectable green color appears and then two additional extractions are made. The last extraction is separated and the funnel is fitted with an

TABLE II
METAL CONTENT OF DITHIZONE-EXTRACTED BUFFER FROM A SOLUTION CONTAINING 500 ppb OF EACH METAL

Metal	ppb (0.1 *M* NaCl)	Molarity (*M*)	Detection Limit (ppb)
Cd^{2+}	<0.2	$<1.7 \times 10^{-9}$	0.2
Co^{2+}	2.9	4.9×10^{-8}	2.0
Cu^{2+}	2.7	4.2×10^{-8}	0.8
Fe^{2+}	7.2	1.3×10^{-7}	3.1
Mn^{2+}	250	4.5×10^{-5}	0.8
Ni^{2+}	5.5	9.4×10^{-8}	2.5
Zn^{2+}	1.6	2.4×10^{-8}	0.3

inverted plastic funnel connected by its spout to a house vacuum supply. The funnel must be void of flutes. When vacuum is applied a seal between the funnel and the neck of the separatory funnel forms and by carefully opening the stopcock a gentle stream of air passes through the solution carrying out the residual organic solvent. The air should be filtered to avoid particulate contamination. With solutions containing less than approximately 1 *M* salt the clearing of solvent will usually take 15–30 minutes of gentle bubbling; visual inspection will indicate when nearly all of the solvent is removed and an additional 10 min of bubbling beyond that time will generally remove the remaining organic solvent. With higher salt concentrations, such as when concentrated stock solutions of buffer are made for later dilution with metal-free water, the solubility of the organic solvent is lower and a more extended bubbling time is required. Small globules of the solvent are visible at the beginning and 10 min of bubbling beyond their disappearance is also required.

As an example of this extraction procedure 100 ml of an aqueous solution of 10 m*M* HEPES, 0.1 *M* NaCl, adjusted to pH 7.5 with saturated NaOH, was made up to 500 ppb in Cd^{2+}, Co^{2+}, Cu^{2+}, Fe^{2+}, Mn^{2+}, Ni^{2+}, and Zn^{2+} by the addition of 100 μl of 1000 ppm stock Fisher certified atomic absorption standards. The solution was extracted with dithizone as detailed above and analyzed for each of the above metals by graphite furnace atomic absorption (Table II).

It can be seen that the method is very effective with all but Mn^{2+} in bringing the concentrations of the metals to a few parts per billion (pbb). Such data are only indicative of the extraction in this particular instance since the efficiency of the process depends on the specific conditions and specific components in the solution extracted.

While such results may seem impressive (one part in 1,000,000,000), contamination is relative and it should be realized that even at these levels of remaining contamination serious misinterpretations of results can occur. A specific case of this is the measurement of the activity of an enzyme that has been demetalated to remove a catalytically essential metal, for example, thermolysin that has been dialyzed against 1,10-phenanthroline to remove the active-site zinc atom.[4] In this case an equivalent of zinc will restore complete activity to the inactive apoenzyme. With a rapidly turned over substrate a typical assay of activity might use 10^{-8} *M* enzyme and were this amount of apoenzyme used, for example, to ensure that without the zinc the enzyme is inactive, the residual zinc in the extracted buffer would be more than sufficient to completely restore all activity leading to the false conclusion that the metal-free enzyme is active. This has actually occurred in early studies of thermolysin. One method to overcome such a problem is to minimize the effects of contamination by raising the enzyme concentration in the measurement, that is to desensitize the assay by, for example, using a substrate with a lower turnover constant. If 10^{-6} *M* enzyme is used, the residual zinc in the buffer can only reconstitute 2.4% of the enzyme based on the data in Table II and assuming that no other metals contribute to the activity.

Metal-Chelating Resins (Chelex 100)

With buffers or aqueous solutions above pH 8 dithizone extraction is difficult if not impossible because of the solubility of dithizonate in the aqueous medium. Hence, above pH 8 such extractions should not be performed. Perhaps the best alternative for removing metals from aqueous media is to use the cation exchangers Chelex 100 or Chelex 20 distributed by Bio-Rad, Richmond, CA. These are resins of styrene divinylbenzene copolymers containing iminodiacetate anions that serve as metal chelating groups to bind most metal ions. Thus, they possess broad chelating properties much like EDTA. They have been widely used for elminating trace metals from reagents, physiological fluids, culture media, soils, and enzymes and to concentrate metals prior to analysis. The circular distributed by Bio-Rad[5] provides a large number of references on the use of Chelex for these purposes.

[4] B. Holmquist and B. L. Vallee, *J. Biol. Chem.* **249,** 4601 (1974).

[5] Bio-Rad Laboratories, "Chelex 100 Chelating Ion Exchange Resin for Analysis, Removal, or Recovery of Trace Metals," Product Information Bulletin 2020. Bio-Rad, Richmond, California.

TABLE III
SELECTIVITY OF CHELEX 100 FOR METALS

			Ba^{2+}					
			Ca^{2+}					
			Mg^{2+}					
			Sr^{2+}		Co^{2+}			
				Cd^{2+}		Pb^{2+}		
Na^{+}			Mn^{2+}	Fe^{2+}	Zn^{2+}	Ni^{2+}	Cu^{2+}	Hg^{2+}
−7	−5	−3		−1		0	1	3
Log stability relative to zinc								

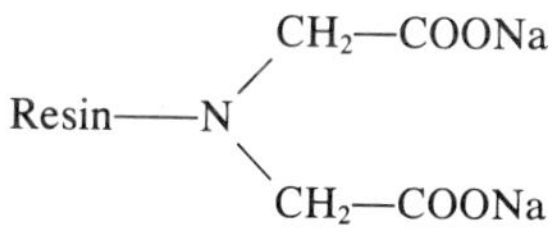

```
              CH2—COONa
             /
Resin——N
             \
              CH2—COONa
```

Table III lists the relative affinity of Chelex 100 with reference to Zn^{2+}. The extremely low value for Na^{+}, the form of the resin most frequently used and supplied by the manufacturer, permits efficient exchange of this ion with all others, particularly those heavy metals most frequently of concern to the enzymologist such as those that inhibit biological reactions or are constituents of enzymes or coenzymes. In addition to those listed the entire series of lanthanides will also bind to Chelex 100.

The actual selectivity of the resin is a function of the composition of the solution treated and the relative affinities can vary. Thus the type of buffer, pH, salts present, and their concentration can compete as ligands for the metal and alter its apparent stability for the resin.

Preparation and Use of Chelex 100

The sodium form of the resin, as supplied by the manufacturer, is frequently used as is without prior treatment other than washing. It is recommended that the resin be cycled to the acid form, then back to the cation form, prior to use. This is accomplished by sequentially equilibrating the resin on a scintered glass funnel with 2 bed volumes of 1 *M* HCl, 5 bed volumes of metal-free water, 2 bed volumes of 1 *M* NaOH, and then 5 bed volumes of water as a final wash. Storage of the resin in the acid form is to be avoided for within a few hours the resin will begin to loose chelating capacity that can only be restored by heating at 60° for 24 hr in 30–50% NaOH.

The resin can be used in either a batch or column technique, the latter being preferred because of its greater efficiency. The column method merely requires that the solution to be demetalized is passed over the resin and after several column volumes of the initial eluate are discarded and the resin equilibrated with the solution, the purified material is collected. Flow rates of as high as 20 ml/min/cm^2 can be used. Examples of the column method include the purification of 0.5 *M* Tris at pH 7.5[6] and of 60 m*M* TES buffer in 150 m*M* NaCl to ensure calcium-free solutions, less than 2×10^{-7} *M*.[7] Such procedures exchange sodium for any metal present.

If another cation is required or if it is necessary to maintain the solution sodium-free an alternate cation form of the resin may be prepared. Thus, the calcium form of the resin is obtained by equilibrating the sodium form of the resin with two bed volumes of 2 *M* calcium chloride. The efficiency of the resin for metal exchange will be considerably altered by this, however, since calcium has a rather high affinity for the resin.

Other forms of the resin may be prepared by treating the acid form with the appropriate salt (buffer). The acid form of the resin is washed with water and then equilibrated with the appropriate buffer salt. Adventitious metals in the solution applied will then be exchanged with the buffer. For example Tris at pH 7.2 was used to neutralize the acid resin to allow sodium-free solutions to be prepared and the potassium form of the resin, prepared using KOH for equilibration, to minimize sodium contamination of HEPES buffers.[8] To prepare 0.5 *M* phosphate buffer for HPLC applications (separation of nucleotides) Reiss *et al.*[9] used a three-column system in which Chelex was used to remove metals. Consecutive columns of 500 g of Dowex AG 1-X8, chloride form, Chelex 100, and activated charcoal were used and the phosphate buffer was continuously cycled through the columns with a circulating pump. The resins were regenerated after approximately 100 liters of buffer were scrubbed.

As a final note, it is stressed that nothing supplants the verification of any procedure used for the elimination of adventitious metal. Hence, actual metal analysis of solutions, using methods such as atomic absorption spectroscopy, is desirable subsequent to metal extraction. It is very easy to recontaminate solutions or to have such procedures go awry; therefore if one or more metals is of particular concern its presence or absence should be verified when possible.

[6] C. Ma and W. J. Ray, *Biochemistry* **19,** 751 (1980).
[7] D. Burger, J. A. Con, M. Compte, and E. A. Stein, *Biochemistry* **23,** 1966 (1984).
[8] J. B. Hunt and A. Ginsberg, *Biochemistry* **20,** 2226 (1981).
[9] P. D. Reiss, P. F. Zuurendonk, and R. L. Veech, *Anal. Biochem.* **140,** 162 (1984).

[3] Metal-Free Dialysis Tubing

By DAVID S. AULD

The degree of care needed in preparing metal-free dialysis tubing is, of course, directly proportional to the concentration level of the metal to be determined. If the concentration level of a metal in an experiment involving dialysis tubing is $\leq 10^{-6}$ *M* contamination from adventitious metal ions, especially zinc, present in trace amounts in water, reagents and dialysis vessels can present formidable problems, requiring care in all phases of the work.

Preparation of Tubing

Commercially obtained dialysis tubing can be contaminated with heavy metals, proteases, and nucleases. Therefore before being used to dialyze protein or DNA solutions, dialysis tubing should be treated in the following manner. A proper amount of tubing (e.g., 10–20 ft of 23 mm flat-width Spectra/Por 1 dialysis tubing), cut to convenient lengths (e.g., 18-in sections) is placed in a 2-liter acid-cleaned glass beaker[1] containing metal-free water. The mixture is heated at 70–80° with occasional gentle stirring with an acid-cleaned glass rod for 2 hr. The water is decanted and this procedure is repeated at least three more times. Most of the metal is removed in the first two washes (Table I). The treated dialysis tubing is stored in metal free water at 4° in a covered plastic container, previously freed of metals. Bacterial growth is usually minimal in a metal-free system, so antibacterial agents such as sodium azide are usually not necessary. However, if azide is used it must be removed before working with metalloenzymes, since it can be an inhibitor. Finally, the tubing is soaked for several hours in the metal-free buffer[1] to be used in the desired experiments.

Handling of Metal-Free Tubing

After the above-described treatment the dialysis tubing should only be handled with metal-free forceps and disposable plastic gloves. Several methods of preparing metal-free water have been described.[2] The choice of gloves can be critical since rubber gloves contain zinc and some gloves

[1] B. Holmquist, this volume [2].

[2] C. Veillon and B. L. Vallee, this series, Vol. 54, p. 446; see also J. F. Riordan and B. L. Valley, this volume [1].

TABLE I
MEASUREMENT OF Zn^{2+} AND Cu^{2+} IN DISTILLED WATER USED FOR WASHING THE DIALYSIS TUBING[a]

Water wash number	$[Zn^{2+}]$ (μg/liter)	$[Cu^{2+}]$ (μg/liter)
0	0.5	1.4
1	12.3	32.6
2	1.7	14.5
3	0.2	4.3
4	0.5	3.8

[a] Eight 18-in. strips of 23-mm flat width dialysis tubing (Spectra POR1) were place in a 1-liter beaker filled with distilled water. Samples were taken at each step of the procedure including a sample of the distilled water before the analysis (0).

are powdered with zinc oxide.[3] The tubing should never be handled with uncovered hands because fingers are a remarkably good source of both metals and degradative enzymes. The tubing should be cut prior to treatment not afterward to reduce the chance of metal contamination from contact with scissors or a razor blade.

Comments

Harsher treatment of the tubing than described may impair the properties of the resulting dialysis bags. Heating the tubing on a steam bath for periods as long as 72 hr has been reported to weaken the tubing to the point where it is permeable to a protein of molecular weight 70,000.[4] All glass and plastic containers in which the dialysis tubing is placed should be free of metal contamination and all water should be deionized and distilled.[1,2] The use of chelators such as ethylenediaminetetraacetic acid (EDTA) to aid in metal removal during the heating is not advisable because residual EDTA can bind to the tubing and to proteins and thus be a source of metal contamination or possibly inhibit the metalloenzyme under investigation.[5] In addition the chelator frequently can contain metals which may result in washing the tubing with a solution containing as much or more metal contamination than is in the tubing itself.

[3] M. Zief and J. W. Mitchell, *Chem. Anal.* **47,** 69 (1976).
[4] T. R. Hughes and I. M. Klotz, *Methods Biochem. Anal.* **3,** 281 (1956).
[5] D. D. Ulmer and B. L. Vallee, *Adv. Chem. Ser.* **100,** 193 (1971).

[4] Metal-Free Chromatographic Media

By MARK T. MARTIN

Introduction

It is often necessary to maintain the integrity of a metal–ligand complex during a chromatographic separation. For example, to demonstrate a biological association of a metal with a protein or enzyme, the intact complex must first be purified, a task which commonly requires some form of chromatography under metal-free conditions. In addition to purification, metal-free chromatography has been useful in the discovery of the proteins and low-molecular-weight ligands that store metals in depot tissues or transport them in biological fluids.[1–3] In general, a metal-free chromatographic medium is necessary in any separation where the metal–ligand relationship under study is susceptible to gain, loss, or substitution of metal ions.

Most, if not all, chromatographic media are capable of binding metals. Thus, in metal-sensitive experiments, precautions should be taken to reduce or obviate this affinity. Highly charged media, such as ion-exchange resins, function by binding ionized solutes and clearly cannot be rendered incapable of binding metal ions or the charged complexes of metal ions. These types of media will not be specifically addressed here, although some of the general concepts described will be useful in reducing their interactions with metals. This chapter will focus primarily on the reduction of metal ion interactions with gel filtration media.

Chemical Nature of the Problem

Gel filtration media are intended to separate molecules solely on the basis of their size and, ideally, the interactions between the gel matrix and the chromatographed sample are strictly steric in nature. However, gels have long been known to bind or repel certain charged or hydrophobic molecules.[4] The specificity and extent of the effects vary with the chemical makeup of the gel.

[1] B. Lönnerdal, B. O. Schneeman, C. L. Keen, and L. C. Hurley, *Biol. Trace Elem. Res.* **2,** 149 (1980).
[2] B. C. Starcher, J. C. Glauber, and J. G. Madaras, *J. Nutr.* **110,** 1391 (1980).
[3] B. Lönnerdal, A. Stanislowski, and L. C. Hurley, *J. Inorg. Biochem.* **12,** 71 (1980).
[4] B. Gelotte, *J. Chromatogr.* **3,** 330 (1960).

Polyacrylamide gels are copolymers of the cross-linking agent *N,N'*-methylenebisacrylamide ($H_2C{=}CHCONHCH_2NHCOCH{=}CH_2$) and allyl dextran (Sephacryl) or acrylamide ($H_2C{=}CHCONH_2$) (BioGel P).[5,6] In addition to metal ions, adsorption of aromatic, basic, or acidic molecules may occur.[6] The copper- and zinc-binding capacities of these gels have been reported and range up to 569 μg zinc or 337 μg copper per gram of dry BioGel P-2.[7] Lesser amounts were found to bind to the sparsely crosslinked BioGel P gels such as P-100 and to Sephadex gels.[7]

Sephadex polydextran gels consist of poly-α-1,6-D-anhydroglucopyranose cross-linked by glyceryl moieties.[8] These gels have been found to contain between 10 and 30 microequivalents of carboxyl groups that are the primary source of the ionic interactions with charged solutes.[4,8,9] Additionally, vicinal hydroxyl groups as well as the glyceryl cross-linkages have been postulated to bind metals.[7,10]

Agarose gels such as BioGel A and Sepharose are derived from agar and are composed of alternating residues of D-galactose and 3,6-anhydro-L-galactose.[11] The purity of the agarose largely determines the extent of interfering ionic interactions since agaropectin, the main contaminent, contains charged carboxyl and sulfate groups.[11] The agarose chains are normally not cross-linked but are held together by hydrogen bonding. An exception is Sepharose CL which is covalently cross-linked and desulfated, resulting in reduced interaction with metal ions.[12,13]

Determination of Metal-Binding Capacity

Previously, the metal-binding capacities of gels have been estimated by quantitating the amount of metal lost from solution during the equilibration of a column with a solvent system of known metal ion concentration.[7,14] This procedure is accurate, but in practice may be rather tedious since the equilibration process may extend over a large elution volume.[14] A simpler but less quantitative method involves the chromatography of a sample containing a known amount of metal and determining the percentage of metal which passes through the column unbound.[13]

[5] S. Hjertén and R. Mosbach, *Anal. Biochem.* **3,** 109 (1962).
[6] M. Belew, J. Porath, J. Fohlman, and J.-C. Jansen, *J. Chromatogr.* **147,** 205 (1978).
[7] P. Johnson and G. Evans, *J. Chromatogr.* **188,** 405 (1980).
[8] J. Reiland, this series, Vol. 22, p. 287.
[9] B. Stenlund, *Adv. Chromatogr.* **14,** 37 (1976).
[10] N. J. Richards and D. G. Williams, *Carbohydr. Res.* **12,** 409 (1970).
[11] C. Araki, *Bull. Chem. Soc. Jpn.* **29,** 543 (1956).
[12] J. Porath, J.-C. Jansen, and T. Låås, *J. Chromatogr.* **60,** 167 (1971).
[13] B. Lönnerdal and B. Hoffman, *Biol. Trace Elem. Res.* **3,** 301 (1981).
[14] G. W. Evans, P. E. Johnson, J. G. Brushmiller, and R. W. Ames, *Anal. Chem.* **51,** 839 (1979).

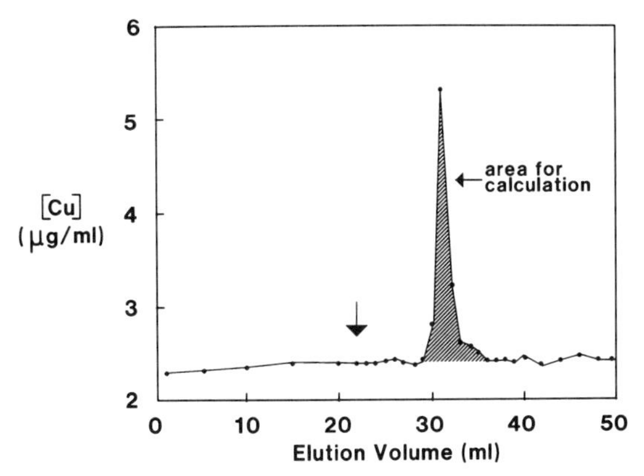

FIG. 1. Quantitation of the copper-binding capacity of Sepharose 4B. An eluent of 10 μg/ml Cu^{2+} as copper acetate, 5 m*M* sodium acetate, 25 m*M* NaCl, pH 5.7 was passed through a 6.0 ml column of gel at a flow rate of 2.0 ml/min until the metal reached equilibrium. The eluting buffer was then changed to the same but containing 50 m*M* HNO_3 (pH 1.7), as indicated by the arrow. One milliliter fractions were collected, diluted 4-fold with 1% HNO_3, and the copper concentrations were measured by atomic absorption spectrophotometry. Further experimental details are provided in the text.

Described here is a simple procedure by which the total metal-binding capacity of a gel can be accurately quantitated. Such a technique is valuable in determining which type of gel and which buffer best resist metal–gel interactions under circumstances where metal-free conditions are required. The method, adapted from that of Porath *et al.*,[12] involves the elution of a small gel column with a metal ion solution until equilibrium (i.e., saturation of metal binding) is reached. The eluent is then changed to the same but containing a sufficient concentration of HNO_3 to lower the pH to approximately 1.5. As a consequence of the decrease in pH a metal ion peak will emerge from the column representing that which was bound to the gel (Fig. 1). The peak can then be quantitated by atomic absorption spectrophotometry such that the amount of metal bound per milliliter of gel can be calculated.

Procedure

1. A known volume (5–6 ml) of the gel of interest is packed in a metal-free plastic, not glass, column.
2. Five column volumes of 0.05 *M* HNO_3 are passed through the column followed by a neutralizing wash of deionized metal-free water.
3. An eluent of chosen pH, buffer, and salt, but also containing 10–15 μg/ml of the metal of interest is passed through the column until the column is saturated with metal (i.e., the concentration of metal leaving

the column equals that of the metal entering the column). This does not have to be rigorously determined at the time but can be simply assumed after 4–5 hr (if equilibrium is not reached, it will be detected later in the experiment). The flow rate is not critical, generally 2.0 ml/min is acceptable. One milliliter fractions are collected.

4. The eluent is changed to the same but containing a concentration of 0.05 *M* HNO_3 (pH approximately 1.5). A greater concentration may be necessary to reach this pH if the eluent is highly buffered.

5. The fractions are diluted with 3.0 ml of 1% HNO_3, mixed, and the metal ion concentrations are measured by atomic absorption spectrophotometry. If the metal ion concentrations before and after the peak are identical, one can be assured that the column was thoroughly equilibrated with metal ions.

6. The weight of metal bound per milliliter gel can be easily calculated by integrating the peak area.

Results

This technique was used to illustrate the copper-binding capacity of four commonly used gels (Table I). The results demonstrate that in unbuffered water the dense polydextran gel Sephadex G-10 binds a great deal of copper; 151.4 μg/ml gel. Sephadex G-100, which contains fewer cross-links and is thus less dense than G-10, binds only 4.0 μg/ml. The other tested gels bind intermediate amounts of copper; 33.0 μg/ml for the polyacrylamide Sephacryl S-200 and 29.0 μg/ml for the agarose Sepharose 4B. This method was also useful in determining the effectiveness of treatments commonly used to decrease metal ion interaction. These treatments, elution with a solvent of substantial ionic strength and alkaline reduction of the gels, are discussed below.

Effect of Eluent Ionic Strength

Including an inert electrolyte in the elution solvent reduces metal–gel interactions. Generally, an ionic strength of at least 0.02 is recommended for all types of media.[5,8] The ionic strength can be provided by a salt such as sodium chloride or simply by using a buffer, as illustrated by Lönnerdal and Hoffman[13] and Johnson and Evans.[7]

Results

The results in Table I show that in all cases, but particularly with Sepharose 4B, a buffered solvent containing an electrolyte reduces metal binding. The extent of the metal–gel interaction is clearly dependent on the gel type, a characteristic that warrants consideration in choosing a gel.

TABLE I
EXAMINATION OF COPPER BINDING TO CHROMATOGRAPHIC MEDIA[a]

Gel type	Treated[b]	Buffered eluent[c]	μg Cu bound/ml gel
Sephadex G-10	−	−	151.4
	−	+	112.6
	+	−	95.8
	+	+	31.8
Sephadex G-100	−	−	4.6
	−	+	1.5
	+	−	3.0
	+	+	2.0
Sepharose 4B	−	−	43.1
	−	+	3.1
	+	−	*d*
	+	+	*d*
Sephacryl S-200	−	−	29.6
	−	+	17.8
	+	−	33.0
	+	+	22.2

[a] Flow rate for these experiments was 2.0 ml/min except with Sephadex G-100 (1.0 ml/min). Experimental procedure is described in the text.

[b] Alkaline reduction was performed according to Lönnerdal and Hoffman.[13]

[c] −, unbuffered eluent which contained 10 μg/ml copper acetate in metal-free distilled water; +, buffered eluent which contained 10 μg/ml copper acetate, 5 m*M* sodium acetate, 24 m*M* Nacl, pH 5.7.

[d] The alkaline reduction treatment destroyed this gel.

Alkaline Reduction of the Gel

It has become fairly common practice to chemically treat chromatographic gel to eliminate the metal-binding carboxyl and sulfate groups.[1-3] A method introduced by Porath *et al.*[12] involves the alkaline hydrolysis of these charged groups in the presence of sodium borohydride or lithium aluminum hydride to prevent subsequent depolymerization. A similar yet simplified procedure was advanced by Lönnerdal and Låås[15] and Lönnerdal and Hoffman.[13] The latter authors examined the method for its effectiveness and found it to be valuable in preventing metal–gel interactions with Sephadex G-15 and G-50 and with the cross-linked agarose

[15] B. Lönnerdal and T. Låås, *Anal. Biochem.* **72,** 527 (1976).

Sepharose CL, particularly when the eluent was well buffered. However, Sepharose 6B (uncross-linked agarose) and BioGel P (polyacrylamide) gels are destroyed by the treatment.

Procedure[13]

1. A volume of swollen gel is taken up in 2.5 volumes of water.
2. The pH of the slurry is adjusted to 10–11 with sodium hydroxide, and $NaBH_4$ is added to a concentration of 2.0 g/liter of the gel slurry.
3. The gel slurry is heated at 80° for 2 hr with gentle stirring.
4. The gel is cooled, then washed with water until neutral in pH.
5. Any fines generated during the procedure are removed.

Results

In this work (Table I) the method was of considerable value with Sephadex G-10, but of little value with Sephadex G-100. Sepharose 4B was destroyed by the treatment (Sepharose CL, which reportedly binds less metal than uncross-linked Sepharose, apparently can be successfully treated). Sephacryl S-200 was still usable after the treatment but the metal-binding capacity of the gel actually increased somewhat.

Addendum: Additional Methods

1. Purchased Sephadex gels have been found to contain zinc,[16] and may contain other metals. Previously, such indigenous metals have been removed by eluting freshly packed gel with a solution of EDTA, followed by large volumes of metal-free water or buffer to remove all EDTA.[17] However, since it has not been shown that EDTA does not bind to gels, it may be safer to elute with 5 column volumes of 0.05 *M* HCl or HNO_3 (a higher concentration of acid or a longer exposure may damage the gel), followed by a neturalizing wash of metal-free water.

2. Oxidizing agents can increase the number of metal-binding carboxyl groups in gels and thus should be avoided when possible.[8]

3. Unconventional solvents have been employed to prevent metal–gel interaction but, due to their harshness, are limited in their applicability. For instance, low pH (0.02 *M* HCl) has been used to fractionate histones.[18]

4. There have been few reports of attempts to reduce metal–media interactions in high pressure liquid chromatography (HPLC). Dissocia-

[16] R. S. Morgan, N. H. Morgan, and R. A. Guinavan, *Anal. Biochem.* **45,** 668 (1972).
[17] B. C. Antanaitis and P. Aisen, *J. Biol. Chem.* **257,** 1855 (1982).
[18] H. J. Cruft, *Biochim. Biophys. Acta* **54,** 611 (1961).

tion of metals in HPLC separations of metallothioneins were avoided by using eluents of neutral to weakly basic pH.[19] Reversed-phase type media appear to be preferable to ion exchange or gel filtration in that they tend to better resist interactions with metal ions.[19]

Summary

The three major considerations in reducing metal-gel binding are (1) the type of gel to be used, (2) the ionic strength of the eluent, and (3) alkaline reduction of the gel. Each of these parameters should be considered in advance of an experiment. A method is offered for determining the optimal gel type, gel treatment, and eluent for a critical metal-free chromatography experiment.

Acknowledgments

The author thanks Drs. D. S. Auld and Barton Holmquist for their critical reading of the manuscript. This work was supported by National Research Service Award 1-F32-HL-06965-01A1(BI-2) from the National Institutes of Health.

[19] K. T. Suzuki, H. Sunaga, Y. Aoki, and M. Yamamura, *J. Chromatogr.* **281,** 159 (1983).

[5] Preparation of Metal-Free Enzymes

By Fred W. Wagner

The need to prepare metal-free enzymes results from the knowledge that bound metal ions are essential to the structure and/or function of enzymes. In most cases, the native ion of metalloenzymes is Zn^{2+} which is spectrally silent. Consequently, methods to study the interaction of protein and metal ion as it affects the structure and function of the enzyme are limited. A common approach has been to substitute the native metal ion with a paramagnetic ion, such as cobalt or cadmium,[1,2] and use their spectral properties to probe the action of the enzyme. The most direct method to substitute enzyme bound metal ions is to prepare the apoen-

[1] K. F. Geoghegan, B. Holmquist, C. A. Spilburg, and B. L. Vallee, *Biochemistry* **22,** 1847 (1983).

[2] P. Gettins and J. E. Coleman, *J. Biol. Chem.* **259,** 4991 (1984).

zyme (metal-free enzyme) and then reconstitute it with stoichiometric amounts of the desired metal ion.

Enzymes that bind metal ions have been divided into two classes: metal-activated enzymes, which associate metal ions loosely ($K_d = 10^{-3}$ to 10^{-8} *M*) and metalloenzymes, which bind metal ions tightly ($K_d < 10^{-8}$ *M*). This chapter will focus on methods used to remove metal ions from metalloenzymes, as they can be readily removed from metal-activated enzymes by dialysis, gel filtration, or dilution.

General Preparation

Metalloenzymes having a high state of purity and well-established metal stoichiometry should be available in good quantities (20–100 mg). It is not advisable to attempt preparing apoenzyme with less than 1 ml of 50–100 μM native enzyme, because smaller amounts are difficult to protect from reaction with aventitious metal ions. The enzyme should be characterized thoroughly with regard to its substrate specificity and stability over a broad range of pH values.

Removing metal ions from metalloenzymes is usually a straightforward procedure and generally does not present problems. The difficult task is controlling the environment of the apoenzyme during and subsequent to its preparation to ensure its stability and ability to be reconstituted with the metal ion of choice. Procedures to control adventitious metal ion concentrations are discussed in previous chapters and should be followed to ensure that reagents and apparatus are "metal free." In attempting to prepare an apoenzyme, be constantly aware that trace levels of metal ions (especially Zn^{2+}) are always present. Zinc ion concentration should not be assumed to be less than the limits of its detection by graphite-furnace atomic absorption spectroscopy (0.05 ppb or approximately 1 n*M*). Generally, solutions should be treated to remove metal contaminants either by extraction with dithizone or Chelex 100.[3] Chromatographic materials should be rendered metal free. Use polystyrene labware when possible as it is usually devoid of detectable metals. Certain types of plastics (Tygon, polyethylene, and polypropylene) can possess significant amounts of contaminating metal ions. It is advisable to rinse plasticware with 20% HNO_3 then metal-free water prior to use. Cuvettes should be soaked in 0.1 *N* NaOH to remove adsorbed protein, then rinsed with 20% HNO_3 and metal-free water. Quartz or glass treated in such a manner possesses ion-exchange properties which may affect concentrations of dilute proteins. Glassware should be avoided because of its indigenous metal content.

[3] C. Veillon and B. L. Vallee, this series, Vol. 54, p. 446.

Special attention should be given to the chemical composition of solutions to be used for preparation and reconstitution of apoenzymes. Many of the buffers commonly employed are also metal-chelating agents at biological pH ranges.[4-6] Notably, organic amines such as Tris and tricine (and other "Good" buffers) should be avoided as they coordinate readily at neutral pH (7.5) with many of the metal ions commonly used to reactivate apoenzymes,[4-6] e.g., cobalt. HEPES, a sulfonic acid derivative, has been successfully used as a buffer salt to prepare a number of apoenzymes. Metal-buffer binding constants of "Good" buffers for Mg^{2+}, Ca^{2+}, Mn^{2+}, and Cu^{2+} have been published.[3]

Enzyme Assays

After being prepared, the apoenzyme should be assayed for residual native enzyme to determine if the loss of activity correlates with the removal of metal ion and to establish that the apoenzyme can be fully reconstituted with the native metal ion. In many cases a special assay procedure must be devised to avoid reactivation of apoenzyme in the substrate solution by adventitious metal ions. The problem centers on the fact that many enzymes are asayed at concentrations comparable to the adventitious metal ions present (0.5–1 n*M*), thus allowing partial or complete reactivation of the apoenzyme during the assay. Assays usually employ substrates most effectively catalyzed by the enzyme, however, when assaying apoenzyme for residual native enzyme a less effective substrate may be more desirable. As an example *Aeromonas* aminopeptidase is routinely assayed at levels of 1 n*M* using 1 m*M* L-leucine-*p*-nitroanilide, a substrate for which the enzyme has a K_m of 19 μM and k_{cat} of 67 min^{-1}.[7] For the apoenzyme, however, the less effective substrate L-alanine-*p*-nitroanilide is used (K_m, 1 m*M*; k_{cat}, 17.3 min^{-1}) at a concentration of about 40 μM (0.04 K_m).[7,8] Consequently, apoenzyme preparations can be assayed at a concentration of about 1 μM, thus avoiding significant reactivation by adventitious ions. Since the substrate concentration is only a minor fraction of its saturation value, tests to reactivate the apoenzyme with different metal ions can be performed in the same solutions without subjecting apoenzyme to further metal ion contamination.

[4] N. E. Good, G. W. Winget, W. Winter, T. N. Connolly, S. Izawa, and R. M. M. Singh, *Biochemistry* **5,** 467 (1966).

[5] J. M. Pope, P. R. Stevens, M. T. Angatti, and W. Wakan, *Anal. Biochem.* **103,** 214 (1980).

[6] E. A. Lance, C. W. Rhodes III, and R. Nakon, *Anal. Biochem.* **33,** 492 (1983).

[7] J. M. Prescott, F. W. Wagner, B. Holmquist, and B. L. Vallee, *Biochem. Biophys. Res. Commun.* **114,** 646 (1983).

[8] J. M. Prescott, F. W. Wagner, B. Holmquist, and B. L. Vallee, *Biochemistry* **24,** 5350 (1985).

Removing Metals from Enzymes

Methods routinely employed to remove metal ions from metalloproteins can be categorized into three groups: those performed at neutral pH, acidic pH, and alkaline pH. The first of these is the most commonly employed as most metalloproteins denature at extremes of pH. Most established procedures are variations of the use of a metal-chelating agent to remove metal ion(s) from the enzyme followed by the removal of chelated metal ions and excess chelating agent. Apoalkaline phosphatase is a notable exception as it can be prepared from the native enzyme by dialysis against 2 *M* $(NH_4)_2SO_4$ at pH 9.0.[2]

The removal of metal ions from enzymes is usually performed from slightly acidic (pH 5.5), neutral, to slightly alkaline (pH 7.5) pH values using water-soluble chelating agents. A single chelating agent cannot be recommended. When developing a procedure to prepare apoenzyme, a variety of chelating agents should be tested for their efficacy. As an example, in the preparation of apothermolysin, Holmquist and Vallee[9] examined a broad selection of potential sequestering agents (Fig. 1) for their utility in removing Zn^{2+} from the active site of thermolysin. Based on the measured equilibrium constants and the stability of apothermolysin in the presence of chelating agent, these workers chose 1,10-phenanthroline for the routine preparation of apoenzyme. Even though 1,10-phenanthroline is the most routinely used chelating agent for preparing apoenzymes at neutral pH, others, such as dipicolinic acid or 8-hydroxyquinoline have also been employed. EDTA is not highly recommended as its strong anionic character allows the possibility for it to bind to proteins and laboratory apparatus at low ionic strengths. If EDTA is used, the procedure of Sabbioni *et al.*[10] is recommended, in which ^{14}C-EDTA is added to the chelating agent to provide a measure of its removal from metal-free enzyme solutions.

Removal of metal ions from enzymes occurs by one of two mechanisms. Either the free metal ion concentration is depleted by the chelating agent to promote dissociation of metal ion from the enzyme or the chelating agent forms a transient ternary complex with the metal and enzyme with subsequent removal of the metal ion from the protein. The former method generally is a time-dependent process, while the latter is usually rapid. Thus in preliminary experiments to determine the choice of chelating agent, it is recommended that the loss of enzymatic activity be monitored with time. In cases such as carbonate dehydratase, Zn^{2+} removal with 1,10-phenanthroline at pH 5.5 occurs slowly with half-lives from 4 to

[9] B. Holmquist and B. L. Vallee, *J. Biol. Chem.* **249,** 4601 (1974).

[10] E. Sabbioni, F. Girardi, and E. Marafante, *Biochemistry* **15,** 271 (1976).

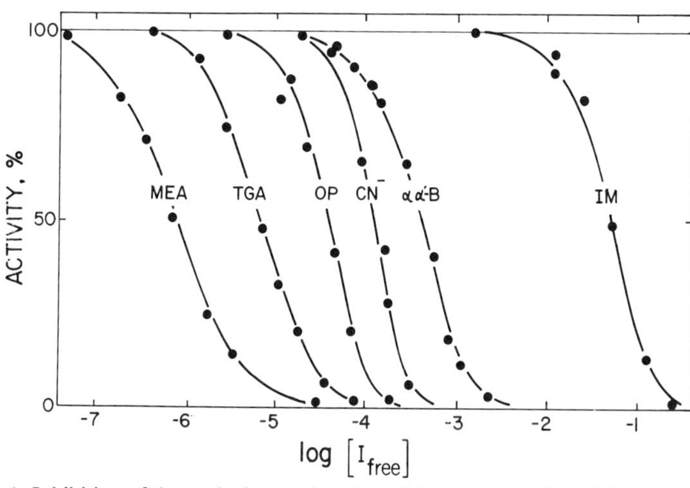

FIG. 1. Inhibition of thermolysin as a function of the concentration of free metal binding agent, I_{free}. Values given for each inhibitor are those of the enzyme concentration employed in experiments with the agent: 2-mercaptoethylamine (MEA), 3.71×10^{-7} *M*; thioglycolic acid (TGA), 3.71×10^{-7} *M*; 1,10-phenanthroline (OP), 9.66×10^{-8} *M*; 2,2′-bipyridine ($\alpha\alpha'$-B), 2.0×10^{-7} *M*; cyanide, 2.4×10^{-7} *M*; imidazole (IM), 2.4×10^{-7} *M*. Conditions: pH 7.5, 0.05 *M* Tris, 0.01 *M* Ca^{2+}, 0.1 *M* NaCl, 25°, 1 m*M* furylacryloylglycyl-L-leucylamide. From Holmquist and Vallee.[9]

10 days depending on the isoenzyme,[11] whereas with dipicolinic acid as the chelating agent, metal ion removal at pH 7.0 is fast ($t_{1/2} \cong 10$ min).[12] Conversely, removal of bound Zn ion from *Aeromonas* aminopeptidase by 1,10-phenanthroline is immediate, while it is time dependent with dipicolinic acid.

The inhibition of human liver alcohol dehydrogenases by 1,10-phenanthroline is not accompanied by loss of metal ion, and is thought to involve a stable ternary complex between enzyme, Zn ion, and chelating agent.[13]

The preparation of apoenzyme useful for reconstitution experiments requires removal of the native metal ion with chelating agent followed by the subsequent removal of the chelating agent without appreciable reactivation of the apoenzyme. It is important to maintain high protein concentrations (0.1–1 m*M*) throughout all manipulative steps in preparing and maintaining apoenzymes. It should be emphasized that 1 ml of 1 μM apoenzyme can be fully reactivated by dialysis against 1 liter of "metal-free" buffer (1 n*M* Zn^{2+}). Protein should be dialyzed, with vigorous stir-

[11] S. Lindskog and B. G. Malmström, *J. Biol. Chem.* **237,** 1129 (1962).

[12] J. B. Hunt, M. J. Rhee, and C. B. Storem, *Anal. Biochem.* **79,** 617 (1977).

[13] F. W. Wagner, X. Páres, B. Holmquist, and B. L. Vallee, *Biochemistry* **23,** 2193 (1984).

ring, against several changes of chelating agent (25–50 volumes each) followed by at least 4 to 5 changes of metal-free buffer (no more than 25 volumes per dialysis). The use of chelating agents with absorptive properties, such as 1,10-phenanthroline or 8-hydroxyquinoline, is recommended as the absorbance of the dialyzate can be monitored to ensure complete removal of the chelating agent. The chelating agent can also be removed by gel filtration; however, this step is accompanied by significant dilution of the apoenzyme and may result in unacceptable levels of reconstituted enzyme in the final apoenzyme preparation. A second and much simpler approach to preparing apoenzyme employs the use of Chelex 100 as a chelating agent.[14] The resin is equilibrated with the metal-free buffer of choice then added to the native enzyme, in metal-free buffer, so that the final volume is about 20% resin. The slurry is gently mixed until loss of enzyme activity is virtually complete. When used to prepare alkaline phosphatase, the loss of activity was time dependent, indicative of Chelex 100 acting to deplete free metal ions from solution. The advantage of this method is that it allows the apoenzyme to be stored in the presence of the chelating agent, thus minimizing contamination by adventitious metal ions. However, certain enzymes (thermolysin) that require Zn^{2+} for catalysis also require Ca^{2+} for stability. In such cases, Zn^{2+} free enzyme must be prepared in the presence of Ca^{2+}, therefore the Chelex 100 method cannot be used. Apoenzyme preparations should be assayed for protein concentration, residual metal ion contents, residual activity, and their ability to be reconstituted upon addition of excess native metal ion. The apoenzyme should be titrated with the native metal ion to establish the stoichiometric relationship of metal binding and enzyme activity.[7–9] This procedure should also be performed for other metal ions to be used for reactivation. Since metal–protein dissociation constants are generally (but not always) below 10 n*M*, reconstitution experiments should employ reasonably large concentrations of apoenzyme if possible (at least 1 μM) to ensure maximum association of added metal ions. In cases where chromophoric metal ions are used to reactivate the apoenzyme, spectral titrations of the reconstitution should collaborate titrations based on activity.[8]

Crystalline apocarboxypeptidase A and metal-depleted horse liver alcohol dehydrogenase have been prepared by dialyzing holoenzyme crystals against chelating agent and subsequently against metal-free buffer (see below).

Removing Residual Enzyme from Apoenzyme

When successfully employed, the methods described above will yield apoenzyme with less than 0.1% of the original activity of the native en-

[14] H. Csopak, *Eur. J. Biochem.* **7**, 186 (1969).

zyme. With both carboxypeptidase A[15] and angiotensin converting enzyme,[16,17] affinity resins used to purify native enzyme can also be used to remove native enzyme from apoenzyme as only the former possesses appreciable affinity for the immobile ligand. While the utility of such a procedure is obvious, the apoenzyme will be diluted somewhat. Precautions must also be taken to render the affinity resin support metal free.

Examples of Apoenzyme Preparation

Enzyme Solutions. Aeromonas aminopeptidase, an unusually stable monomeric enzyme of 29,500 Da, binds 2 equivalents of zinc ion. The enzyme can be obtained in large quantities,[18] thus millimolar concentrations can be prepared with some effort. Concentrations of enzyme above 0.2 m*M* are used to prepare apoenzyme.[7,8]

The chelating agent 1,10-phenanthroline was chosen to prepare apoenzyme by comparing its effectiveness with that of dipicolinic acid.[7] For the experiment, samples of enzyme (3.26×10^{-9} *M*) are added to concentrations of chelating agent which vary from 0.1 μM to 20 m*M*. The resulting solutions are assayed immediately or between 15 and 30 min using L-leucine-*p*-nitroanilide as the substrate. As shown in Fig. 2, inhibition by 1,10-phenanthroline is immediate, while complete inhibition by dipicolinic acid is time dependent.

Apoaminopeptidase is routinely prepared by dialyzing approximately 0.5 m*M* native enzyme against 3 changes of 25 volumes of 2 m*M* 1,10-phenanthroline in 50 m*M* tricine containing 50 m*M* KCl, pH 7.5. Apoenzyme is freed from the chelating agent by dialysis against a minimum of 7 changes of 25 volumes of 50 m*M* HEPES (pH 7.5) over a 48-hr period. An absorption spectrum (400–200 nm) of the final dialyzing fluid is routinely measured to confirm the complete removal of residual 1,10-phenanthroline ($\varepsilon_{265} = 3.15 \times 10^4\ M^{-1}\ \text{cm}^{-1}$). Zinc content of the preparation, measured by atomic absorption spectroscopy, is usually 0.05–0.1 g-atom per mole of protein; however, this value is somewhat higher (0.12 g-atom/mole) when lower concentrations of enzyme are used. The residual enzyme activity is assessed using 37.5 μM L-alanine-*p*-nitroanilide in 50 m*M* HEPES, pH 7.5, as the substrate. As mentioned earlier, under these conditions substrate concentration is about 0.04 K_m and reaction rates are

[15] L. B. Cueni, T. A. Bazzone, J. F. Riordan, and B. L. Vallee, *Anal. Biochem.* **107,** 341 (1980).

[16] M. W. Pantoliano, B. Holmquist, and J. F. Riordan, *Biochemistry* **23,** 1037 (1984).

[17] J. F. Riordan, F. Lee, M. Pantoliano, and B. Holmquist, *in* "Frontiers of Bioinorganic Chemistry" (A. Xavier, ed.), p. 612. VCH, Weinheim, Federal Republic of Germany, 1986.

[18] J. M. Prescott and S. H. Wilkes, this series, Vol. 44, p. 530.

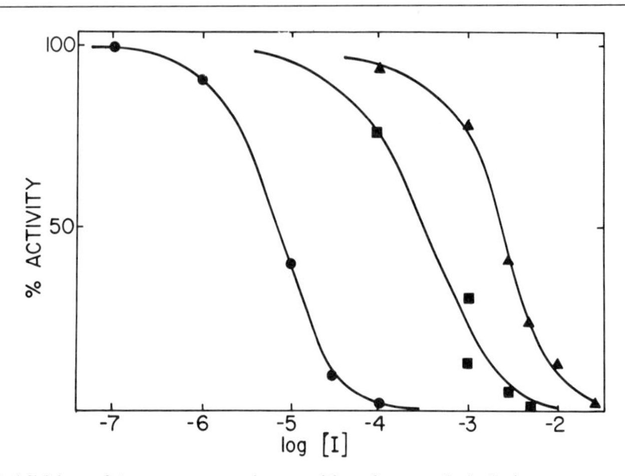

FIG. 2. Inhibition of *Aeromonas* aminopeptidase by metal-chelating reagents. Concentration dependence of inhibition by 1,10-phenanthroline (●) and by dipicolinic acid at zero (▲) and 17–28 min (■) incubation. In the experiments with 1,10-phenanthroline, samples of enzyme (3.26×10^{-9} *M*) were allowed to stand for 30 min in the concentrations of reagent shown. The enzyme concentration in the dipicolinic acid experiments was 1.43×10^{-9} *M*. First-order assays were started by adding L-leucine-*p*-nitroanilide to yield a final concentration of 5×10^{-6} *M* in 10 m*M* Tricine, pH 7.5. From Prescott *et al.*[7]

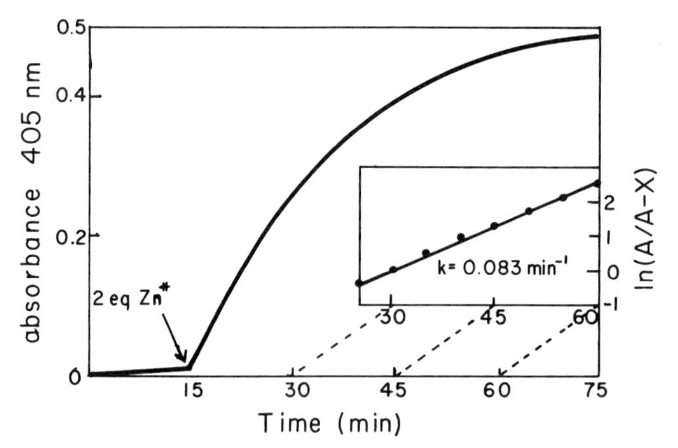

FIG. 3. Assay of apoaminopeptidase. The assay mixture contained 1 ml of 5×10^{-4} *M* L-alanine-*p*-nitroanilide and 3.57×10^{-6} *M* apoaminopeptidase in 10 m*M* HEPES, pH 7.5. The residual activity of the apoenzyme was monitored for 15 min then the reaction mixture was rendered 7.14×10^{-6} *M* with $ZnSO_4$. The limiting absorbance was approximately 0.5. The inset figure is a first-order plot of the data collected from 15 min until completion of the experiment. In the expression $\ln(A/A - X)$, A was assumed to be 0.5 and X was the change in absorbance at a given time, measured relative to the absorbance at 15 min. The first-order rate constant, 0.083 min^{-1} (measured as the slope), was divided by the enzyme concentration to calculate k_{cat}/K_m, which was 2.32×10^{-4} M^{-1} min^{-1}.

first order; however, enzyme can be assayed at a concentration of 1 μM. After residual rates are determined (Fig. 3), an amount of $ZnSO_4$ equivalent to the apoenzyme concentration is added to the assay mixture and the velocity of the reconstituted enzyme is measured. First-order rate constants are obtained by conventional plotting methods and values of k_{cat}/K_m can be estimated (Fig. 3). Using the same assay method and by adding successive increments of $ZnSO_4$, apoenzyme can be titrated to determine the Zn^{2+} stoichiometry for reactivation (Fig. 4).[7,8] Also shown in Fig. 4 is a spectral titration of apoaminopeptidase plotted with an activity titration using Co^{2+} to reconstitute apoenzyme.

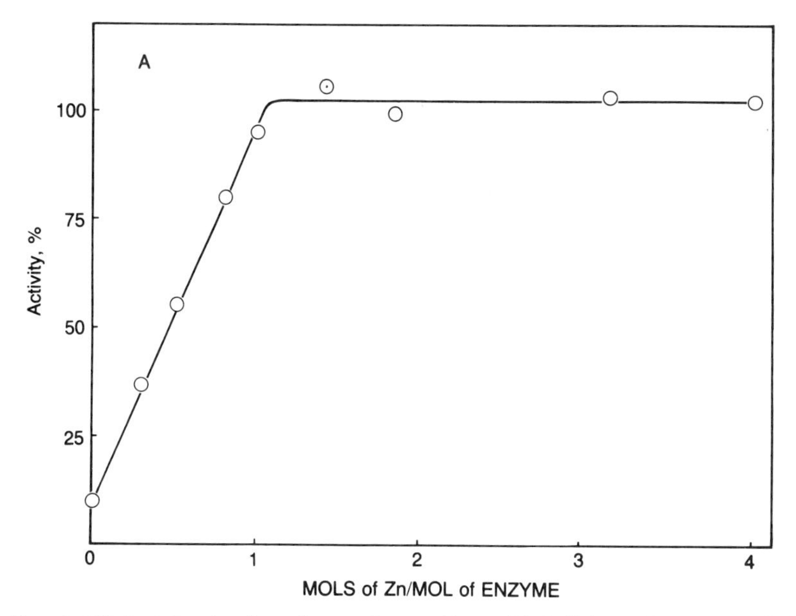

FIG. 4. (A) Activity titration of apoaminopeptidase with Zn^{2+} ions. Apoenzyme (8.5×10^{-7} M) was incubated for 1 hr at room temperature in 50 mM HEPES, pH 7.5, with Zn^{2+} ions at the molar ratios indicated. Reactions were started by adding sufficient L-alanine-*p*-nitroanilide to yield a concentration of 43 μM in the reaction mixture. Activities are expressed relative to that of the sample containing 2 mol/mol. (B) Spectroscopic (○) and activity (△) titrations of apoaminopeptidase by Co^{2+}. In the spectroscopic titration, aliquots of Co^{2+} were added sequentially to apoenzyme (2.39×10^{-4} M), and the visible region of the spectrum was scanned after each addition. Corrections were made for absorption by unbound Co^{2+} ($\varepsilon = 4.8$ M^{-1} cm^{-1}) above 2 mol/mol. The left-hand ordinate is the molar absorptivity at 527 nm. Enzyme activity was titrated by mixing apoenzyme (1×10^{-6} M) and substrate (43 μM L-alanine-*p*-nitroanilide) and then adding Co^{2+} to yield the molar ratios shown. Activities (right-hand ordinate) are expressed as percentage of the activity of the enzyme fully reconstituted with Co^{2+}. From Prescott *et al.*[8]

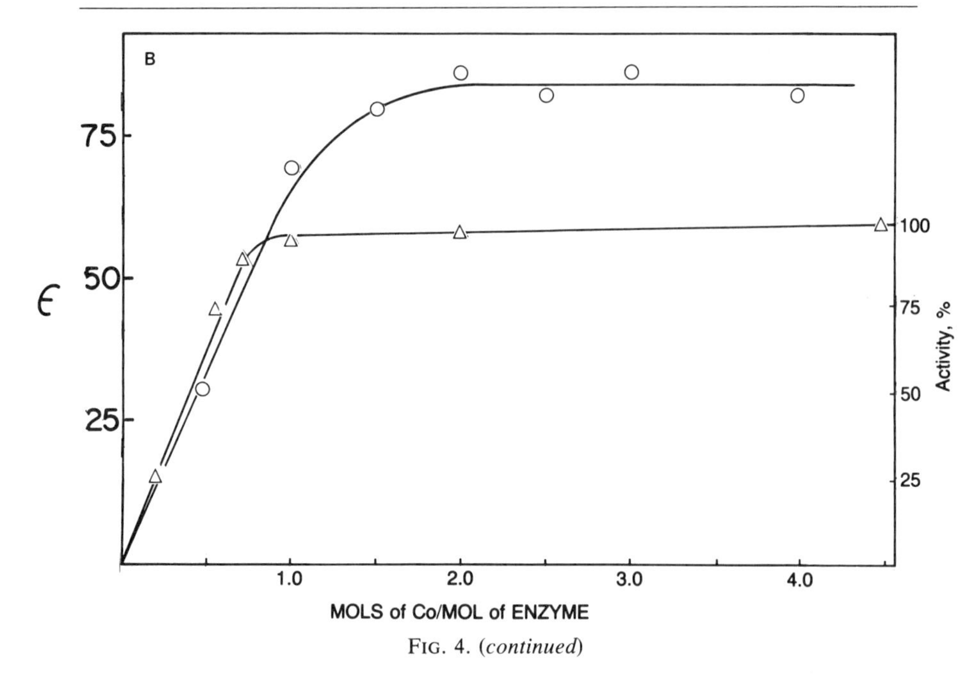

FIG. 4. (*continued*)

Crystalline Enzymes

Apocarboxypeptidase A. Apocarboxypeptidase A was originally prepared in the aqueous state by a procedure[19] analogous to the one described for *Aeromonas* aminopeptidase. Currently, the preferred method involves removing native metal ion from the crystalline enzyme.[20] Additionally, metal substitutions can be performed with the crystalline apoenzyme. The advantage of this procedure is that it provides maximally concentrated enzyme to be treated with minimum exposure to adventitious metal ions.

A suspension of carboxypeptidase A crystals (5 mg/ml), prepared either by the method of Anson[21] or Cox *et al.*,[22] is suspended in 10 m*M* 1,10-phenanthroline, dissolved in 1 m*M* MES at pH 7.0 for 1 hr. The chelation treatment is repeated 3 times, then the crystals of apocarboxypeptidase A are washed for 30 min with at least 4 changes of metal-free MES (1 m*M*) pH 7.0. Apoenzyme dissolved in 50 m*M* HEPES containing 1 *M* Nacl and assayed, using furylacryloylglycyl-L-phenylalanine,[23] should possess 3%

[19] J. E. Coleman and B. L. Vallee, *J. Biol. Chem.* **235,** 390 (1960).

[20] D. W. Auld and B. Holmquist, *Biochemistry* **13,** 4355 (1974).

[21] M. L. Anson, *J. Gen. Physiol.* **20,** 633 (1937).

[22] D. J. Cox, F. C. Bovard, J.-P. Borgetzi, K. A. Walsh, and H. Nemath, *Biochemistry* **3,** 44 (1964).

[23] L. M. Peterson, B. Holmquist, and J. L. Bethune, *Anal. Biochem.* **125,** 420 (1982).

TABLE I
METALLOENZYMES FROM WHICH APOENZYMES HAVE BEEN PREPARED

Enzyme	Number of subunits	Metal/subunit	Chelating agent used to prepare apoenzyme	Refs.
Aeromonas aminopeptidase	1	2 Zn^{2+}	OP[b]	*a*
Yeast (Class II) aldolase	2	1 Zn^{2+}	EDTA	*c*
Alkaline phospatase (*E. coli*)	2	2 Zn^{2+}, 1 Mg^{2+}	OP, EDTA Chelex 100[g]	*d–f*
Angiotensin converting enzyme	1	1 Zn^{2+}	OP	*h*
Bacillus subtilis neutral protease	1	1 Zn^{2+}	EDTA	*i*
Carbonate dehydratase	1	1 Zn^{2+}	OP, DPA	*j, k*
Carboxypeptidase A	1	1 Zn^{2+}	OP	*l, m*
Carboxypeptidase B	1	1 Zn^{2+}	OP	*n*
Horse liver alcohol dehydrogenase	2	2 Zn^{2+}	DPA	*o*
Bovine lens aminopeptidase	6	2 Zn^{2+}	OP	*p*
Procarboxypeptidase A	1	1 Zn^{2+}	OP	*q*
Steptomyces griseus carboxypeptidase	1	1 Zn^{2+}	OP	*r*
Superoxide dimutase	2	2 Cu^{2+}, 1 Zn^{2+}	EDTA	*s*
Thermolysin	1	1 Zn^{2+}, 1 Ca^{2+}	OP	*t*

[a] J. M. Prescott, F. W. Wagner, B. Holmquist, and B. L. Vallee, *Biochem. Biophys. Res. Commun.* **114,** 646 (1983).

[b] OP, 1,10-phenanthroline; DPA, dipicolinic acid.

[c] R. D. Kobes, R. T. Simpson, B. L. Vallee, and W. J. Rutter, *Biochemistry* **8,** 585 (1969).

[d] W. F. Bosron, R. A. Anderson, M. C. Falk, F. S. Kennedy, and B. L. Vallee, *Biochemistry* **16,** 610 (1977).

[e] E. Sabbioni, F. Girardi, and E. Marafante, *Biochemistry* **15,** 271 (1976).

[f] H. Csopak, *Eur. J. Biochem.* **7,** 186 (1969).

[g] Metals can also be removed by dialysis against $(NH_4)_2SO_4$ [P. Gettins and J. E. Coleman, *J. Biol. Chem.* **259,** 4991 (1984)].

[h] P. Buenning and J. F. Riordan, *Bioinorg. Chem.* **24,** 183 (1985).

[i] J. D. McConn, P. Tsuru, and K. T. Yasunobu, *J. Biol. Chem.* **239,** 3706 (1964).

[j] S. Lindskog and B. G. Malmström, *J. Biol. Chem.* **237,** 1129 (1962).

[k] J. B. Hunt, M. J. Rhee, and C. B. Storm, *Anal. Biochem.* **79,** 617 (1977).

[l] J. E. Coleman and B. L. Vallee, *J. Biol. Chem.* **235,** 390 (1960).

[m] D. W. Auld and B. Holmquist, *Biochemistry* **13,** 4355 (1974).

[n] E. Wintersberger, H. Neurath, T. L. Coombs, and B. L. Vallee, *Biochemistry* **4,** 1526 (1965).

[o] W. Maret, I. Anderson, H. Dietrich, H. Schneider-Bernlohr, R. Einarsson, and M. Zeppezauer, *Eur. J. Biochem.* **98,** 501 (1979).

[p] H. Hanson and M. Frohne, this series, Vol. 44, p. 504.

[q] W. D. Behnke and B. L. Vallee, *Proc. Natl. Acad. Sci. U.S.A.* **69,** 2442 (1972).

[r] K. Breddam, T. J. Bazzone, B. Holmquist, and B. L. Vallee, *Biochemistry* **18,** 1563 (1979).

[s] J. M. McCord and I. Fridovich, *J. Biol. Chem.* **244,** 6049 (1969).

[t] B. Holmquist and B. L. Vallee, *J. Biol. Chem.* **249,** 4601 (1974).

of its original activity and no greater than 0.003 g-atom Zn^{2+}/mol of protein. Apoenzyme can be reconstituted by the addition of a 3-fold molar excess of the desired metal ion followed by at elast 3 washings with metal-free buffer.

Metal-Depleted Horse Liver Alcohol Dehydrogenase. Horse liver alcohol dehydrogenase is a dimeric enzyme that binds 4 g-atom Zn^{2+}/mol of protein. Each subunit contains one catalytically functional metal ion chelated to the protein by three cysteine residues. The second Zn^{2+} is 2 nm from the active site and is thought to confer structural stability to the protein.[24] Completely metal-free alcohol dehdyrogenase cannot be prepared, but the catalytically essential Zn ions can be removed from alcohol dehydrogenase solutions or crystals under anaerobic conditions using dipicolinic acid.

Commercially available enzyme is crystallized at pH 8.4 using 2-methyl-2,4-pentanediol as precipitant or at pH 6.9 in 50 m*M* TES containing 25% *tert*-butanol. Metal depleted enzyme is prepared by adding 5 m*M* dipicolinic acid to the mother liquor and dialyzing the crystals until the specific activity of the enzyme is less than 0.5% of the original value. Chelating agent is removed by dialysis against buffer with the appropriate precipitant. Apoenzyme prepared in such a fashion contains a full complement of structural but not catalytic Zn^{2+}. Reconstitution of the "apoenzyme" with Co^{2+} yields a hybrid metal enzyme containing two catalytically functional Co^{2+} ions and two structural Zn^{2+} ions. Methods to prepare completely metal-free apoalcohol dehydrogenase capable of being reconstituted with metal ions have not been reported, however, structural Zn ions may be exchanged for other ions by procedures which do not involve preparing apoenzymes.[25]

Other Apoenzymes

Table I gives a list of metalloenzymes from which apoenzymes have been prepared. Also included are abbreviated comments on general methods used to prepare the apoenzyme and metal ions used to reconstitute the apoenzymes.

Acknowledgments

The author is indebted to his colleagues, J. M. Prescott, B. Holmquist, S. W. Wilkes, M. E. Bayliss, and R. de la Motte, for their assistance during the preparation of this manuscript.

[24] W. Maret, I. Andersson, H. Dietrich, H. Schneider-Bernlohr, R. Einarsson, and M. Zeppezauer, *Eur. J. Biochem.* **98,** 501 (1979).

[25] A. J. Sytkowski and B. L. Vallee, *Biochemistry* **17,** 2850 (1978).

[6] Metal-Buffered Systems

By JOHN O. BAKER

A number of enzymes bind catalytically essential or regulatory divalent metal ions tightly enough that the metal ions remain with the enzyme throughout the course of purification. These metal-binding enzymes, termed metalloenzymes[1,2] to distinguish them from the less-stable metal–enzyme complexes which dissociate during purification, may have metal dissociation constants of 10^{-8} to 10^{-10} *M* or even smaller. The measurement of binding constants of this order of magnitude will require the maintenance in the solution of accurately known "free" metal ion concentrations well below the total concentration of the ion in question even in carefully extracted solutions, and below the limit of detection for a number of biochemically important metal ions. Maintenance of such low metal ion concentrations is for all practical purposes impossible without the use of metal ion buffer systems, in which the vast majority of metal ions are present as complexes with an organic sequestering agent (ligand). The small proportion of metal ion present as the "free" ion is calculated from the known total concentration of ligand, the total concentration of metal ion in all species, and the stability constants for interaction of the metal ion and ligand under the conditions of the experiment.

In addition to controlling the "free" concentration of a particular metal ion, a well-designed metal ion buffer system can provide other advantages. For instance, provided the ligand chosen has significant affinity for a wide variety of metal ions, the free concentrations of adventitious, contaminating metal ions will be reduced far below the "free" concentration of the desired metal ion, which is added in substantial quantities in construction of the buffer system.[3] This feature of metal ion buffer systems should not be seen as removing the necessity for taking the standard precautions against metal ion contamination,[4] but it can reduce the effects of any contamination occurring despite such precautions.

Use of a metal ion buffer system of fairly high capacity can also guard against other types of interferences. Many of the molecules used as hy-

[1] B. L. Vallee and W. E. C. Wacker, *in* "The Proteins" (H. Neurath, ed.), Vol. 5. Academic Press, New York, 1971.

[2] J. F. Morrison, this series, Vol. 63, p. 257.

[3] L. L. Hendrickson, M. A. Turner, and R. B. Corey, *Anal. Chem.* **54,** 1633 (1982).

[4] R. T. Thiers, *Methods Biochem. Anal.* **5,** 273 (1957).

drogen ion buffers also have significant affinity for metal ions,[2,5–8] as do many enzyme substrates and inhibitors. Even inorganic salts such as chloride, if present in sufficient quantities, can form complexes with metal ions to an extent sufficient to perturb the "free" metal ion concentration.[6] If, however, a metal ion buffer system of sufficiently high capacity is used, the quantity of a given metal complexed with the metal ion buffer ligand will be much larger than the total of the quantities of metal sequestered by other molecules present; the effects on the free metal ion concentration of the hydrogen ion buffers, substrates, inhibitors, inorganic salts, etc., will then be neligibile. The effect of the chosen metal ion buffer on free metal ion concentration will still have to be calculated, but the investigator now has only to make calculations based on a single, presumably well-characterized set of metal–ligand interactions, rather than being confronted with the necessity of evaluating the effects of a number of different interactions.

One type of experiment in which metal ion buffers are especially useful is that in which metal–protein interactions are studied by means of activity measurements in the presence of different levels of substrate and metal ion. If the enzyme–metal dissociation constants involved are such that the required metal ion concentrations are 10^{-5} *M* or higher, such experiments can be performed without metal ion buffers.[2,9,10] If, on the other hand, the required free metal ion concentrations are on the order of 10^{-8} to 10^{-10} *M*, there is essentially no way in which the experiments can be run meaningfully without the use of metal ion buffers.

Other applications include maintaining optimal metal ion concentrations in standard assays for a given enzyme, in the face of widely varying contents of metal ion and potential chelating agents in the solutions to be tested,[11] and the provision of higher total solution content of a given metal ion that would otherwise be permitted by the relative insolubility of a hydroxo complex.

The choice of sequestering agent for use as a metal ion buffer will depend on two principal criteria. First, the stability constants for complexation with the metal ion of interest, and the solubilities of the free ligand and its complexes with the metal, should be such that one can obtain both the desired range of free metal ion concentration and ade-

[5] C. Veillon, this volume [7].

[6] J. E. Coleman and B. L. Vallee, *J. Biol. Chem.* **236,** 2244 (1961).

[7] P. Vieles, C. Frezou, J. Galsomias, and A. Bonniol, *J. Chim. Phys. Phys.-Chim. Biol.* **69,** 869 (1972).

[8] J. M. Prescott, F. W. Wagner, B. Holmquist, and B. L. Vallee, *Biochemistry* **24,** 5350 (1985).

[9] S. Lindskog and P. O. Nyman, *Biochim. Biophys. Acta* **85,** 462 (1964).

[10] J. O. Baker and J. M. Prescott, *Biochemistry* **22,** 5322 (1983).

[11] J. Raaflaub, *Methods Biochem. Anal.* **3,** 301 (1956).

quate buffering capacity. Second, the metal ion buffer ideally should exert its effect *only* by controlling the free metal ion concentration in the solution, and should not interact with the enzyme in any way that will perturb the equilibrium between enzyme and metal, or between enzyme and inhibitor or substrate.

In chemical terms, most substances used as soluble metal ion buffers fall into one of two major categories: the carboxylic acid buffers and the polynuclear aromatic hydrocarbon buffers. The carboxylic acid buffers include compounds such as ethylenediaminetetracetic acid (EDTA), ethylenebis(oxyethylenenitrilo)tetraacetic acid (EGTA), 2-*N*-hydroxyethylethylenediaminetriacetic acid (HEDTA), and nitrilotriacetic acid (NTA). Prominent among the polynuclear heteroaromatic hydrocarbon buffers are 1,10-phenanthroline (OP), 8-hydroxyquinoline-5-sulfonic acid (HQSA), 2,2′-dipyridyl, and 2,2′ : 6′,2′-terpyridine (terpy). In addition to the ability to complex with metal ions, the carboxylic acid–metal ion buffers have considerable potential for both ionic and hydrogen bonding, and the aromatic hydrocarbon buffers are capable of hydrophobic interactions. The meeting of the second criterion listed above should therefore not be taken for granted, but should be tested by procedures to be detailed later in this chapter.

A number of procedures have been developed for the use of insoluble metal-chelating agents as metal ion buffers.[3,12–15] Most notable are those in which calcium ion concentrations in the 10^{-3} to 10^{-6} *M* range are controlled by use of Chelex 100, a cross-linked polystyrene resin having iminodiacetic acid functional groups. Insoluble buffers have certain advantages; in particular, in equilibrium dialysis experiments direct contact between enzyme and metal ion buffer can be prevented simply by placing enzyme and metal ion buffer on opposite sides of the membrane. The present discussion will be confined, however, to the use of the soluble buffers, for which quantitation is more straightforward, the necessary physicochemical parameters have been worked out more precisely, and the metal/ligand stability constants are more appropriate for work with metal ions "tightly bound"[1,2] to proteins.

Equilibrium Constant Nomenclature

Both dissociation constants and stability (formation) constants appear in this chapter. For all interactions involving enzymes the equilibrium

[12] R. H. Wasserman, R. A. Corradino, and A. N. Taylor, *J. Biol. Chem.* **243,** 3978 (1968).
[13] N. Briggs and M. Fleischman, *J. Gen. Physiol.* **49,** 131 (1965).
[14] L. L. Hendrickson and R. B. Corey, *Soil Sci. Soc. Am. J.* **47,** 467 (1983).
[15] D. M. Waisman and H. Rasmussen, *Cell Calcium* **4,** 89 (1983).

constants are written as dissociation constants, after the convention almost invariably followed for interactions between enzymes and substrates or inhibitors. Protonation/deprotonation equilibria are likewise described by dissociation constants (K_a). Interactions between metal ions and nonprotein ligands (metal ion buffer ligands in the present case) are on the other hand given as stability (formation) constants (K_1, K_2, K_3, β_2, β_3), in accordance with the usual literature practice in describing these interactions. The dissociation constants for these latter interactions can be obtained simply by taking the reciprocal of the corresponding stability constant.

Calculation of Free Metal Ion Concentration

Of the carboxylic acid buffers, a number are hexadentate and thus tend to form only 1 : 1 complexes with metal ions since they are capable of filling all coordination positions for a metal ion prone to octahedral coordination geometry.[1] For these buffer systems, calculation of free metal ion concentration is straightforward. The concentration of free metal ion (denoted [M] in this chapter) can be obtained from Eq. (1a), which in its logarithmic form [Eq. (1b)] is analogous to the familiar Henderson–Hasselbalch equation for hydrogen ion buffers.

$$[\mathrm{M}] = \left(\frac{1}{K_1}\right)\left(\frac{[\mathrm{ML}]}{[\mathrm{L}]}\right) \tag{1a}$$

$$\mathrm{pM} = \log K_1 + \log([\mathrm{L}]/[\mathrm{ML}]) \tag{1b}$$

In Eqs. (1a) and (1b), K_1 is the formation, or stability constant for formation of the 1 : 1 complex ML from metal and ligand (L), and [L] refers to the concentration of free ligand in the solution.

Nitrilotriacetic acid resembles EDTA chemically, but is only tetradentate. Terpyridine is tridentate, and the rest of the aromatic hydrocarbon buffers listed earlier are bidentate. For these buffers, calculation of free metal ion concentration is more complicated. Because more than one ligand molecule can bind simultaneously to a single metal ion in these systems, the expression for free metal concentration will involve terms second order (in the case of tetra- and tridentate ligands) and third order (for bidentate ligands) in free ligand concentration. For a bidentate ligand capable of forming complexes with up to 3 : 1 ligand : metal stoichiometry [Eq. (2)],

$$\mathrm{M} \underset{}{\overset{K_1}{\rightleftharpoons}} \mathrm{ML} \overset{K_2}{\rightleftharpoons} \mathrm{ML_2} \overset{K_3}{\rightleftharpoons} \mathrm{ML_3} \tag{2}$$

the conservation equation for total metal (M_t) can be written

$$[M_t] = [M] + [ML] + [ML_2] + [ML_3] \quad (3)$$

Given the relationships in Eqs. (4a)–(4c)

$$K_1 = \frac{[ML]}{[M] \times [L]} \quad (4a)$$

$$K_2 = \frac{[ML_2]}{[ML] \times [L]} \quad (4b)$$

$$K_3 = \frac{[ML_3]}{[ML_2]\,[L]} \quad (4c)$$

Eq. (3) can be rewritten

$$[M_t] = [M] + K_1[M][L] + K_1K_2[M][L]^2 + K_1K_2K_3[M][L]^3 \quad (5)$$

and the "free" metal concentration, [M], will be given by

$$[M] = \frac{[M_t]}{1 + K_1[L] + K_1K_2[L]^2 + K_1K_2K_3[L]^3} \quad (6a)$$

When the substitutions $\beta_2 = K_1K_2$ and $\beta_3 = K_1K_2K_3$ are made, Eq. (6a) becomes

$$[M] = [M_t]/(1 + K_1[L] + \beta_2[L]^2 + \beta_3[L]^3) \quad (6b)$$

Utility of this equation is limited by the fact that the ligand concentration [L] used in the equation is the equilibrium free ligand concentration and not the concentration of ligand originally added ($[L_t]$).

Rewritten in terms of $[L_t]$, Eq. (6b) becomes

$$[M] = \frac{[M_t]}{1 + K_1([L_t] - [L_{bound}]) + \beta_2([L_t] - [L_{bound}])^2 + \beta_3([L_t] - L_{bound})^3} \quad (6c)$$

For systems in which the highest ligand : metal binding stoichiometry is 2 : 1, the term third order in [L] is absent from Eq. (6c) and the remaining quadratic equation can be solved for [M].[16] For systems that do form the complex ML_3, and thus require the solution of a cubic equation, [M] is most conveniently calculated by means of an iterative computer routine. Table I presents the essential elements of a BASIC program which carries out the required successive approximations, for all values of $[M_t]$ and $[L_t]$ such that $[M_t] < [L_t]$. The program is written in Tektronix BASIC for the

[16] P. Aisen, A. Leibman, and J. Zweier, *J. Biol. Chem.* **253,** 1930 (1978).

TABLE I

BASIC PROGRAM FOR VALUES OF $[M_t]$ AND $[L_t]$ FOR $[M_t] < [L_t]$

```
95 PRINT "PROGRAM TO CALCULATE Mfree ";
96 PRINT "FROM Mtotal AND Ltotal (Lt)"
97 PRINT "(Mtotal MUST BE SMALLER THAN Ltotal)"
98 PRINT "   "
100 CALL "CMFLAG",2,3
110 PRINT @37,0:13,28,255
120 CALL "RATE",9600,0,2
130 PRINT "ENTER LOGS OF STABILITY CONSTANTS";
140 PRINT " LOG K1, LOG BETA-2, LOG BETA-3)J"
150 INPUT Lsk1,Lsb2,Lsb3
160 PRINT @40:"J"
170 PRINT @40:"LOG K1 = ";Lsk1;"   LOG BETA-2 = ";Lsb2;
180 PRINT @40:"   LOG BETA-3 = ";Lsb3,"J"
190 K1=10^Lsk1
200 B2=10^Lsb2
210 B3=10^Lsb3
220 PRINT @40:"   Ltotal","   Mtotal","  Lfree","   Mfree"
230 PRINT "ENTER LTOTAL (MOLAR)"
240 INPUT Lt
250 PRINT "ENTER MTOTAL (MOLAR)"
260 INPUT Mt
270 IF Mt/Lt>0.33 THEN 330
280 Lprev=Lt
290 Bite=1
300 Diffcrit=1.0E-4
310 Mprev=Mt
320 GO TO 360
330 Lprev=1.0E-6
340 Diffcrit=1.0E-7
350 Bite=0.005
360 Mprev=Mt
370 REM *** ITERATION LOOP BEGINS ***
380 Mfree=Mt/(1+K1*Lprev+B2*Lprev^2+B3*Lprev^3)
390 M1=Mfree*K1*Lprev
400 M12=Mfree*B2*Lprev^2
410 M13=Mfree*B3*Lprev^3
420 Lfree=Lt-M1-2*M12-3*M13
430 Ldiff=Lfree-Lprev
440 Lprev=Lprev+Bite*Ldiff
450 REM ** PRINT @40:Lprev,Lfree,Ldiff,Mfree
460 Mchange=ABS((Mfree-Mprev)/Mprev)
470 IF Mchange<Diffcrit THEN 510
480 Mprev=Mfree
490 GO TO 370
500 REM *** LINE ABOVE ENDS ITERATION LOOP ***
510 PRINT @40:Lt,Mt,Lfree,Mfree
520 PRINT "DO ANOTHER CALCULATION? (1=YES,0=NO)"
530 INPUT Yesorno
540 IF Yesorno=0 THEN 590
550 PRINT "NEW STABILITY CONSTANTS?(1=YES,0=NO)"
560 INPUT Yesorno
570 IF Yesorno=1 THEN 130
580 GO TO 230
590 END
```

Tektronix 4052 and should be readily adaptable to other BASIC languages. The program initially sets the value of [L] equal to [L_t] (i.e., [L_{bound}] is taken as 0), and calculates first a value of [M] (= free metal ion). Then, using this value of [M] and the previously assumed value of [L], the program calculates [L_{bound}]. A new provisional value of [L] is then calculated as [L] = [L_t] − [L_{bound}], and [M_{free}] is recalculated using this value of [L]. The process is repeated until the fractional difference between [M], the most recently calculated value of free metal, and [M_{prev}], the value of [M_{free}] obtained in the preceding cycle, is less than the fractional difference criterion, DIFFCRIT. With the provision of a line printer (address "40" in this program), this program produces a table with columns for [L_t], [M_t], [L], and [M], and identifies the stability constants used for each batch of calculations.

Buffer Capacity

In buffer systems in general, the buffer capacity can be evaluated by comparing the concentration of the "minority" constituent of the buffer system with the concentrations in the solution of species that may challenge control by the buffer. In metal ion buffer systems the lower limit for [M_t] is set by the requirement that the concentration of added metal ion be sufficiently larger than the sum of concentrations of adventitious metal ions and of adventitious strong chelators, that the effect of these interfering species on the "free" concentration of M can be neglected. The upper limit for the ratio of [M_t] to [L_t] is set by the need to leave free in solution a concentration of ligand that is also substantially larger than adventitious ions and other challenging species. In this regard, systems utilizing ligands capable of forming 2 : 1, 3 : 1, or higher ligand : metal complexes operate under constraints not encountered with hydrogen ion buffers or "1 : 1" metal ion buffers. With the "1 : 1" buffers the investigator has the option of simply increasing the total concentration of the buffer in order to increase the capacity at a value of pH or pM somewhat removed from the pH or pM of maximum buffering capacity. An attempt to make use of this simple expedient with a 2 : 1 or 3 : 1 metal ion buffer system encounters a complication due to the ability of ML (and ML_2 as well for a 3 : 1 buffer) to form complexes with additional ligand molecule(s)/ion(s) [Eq. (2)]. As the concentrations of ML and ML_2 approach and exceed the reciprocals of the values of K_2 and K_3, respectively [Eqs. (4a–c)], these species will begin to scavenge the remaining free ligand from the solution. Owing to the dependence of free metal ion concentration on the square and cube of free ligand concentration, an attempt to maintain the free metal ion concentration constant while increasing M_t and L_t will result in a dramatic increase in the ratio of bound to free ligand. This in turn will cause the free metal ion concentration to become extremely sensitive to uncertainities in the values of the metal–ligand stability constants used (see the following section). Thus, although the computer program described above and shown in Table I will calculate free metal ion concentrations based on combinations in which the free ligand concentration is quite small, it is prudent to keep the free ligand concentration at least as high as $0.1 \times [M_t]$ and preferably considerably higher, for reasons outlined in the following section.

Sources of Uncertainty in Calculated Free Metal Ion Concentrations

Stability Constants

The stability constants used in Eqs. (6a) through (6c) are generally reported in logarithmic form with one- or two-digit mantissas; the uncertainties in the values, where given, are rarely much better than a factor of 1.259, or $\text{antilog}_{10}(\pm 0.1)$.[17] As Eq. (6c) makes clear, there are two ways in which the uncertainty in the values of stability constants can lead to uncertainty in the calculated free metal ion concentrations. First, for a given free ligand concentration [L], any error in the values of K_1, β_2, or β_3 will naturally result in an error in the calculated free metal ion concentration [M]. Second, for given values of $[M_t]$ and $[L_t]$, errors in the stability constants will also result in errors in the concentration of free ligand, [L]. The situation is ameliorated somewhat by the fact that these errors tend to have opposite effects on the calculated free metal concentration. The best strategy is to minimize the effect of stability constant uncertainties on free ligand concentration by making the ratio of $[M_t]$ to $[L_t]$ as small as is possible under the other design constraints on the system. If the magnitude of $[L_{bound}]$ can be made negligible with respect to $[L_t]$, one can establish a range of free metal ion concentrations over which the free metal ion concentration is a virtually linear function of $[M_t]$. Even if the exact proportionality between [M] and $[M_t]$ is not known, the establishment of such a linear range can be of tremendous value.

An example of the use of such a linear range can be found in the study of inhibition of phosphoglucomutase by Zn^{2+} ion.[18] By using a rather high concentration of imidazole (32 m*M*) as a metal ion buffer, along with total added zinc concentrations ranging from 2.5×10^{-5} to 2.5×10^{-4} *M*, Ray[18] was able to establish a range of free Zn^{2+} concentrations directly propor-

[17] R. M. Smith and A. E. Martell, "Critical Stability Constants," Vol. 2. Plenum Press New York, 1975.

[18] W. J. Ray, Jr., *J. Biol. Chem.* **242,** 3737 (1967).

tional to total zinc, with $[Zn^{2+}_{free}] = 4 \times 10^{-6}\ [Zn_t]$. Using this metal ion buffer system, the effect of substrate on the inhibitory binding of Zn^{2+} ion to phosphoglucomutase was demonstrated over a range of free Zn^{2+} concentrations from 1.0×10^{-10} to 1.0×10^{-9} *M*.

Effects of pH

Since over the pH range of biochemical interest, most metal ion buffers are also capable of binding protons in place of metal ions, account must be taken of the pH dependence of the equilibria discussed previously. Terms for the complexes involving protons could be written into the conservation equations, but the effects of pH are most conveniently handled by adjusting the values of the stability constants to give "conditional," pH-dependent constants.

Where protonation of a ligand atom prevents complexation with the metal ion, the values of the conditional, pH-dependent constants may be obtained from the absolute constants found in standard tables[14] by means of Eqs. (7a–c),

$$K_1' = K_1(\text{abs})/(1 + [\text{H}]/K_a) \tag{7a}$$
$$\beta_2' = \beta_2(\text{abs})/(1 + [\text{H}]/K_a)^2 \tag{7b}$$
$$\beta_3' = \beta_3(\text{abs})/(1 + [\text{H}]/K_a)^3 \tag{7c}$$

in which the factor $1/(1 + [\text{H}]/K_a)$ represents the fraction of total ligand that is in the form (i.e., unprotonated) that is capable of binding to the metal ion.

Effects of Temperature

Both the metal–ligand stability constants and the acid dissociation constants for protonated ligand are temperature dependent, so temperature effects must also be calculated. One of the advantages of using a physicochemically well-characterized compound, such as 1,10-phenanthroline, is that ionization–enthalpy values have been tabulated, both for H^+ ion and for a large number of metal ions, over the temperature range of greatest interest to enzymologists.[17,19]

In addition to the influences of pH and temperature, the effective values of metal–ligand stability constants are also functions of the ionic strength of the experimental mixture and of the nature of the electrolyte used to support the ionic strength. A detailed discussion of all four of these factors can be found in Ref. 7.

[19] D. J. Eatough, *Anal. Chem.* **42,** 635 (1970).

Complications Due to Binding of Metal–Buffer Ligands to Active Sites of Enzymes

The objective of binding studies utilizing metal ion buffers is to measure the equilibrium between metalloenzyme, apoenzyme, and metal ion, and the relationship of this equilibrium to activity and to enzyme–substrate and enzyme–inhibitor equilibria. Ideally, the metal ion buffer will exert its effect on these equilibria solely through control of the free metal ion concentration, without interacting in any other way with the enzyme or substrates/inhibitors, as shown in Eq. (8a). In actuality, the ability of the ligand species to bind metal ions means that, steric conditions permitting, it may also bind to the enzyme-bound active-site metal ion. This additional interaction is shown in Eq. (8b).

$$\mathrm{EM} \underset{}{\overset{K_{\mathrm{Me}}}{\rightleftharpoons}} \mathrm{E} + \mathrm{M} \rightleftharpoons \mathrm{E} + \mathrm{ML}_n \tag{8a}$$

$$\begin{array}{ccccc} \mathrm{EM} & \overset{K_{\mathrm{Me}}}{\rightleftharpoons} & \mathrm{E} + \mathrm{M} & \rightleftharpoons & \mathrm{E} + \mathrm{ML}_n \\ K_{\mathrm{I}} \Big\updownarrow & & \Big\updownarrow & & \\ \mathrm{EML} & \overset{K_{\mathrm{MeL}}}{\rightleftharpoons} & \mathrm{E} + \mathrm{ML} & \rightleftharpoons & \mathrm{E} + \mathrm{ML}_n \end{array} \tag{8b}$$

Here we consider the case in which the buffer ligand in question is bound to the metal ion in the active site of the enzyme, and prevents binding of at least one of the substrates.[1] The equilibrium constant describing the interaction of buffer ligand and enzyme-bound metal is written as K_{I}, the equilibrium constant for dissociation of L from the enzymatically inactive ternary complex EML. The formation of EML may perturb the equilibrium between enzyme and metal ion, and will certainly perturb substrate kinetics in experiments designed to measure the effect of free metal ion concentration upon activity. The occurrence or nonoccurrence of such perturbations in a given system of metal ion buffer and enzyme will affect both the choice of chelator used as buffer in a given system and the interpretation of the results of the experiments.

An entire range of behavior can be postulated from the scheme of Eq. (8b), depending on the values assigned to the various equilibrium constants. One combination, in which K_{I} is extremely large ($K_{\mathrm{I}} \gg [\mathrm{L}]$) and K_{MeL} is extremely small, approaches the situation in Eq. (8a), in which the metal ion buffer is without additional effect on the enzyme. Another possibility is that K_{I} is *not* large with respect to the concentration of free L used in the experiment, and EML is formed to a sufficient extent that dissociation of the binary complex, ML, from this ternary complex constitutes a significant pathway for removal of metal from the enzyme.[20,21]

[20] M. F. Dunn, S. E. Pattison, M. C. Storm, and E. Quiel, *Biochemistry* **19,** 718 (1980).
[21] P. Bünning and J. F. Riordan, *J. Inorg. Biochem.* **24,** 183 (1985).

In still another scenario, the EML complex may be formed, but may not dissociate to form E + ML. In this latter case, the binding of L to the enzyme-bound metal will not promote the removal of metal from the enzyme, but may actually perturb the enzyme–metal equilibrium in the direction of holoenzyme, if other interactions of ligand with the active site can add to the metal–ligand interaction to make the binding of metal and ligand to enzyme significantly cooperative.

From the foregoing, it is obviously essential that an investigator understand whether or not a ternary complex is formed, before confidence can be placed in enzyme/metal stability constants measured in experiments using metal ion buffers. This distinction can be made by several procedures discussed below.

Ternary Complex Formation: Diagnostic Procedures under Conditions of Enzyme/Metal Equilibrium

A number of diagnostic approaches are available for determining the mode(s) of inhibition and/or inactivation of metalloenzymes by metal ion-sequestering agents. Perhaps the most direct is an initial rate method in which enzyme is maintained under conditions such that it initially has a full complement of tightly bound metal ion, before being mixed with substrate solution containing the metal ion buffer. If the metal ion is removed from the enzyme, and if removal is not so fast as to appear instantaneous on the time scale of the assay, then one may be able to distinguish between an initial velocity measured before significant metal removal has occurred, and the velocity reached after enzyme, metal ion buffer, and metal ion have equilibrated. The type of primary data to be expected in such experiments is shown in Fig. 1. Comparison of the initial velocities obtained in the presence and absence of metal ion buffer but under otherwise identical conditions will indicate whether there is formation of ternary complex to result in "instantaneous, reversible inhibition" of the enzyme reaction. In especially favorable cases, it may be possible to determine the value of K_I for the interaction from these initial velocities.

The latter, equilibrated portions of the progress curves in Fig. 1 can also be used to obtain information about the mode of inhibition of metalloenzymes by sequestering agents.[22] This approach (Fig. 2) utilizes plots of $\log[(k_0/k_I) - 1]$ vs log[free ligand], where k_I is the "equilibrated" value of the pseudo-first-order rate constant for disappearance of substrate, measured after equilibration in the presence of metal-chelating inhibitor, and k_0 is the value measured in the absence of inhibitor. According to Eqs.

[22] B. Holmquist and B. L. Vallee, *J. Biol. Chem.* **249,** 4601 (1974).

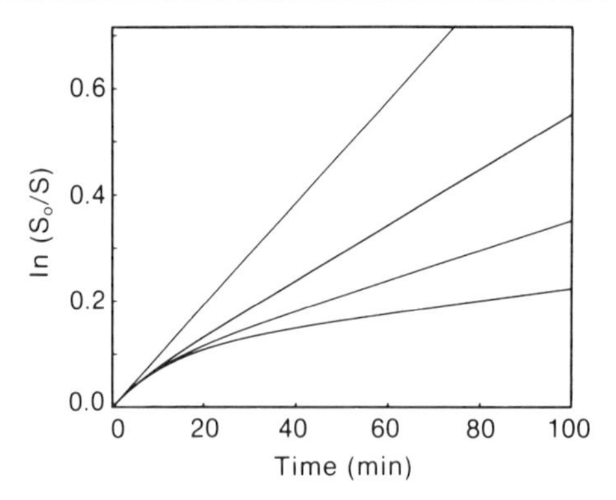

FIG. 1. Equilibration of a metalloenzyme with different levels of a catalytically essential metal ion, during the course of an assay run under conditions such that the reaction is pseudo-first order in substrate ($[S] \ll K_m$). Computer simulation for a zinc metalloenzyme with $K_m = 2.0 \times 10^{-3}$ M, $k_{cat} = 380$ min^{-1}, $K_{Me} = 2.0 \times 10^{-10}$ M, $k_{in} = 3.46 \times 10^{8}$ M^{-1} and $k_{out} = 0.0696$ min^{-1}, where k_{in} and k_{out} are, respectively, the rate constants for formation and dissociation of the active complex (holoenzyme) between enzyme and zinc. For all assays, initial substrate concentration is 2×10^{-5} M, $[E_t]$ is 1.0×10^{-8} M, and $[M_t]$ is 1.0×10^{-5} M. The curves were calculated by means of Eq. (13), and represent, from top to bottom, the course of reaction in assay solutions containing 0, 1.0, 1.2, and 1.6×10^{-4} M 1,10-phenanthroline, with the resultant free Zn^{2+} ion concentrations of 1.0×10^{-5}, 1.84×10^{-10}, 9.2×10^{-11}, and 3.2×10^{-11} M, respectively. "Equilibrated" first-order rate constants are obtained from the slopes of the latter, linear portions of the curves.

(9a) and (9b), the slope of the plot is a measure of $\bar{n}$, the average number of molecules of ligand involved in inhibition of a molecule of enzyme.

$$(k_o/k_I) - 1 = K'[L]^{\bar{n}} \tag{9a}$$
$$\log[(k_o/k_I) - 1] = \log K' + \bar{n} \log[L] \tag{9b}$$

The equilibrium "constant," K', for the observed reduction in activity in the presence of the ligand is a conditional constant that is applicable only over a certain range of large values of [L], and is a function of the total concentration of metal ion in the solution. K' can be determined from the value of $\log[(k_0/k_I) - 1]$ when $[L] = 1$, and has the units of molarity, cubed. The maximum value of $\bar{n}$ that would be expected from interaction of a bidentate ligand molecule with a metal ion bound in an enzyme active site (EML formation) is 1.0. Values of $\bar{n}$ greater than 1.0 are taken as an indication that the metal ion has been removed from the enzyme and is free to interact with more than one sequestering molecule in solution.[1,21,22]

The assumptions required for the derivation of Eqs. (9a) and (9b), however, impose certain restrictions on the applicability of these equa-

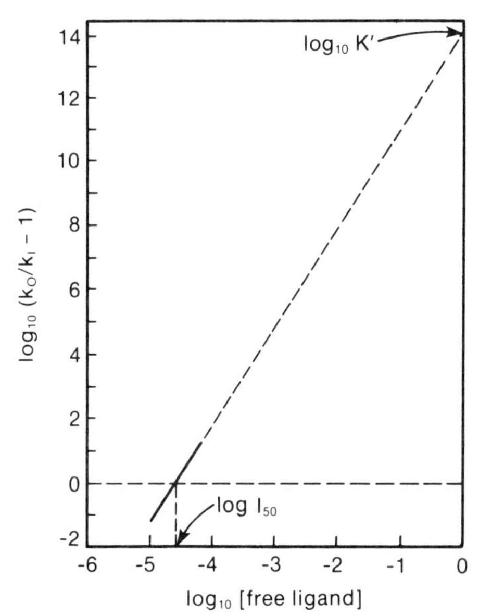

FIG. 2. Dependence of the degree of inhibition/inactivation of a metalloenzyme by a chelating agent (here referred to as ligand) on the concentration of the ligand. The values of k_0 and k_I are obtained from first-order plots like those in Fig. 1, with k_0 being the "equilibrated" rate constant measured in the absence of chelator, and k_I representing the value measured in the presence of a particular concentration of chelating ligand. I_{50} is the concentration of chelator required to reduce the rate constant to one-half the value measured in the absence of chelator, and K' is the equilibrium constant in Eqs. (9a,b). The slope of the plot gives a measure of the value of $\bar{n}$, the average number of chelator molecules involved in the inhibition/inactivation of each molecule of the metalloenzyme. Equation (9b), upon which this type of linear plot is based, is valid only at ligand concentrations sufficiently high that essentially all of the metal ion present in solution exists as the complex ML_3 (see text).

tions and of the plots[23,24] based on the equations. If one assumes the general case described by Eq. (8b), in which the metal ion buffer species can both scavenge free metal ion from the solution and bind to enzyme-bound metal ion to form the ternary complex, EML, the conservation equation for enzyme is written as

$$[E_{total}] = [EM] + [EML] + [E] \tag{10a}$$

Here E represents enzymatically inactive apoenzyme. When the concentrations of the other two enzyme species are written in terms of [EM],

[23] J. F. Ambrose, G. B. Kistiakowsky, and A. G. Kridl, *J. Am. Chem. Soc.* **72,** 317 (1950).

[24] F. H. Johnson, H. Eyring, R. Steblay, H. Chaplin, C. Huber, and G. Gherardi, *J. Gen. Physiol.* **28,** 463 (1945).

making use of the relationships denoted by the dissociation constants K_{Me} and K_I [Eq. 8b], the conservation equation becomes

$$[E_{total}] = [EM] + ([L]/K_I)[EM] + (K_{Me}/[M])[EM] \quad (10b)$$
$$= (1 + [L]/K_I + K_{Me}/[M])[EM]$$

Rearranging, we have

$$\frac{[E_{total}]}{[EM]} - 1 = \frac{[L]}{K_I} + \frac{K_{Me}}{[M]} \quad (10c)$$

If one assumes further that the metal ion buffer, L, is a bidentate ligand such as 1,10-phenanthroline, then as many as three molecules of a bidentate ligand will be capable of binding to a single metal ion (zinc, manganese, cobalt, etc.) in solution. The concentration of free metal ion, M, will then be given by Eq. (6b), where [L] is the concentration of free buffer ligand. Substituting this expression for [M] in Eq. (10c), we have

$$\frac{[E_{total}]}{[EM]} - 1 = \frac{[L]}{K_I} + \frac{K_{Me}}{[M_t]}(1 + K_1[L] + \beta_2[L]^2 + \beta_3[L]^3) \quad (11)$$

In order to determine a value for K_{Me}, it is necessary to measure the interactions between protein and metal ion, unperturbed by any mutual dependence of substrate and metal ion binding. Accordingly, in using activity measurements to monitor such interactions one should choose a sensitive assay utilizing a substrate for which the enzyme has a reasonably large k_{cat}, but also a rather large K_m, and then conduct the asays with $[S] \ll K_m$.[22] Under these conditions, with a fixed standard concentration of substrate and with the same amount of enzyme added in each assay, the measured velocity in a given experiment will be proportional to the fraction of total enzyme that is present as EM, or free, enzymatically active holoenzyme. In the absence of metal ion sequestering agents and with $[M_t] \gg K_{Me}$, the enzyme can be assumed to exist almost entirely as holoenzyme, EM; thus either the velocity or the apparent first-order rate constant for substrate measured under these conditions can be taken as a measure of $[E_{total}]$. The quotient $[E_{total}]/[EM]$ in Eq. (10c) can therefore be replaced by k_0/k_I, with k_0 equal to the rate constant in the absence of sequestering agent and k_I equal to the rate constant in the presence of a given concentration of sequestering agent, L. The result of this substitution is Eq. (12).

$$\frac{k_0}{k_I} - 1 = \frac{[L]}{K_I} + \frac{K_{Me}}{[M_t]}(1 + K_1[L] + \beta_2[L]^2 + \beta_3[L]^3) \quad (12)$$

Taking the logarithm of both sides of the equation yields Eq. (13).

$$\log\left[\frac{k_o}{k_I} - 1\right] = \log\left\{\frac{[L]}{K_I} + \frac{K_{Me}}{[M_t]}(1 + K_1[L] + \beta_2[L]^2 \times \beta_3[L]^3)\right\} \quad (13)$$

Equation (13) does not have quite the form of Eq. (9b), but if K_I is very large (i.e., EML is not formed), and [L] is much larger than the largest of K_1, K_2, and K_3, so that $\beta_3[L]^3$ is much larger than the sum of $\beta_2[L]^2$, K_1L, and 1, then Eq. (13) will approximate Eq. (14), which is identical in form to Eq. (9b) (with $\bar{n}$ set equal to 3), and is compatible with the plotting method of Fig. 2.

$$\log\left(\frac{k_o}{k_I} - 1\right) = \log\left(\frac{K_{Me}\beta_3}{[M_t]}\right) + 3\log[L] \quad (14)$$

The determination of $\bar{n}$ from the slopes of linear plots such as that in Fig. 2 is thus valid only in the range of ligand concentration sufficiently high that essentially all of the metal ion not bound to enzyme has been converted to the metal–ligand complex of highest possible ligand : metal stoichiometry (ML_3 in the case of a bidentate ligand such as 1,10-phenanthroline). More important, while a linear plot with slope significantly greater than 1.0 does imply that removal of metal ion from the enzyme plays a major role in "inhibition" of enzyme activity by sequestering agents, the mere observation at high [L] of a linear segment of slope greater than 1.0 does not necessarily indicate that the ternary complex EML is not present in amounts significant in comparison with [EM]. To the contrary, even systems in which there is substantial ternary complex formation can be expected to yield linear plots with slope > 1.0, provided the free ligand concentration is high enough.

Figure 3 illustrates a plot of expected values of $\log[(k_0/k_I) - 1]$ versus log [free ligand], for varying levels of inhibitory binding of L to the enzyme-bound metal ion. In Fig. 3 the values of $\log[(k_0/k_I) - 1]$ are calculated not by use of Eq. (14), but by means of Eq. (13), the more general equation which takes into account both formation of ternary complex (EML) and the distribution of metal ion among metal–ligand complexes other than ML_3. In Fig. 3 the metal–ligand formation constants are those for zinc and unprotonated 1,10-phenanthroline at 25° and ionic strength 0.1, i.e., $\log K_1 = 6.4$, $\log \beta_2 = 12.2$, and $\log \beta_3 = 17.1$.[17] The value chosen for K_{Me} ($= 2 \times 10^{-10}$ *M*) is arbitrary, but is near the upper end of the range of values found in near-neutral media for the protein/metal ion dissociation constants for zinc–carboxypeptidase A,[6] for the more tightly bound Zn^{2+} ion of alkaline phosphatase,[25] for zinc–carbonate dehydratase,[9] for zinc–phosphoglucomutase,[18] and for the more tightly bound Zn^{2+} ion of

[25] S. R. Cohen and I. B. Wilson, *Biochemistry* **5**, 904 (1966).

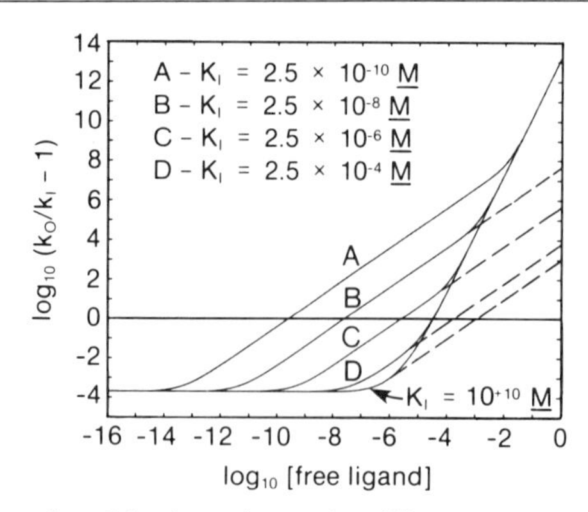

FIG. 3. Wide-range plot of the dependence of equilibrated enzyme activity upon concentration of chelating ligand. The ordinate values in this figure are based on the scheme of Eq. (8b), and are calculated using Eq. (13), which is valid at all concentrations of free ligand, rather than only at high [L], as in the case of Eq. (9b) and Fig. 2. The chelating ligand is assumed to have the characteristics of 1,10-phenanthroline, in that it is bidentate and capable of forming the 3 : 1 ML_3 complex with metal ion free in solution. Stability constants for binding to the metal ion are given by $\log(K_1) = 6.4$, $\log(\beta_2) = 12.2$, $\log(\beta_3) = 17.1$. $[M_t]$ is assumed to be 1.0×10^{-6} *M*, K_{Me} is $2.0 \times 10^{+10}$ *M*, and the values of K_I [see the scheme of Eq. (8b)] are as given on the figure. Note; The lower curve, generated using an extremely large value of K_I (= 1.0×10^{-10} *M*), represents an approximation of the situation in which EML is not formed.

Aeromonas aminopeptidase.[26] One of the values of K_I, that at 2.5×10^{-4} *M*, is near that found for dissociation (EML → EM + L) of the ternary complex formed between 1,10-phenanthroline and horse liver alcohol dehydrogenase.[27] As an illustration of the dependence of curve shape on the values of the various constants, the function has been plotted over a much wider range of ligand concentration than could be meaningfully covered in practice by even the most sensitive assays available. The lower line, in which K_I is given the extremely large value of $1.0 \times 10^{+10}$ *M*, corresponds to the case in which EML is not formed, and consists of three segments: (1) an essentially linear (and horizontal) segment at extremely low [L], over which the ligand concentration is simply too small to perturb the free zinc ion concentration significantly, and over which the value of $\log[(k_0/k_I) - 1]$ is given by K_{Me}/M_t, (2) a curved segment over which the solution free Zn^{2+} concentration is sensitive to the formation of all three zinc–ligand complexes, ML, ML_2, and ML_3, and (3) a linear segment at high ligand concentration over which essentially all of the solution metal is present as ML_3, with $[M] \simeq [M_{total}]/(\beta_3 L^3)$, and $\log[(k_0/k_I) - 1] \simeq \log$

[26] J. O. Baker and J. M. Prescott, *Biochem. Biophys. Res. Commun.* **130,** 1154 (1985).
[27] A. J. Sytkowski and B. L. Vallee, *Biochemistry* **18,** 4095 (1979).

$(K_{Me}\beta_3/M_{total}) + 3\log[L]$ (i.e., $\bar{n} = 3$). As the value of K_I is decreased, and formation of the ternary complex EML becomes significant, a third linear segment appears between the first two, representing a range of ligand concentrations over which the dependence of $\log[(k_0/k_I) - 1]$ on [L] is dominated by a process that is first order in [L] (the formation of ternary complex EML). Over this range the plot has a slope of 1.0. Between the central linear portion of the curve, dominated by formation of EML, and the right-hand linear portion, dominated by removal of metal ion from the enzyme, the slope of the curve will have a continuum of values intermediate between 1.0 and 3.0. Since in reality the range of relative enzyme activities accessible to a given experimental design will tend to be a small portion of the range shown in Fig. 3, the curvature of the plot in this transitional region may not be obvious.

The straight broken lines in Fig. 3 illustrate the patterns expected of a system utilizing a metal ion buffer ligand capable of forming only a 1 : 1 complex with the metal. For comparison with the behavior of the 3 : 1 complex-forming ligand represented by the solid lines, the value of K_1 for this 1 : 1 complex-forming ligand has been set equal to K_1 for the 3 : 1 buffer.

The important point to be taken from Fig. 3 is that even in cases in which the ternary complex is formed to a significant extent along with scavenging by the chelator of metal ion dissociating from the enzyme, there are still straight line segments in the plot of $\log[(k_0/k_I) - 1]$ vs $\log[L]$, with slope equal to the maximum number of ligand molecules capable of binding to the metal ion free in solution. The simple observation of linear plots at high [L] cannot, therefore, be taken as indicating that the ternary complex is not formed and that the chosen sequestering agent is acting only on the free metal ion concentration.

In actual experimental work, the readily accessible parameter is the total concentration of added ligand, rather than the equilibrium concentration of free ligand. Figure 4 shows the effect of variation in total added ligand upon the "equilibrated" reaction rate (Fig. 1). The values (solid lines) have been calculated as in Fig. 3, with the same values and ranges of values taken for K_{Me}, $[M_t]$, and K_I but with the effects of ligand depletion by binding to M and EM taken into account. For purposes of comparison, the broken lines indicate values calculated as though the concentration of free ligand were equal to the concentration of added ligand, i.e, with ligand depletion by binding to M and EM considered negligible. The right-hand portion of Fig. 4, at high total ligand concentration, is essentially identical to the corresponding region of Fig. 3. The deviations seen beginning in the region from $[L_t] = 10^{-5}$ *M* down to 10^{-6} *M* are due to depletion of free L by binding to M; those observed near $[L_t] = 10^{-7}$ *M* (in

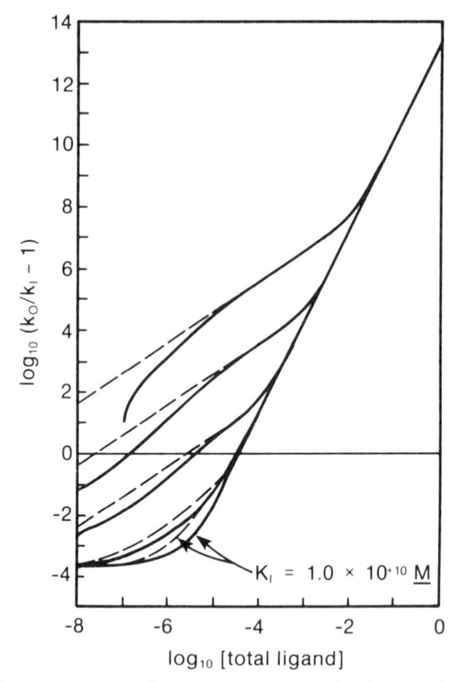

FIG. 4. Plot analogous to that of Fig. 3, but with calculated values of log $[(k_0/k_I) - 1]$ (solid lines) based on total added ligand, rather than on [free ligand]. Depletion of ligand by binding to EM and to free metal-ion is taken into account, with $[E_t] = 1.0 \times 10^{-7}$ *M*. Other constants and concentrations are the same as in Fig. 3. For comparison purposes, the values calculated assuming that free ligand concentration is equal to $[L_t]$ are shown as broken lines. Values of K_I for the solid curves and associated broken curves are, from top to bottom, 2.5 $\times 10^{-10}$, 2.5×10^{-8}, 2.5×10^{-6}, 2.5×10^{-4}, and 1.0×10^{-10} *M*.

the plots for very small values of K_I) reflect depletion of free L by binding to EM. ($[E_t]$ is assumed to be 10^{-7} *M* in these calculations.)

Ternary Complex Formation: Diagnostic Procedures under Preequilibrium or "Pseudo-Irreversible" Conditions

Further information concerning the contribution of ternary complexes to the overall metal ion buffer/enzyme equilibria can be obtained by studying the *rate* of removal of metal ion from the enzyme as a function of chelator concentration. If the inhibition of the metalloenzyme is due solely to scavenging by the chelator of free metal ion after it has dissociated from the enzyme, then at chelator concentrations above the concentration required for complete inactivation at equilibrium, the observed rate of inactivation/metal removal will reach a constant value reflecting the rate of the first-order dissociation of the metal M from the holoenzyme

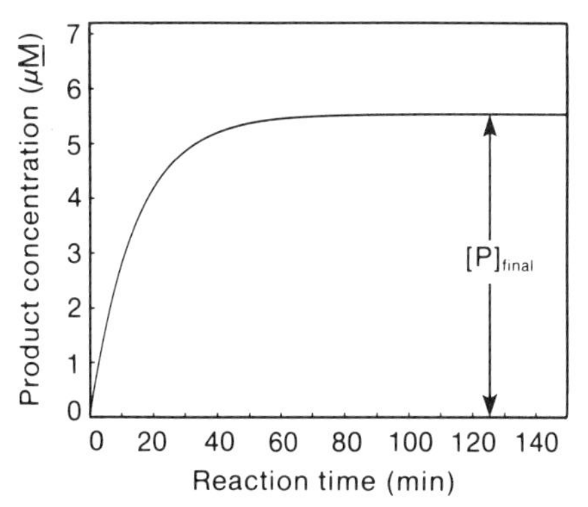

FIG. 5. Generation of product, P, by a metalloenzyme concurrently with the quasi-irreversible loss of catalytically essential metal ion from the enzyme due to the presence in the assay mixture of high concentrations of chelating ligand. It is shown in the text that, provided the concentration of chelator is sufficient to produce essentially complete inactivation at equilibrium, the value of $[P]_{final}$ is proportional to k_{cat}/k_{out}, the constant k_{out} being the effective rate constant for removal of metal ion from the enzyme. For the plot shown, $k_{cat} = 3880\ min^{-1}$, $k_{out} = 0.0693\ min^{-1}$, $K_m = 1.0 \times 10^{-2}\ M$, $[S] = 1.0 \times 10^{-4}\ M$, and $[E_t] = 1.0 \times 10^{-8}\ M$.

EM. Figure 5 illustrates an experiment in which a metalloenzyme is incubated with a solution containing such a totally inactivating concentration of metal ion buffer ligand, along with a nonperturbing ($\ll K_m$) concentration of substrate. The "pseudo-irreversible" removal of the catalytically essential metal ion is followed by monitoring the generation of product by the decreasing fraction of active holoenzyme, EM. The scheme of Eq. (15) illustrates the most general case of the process:

$$\begin{array}{ccc} EM & \xrightarrow{k_{diss}} & E + M \longrightarrow ML_n \\ K_I \Big\updownarrow & & \\ EML & \xrightarrow{k_{ext}} & E + ML \longrightarrow ML_n \end{array} \tag{15}$$

Here k_{diss} is the first-order rate constant for simple dissociation of metal ion from the enzyme, with the reaction rendered essentially irreversible due to the removal of the dissociated metal ion by excess L (formation of ML_n), and k_{ext} is the rate constant for "extraction" of the metal ion from the metalloenzyme through formation of EML and subsequent breakdown to E + ML. The rate of formation of product is given by Eq. (16),

$$\frac{dP}{dt} = \frac{k_{cat}[S]}{K_m(1 + [L]/K_I)} [EM]_0 e^{-k_{out}t} \tag{16}$$

in which k_{out} is the effective rate constant for conversion of holoenzyme to apoenzyme under a particular set of conditions, and $[EM]_0$ is the concentration of holoenzyme, EM, added to the assay at $t = 0$. According to the scheme of Eq. (15), the general expression for k_{out} is given by Eq. (17).

$$k_{out} = k_{diss}\left(\frac{K_I}{K_I + [L]}\right) + k_{ext}\left(\frac{[L]}{K_I + [L]}\right) \tag{17}$$

Provided only a small fraction of the substrate present is consumed during the inactivation of the enzyme, the substrate concentration, [S], can be treated as a constant in Eq. (16). The progress curve for product formation under these inactivating conditions will then be described by Eq. (18), which is obtained by integrating Eq. (16) with [S] treated as a constant.

$$P(t) = \left(\frac{k_{cat}}{k_{out}}\right)\left(\frac{[S]}{K_m}\right)\left(\frac{[EM]_0}{1 + [L]/K_I}\right)(1 - e^{-k_{out}t}) \tag{18}$$

Thus, if the inactivation can be run to completion (reaction time 8–10 times as long as $0.693/k_{out}$) without significantly changing the substrate concentration, the influence of EML formation on metal ion removal can be evaluated simply by measuring the final amount of product present after the enzyme has been completely inactivated. At such long reaction times, t, the term $e^{-k_{out}t}$ will be negligible with respect to 1, and Eq. (18) will be approximated by Eq. (19).

$$P(t) = \left(\frac{k_{cat}}{k_{out}}\right)\left(\frac{[S]}{K_m}\right)\left(\frac{[EM]_0}{1 + [L]/K_I}\right) \tag{19}$$

Examination of Eqs. (17) and (19) reveals that if the formation of EML is significant (K_I *not* much larger than the range of [L] employed) then the total amount of product generated before complete inactivation of the enzyme will be a function of [L] at high [L]. This is true even in the case in which $k_{ext} = k_{diss}$ and k_{out} is therefore independent of [L]; in this case the extent of product generation will still depend on [L] because of simple competitive inhibition of the enzyme reaction by L, as indicated by the factor $(1 + [L]/K_I)$ in Eqs. (18) and (19). If EML is not formed to any significant extent, and the desired condition that L acts purely as an "external" metal ion buffer therefore occurs, the total product generated will be independent of L above a certain concentration. This "total product" approach should be especially advantageous in studies of enzymes for which a continuously monitorable assay is not available.

A cautionary note should be injected here, to the effect that this "pseudo-irreversible" method measures effects on only one side of the holoenzyme/apoenzyme equilibrium. The apparent absence of EML formation, as tested for by this procedure, cannot by itself be taken as a guarantee that a given metal ion buffer system is free of unwanted side interactions with the enzyme. Instead, this method is best employed in rapid screening of prospective buffer systems in order to identify those that produce the least complications on the dissociation side of a given enzyme/metal equilibrium. The final test of the innocuous nature of a metal ion buffer system (with respect to a given enzyme/metal system) is met when the same values are obtained for the enzyme/metal binding constant in experiments utilizing different concentrations of a given buffer to achieve the same range of calculated free metal ion concentrations,[28] or utilizing chemically different metal ion buffers to control free metal.[29]

Analytical Complications in Use of Metal Ion Buffers

In graphite-furnace atomic absorption spectrophotometry of neutral or only weakly acidic samples containing zinc and 1,10-phenanthroline, the 1,10-phenanthroline has been found to cause significant supression of the zinc signal.[26] This interference can readily be eliminated either by acidification with high-purity nitric acid or by addition of another metal ion, such as nickel, which displaces zinc from its 1,10-phenanthroline complexes by virtue of its higher affinity for the chelator, and which does not interfere with the subsequent atomic absorption determination of zinc.

Another potential complication is an apparent interaction between zinc–1,10-phenanthroline complexes and polystyrene laboratory ware. In a recent study of zinc binding to *Aeromonas* aminopeptidase,[30] disposable polystyrene centrifuge tubes (Corning 25310-15) were used in making dilutions for atomic absorption measurements. In terms of cleanliness, these tubes are excellent for the purpose, a simple rinsing with high-purity water (10^{-9} *M* or less zinc) sufficing to render the tubes useable with zinc solutions on the order of 10^{-8} *M*. In neutral or only weakly acidic dilutions of zinc–1,10-phenanthroline buffer systems, however, a time-dependent decrease in zinc signal was observed, amounting to as much as 35–40% of the initial signal over a 2-hr period. Since this time-dependent decrease did not occur either in zinc solutions without 1,10-phenanthroline in the polystyrene tubes, or in zinc–1,10-phenanthroline solutions in acid-cleaned glass test tubes, it would appear to be a result of the adsorption of a zinc–1,10-phenanthroline complex on the walls of the polystyrene tubes. Immediate acidification of the dilutions, or addition of sufficient nickel to displace the zinc from its 1,10-phenanthroline complex effectively eliminated this complication as well.

[28] S. Sillen and B. Mannervik, *J. Biol. Chem.* **259,** 11426 (1984).
[29] W. R. Harris, *Biochemistry* **22,** 3920 (1983).
[30] J. O. Baker, unpublished results (1983).

In addition to being pH dependent, the values of the apparent, or "conditional" stability constants for interaction of a metal ion buffer with a given metal ion may also depend on the choice of cation used to support the ionic strength of the solution. In solutions at 20°, at an ionic strength of 0.1, the affinity of EDTA for Na^+ is approximately 7 times as great as that for K^+ ($\log K_{Na^+} = 1.66$, $\log K_{K^+} = 0.8$).[17] In an experimental solution, in which an ionic strength on the order of 0.1 is supported primarily by sodium salts, the apparent affinity of EDTA for other metals will therefore differ significantly from values determined in solutions of potassium salts.[31]

Possible Future Directions: Crown Ethers and Cryptands as Metal Ion Buffer Ligands

During the past 20 years a large number of cyclic polyethers (crown ethers) and polyoxa macrobicyclic diamines (cryptands) have been synthesized,[32–34] and their varying abilities to form complexes with a variety of metal ions have been studied both thermodynamically and kinetically.[35–38] Both of these classes of molecules form 1 : 1 complexes with metal ions, with the stability of the complexes a sensitive function of ligand cavity size and metal ion size. These properties permit fine-tuning of the affinity for a given metal ion by adjustment of the size of the ligand rings,[35,38] and both of these classes of compounds would seem, therefore, to be excellent prospects for the construction of new metal ion buffer systems.

In a study of the interaction of *E. coli* tryptophanase with K^+ ions, addition of the crown ether, 18-crown-6 (1,4,7,10,13,16-hexaoxacyclooctadecane), was used to drop the free K^+ ion concentration of the solution suddenly from 17 to 1.06 m*M*.[39] The metal complex stability constants and solubilities reported for another compound, the bicyclic [2.2.1]cryptand,[35] indicate that this compound should be useful as a calcium ion buffer in the range of 10^{-6} to 10^{-9} *M* and perhaps slightly below. Since the crown ethers are neutral molecules in the pH range of enzymological interest, buffer systems constructed using these molecules should be very insensitive to pH, at least in acidic and neutral media.

[31] A. C. H. Durham, *Cell Calcium* **4,** 33 (1983).

[32] C. J. Pedersen, *J. Am. Chem. Soc.* **89,** 7017 (1967).

[33] J. J. Christensen, J. O. Hill, and R. M. Izatt, *Science* **174,** 459 (1971).

[34] J. J. Christensen, D. J. Eatough, and R. M. Izatt, *Chem. Rev.* **74,** 351 (1974).

[35] J. M. Lehn and J. P. Sauvage, *Chem. Commun.* **1971,** 440 (1971).

[36] E. L. Yee, O. A. Gansow, and M. J. Weaver, *J. Am. Chem. Soc.* **102,** 2278 (1980).

[37] B. G. Cox, J. Garcia-Rosas, and H. Schneider, *J. Am. Chem. Soc.* **103,** 1054 (1981).

[38] R. M. Izatt, D. J. Eatough, and J. J. Christensen, *Struct. Bonding* **11,** 161 (1973).

[39] I. Behbahani-Nejad, C. H. Suelter, and J. L. Dye, *Curr. Top. Cell. Regul.* **24,** 219 (1984).

The crown ethers and cryptands are not expected to be without their problems in such applications. Although the 18-crown-6 complex with K^+ ion was found to be noninhibitory in the tryptophanase studies cited above,[39] the corresponding complex of K^+ ion with [2.2.1] cryptand decreased enzyme activity by about one-third even in the presence of excess K^+ ion. In addition, the rate constants for formation of metal ion/cryptand complexes tend to be several orders of magnitude lower than those for simple ligands,[40] and this fact may need to be taken into account in fast-kinetic studies of enzyme/metal association and dissociation. These possible difficulties notwithstanding, the addition of two more classes of metal ion buffer systems will increase the options available to the investigator in the search for buffer systems without undesirable side-effects on a given enzyme system.

Summary and Conclusions

The design of a metal ion buffer system useful in a given enzymological application is subject to a number of different requirements.

1. The total concentration of added metal ion, M_t, should be large enough to damp out the effect of any adventitious quantities of the same metal ion and to overwhelm adventitious quantities of other metal ions.
2. The ratio of free to bound ligand should be high enough that the calculated ratio between the concentrations of free metal ion M and M_t will not be unduly sensitive to uncertainties in the values of metal–ligand stability constants. If possible, $[L_t]/[M_t]$ should be large enough that the variation of free metal ion concentration, [M], with $[M_t]$ will be effectively linear in the range of interest.
3. The concentrations of metal ion buffer species, both the free ligand and metal–ligand complexes, should be kept reasonably low in order to minimize the possibility of perturbation of the enzyme/metal ion equilibrium.

The best design will be that which most successfully balances these sometimes opposing requirements.

Acknowledgments

Development of some of the ideas expressed herein was supported by grants to John M. Prescott from the Robert A. Welch Foundation (Grant A-003) and from the National Institute of General Medical Sciences (Grant GM 32181). This work was also funded by the Department of Energy Office of Alcohol Fuels under WPA number 349.

[40] V. M. Loyola, R. Pizer, and R. G. Wilkins, *J. Am. Chem. Soc.* **99,** 7185 (1977).

[7] Standards for Metal Analysis

By CLAUDE VEILLON

Most analytical methods for elemental analysis fall into three general categories: reference methods, working methods, and routine methods. A reference method is one in which the inherent accuracy is high or well established and the sources of error are known and can be measured and controlled. Working methods are those that are more widely available and which have had their accuracy in a particular determination established either by reference methods or by independent means. Routine methods are those that are widely available, rapid and/or convenient, and generally need to have their accuracy in a particular determination established by independent means or by analyzing a suitable reference material of established analyte content. For example, the determination of selenium in blood plasma could be established to a high degree of certainty by a reference method like stable isotope dilution, isotope ratio mass spectrometry, or by neutron activation analysis. This material could then be used to help establish the accuracy of a more readily available working method like fluorometry, following perchloric acid digestion and formation of the fluorophor with diaminonaphthalene. However, for rapid routine plasma selenium determinations on a day-to-day basis or for large numbers of samples, direct measurement by graphite furnace atomic absorption spectrometry might be a better choice. It would be necessary to establish the accuracy of such a method by comparison with a reference or working method, or by using reference plasmas with established selenium content as a quality control material.

Far too often in the literature one sees analytical data obtained by unvalidated methods, or by insufficiently validated methods. Examples of the latter include "recovery" experiments, and using the method of additions. If one adds analyte to a sample, analyzes it, and obtains 100% "recovery," all that can be said with complete certainty is that the method is capable of finding the added analyte. In the method of additions, the assumption (actually, a hope) is made that any matrix effects on the determination are offset. However, in the earlier examples, the selenium in plasma probably exists as several organoselenium compounds, and it is perhaps a lot to hope for that added inorganic selenium will behave in the same way as the endogenous analyte during the determination. Even when the slope of the standard additions curve and that of aqueous inorganic standards is the same, this does not necessarily nor completely establish the accuracy of the determination. This can be achieved only by simultaneous measurement of a reference material, as nearly identical to the sample as possible, of established analyte content, or by analysis of the same sample by an additional, independent method.

Once a routine methods' accuracy is established for a particular determination in a particular sample matrix and suitable quality control materials are available, it is usually necessary to have available "standards" to establish the analytical response (i.e., calibrate) of the instrument being used. For many determinations of elements in biological samples, aqueous inorganic standards can be used. If the analysis procedure involves destruction of the organic matrix by dry ashing or acid digestion, the analyte is then in an aqueous inorganic matrix, and aqueous standard solutions can often be used. However, this must still be established by acceptable procedures as before.

From the above discussion, the topic of standards for metal analysis, in the context of elemental analysis of biological materials, can be divided into two areas: aqueous inorganic standard solutions for calibration of instrument response, standard additions curves, use where acceptability established, etc.; and reference materials, i.e., biological materials as nearly identical to the samples as possible and with established analyte content. Only recently have the latter been available, and many more are needed. A third area to be discussed in this context is quality control procedures and materials.

Aqueous Standards

These are invariably needed for the purposes stated above and can either be made by dissolution of a known weight of the pure metal in appropriate acids and diluted to a known volume, or they can be purchased. Several suppliers sell standard solutions of most of the elements at a known concentration, and we have experienced no difficulties or errors in concentration with the several we have used. It is good laboratory practice however to check these against standards quantitatively prepared from the metal and to check their concentration periodically, because changes caused by evaporation, adsorption, and contamination are always possible. Some of these also have "expiration dates," although how (or why) these are established is not clear.

Even definitive methods like stable isotope dilution mass spectrometry can require the use of accurate standards. Isotope dilution procedures employ the addition of a known quantity of an enriched stable isotope (spike) of the analyte element to the sample. Thus, the accuracy of the determination is highly dependent on knowing the amount of spike added.

The accuracy of the spike solution concentration can be established against a known amount of unenriched analyte (i.e., a known amount of an accurate standard) by reverse isotope dilution.

Instrument manufacturers often encourage, either in their operating instructions or in the way the instrument is set up, the use of only one or perhaps two standards for calibration. This is to be seriously discouraged. At least three, and preferably four, standards (concentrations) should be used for calibration curves or standard additions. With a single standard, any error in its preparation renders all of the measurements in error. With only two standards, linearity is assumed (and achieved, by definition) and an error in one of the standards is not evident. This is not to say that more standards eliminate the possibility of errors, but they do reduce the number of opportunities. The use of three or four standards, coupled with a conscientious quality control program (vide infra), will greatly help eliminate errors caused by serial dilutions.

Reference Materials

When determining elements present in biological samples at a fairly high level using sensitive methods, often the sample can be, and in fact needs to be, diluted considerably. This also dilutes the sample matrix, so the determination approaches that of simple aqueous solutions, and calibration curves obtained from standard solutions are often sufficient to obtain accurate results. Interest has increased in recent years in trace elements present in biological samples at very low concentration levels. Here, samples often cannot be diluted, and may even need to be concentrated prior to analysis. In these situations, the sample matrix and chemical form of the analyte element in the sample can effect substantial deviations in response compared to aqueous standards. Likewise, as concentration levels go down, sample contamination and blank levels become increasingly important, occasionally leading to serious errors in the data.

The answer lies in analyzing uncontaminated samples and acceptably low blanks by a method of established accuracy—established for that elemental determination in that sample matrix. As pointed out earlier, validation of the accuracy of a method can be achieved in only two ways: analysis of the same sample(s) by one or more additional, *independent* methods (it goes without saying that the answer by the two methods should agree). The second way is to analyze reference materials of identical (or as nearly identical as possible) matrix to that of the samples, the reference materials having had their analyte content established by acceptable means.

Until recently, determinations with validated methods posed serious problems, contributing much erroneous data (and conclusions therefrom) to the literature. First, few laboratories have, or have access to two independent methods. Second, appropriate biological reference materials were almost nonexistent a few years ago. This situation has improved recently, but a great distance is left to go. Fortunately, there is much interest now in the problem, and several organizations are currently working on it. The most recent summary of available biological reference materials is that of Muramatsu and Parr[1] of the International Atomic Energy Agency (IAEA).

Perhaps the first biological reference material for trace elements was produced by Dr. H. J. M. Bowen of the University of Reading in the United Kingdom. This is a kale material, and has become known as "Bowen's Kale." Over the years, a number of laboratories have analyzed this plant material allowing the most probable average concentrations to be assessed.

The primary organization producing reference materials in the United States is the National Bureau of Standards (NBS). Until recently, their biological reference materials were mostly of plant origin, but also included bovine liver and oyster tissue. These are not particularly well-suited matrices for many biological samples, particularly those wherein the analyte concentrations are very low. More recently, materials like urine, milk, and serum have become available, with more on the way.

The organizations providing biological reference materials are listed in Table I[1a] and an overview of their products is given in Table II.[1a] In addition to these, a new pool of bovine serum has been prepared by the U.S. Department of Agriculture under the direction of the author. This will replace the current NBS reference material 8419, and will be issued as a certified standard reference material (SRM), designated as SRM-1598. Reference material 8419 (and the new SRM) represents one of the first biological materials with analytes at the very low levels encountered in samples of this kind. An additional material of which I am aware is a large pool of human serum, collected by Dr. J. Versieck of the University Hospital in Ghent, Belgium. It is currently being analyzed by selected laboratories worldwide.

These newer biological reference materials, being of matrix compositions and analyte concentration levels more like the samples being analyzed today, will go a long way toward validating analytical methods used and, hopefully, thus remove one of the major barriers to meaningful inter-

[1] Y. Muramatsu and R. M. Parr, *I. A. E. A. [Publ.] RL/128* **December** (1985). (Copies may be requested from Dr. R. M. Parr, IAEA, P.O. Box 100, A-1400 Vienna, Austria.)

[1a] This material became available in late 1987.

TABLE I
SUPPLIERS OF BIOLOGICAL AND ENVIRONMENTAL REFERENCE MATERIALS[a]

Abbreviated name	Full name and address
BCR	Community Bureau of Reference (BCR) Commission of the European Communities 200 Rue de la Loi B-1049 Brussels Belgium
BI	Behring Institute P.O. Box 1140 D-3550 Marburg 1 Federal Republic of Germany
BOWEN	Dr. H. J. M. Bowen Department of Chemistry The University of Reading Whiteknights P.O. Box 224 Reading RG6 2AD United Kingdom
IAEA	International Atomic Energy Agency Analytical Quality Control Services Laboratory Seibersdorf P.O. Box 100 A-1400 Vienna Austria
KL	Kaulson Laboratories, Inc. 691 Bloomfield Avenue Caldwell New Jersey 07006 USA
IRANT	Institute of Radioecology and Applied Nuclear Techniques Komenského 9 P.O. Box A-41 040 61 Kosice Czechoslovakia PZO Služba výskumu Koněvova 131 130 86 Prague 3-Žižkov Czechoslovakia
NBS	Office of Standard Reference Materials Room B311, Chemistry Building National Bureau of Standards Gaithersburg, MD 20899 USA

TABLE I (*continued*)

Abbreviated name	Full name and address
NIES	National Institute for Environmental Studies Japan Environment Agency P.O. Yatabe Tsukuba Ibaraki 300-21 Japan
NRCC	National Research Council Canada Division of Chemistry Ottawa K1A OR6 Canada
NYE	Nyegaard & Co. AS Diagnostic Division Postbox 4220 Torshov N-0401 Oslo 4 Norway
SABS	South African Bureau of Standards Private Bag X191 Pretoria 0001 Republic of South Africa

[a] From Muramatsu and Parr.[1]

laboratory comparisons and data interpretation. I implore everyone to use these whenever possible, as journal editors should *require* method validation whenever a suitable reference material is available.

Quality Control

It is important that analyses, particularly those of a recurring nature, be monitored by means of a quality control program. Naturally, standards need to be checked periodically, and reference materials need to be analyzed to validate the accuracy of the method. Ideally, one would run suitable reference materials with each batch of samples, thus validating (and sometimes invalidating) each run. However, if large numbers of samples of a particular type are analyzed frequently, the cost of using reference materials for each run could become considerable. In a situation such as this, one should consider pool samples for quality control purposes.

For example, suppose a laboratory is analyzing large numbers of samples of, say, serum, or urine, or buffer solutions from a fraction collector.

TABLE II

OVERVIEW OF BIOLOGICAL AND ENVIRONMENTAL REFERENCE MATERIALS AND ELEMENTS QUOTED

Supplier	Material	Name or code number (supplier)	Code number	Unit weight or volume	Cost ($)	Quoted elements[a]
BCR (Community Bureau of Reference, CEC)	Biological					
	Aquatic plant	CRM-060	BCR-CRM-060	25 g	25	Al Ca *Cd* Cl *Cu* Fe *Hg* K Mg *Mn* N Na P *Pb* S Si Ti *Zn*
	Aquatic plant	CRM-061	BCR-CRM-061	25 g	25	Al Ca *Cd* Cl *Cu* Fe *Hg* K Mg *Mn* N Na P *Pb* S Si Ti *Zn*
	Olive leaves	CRM-062	BCR-CRM-062	25 g	25	Al Ca *Cd* Cl *Cu* Fe *Hg* K Mg *Mn* N Na P *Pb* S Si Ti *Zn*
	Skim milk powder	CRM-063	BCR-CRM-063	30 g	45	*Ca Cd Cl* Co *Cu Fe Hg K Mg* Mn *N Na* Ni *P Pb* Se Tl Zn
	Skim milk powder (lower level spiked)	CRM-150	BCR-CRM-150	30 g	30	*Cd* Co *Cu Fe Hg I* Mn Ni *Pb* Se Tl Zn
	Skim milk powder (higher level spiked)	CRM-151	BCR-CRM-151	30 g	30	*Cd* Co *Cu Fe Hg I* Mn Ni *Pb* Se Tl Zn
	Blood	CRM-194	BCR-CRM-194	?	?	*Cd Pb*
	Blood	CRM-195	BCR-CRM-195	?	?	*Cd Pb*
	Blood	CRM-196	BCR-CRM-196	?	?	*Cd Pb*
	Single cell protein	CRM-273	BCR-CRM-273	?	?	*Ca Fe K Mg N P*
	Environmental (nonbiological)					
	Fly ash	CRM-038	BCR-CRM-038	6 g	20	*As Cd Co* Cr *Cu Fe Hg Mn Na* Ni *Pb* Th V *Zn*
	Coal	CRM-040	BCR-CRM-040	50 g	56	*As Cd Co Cr F Hg Mn Ni Pb* Tl *Zn*
	Soil (calcareous loam)	CRM-141	BCR-CRM-141	40 g	25	Al Ca *Cd* Co Cr *Cu* Fe *Hg* K Mg Mn Na Ni P *Pb* Se Si Ti *Zn*
	Soil (light sandy)	CRM-142	BCR-CRM-142	40 g	25	Al Ca *Cd* Co Cr *Cu* Fe *Hg* K Mg Mn Na *Ni* P *Pb* Se Si Ti *Zn*
	Soil (amended sewage sludge)	CRM-143	BCR-CRM-143	40 g	25	Al Ca *Cd* Co Cr *Cu* Fe *Hg* K Mg Mn Na *Ni* P *Pb* Se Si Ti *Zn*

	Sewage sludge (domestic origin)	CRM-144	BCR-CRM-144	40 g	25	Al As Ca *Cd Co* Cr *Cu* Fe *Hg* K Mg *Mn* Na *Ni* P *Pb* Se Si Ti *Zn*
	Sewage sludge	CRM-145	BCR-CRM-145	40 g	25	Al Ca *Cd Co* Cr *Cu* Fe *Hg* K Mg *Mn* Na *Ni* P *Pb* Se Si Ti *Zn*
	Sewage sludge (industrial origin)	CRM-146	BCR-CRM-146	40 g	25	Al Ca *Cd Co* Cr *Cu* Fe *Hg* K Mg *Mn* Na *Ni* P *Pb* Se Si Ti *Zn*
	City waste incineration ash	CRM-176	BCR-CRM-176	30 g	25	Al As Ca *Cd Co Cr Cu Fe Hg* K Mg Mn Na *Ni* P *Pb* S *Sb Se* Si Ti *Tl Zn*
	Gas coal	CRM-180	BCR-CRM-180	?	?	*As C Cd Cl H Hg Mn N Pb Se V Zn*
	Coking coal	CRM-181	BCR-CRM-181	?	?	*As C Cd Cl H Hg Mn N Pb Se V Zn*
	Steam coal	CRM-182	BCR-CRM-182	?	?	*As C Cd Cl H Hg Mn N Pb Se V Zn*
BI (Behring Institute, FRG)	Biological					
	Blood	Control Blood for Metals 1 (OSSD)	BI-CBM-1	4 × 5 ml	22	*Cd Cr Hg Pb*
	Blood	Control Blood for Metals 2 (OSSE)	BI-CBM-2	4 × 5 ml	22	*Cd Hg Pb*
	Urine	Lanonorm Metals 1 (OSSA)	BI-CUM-1	12 × 50 ml	74	*As* Cd *Co Cr Cu F Hg Ni Pb*
	Urine	Lanonorm Metals 2 (OSSB)	BI-CUM-2	12 × 50 ml	74	*As* Cd *Co Cr Cu F Hg Ni Pb Tl*
	Urine	Lanonorm Metals 3 (OSSC)	BI-CUM-3	12 × 50 ml	74	*As Cd Co Cr Cu F Hg Ni Pb Tl*
BOWEN	Biological					
	Kale	Kale	BOWEN's Kale	100 g	15	Ag *Al As Au B Ba Br C Ca Cd* Ce *Cl Co Cr Cs Cu* Eu *F Fe Ga* H Hf *Hg* I *In K La* Li Lu *Mg Mn Mo N Na Ni* O *P Pb Rb* Ru *S Sb* Sc *Se Si* Sm Sn *Sr Th U V* W *Zn*

(continued)

TABLE II (*continued*)

Supplier	Material	Name or code number (supplier)	Code number	Unit weight or volume	Cost ($)	Quoted elements[a]
IAEA (International Atomic Energy Agency)	Biological					
	Milk powder	A-11	IAEA-A-11	25 g	40	Al As Au B Ba Br *Ca* Cd *Cl* *Co* Cr Cs *Cu* F *Fe* *Hg* I *K* Li *Mg* *Mn* Mo *Na* Ni *P* Pb *Rb* Sb *Se* Si Sn Sr V *Zn*
	Animal blood	A-13	IAEA-A-13	25 g	80	*Br* *Ca* *Cu* *Fe* *K* Mg *Na* Ni P Pb *Rb* *S* *Se* *Zn*
	Animal muscle	H-4	IAEA-H-4	2 × 10 g	80	Al As *Br* *Ca* Ce *Cl* Co Cr *Cs* *Cu* *Fe* *Hg* *K* *Mg* *Mn* Mo *Na* *Rb* S *Se* V W *Zn*
	Animal bone	H-5	IAEA-H-5	2 × 15 g	40	*Ba* *Br* *Ca* *Cl* *Fe* *K* *Mg* *Na* *P* *Pb* *Sr* *Zn*
	Horse kidney	H-8	IAEA-H-8	30 g	40	*Br* *Ca* *Cd* *Cl* Co Cs *Cu* *Fe* *Hg* *K* *Mg* *Mn* *Mo* *Na* *P* *Rb* S *Se* Sr *Zn*
	Copepod	MA-A-1(TM)	IAEA-MA-A-1	30 g	40	*Ag* *As* *Cd* *Co* *Cr* *Cu* *Fe* *Hg* *Mn* *Ni* *Pb* *Sb* *Se* *Zn*
	Fish flesh	MA-A-2(TM)	IAEA-MA-A-2	30 g	40	*Ag* *As* *Cd* *Co* *Cr* *Cu* *Fe* *Hg* *Mn* *Ni* *Pb* *Sb* *Se* *Zn*
	Mussel tissue	MA-M-2(TM)	IAEA-MA-M-2	25 g	40	Ag *As* Au *Br* *Ca* *Cd* Cl *Co* *Cr* *Cu* *Fe* *Hg* *Mg* *Mn* *Na* Pb *Rb* Sb Sc *Se* *Sr* *Zn*
	Rye flour	V-8	IAEA-V-8	50 g	40	Al Au Ba *Br* *Ca* Cd *Cl* Co Cs *Cu* *Fe* *K* *Mg* *Mn* Mo Na *P* *Rb* S Sb *Zn*
	Cotton cellulose	V-9	IAEA-V-9	25 g	80	Al *Ba* Br *Ca* Cd *Cl* *Cr* *Cu* Fe Ga Hf *Hg* Li *Mg* *Mn* *Mo* *Na* *Ni* *Pb* S Sc Se Sm Sn *Sr* Th U V
	Hay powder	V-10	IAEA-V-10	50 g	80	Al *Ba* *Br* *Ca* *Cd* *Co* *Cr* Cs *Cu* Eu *Fe* *Hg* K La *Mg* Mn *Mo* Na *Ni* *P* *Pb* *Rb* Sb *Sc* Se *Sr* *Zn*

Environmental (nonbiological)					
Air filter	Air-3/1	IAEA-Air-3/1	6 Filters (+ 6 blanks)	80	*As Au* Ba *Cd Co Cr* Cu *Fe* Hg *Mn* Mo *Ni Pb Se U V Zn*
Marine sediment	SD-N-1/2(TM)	IAEA-SD-N-1/2	25 g	80	*Ag* Al *As* Au *Ba* Be *Br* Ca *Cd Ce* Cl *Co Cr Cs Cu* Dy *Eu* Fe Hf Hg I K *La* Li *Lu* Mg *Mn* Mo Na *Nd Ni* P *Pb Rb Sb Sc* Se Si *Sm Sr* Ta *Tb Th* Ti *U V* W Y Yb *Zn* Zr
Lake sediment	SL-1	IAEA-SL-1	25 g	80	*As Ba Br* Ca *Cd Ce Co Cr Cs Cu Dy* Eu *Fe* Ga Ge *Hf* Hg K *La* Li Lu Mg *Mn* Mo *Na Nd Ni Pb Rb* S *Sb Sc* Se *Sm* Sr Ta Tb *Th Ti U V* W Y *Yb Zn* Zr
Soil	SOIL-7	IAEA-SOIL-7	25 g	80	Al *As* Ba Br Ca Cd *Ce Co Cr Cs Cu Dy Eu* F Fe Ga *Hf* Hg *Ho* K *La* Li Lu Mg *Mn* Mo Na Nb *Nd* Ni P *Pb Rb Sb Sc* Se Si *Sm Sr Ta Tb Th* Ti *U V Y Yb Zn Zr*
Fresh water	W-4	IAEA-W-4	Concentrate in quartz ampoule	80	*Al As B Ba Be Ca Cd Co Cr Cu Fe Hg K Mg Mn Mo Na Ni Pb Se Sr U V Zn*
Fresh water	W-5	IAEA-W-5	Concentrate in plastic bottle	40	(Same as W-4)
IRANT Environmental (nonbiological) (Institute of Radioecology and Applied Nuclear Techniques, CSSR)					
Coal fly ash	ECH	IRANT-ECH	50 g	80	*Al As Ba Be Ca Cd Ce Co Cr Cs Cu Eu Fe Ga Hf K La Lu Mg Mn Na Ni Pb Rb Sb Sc Si Sm Sr Ta Tb Th Ti U V Yb Zn Zr*

(*continued*)

TABLE II (*continued*)

Supplier	Material	Name or code number (supplier)	Code number	Unit weight or volume	Cost ($)	Quoted elements[a]
	Coal fly ash	ENO	IRANT-ENO	50 g	80	*Al As Ba Ca Ce Co Cr Cs Cu Eu Fe Ga Hf K La Lu Mg Mn Na Ni Pb Rb Sb Sc Si Sm Sr Ta Th Ti U V Yb Zn Zr*
	Coal fly ash	EOP	IRANT-EOP	50 g	80	*Al As Ba Be Ca Cd Ce Co Cr Cs Cu Eu Fe Ga Hf K La Lu Mg Mn Na Ni Pb Rb Sb Sc Si Sm Sr Ta Tb Th Ti U V Yb Zn Zr*
KL (Kaulson Laboratories Inc., USA)	Biological					
	Blood (lead control)	Contox No.0100 (high)	KL-100-H	4 × 5 ml	40	*Pb*
	Blood (lead control)	Contox No.0100 (low)	KL-100-L	4 × 5 ml	30	*Pb*
	Blood (lead control)	Contox No.0100 (med.)	KL-100-M	4 × 5 ml	30	*Pb*
	Urine (lead control)	Contox No.0110 (high)	KL-110-H	4 × 5 ml	40	*Pb*
	Urine (lead control)	Contox No.0110 (low)	KL-110-L	4 × 5 ml	30	*Pb*
	Urine (lead control)	Contox No.0110 (med.)	KL-110-M	4 × 5 ml	40	*Pb*
	Urine (heavy metal control)	Contox No.0140 (I)	KL-140-I	4 × 5 ml	50	*As Cd Hg*

	Urine (heavy metal control)	Contox No.0140 (II)	KL-140-II	4 × 5 ml	50	*As Cd Hg*
	Blood (heavy metal control)	Contox No.0141 (I)	KL-141-I	4 × 5 ml	50	*As Cd Hg*
	Blood (heavy metal control)	Contox No.0141 (II)	KL-141-II	4 × 5 ml	50	*As Cd Hg*
	Serum (trace metal control)	Contox No.0146 (I)	KL-146-I	4 × 5 ml	50	*Cu Fe Zn*
	Serum (trace metal control)	Contox No.0146 (II)	KL-146-II	4 × 5 ml	50	*Cu Fe Zn*
NBS (National Bureau of Standards, USA)	Biological					
	Albacore tuna	RM-50	NBS-RM-50	2 × 35 g	81	*As Hg* K Mn Na Pb *Se Zn*
	Bovine serum	RM-8419	NBS-RM-8419	3 × 4 ml	52	*Al Ca Co Cr Cu Fe K Mg Mn Mo Na Ni Se* V *Zn*
	Milk powder (nonfat)	SRM-1549	NBS-SRM-1549	100 g	138	Ag Al Al As Br *Ca Cd Cl* Co *Cr Cu* F *Fe Hg I K Mg Mn* Mo *Na P Pb* Rb *S* Sb *Se* Si Sn *Zn*
	Oyster tissue	SRM-1566	NBS-SRM-1566	30 g	89	*Ag As* Br *Ca Cd* Cl Co *Cr Cu* F *Fe Hg* I *K Mg Mn* Mo *Na Ni* P *Pb Rb* S *Se Sr* Th Ti *U V Zn*
	Wheat flour	SRM-1567	NBS-SRM-1567	80 g	105	As Br *Ca Cd Cu Fe Hg K Mn* Mo *Na* Ni Rb *Se* Te *Zn*
	Rice flour	SRM-1568	NBS-SRM-1568	80 g	105	*As* Br *Ca Cd Co Cu Fe Hg K Mn* Mo *Na* Ni *Pb* Rb *Se* Te
	Brewers yeast	SRM-1569	NBS-SRM-1569	50 g	88	*Cr*
	Citrus leaves	SRM-1572	NBS-SRM-1572	70 g	101	*Al As Ba* Br *Ca Cd* Ce Cl Co *Cr* Cs *Cu* Eu *Fe Hg I K* La *Mg Mn Mo* N *Na Ni P Pb Rb S* Sb Sc Se Sm Sn *Sr* Te Tl U *Zn*
	Tomato leaves	SRM-1573	NBS-SRM-1573	70 g	102	Al *As* B Br *Ca* Cd Ce Co *Cr Cu* Eu *Fe* Hg *K* La Mg *Mn* N *P Pb Rb* Sc *Sr Th* Tl *U Zn*

(*continued*)

TABLE II (*continued*)

Supplier	Material	Name or code number (supplier)	Code number	Unit weight or volume	Cost ($)	Quoted elements[a]
	Pine needles	SRM-1575	NBS-SRM-1575	70 g	102	*Al As* Br *Ca* Ce Co Cr *Cr Cu* Eu *Fe Hg K* La *Mn* N Ni *P Pb Rb* Sb Sc *Sr Th* Tl *U*
	Bovine liver	SRM-1577a	NBS-SRM-1577a	50 g	116	*Ag* Al *As* Br *Ca Cd Cl Co Cu Fe Hg K Mg Mn Mo* N *Na P Pb Rb S* Sb *Se Sr* Tl *U Zn*
	Urine (normal) (Freeze-dried)	SRM-2670	NBS-SRM-2670	2 × 20 ml	184	Al As Be *Ca* Cd Cl Cr *Cu* Hg K *Mg* Mn *Na* Ni Pb Pt S *Se*
	Urine (spiked) (Freeze-dried)	SRM-2670	NBS-SRM-2670			Al *As* Be *Ca Cd* Cl *Cr Cu Hg* K *Mg* Mn *Na* Ni *Pb* Pt S *Se*
	Environmental (nonbiological)					
	Crude oil	RM-8505	NBS-RM-8505	275 ml	52	V
	Coal fly ash	SRM-1633a	NBS-SRM-1633a	75 g	125	Al *As* Ba Be *Ca Cd* Ce Co *Cr* Cs *Cu* Eu *Fe* Ga Hf *Hg K Mg* Mn Mo *Na Ni Pb Rb* Sb Sc *Se Si Sr Th* Ti *Tl U* V *Zn*
	Fuel oil	SRM-1634a	NBS-SRM-1634a	100 ml	153	As Be Br Ca Cd Cl Co Cr Fe Hg *Mn* Mo *Na Ni Pb S Se V Zn*
	Coal (subbituminous)	SRM-1635	NBS-SRM-1635	75 g	108	Al *As Cd* Ce Co *Cr Cu* Eu *Fe* Ga Hf *Mn* Na *Ni Pb S* Sb Sc *Se Th* Ti *U V Zn*
	Fuel	SRM-1636a	NBS-SRM-1636a	set (12)	111	*Pb*
	Water	SRM-1641b	NBS-SRM-1641b	6 × 20 ml	126	*Hg*
	Water	SRM-1642b	NBS-SRM-1642b	950 ml	154	*Hg*
	Water	SRM-1643b	NBS-SRM-1643b	950 ml	163	*Ag* As B *Ba Be* Bi *Cd Co Cr Cu Fe Mn Mo Ni Pb Se Sr Tl V Zn*

	River sediment	SRM-1645	NBS-SRM-1645	70 g	135	*Al* As Ca *Cd Co Cr Cu* F *Fe Hg K* La *Mg Mn Na Ni Pb* S Sb Sc Se *Th Tl U V Zn*
	Estuarine sediment	SRM-1646	NBS-SRM-1646	75 g	116	*Al As* Be *Ca Cd* Ce *Co Cr* Cs *Cu* Eu *Fe* Ge *Hg* K Li *Mg Mn* Mo Na *Ni P Pb* Rb S Sb Sc Se Si Te Th Ti Tl *V Zn*
	Urban particulate	SRM-1648	NBS-SRM-1648	2 g	127	Ag *Al As* Ba Br *Cd* Ce Cl Co *Cr* Cs *Cu* Eu *Fe* Hf I In *K* La Mg Mn *Na Ni Pb* Rb S Sb Sc *Se* Sm Th Ti *U V* W *Zn*
NIES	Biological (National Institute for Environmental Studies, Japan)					
	Pepperbush	CRM-1	NIES-CRM-1	14 g	Free	*As Ba Ca Cd Co* Cr Cs *Cu Fe* Hg *K Mg Mn Na Ni* P *Pb Rb Sr* Tl *Zn*
	Chlorella	CRM-3	NIES-CRM-3	36 g	Free	*Ca* Cd *Co Cu Fe K Mg Mn* P Pb Sc *Sr Zn*
	Human hair	CRM-5	NIES-CRM-5	2 g	Free	Al Ba Br *Ca Cd* Cl Co *Cr Cu Fe Hg K Mg Mn Na Ni* P Pb Rb Sb Sc Se *Sr* Ti *Zn*
	Mussel	CRM-6	NIES-CRM-6	10 g	Free	*Ag* Al *As Ca Cd* Co *Cr Cu Fe* Hg *K Mg Mn Na Ni* P *Pb* Se Sr *Zn*
	Environmental (nonbiological)					
	Pond sediment	CRM-2	NIES-CRM-2	20 g	Free	*Al As* Br *Ca Cd Co Cr Cu Fe* Hg *K* La Mn *Na Ni* P *Pb* Rb Sb Sc Si Sr Ti V *Zn*
NRCC	Biological (National Research Council, Canada)					
	Lobster hepatopancreas	TORT-1	NRCC-TORT-1	30 g	62	*As Ca Cd Cl Co Cr Cu Fe Hg K Mg Mn Mo Na Ni P Pb S Se Sr V Zn*
	Environmental (nonbiological)					
	Marine sediment	BCSS-1	NRCC-BCSS-1	80 g	87	*Al As Be C Ca Cd Cl Co Cr Cu Fe Hg K Mg Mn Na Ni P Pb S Sb Si Ti V Zn*

(continued)

TABLE II (*continued*)

Supplier	Material	Name or code number (supplier)	Code number	Unit weight or volume	Cost ($)	Quoted elements[a]
	Seawater	CASS-1	NRCC-CASS-1	2 L	109	*As Cd Co Cr Cu Fe Mn Ni Pb Zn*
	Marine sediment	MESS-1	NRCC-MESS-1	80 g	87	*Al As Be C Ca Cd Cl Co Cr Cu Fe Hg K Mg Mn Na Ni P Pb S Sb Si Ti V Zn*
	Seawater	NASS-1	NRCC-NASS-1	2 L	109	*As Cd Co Cr Cu Fe Mn Mo Ni Pb Zn*
NYE (Nyegard & Co., Norway)	Biological					
	Serum	SERONORM(105)	NYE-105	6 × 3 ml	?	*Se*
	Urine	SERONORM(108)	NYE-108	6 × 10 ml	?	*Se*
	Serum	SERONORM(164)	NYE-164	10 × 5 ml	?	*Ca Cl Cu Fe K Mg N Na P Zn*
SABS (South African Bureau of Standards)	Environmental (nonbiological)					
	Coal (Witbank)	SARM-18	SABS-SARM-18	120 g	?	*Al* B *Ba Be* Br *Ca Ce Co Cr* Cs *Cu* Eu *Fe* Ga Ge *Hf* Hg *K La* Li *Mg Mn* Mo Na Nb *Ni P* Pb *Rb S* Sb *Sc Si Sm* Sn *Sr* Ta Tb *Th Ti U V* W Y *Zn Zr*
	Coal (O.F.S)	SARM-19	SABS-SARM-19	120 g	?	*Al As* B *Ba Be* Br *Ca Ce* Cl *Co Cr Cs Cu* Eu *Fe Ga Ge Hf* Hg *K La* Li *Mg Mn* Mo *Na* Nb *Ni P Pb Rb S* Sb *Sc* Se *Si Sm* Sn *Sr* Ta Tb *Th Ti U V* W Y Yb *Zn Zr*
	Coal (Sasolburg)	SARM-20	SABS-SARM-20	120 g	?	*Al As* B *Ba Be* Br *Ca Ce Co* Cr Cs *Cu* Eu *Fe Ga Hf Hg K La* Li *Mg Mn Na* Nb *Ni P Pb Rb S* Sb *Sc Se Si Sm* Sn *Sr Ta* Tb *Th Ti U V* W *Y* Yb *Zn* Zr

[a] If the element symbol is in italics, this indicates that a certified or recommended value is available; if not, only an information value is available.

One could then put together a large pool, or even several pools, of the material of interest. This would then be carefully analyzed against appropriate reference materials or by an independent method(s). Running an aliquot of the pool with each batch of samples, and knowing what the value is supposed to be, would serve as a quality control measure for each sample batch. The "correct" reading on the pool sample would serve to indicate that all is well regarding standards, additions, instrument response, etc. It would also quickly indicate when one of these parameters is amiss, though not necessarily which one. Pool samples used in this fashion also provide a measure of long-term continuity of the results, as well as indicating any slow "drift" in the results with time. As a pool is nearing exhaustion, it can be used to transfer accuracy to the next new pool.

Implementing all of these standardization procedures will result in improved accuracy of the methods and allow more meaningful interlaboratory comparisons of data on similar samples. Most important is the use of *appropriate* reference materials with the proper analyte concentration. These are only now becoming available and the situation will continue to improve rapidly.

[8] Methods for Metal Substitution

By DAVID S. AULD

Metal substitution can be performed by three conceptually different approaches: (1) removal of the native metal by a metal binding agent and/or competition with hydrogen ions at low pH, followed by insertion of the new metal, (2) direct displacement of the first metal by a second metal, and (3) biosynthesis of the metalloprotein under enriched conditions of the metal of choice. The last approach will not be described in this chapter although it has been used successfully to replace zinc by cobalt and other metals in the multisubunit enzyme RNA polymerase.[1] The first two methods can be applied in either the crystalline or solution states. The choice of the method and physical state of the protein will depend on the protein solubility, ease of metal removal, and stability of the apoprotein. Since metals frequently are involved in structural as well as functional roles the stability of the apoprotein is a particularly important consideration. If the apoprotein is unstable and/or metal removal is a very slow process direct

[1] D. C. Speckhard, F. Y.-H. Wu, and C.-W. Wu, *Biochemistry* **16**, 5228 (1977).

displacement may be the method of choice. However, if these factors are not pertinent the use of a metal-binding agent and/or pH is a convenient method to produce a large amount of apoenzyme in a short time. In addition removal of the metal in the crystalline state can aid in reducing the risk of contamination from adventitious metal ions and help to stabilize the apoprotein through the restricted motion of the protein in the crystalline state. The methodology involved in the first two approaches for both physical states of the enzyme will be illustrated in detail for carboxypeptidase A. A few exemplary references will be given to where these approaches have been used in other metalloproteins.

Three forms of bovine carboxypeptidase A are available, α, β, and γ, which differ in their length at the amino-terminal portion of the protein. The length and amino-terminal residues are for α, 307, Asn, β, 305, Val, and γ, 300, Ile.[2] Three methods of preparing the enzyme known as the Cox,[3] Allan,[4] and Anson[5] preparations yield predominantly the α, β, and γ forms, respectively. The apoenzymes of α and β are stable in solution but the γ enzyme looses activity very rapidly. If commercial Cox enzyme (Sigma, C 0261 or C6393) is used for the preparation of metallo derivatives, it should first be purified by affinity chromatography[6] since a contaminating endoprotease in combination with residual zinc carboxypeptidase will digest the apoenzyme slowly.[7] This contaminating protease is not removed by washing the crystals with water.

Preparation of Metallocarboxypeptidases in Solution

Equilibrium Dialysis in Presence of a Chelator

The Cox and Allan but not the Anson preparation of carboxypeptidase A can be used to prepare their respective apoenzymes by equilibrium dialysis versus metal-binding agents.[8–11]

[2] H. Neurath, R. A. Bradshaw, P. H. Petra, and K. A. Walsh, *Philos. Trans. R. Soc. London, B* **257,** 159 (1970).

[3] D. J. Cox, F. C. Bovard, J.-P. Bargetzi, K. A. Walsh, and H. Neurath, *Biochemistry* **3,** 44 (1964).

[4] B. J. Allan, P. J. Keller, and H. Neurath, *Biochemistry* **3,** 40 (1964).

[5] M. L. Anson, *J. Gen. Physiol.* **20,** 633 (1937).

[6] T. J. Bazzone, M. Sokolovsky, L. B. Cueni, and B. L. Vallee, *Biochemistry* **18,** 4362 (1979).

[7] R. Bicknell, A. Schäffer, D. S. Auld, J. F. Riordan, R. Monnanni, and I. Bertini, *Biochem. Biophys. Res. Commun.* **133,** 787 (1985).

[8] J. E. Coleman and B. L. Vallee, *J. Biol. Chem.* **235,** 390 (1960).

[9] R. C. Davies, J. F. Riordan, D. S. Auld, and B. L. Vallee, *Biochemistry* **7,** 1090 (1968).

[10] D. S. Auld and B. L. Vallee, *Biochemistry* **9,** 602 (1970).

[11] D. S. Auld and B. Holmquist, *Biochemistry* **13,** 4355 (1974).

Procedure. Apocarboxypeptidase Aα is prepared by dialysis of 2 ml of 2×10^{-4} *M* carboxypeptidase Aα in Visking-Nojax dialysis bags against four changes of a 100-fold volume excess of 2 m*M* 1,10-phenanthroline (OP) (Aldrich Chemical Co.) in 1.0 *M* NaCl, 0.05 *M* Tris buffer (pH 7.5), at 4° in a 250-ml plastic graduated cylinder. The OP is removed by dialysis vs four changes of dithizone-extracted Tris buffer.[8-10] In each case the first three steps should be 4–8 hr and the last one is an overnight dialysis.

Apoenzyme samples (10^{-4} *M*) are dialyzed at 4° with a 100-fold volume excess of 1 m*M* metal ion in 1 *M* NaCl, 0.05 *M* Tris (pH 7.5). The enzyme is stored in this manner until use. Enzyme is diluted to 10^{-5} *M* with 10^{-3} *M* metal ion solutions in order to ensure better than 99% formation of the metalloenzyme.[8] Each metalloenzyme prepared in this fashion displayed activities characteristic of the particular metal substituted for zinc.[9,10] The amount of zinc present in the apoenzyme and all of the metalloenzymes, as determined by atomic absorption spectrometry, was always in the range of 5×10^{-3} to 2.0×10^{-2} g-atom of zinc per mole of enzyme.[8-10]

Comments. A Fiberglas sewing thread is attached to the dialysis bag so that the bags can be conveniently held while changing the dialysis solution. The thread is taped to the outside of the cyclinder. Metal-free plastic gloves should always be used in handling the thread or touching the graduate cyclinder. A piece of clean Parafilm is placed over the top of the graduate cyclinder to prevent exposure of the water to airborne contaminates. This parafilm cover should never be reused. The preparation and handling of a buffer are more important than an accurate concentration. The preparation of 2 m*M* OP in 1 *M* NaCl, 0.05 *M* Tris will be used as an illustration. A stock solution of 40 m*M* OP adjusted to pH 7 is prepared in an acid-cleaned vessel using metal-free water[12] and stored in a closed metal-free container. The solution is poured directly into the graduate cyclinder to a level of 10 ml and then diluted to 200 ml with the dithizone-extracted[13] 1 *M* NaCl, 0.05 *M* Tris buffer with a Teflon-coated magnetic stirrer in operation.

Particular care must be taken during the removal of OP by dialysis. The dialysis bag should never be handled with uncovered hands since they are a good source of metal contamination, particularly zinc. The metallic salts added should be of the highest purity (e.g., Johnson-Matthey spec pure). A 1% contamination of zinc of the dialysis solution could easily prevent formation of a metalloenzyme which has a 1000-fold or less weaker stability constant than zinc. The new metalloenzyme is released from the dialysis bag by hanging it by the thread in a metal-free centrifuge

[12] J. F. Riordan and B. L. Vallee, this volume [1].

[13] B. Holmquist, this volume [2].

tube and piercing it with a sharp piece of acid cleaned glass rod drawn to a sharp point.

Alternatively, the new metalloenzyme can be crystallized by reducing the salt concentration to below 100 mM NaCl, keeping the metal ion constant at 10^{-3} M, and harvesting the crystals in the above described manner.

The use of OP to remove zinc by equilibrium dialysis has been used successfully with slight modifications of the above described procedure for a number of zinc enzymes, e.g., thermolysin,[14] angiotensin converting enzyme,[15] *Streptomyces griseus* carboxypeptidase,[16] and *Aeromonas* aminopeptidase.[17] Similar procedures have been described for alkaline phosphatase using 8-hydroxyquinoline-5-sulfonic acid[18] and for carbonic anhydrase using pyridine-2,6-dicarboxylic acid[19] as the metal-binding agents.

Preparation by Gel Permeation Chromatography

This procedure can be used with all forms of carboxypeptidase A.

Procedure. Gel filtration is carried out on a 2 × 35 cm column of Sephadex G-25.[20] The column is first washed at 4° with 3 column volumes of 2 mM OP in metal-free 1 M NaCl–0.05 M Tris or HEPES (pH 7.5) buffer to remove metals from the column. The column is eluted with metal-free buffer until the absorbance at 320 μm, characteristic of the presence of OP, is negligible.[21] The metal-free column is equilibrated with 1 column volume of the buffer of choice made 1 mM in the new metal. Approximately 10 ml of metal-free buffer followed by 20 ml of buffered 2 mM OP is placed on the top of the column. The apoenzyme is prepared by adding 0.2 ml of 20 mM OP to 2 ml of 2×10^{-4} M carboxypeptidase A in 1 M NaCl–0.05 M Tris or HEPES, pH 7.5, on ice. After an approximately 15 min preincubation the enzyme–OP mixture is applied to the column and the enzyme is eluted with buffer containing the new metal ion at 1 mM. The metallocarboxypeptidase prepared by this manner normally contained less than 0.10 g-atoms of residual zinc.

[14] B. Holmquist and B. L. Vallee, *J. Biol. Chem.* **249,** 4601 (1974).

[15] P. Bünning and J. F. Riordan, *J. Inorg. Biochem.* **24,** 183 (1985).

[16] K. Breddam, T. J. Bazzone, B. Holmquist, and B. L. Vallee, *Biochemistry* **18,** 1563 (1979).

[17] J. M. Prescott, F. W. Wagner, B. Holmquist, and B. L. Vallee, *Biochemistry* **24,** 5350 (1985).

[18] R. T. Simpson and B. L. Vallee, *Biochemistry* **7,** 4343 (1968).

[19] J. B. Hunt, M. J. Rhee, and C. B. Storm, *Anal. Biochem.* **79,** 614 (1977).

[20] T. L. Coombs, Y. Omote, and B. L. Vallee, *Biochemistry* **3,** 653 (1964).

[21] D. E. Drum and B. L. Vallee, *Biochemistry* **9,** 4078 (1970).

Direct Exchange by Equilibrium Dialysis

This approach has been described for the preparation of the Allan (β form) metallocarboxypeptidases[8,22] and should be useful for the α and γ forms of carboxypeptidase A. The procedure will be described for the preparation of the cobalt and cadmium enzymes.

Procedure. Samples of 1 ml of a solution of zinc carboxypeptidase, 1.0×10^{-4} *M*, in 1 *M* NaCl, 0.05 *M* Tris, pH 8.0, are placed in closed dialysis bags at 4°. These samples are dialyzed against 100-ml volume excess of buffer containing either 1.0×10^{-2} *M* Co^{2+} or 10^{-4} *M* Cd^{2+} for 2 and 4 days, respectively.[22] The metal ion stoichiometry can be determined either by atomic absorption or radioactive isotope techniques. If the latter approach is used, tracer quantities of ^{60}Co or ^{115}Cd (Oak Ridge National Laboratories) can be added as the chlorides to the dialyzates. The quantities of total metal added with the isotope is negligible, and the addition of 1.0×10^{-8} *M* labeled cobalt or cadmium to the stable ion is sufficient to yield a convenient counting level of 5000 counts per minute per ml in a well-type scintillation counter.[22] The radioactivity bound to the enzyme is calculated by subtracting the radioactivity of a control (no protein present) from that of the protein bag. The molarity of the metal bound to the protein is calculated from the radioactivity bound to the enzyme and the specific radioactivity of the original dialyzate. This value divided by the known molarity of the protein gives the g-atoms of metal per mole of protein.

Comments. The concentration of metal ion needed to displace the zinc is of course related to their relative stability constants. Since cobalt binding to carboxypeptidase is several orders of magnitude weaker than zinc, a 10^{-4} *M* Co^{2+} solution will not completely displace the Zn^{2+} from the native enzyme even at a 100-fold excess of Co^{2+}.[8,22] Multiple changes of the dialysis solution would of course help to shift the equilibrium in favor of the cobalt by removing the zinc. Ultrafiltration devices may be useful for this procedure.

If the rate of exchange is not the limiting step, multiple changes of a buffered 10^{-3} *M* metal ion dialyzate after 4–6 hr of dialysis will speed up the above described process, likely decreasing the time of preparation of a new metalloenzyme into the 24–36 hr time range. The rate of exchange is fast with cobalt but much slower with cadmium[22] or copper.[23] The counteranion used in these studies can also influence the results. In the case

[22] J. E. Coleman and B. L. Vallee, *J. Biol. Chem.* **236,** 2244 (1961).

[23] A. Schäffer and D. S. Auld, *Biochemistry* **25,** 2476 (1986).

Hg^{2+} and Cd^{2+}, the presence of Cl^- ions decreases the free concentration of metal ions because of their high chloride stability constants.[24]

Direct exchange of zinc for another metal has been used successfully with liver[25,26] and yeast[27] alcohol dehydrogenases, leucine aminopeptidase,[28] and hemorrhagic toxin e.[29]

Use of Ultrafiltration Device

This procedure should be useful for the Cox[3] and Allan[4] forms of the enzyme.

Procedure. Apocarboxypeptidase Aα is prepared in the following manner. Zinc enzyme, 0.2 ml of 2×10^{-4} *M* in 1.0 *M* NaCl–0.05 *M* HEPES, pH 7.5 buffer is added to a Centricon-30 microconcentrator (Amicon). Then 1.8 ml of 2 m*M* 1,10-phenanthroline (Aldrich Chemical Co.) in 1.0 *M* NaCl–0.05 *M* HEPES, pH 7.5 buffer is added and gently vortexed. Parafilm is placed on the top of the tube. A small hole is made in the parafilm to increase the rate of filtration during centrifugation. After a 5 min incubation at 22° the Centricon-30 device is placed in a plastic 40-ml centrifuge tube, inserted in a fixed-angle rotor, and centrifuged at 3000 *g* at 10° for 30 min. After centrifugation the device is removed from the rotor and the filtrate emptied into a beaker. The sample should be concentrated to approximately 0.2 ml during the centrifugation. The sample is again diluted to 2 ml with buffered 2 m*M* 1,10-phenahthroline solution and the entire procedure repeated 2 more times. The zinc is removed during this repeated incubation with 1,10-phenanthroline. The 1,10-phenanthroline solution is then removed by successive dilution and concentration of the sample with metal free buffer. This procedure is repeated until the filtrate shows negligible (<0.01 absorbance units) absorbance at 320 nm (usually 3 washes). The apoenzyme is then mixed with 1.8 ml of 1.0 *M* NaCl–0.05 *M* HEPES containing 5×10^{-5} *M* of the desired metal ion. After a 10 min incubation the sample is centrifuged and concentrated to 0.2 ml as described above. The new metalloenzyme is then diluted and concentrated again with metal free buffer to remove excess metal ion if excess metal ion inhibits the enzyme.

[24] L. G. Sillen and A. E. Martell, *in* "Stability Constants of Metal-Ion Complexes," Special Publ. No. 17. The Chemical Society, London, 1964.

[25] A. J. Sytkowski and B. L. Vallee, *Biochemistry* **17,** 2850 (1978).

[26] W. Maret, I. Andersson, H. Dietrich, H. Schneider-Bernlöhr, R. Einarsson, and M. Zeppezauer, *Eur. J. Biochem.* **98,** 501 (1979).

[27] A. J. Sytkowski, *Arch. Biochem. Biophys.* **84,** 505 (1977).

[28] G. A. Thompson and F. M. Carpenter, *J. Biol. Chem.* **25,** 53 and 1618 (1976).

[29] J. B. Bjarnason and J. W. Fox, *Biochemistry* **22,** 3770 (1983).

Comments. The Centricon device can be freed of metal ions prior to use by centrifuging with buffered 2 mM 1,10-phenanthroline solution. Care must be taken throughout the procedure not to touch the membrane of the sample reservoir with a pipet tip in order to prevent damage to the membrane and subsequent leaking of the sample. The most important step in the procedure is the care in handling the Centricon device as the 1,10-phenanthroline is removed from the system. It should never be touched along the rim with the fingers since they are a ready source of metal contamination. Disposable plastic pipet tips should be used to add metal free buffer or metal ion to the apoenzyme.

In the above described procedure the time of incubation of the 1,10-phenanthroline with the enzyme is very short (5 min) since the metal can be rapidly removed from Carboxypeptidase A. In addition centrifugation time must also be considered as further incubation time. For some proteins, e.g., carbonic anhydrase, the time of incubation would likely need to be much longer. The use of Centriprep concentrators (15 ml) should allow the processing of several milligram quantities of enzyme per centrifuge tube. In many cases this procedure maybe the preferred method of metal substitution since metal contamination is being minimized by the removal and replacement being performed in the same container and only small volumes of buffers are needed. In addition, the entire procedure can be completed in less than 1 day if the enzyme releases its metal rapidly to the metal-binding agent. The ultrafiltration devices may also be useful for direct exchange of metals through successive dilution and concentration of the metalloenzyme with an excess of the new metal ion (see section above).

This procedure produced apocarboxypeptidase A which contained <0.006 g-atom of metal per mole of protein.

Preparation of Metallocarboxypeptidase Derivatives from Crystals

This procedure has been used on the Anson and Cox preparations of the enzyme and should also be applicable to Allan enzyme. While the Anson apoenzyme looses activity within minutes in solution the crystalline apoenzyme can be converted to a new metallo derivative, with little loss in activity if the conversion is made within a 8-hr period.[11] Storing the Anson crystalline apoenzyme for periods of 24 hr or longer will result in progressively greater amounts of inactive enzyme. The following procedure can readily convert gram quantities of the zinc enzyme into new metallo derivatives within an 8-hr period.

Procedure. Add 200–250 mg of carboxypeptidase A as an aqueous suspension to a 50-ml round-bottom polypropylene centrifuge tube with a magnetic spin bar in the bottom. Centrifuge and decant liquid. Resuspend and stir the crystals (using a metal free glass rod) with 40 ml of 10 m*M* 1,10-phenanthroline (Aldrich Chemical Co.) in 10 m*M* MES (Sigma) at pH 7.0 ± 0.1 at room temperature (20 ± 5°) for 1–1.5 hr. Spin down the crystals in the tube at 10,000 rpm (1 or 2 min). A thin piece of parafilm is placed over the tube during this entire process to prevent contamination from dust in the air. When it is removed, it is discarded and a fresh piece used for the subsequent step, etc. Centrifugation is performed with the spin bar in the tube. Decant and repeat the above procedure at least 3 more times.

Wash the crystals with a 40 ml solution of 10 m*M* MES, pH 7.0 ± 0.1, stirring for 20–30 min in order to remove the 1,10-phenanthroline. Centrifuge, decant the solution, and repeat the procedure until the absorbance at 320 nm is neligible (<0.01 absorbance units) usually at least 4 times. Resuspend the crystals in 20 ml of 10 m*M* MES at pH 7.0 containing 5×10^{-4} *M* of the desired metal ion for 30 min. Centrifuge, decant the solution, and repeat the procedure. Before dissolving crystals, wash 1 or 2 times with 10 m*M* MES, pH 7.0 buffer. Crystals should dissolve easily in pH 7.5 buffered 2 *M* NaCl to at least 1 m*M* and remain in solution when diluted down to 1 *M* NaCl. This procedure should yield a metallo derivative which has less than 0.01 g-atoms zinc remaining. Zinc analysis is most easily determined by dissolving the crystals in 1 m*M* metal free HNO_3 to give $\sim 2 \times 10^{-4}$ *M* enzyme then diluting 10-fold in 10 m*M* HNO_3 before performing atomic absorption measurements. The enzyme concentration is determined using the molar absorbtivity of 6.42×10^5 M cm^{-1}.[30]

Comments. The key to the success of the above method is minimal contamination from adventitious metal ions as the 1,10-phenanthroline is removed from the system. Thus particular care must be taken when handling the centrifuge tube. It should never be touched on the rim with the fingers which are ready sources of metal ions. Parafilm covers should never be reused. Only the section of the parafilm next to the paper should touch the rim of the tube. Disposable plastic tips should be used to add buffer or metal ion to the apoenzyme. The glass rod used to loosen the crystals after centrifugation and plastic tips should be acid cleaned with a nitric acid solution (1 volume nitric to 2 volumes deionized distilled water) before using. These items should always be placed on metal-free surfaces or in clean storage containers when not in use. The accuracy of the concentration of a reagent is less important than its handling. The preparation of 10 m*M* MES will be used as an illustration. For the above experiments a 100 ml volume of 0.40 *M* MES buffer is prepared, adjusted

[30] R. T. Simpson, J. F. Riordan, and B. L. Vallee, *Biochemistry* **2,** 616 (1963).

to pH 7.0. This solution is extracted with dithizone and placed in a metal-free plastic bottle.[12,13] An acid cleaned 1-ml Teflon beaker is then used to transfer the 0.40 *M* MES buffer to the centrifuge tube. Deionized distilled water is added to the crystals to a predetermined level of ~39 ml and 1 ml of MES is then poured in. This procedure of preparing the 10 m*M* buffer is preferred to using a dithizone-extracted solution of 10 m*M* MES buffer for the following reasons. Much larger containers are needed for the storage and transfer of the dilute buffer solution (i.e., the quantity needed is 500 ml of 0.01 *M* vs 12 ml of 0.40 *M* MES buffer and the aliquot transferred is 40 ml vs 1 ml) thus providing a greater chance for surface contamination. Since deionized distilled water is the purest source of metal free water, dithizone extraction of the 0.40 *M* MES buffer and dilution with metal-free water will reduce the level of metal contamination further than dithizone extraction of the dilute solution. Only the purest of metal salts should be used in reconstitution of the apocrystals. Johnson-Matthey "Spec-pure" grade is an excellent source of such salts. If the metal salts are made by dissolving the pure metal in 6 *N* HCl or nitric acid (Aristar, BDH Chemicals) precautions must be taken in adding the dilute metal solutions to the apocrystals to prevent large pH changes. Metal sulfates are usually the best source since they do not cause a pH change upon preparation of a 0.1 *M* stock solution. If anions inhibit the metalloenzyme the least inhibitory one would dictate the choice of the metal salt.

[9] Preparation of Metal-Hybrid Enzymes

By WOLFGANG MARET and MICHAEL ZEPPEZAUER

General Principles

Metalloproteins containing one metal ion constitute only a small fraction as compared to the large number of those metalloproteins showing higher structural complexity with several subunits and/or metal-binding sites. Representative multimetal proteins, for which metal exchange studies have been reported, are listed in Table I. They greatly differ in the type(s) of metal ions present and the distribution of metal ions among subunits. The abundance and hence importance in many biological processes of metalloproteins call for a review of available strategies to exchange intrinsic metal ions. Such metal replacements are desirable and often necessary in metalloproteins to define the function of individual

TABLE I
MULTIMETAL PROTEINS[a]

Protein	Metal site(s) and subunit composition
Leucine aminopeptidase (*Aeromonas proteolytica*)	Two types of zinc ions in the monomer
Alcohol dehydrogenase (horse liver)	Two types of zinc ions in each subunit of the dimer
RNA polymerase (*Escherichia coli*)	Two types of zinc ions in two different subunits
Alkaline phosphatase (*Escherichia coli*)	Two types of zinc ions and one magnesium ion in each subunit of the dimer
Thermolysin (*Bacillus thermoproteolyticus*)	One zinc ion and four calcium ions in the monomer
Laccase (*Rhus vernicifera*, *Polyporus versicolor*)	Three types of copper centers in the monomer
Homobinuclear sites	
Acid phosphatase (pig allantoic fluid)	Binuclear iron center in the monomer
Hemocyanin (*Panulirus interruptus*)	Binuclear copper center in each subunit of the hexamer
Heterobinuclear sites	
Superoxide dismutase (bovine erythrocytes)	Binuclear copper/zinc site in each subunit of the dimer
Concanavalin A (jack bean)	Binuclear calcium/manganese site in the monomer

[a] Only examples are given in the table that are treated with selected references in the text.

metal centers, which would otherwise be buried under their overlapping features. This is achieved either by making visible or by masking one metal center by use of a particular probe. In any case the result of the metal substitution is a metal-hybrid enzyme[1] in which the properties of the metal center to be studied are enhanced. These metal-hybrid enzymes comprise not only derivatives with another type of metal ion as a more suitable spectroscopic or kinetic marker, but also derivatives with a particular isotope of the native metal ion in a selected site, e.g., ^{57}Fe for Mössbauer studies of iron proteins and ^{61}Ni to resolve metal hyperfine structure in EPR studies of nickel proteins. The complete retention of enzymatic function is not an absolute requirement for the usefulness of

[1] Enzymes such as alkaline phosphatase (*Escherichia coli*) containing zinc and magnesium must be considered *native* metal-hybrid enzymes.

the metal-substituted enzyme. Frequently an enzyme exhibits varying degrees of enzymatic activity when substituted with a series of metal ions. The comparison of these species with the native enzyme reveals important structural features of the active site pertinent to the following question: Why during evolution was a particular metal ion chosen for the proper function of the enzyme?[2] This by no means indicates that the native metal ion is optimally adapted to fulfill the enzyme's function, since hyperactivation has been found in a metal-substituted leucine aminopeptidase.[3]

It is not our intention to review the techniques of metal exchange, since this has already been done,[4,5] but rather to treat the inherent problems in metal exchange studies of multimetal enzymes. Also not considered in this chapter are metal exchange studies in metal-containing prosthetic groups or in proteins containing metal clusters, for which a specific chemistry has developed.

For the preparation of metal-hybrid enzymes the crucial question is: How can the metal exchange be designed to be selective for one metal center leaving the other metal centers unaffected?

Three approaches exist for metal exchange: (1) preparation of an apoenzyme with subsequent insertion of a metal ion, (2) displacement of the intrinsic metal ion by another metal ion (sometimes referred to as interchange or direct exchange), and (3) growth of a microorganism producing the metalloenzyme in a medium, in which the concentration of the native metal ion is very low and the concentration of the metal ion to be incorporated is relatively high. This biosynthetic incorporation of metal ions does not necessarily provide selectivity, since it will most probably produce a fully metal-substituted species.[6] Only the remaining two approaches are suited for the preparation of metal-hybrid enzymes. The first approach (1)

[2] M. Zeppezauer and W. Maret, *in* "The Importance of Chemical Speciation in Environmental Processes" (M. Bernhard, F. E. Brinckman, and P. J. Sadler, eds.), p. 99. Springer-Verlag, Berlin, 1986.

[3] J. M. Prescott, F. W. Wagner, B. Holmquist, and B. L. Vallee, *Biochemistry* **24,** 5350 (1985).

[4] W. Maret, *in* "Zinc Enzymes" (I. Bertini, C. Luchinat, W. Maret, and M. Zeppezauer, eds.), p. 17. Birkhäuser-Verlag, Basel, 1986.

[5] D. Auld, this volume [8].

[6] However, one exception is worth mentioning in this context. In a recent study, P. G. Nettersheim, H. G. Engeseth, and J. D. Otvos [*Biochemsitry* **24,** 6744 (1985)] have isolated Zn,Cd-hybrid metallothioneins from the livers of rabbits which were exposed to cadmium. In these proteins four to five cadmium atoms are distributed among the seven metal-binding sites. Exactly the same Zn,Cd-hybrid metallothionein has been obtained through an intermolecular metal exchange by simply mixing together Zn_7- and Cd_7-metallothionein.

describes a two-step procedure. Therefore, it offers the advantage of being site-specific during either metal removal or metal insertion.

Apoproteins

Upon selective removal of a fraction of metal ions from metalloproteins partially metal-depleted apoenzymes are formed. When two metal ions per protein are present the latter have been designated as half-apoenzymes. For instance, in the characterization of metalloproteins which already contain useful spectroscopic probes such as copper or iron one does not necessarily aim at replacing the metal ion, but instead, specifically depleting one metal site. Hence, the partial apoenzymes are the important hybrid species and, therefore, they will also be considered metal-hybrid enzymes in this chapter. In these partial apoenzymes migration of metal ions from occupied to vacant sites has to be considered. Such a migration has been described in a derivative of superoxide dismutase (SOD) from bovine erythrocytes, a dimeric protein containing a binuclear copper/zinc site in each subunit. In Cu_2-SOD, the half-apoenzyme lacking zinc, copper was found to migrate as a function of pH, resulting in a species having the copper and zinc site filled by copper in one subunit and both sites empty in the other subunit.[7] In contrast, metal ions do not migrate in half-apo alcohol dehydrogenase from horse liver, in which the catalytic zinc ions have been removed without affecting the noncatalytic zinc ions.[8] Since affinity constants for metal ions in metalloproteins are usually high as compared to metal-activated enzymes, empty metal sites are avid chelating agents. Therefore the partial apoenzymes are stable toward partitioning of metal ions between different sites only within a narrow pH range.

It is known from the X-ray structure determination of half-apo alcohol dehydrogenase that this derivative is a stable species and that the metal-binding site is preformed in the absence of the metal ion.[9] However, if the metal ion plays a role in organizing the protein structure or a part of it, the apoenzyme may not be stable or may not be a species in which the structures of the other metal centers correspond to their native states. Then it is necessary to insert a "silent" metal ion, i.e., a metal ion that does not perturb spectroscopic measurements at other metal centers, to

[7] J. S. Valentine, M. W. Pantoliano, P. J. McDonnell, A. R. Burger, and S. J. Lippard, *Proc. Natl. Acad. Sci. U.S.A.* **76,** 4245 (1979).

[8] W. Maret, I. Andersson, H. Dietrich, H. Schneider-Bernlöhr, R. Einarsson, and M. Zeppezauer, *Eur. J. Biochem.* **98,** 501 (1979).

[9] G. Schneider, H. Eklund, E. Cedergren-Zeppezauer, and M. Zeppezauer, *Proc. Natl. Acad. Sci. U.S.A.* **80,** 5289 (1983).

restore the structure or function of the enzyme. Instability of an apoenzyme can also limit the choice of techniques for metal exchange. Even in metal sites seemingly very similar a technique which has been successfully applied in one system may fail in another. The following two examples illustrate that point. Both liver alcohol dehydrogenase and *Escherichia coli* aspartate carbamoyl transferase contain a zinc ion tetrahedrally surrounded by four sulfur atoms from cysteines. Yet, the removal of the metal ion and subsequent metal insertion has been possible only in the latter enzyme.[10] Pea and lentil lectins have a binuclear calcium/manganese site as in concanavalin A. In contrast to the behavior of concanavalin A, apoproteins could not be prepared from the lectins and one had to resort to the direct metal exchange procedure.[11]

Experimental Design: Factors Affecting Selectivity

The general pathways for metal exchange in multimetal enzymes are summarized in Scheme I. In this scheme steps eliciting selectivity are indicated. The asterisk denotes steps allowing for a specific blocking of a particular site. Often a combination of methods is rewarding, especially with increasing complexity of the system. After the right selectivity parameters have been determined according to considerations stressed in the following discussion, they have to be optimized as a function of temperature, time, and concentration of reactants, which can be difficult and time consuming.

Metal Removal

The parameters affording selectivity during metal removal are pH, type of chelating agent, physical state of the protein, and oxidation state of the metal ion. Lowering the pH leads to protonation of metal ligands and, if the properties of the metal sites are sufficiently different, to selective loss of one type of metal ion. Increasing the pH may have a similar effect. However, metal removal at high pH will proceed through a different mechanism, in which local conformational changes of the protein are thought to loosen the metal–protein interaction. If a change of pH alone does not provide the right selectivity for metal removal, a combination of pH variation with the use of a chelating agent may be helpful. The effect of pH is then 2-fold: it may assist the chelating agent by controlling the dissociation constant of the metal–protein complex. Moreover, it defines

[10] R. S. Johnson and H. K. Schachmann, *Proc. Natl. Acad. Sci. U.S.A.* **77,** 1995 (1980).

[11] L. Bhattacharyya, C. F. Brewer, R. D. Brown III, and S. H. Koenig, *Biochemistry* **24,** 4974 (1985).

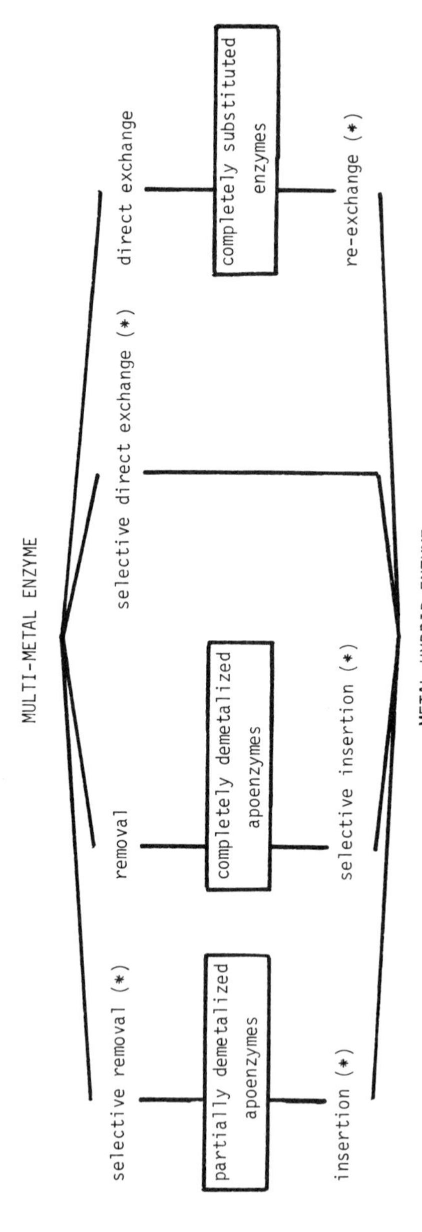

SCHEME I. Metal exchange pathways in multimetal proteins.

the correct charges of the protein and of the chelating agent. The chelating agent can be tailored to suit the particular purpose in the following way: the type, charge, and size of the chelating agent influence its approach to the metal ion and its preference for a site or type of metal ion. The denticity of the chelating agent determines the velocity of metal removal. For instance, chelating agents have been developed which show site specificity for iron either in the C-terminal or in the N-terminal part of transferrin.[12] This preference has been attributed to the charge and size of the chelating agent. In equine liver alcohol dehydrogenase the physical state of the protein turned out to be an important factor for metal exchange, since the catalytic pair of metal ions could be specifically removed only when the protein was kept in the crystalline state.[8]

The oxidation state of the redox-active metal ions determines the thermodynamic and kinetic stability of a complex. Removal of iron(III) from iron proteins is usually preceded by reduction of iron(III) to iron(II). On the other side, iron(II) is used for metal insertion. Therefore, the choice of the reducing or oxidizing agent is another means to modify selectively one metal site.

Metal Insertion

Site specificity in metal insertion has been achieved only by pH adjustments. Two counteracting effects of pH influence metal insertion. With increasing pH the solubility of the metal ion decreases whereas the stability of the metal–protein complex increases. Given this restriction, reconstitution may be successful only if a carrier for the metal ion is used. Whereas a chelating agent used for metal removal should have high kinetic and thermodynamic stability, sufficient lability is required for the carrier. A carrier keeps the metal ion soluble and helps to overcome barriers of charge and hydrophobicity posed by the protein. These factors dictate the choice of the carrier. So far, the design of carriers optimized to direct metal ions in various binding sites has not yet been pioneered. Instead, metalloproteins have been employed as biological carriers to deliver their metal ion(s) to an apoprotein. Examples are the reconstitution of native copper or zinc proteins by copper or zinc thioneins.[13,14]

Metal Displacement

The basis of this procedure is the relative magnitude of the affinity constants of the foreign versus the native metal ion for the binding site in

[12] G. J. Kontoghiorghes and R. W. Evans, *FEBS Lett.* **189,** 141 (1985).

[13] A. O. Udom and F. O. Brady, *Biochem. J.* **187,** 329 (1980).

[14] H.-J. Hartmann, L. Morpurgo, A. Desideri, G. Rotilio, and U. Weser, *FEBS Lett.* **152,** 94 (1983).

the protein. Thus, the exchange is mainly controlled by the concentration of reactants, pH, and the type of complexation of the foreign metal ion. In metal exchange studies of alcohol dehydrogenase it was proposed that the noncatalytic pair of zinc ions, which is easily accessible on the surface of the protein, is able to form a transient binuclear complex with the entering metal ion. Such a kinetic mechanism is thought to favor metal exchange of the noncatalytic metal ion in this enzyme.[15]

Blocking Reactions

An additional method encountered is the blocking of one or several sites in such a way that the blocked site will not interfere with metal exchange at other sites. This may be accomplished in different ways. For instance, the oxidation of cobalt(II) to cobalt(III) creates a kinetically inert metal ion that does not exchange over the time scale of typical metal exchange studies.[16] Alternatively, the ligands of a metal-binding site are modified in order to prevent them from binding a metal ion, or the metal–protein complex is made thermodynamically more stable by site-directed mutagenesis. Finally, a catalytic metal site is susceptible to blocking by binding of a molecule that makes this site inaccessible for a chelating agent or for a metal ion used in a displacement experiment.

Assignment of Metal Ions in a Metal-Substituted Enzyme

It is not sufficient to assume that the established metal–protein stoichiometry in a metalloprotein remains the same in the metal-substituted enzyme. Due to the different affinity of the foreign metal ion to the protein, the metal ion may occupy unspecific binding sites and/or leave binding sites unoccupied. As a consequence the metal–protein stoichiometry has to be checked and binding sites have to be assigned. The more difficult task is to identify individual binding sites and to decide which metal ion has been exchanged. To this end the metal ions can be operationally defined on the basis of their exchange kinetics (fast/slow) or spectroscopic properties (symmetry, metal environment). Usually a spectroscopic parameter is measured as a function of metal–protein ratios. The correlation of such binding data with measurements of the enzymatic activity helps to identify catalytic metal ions. Regulatory metal ions are characterized by their influence on the catalytic performance of the enzyme. Finally, ligand-sensitive spectroscopic features will be used to characterize metal-

[15] A. J. Sytkowski and B. L. Vallee, *Biochemistry* **17,** 2850 (1978).
[16] J. I. Legg, *Coord. Chem. Rev.* **25,** 103 (1978).

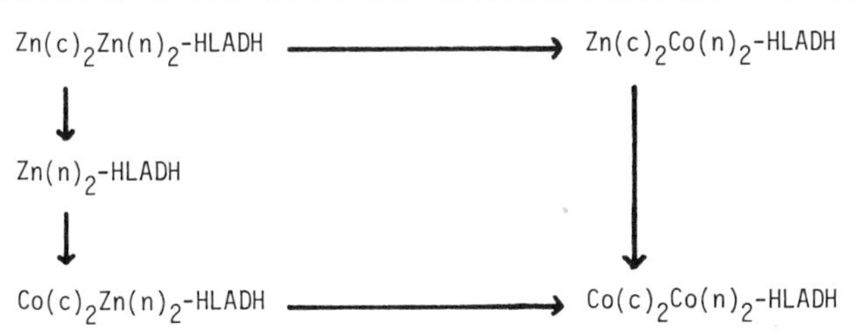

SCHEME II. Metal exchange in horse liver alcohol dehydrogenase.

binding sites, e.g., hyperfine interactions in magnetic resonance and charge transfer spectra in absorption spectroscopy.

Examples

Presently, it is impossible to predict the function of metal ions in organizing and stabilizing a protein structure. The very nature of the latter may limit the applicability of some of the outlined principles for metal exchange. Therefore, we want to deepen the subject with the following instructive examples. Metal exchange studies in three types of proteins containing two metal ions will be presented: liver alcohol dehydrogenase with two zinc-binding sites far from each other in one subunit, *E. coli* RNA polymerase with two zinc ions in different subunits, and superoxide dismutase from bovine erythrocytes with a binuclear copper/zinc site in each subunit. A detailed description of the three interacting metal-binding sites in *E. coli* alkaline phosphatase follows. The last examples summarize metal exchange experiments in laccase, an oxidase containing four copper ions, and in thermolysin with one zinc and four calcium ions.

Horse Liver Alcohol Dehydrogenase (HLADH, EC 1.1.1.1)

The two hybrid enzymes with Co, Cd, or ^{65}Zn in either the noncatalytic (n) or catalytic (c) site are prepared by two different procedures (Scheme II). Substitution of the noncatalytic zinc ions to prepare Me(n)$_2$Zn(c)$_2$-HLADH was carried out by the metal displacement technique in solution at pH below 6.[15] Under these conditions the noncatalytic zinc ions are exchanging fast against other metal ions whereas the exchange of the catalytic metal ions is slow and becomes significant only if the pH is lowered further. Conditions had to be worked out for each metal ion.[15,17] Both ^{65}Zn(n)$_2$Zn(c)$_2$-HLADH and ^{65}Zn(n)$_2$^{65}Zn(c)$_2$-HLADH have been prepared according to these procedures. In the latter derivative a

[17] A. J. Sytkowski and B. L. Vallee, *Biochemistry* **18,** 4095 (1979).

reexchange of the fast-exchanging noncatalytic zinc ions against natural abundance zinc was possible, since the use of acetate buffer leads to a selective blocking of the catalytic zinc site. The result of the reexchange is the second hybrid enzyme $Zn(n)_2{}^{65}Zn(c)_2$-HLADH. The transfer of this reexchange to metal ions other than zinc has not been successful. To introduce other metal ions specifically in the active site, a completely different route was described. In crystal suspensions of the enzyme the catalytic metal ions are specifically removed by the chelating agent dipicolinic acid at pH 7.[8] It was concluded that the physical state of the protein is responsible for the specificity of this procedure. The resulting half-apoenzyme, $Zn(n)_2$-HLADH, can be reconstituted with various metal ions in solution or in crystal suspension. The kinetics and mechanisms of reconstituting $Zn(n)_2$-HLADH with various metal ions have recently been reviewed.[18] As noted for the exchange of the noncatalytic zinc ions, the conditions for metal insertion had to be newly defined for each metal ion. The second half-apoenzyme, $Zn(c)_2$-HLADH, has not been prepared. The hybrid enzymes $Co(n)_2Zn(c)_2$-HLADH and $Zn(n)_2Co(c)_2$-HLADH are stable species and interchange of the two metal ions between the two binding sites is extremely slow if it occurs at all. In these species the binding site of cobalt has been assigned on the basis of activity measurements, electronic absorption spectra of the cobalt ion in each site,[8] and X-ray structure analyses of both the half-apoenzyme $Zn(n)_2$-HLADH and the hybrid enzyme $Zn(n)_2Co(c)_2$-HLADH.[9]

Bovine Erythrocyte Superoxide Dismutase (SOD, EC 1.15.1.1)

The dimeric enzyme has a binuclear copper/zinc site in each subunit. Such heterobinuclear metal sites are easier to deal with in a metal exchange experiment than homobinuclear sites (see below). The starting point for the preparation of metal-hybrid enzymes of superoxide dismutase are the two half-apoenzymes Cu_2-SOD and Zn_2-SOD, i.e., species either lacking zinc or copper. They are prepared by removing both metal ions at pH 3.8 with EDTA followed by incubation with copper at pH 3.8 or with stoichiometric amounts of zinc at pH 5.9.[19] This *selective insertion* is possible, since zinc binds about 10^3 times tighter to its native site than to the copper site. In addition, the affinity of copper for its native site is about 10^6 times higher than for the zinc site.[20] An alternative pathway to

[18] M. Zeppezauer, I. Andersson, H. Dietrich, M. Gerber, W. Maret, G. Schneider, and H. Schneider-Bernlöhr, *J. Mol. Catal.* **23,** 377 (1984).

[19] J. S. Valentine and M. W. Pantoliano, *in* "Copper Proteins" (T. G. Spiro, ed.), p. 291. Wiley, New York, 1981.

[20] J. Hirose, M. Yamada, C. Hayakawa, H. Nagao, M. Noji, and Y. Kidani, *Biochem. Int.* **8,** 401 (1984).

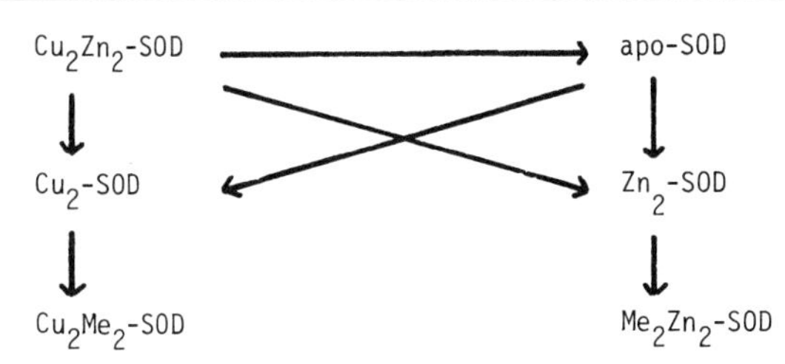

SCHEME III. Metal exchange in bovine erythrocyte superoxide dismutase.

the half-apoenzymes is a *selective removal* of copper with cyanide at pH 6 or of zinc at pH 3.6 in the absence of EDTA. These metal exchange procedures are summarized in Scheme III. It is conceivable to hybridize subunits of different metal composition in dimeric or oligomeric enzymes. Such hybridization studies have been carried out between iron and manganese superoxide dismutases.[21] Metal-hybrid enzymes of this kind will certainly be important tools in the future for studying subunit interactions in analogy to the existing investigations on iron/cobalt hybrid hemoglobins.

RNA Polymerase from E. coli (RPase, EC 2.7.7.6)

The core enzyme, lacking the σ-subunit, has the subunit composition $\alpha_2\beta\beta'$. The β- and β'-subunit each bind one zinc ion. This subunit location of the metal ions was established by oxidizing the cobalt(II) ions in the fully substituted and the hybrid enzyme to cobalt(III) and then demonstrating the association of cobalt(III) ions with the β- and β'-subunit. The complete substitution of zinc with cobalt was achieved by growing *E. coli* on a Co^{2+}-enriched and Zn^{2+}-depleted medium.[22] In contrast to this *in vivo* procedure, hybrid RPase with selective substitution of zinc in the β-subunit by cobalt was prepared by a denaturation–reconstitution procedure *in vitro*.[23] The denaturation with 7 *M* urea was found to be a compulsory step demonstrating that the exposure of the enzyme to a chelating agent at physiological pH alone is not sufficient for metal removal.

Alkaline Phosphatase from E. coli (AP, EC 3.1.3.1)

The history of establishing the correct metal–protein stoichiometry in this enzyme[24] illustrates several crucial aspects: (1) the quality of the

[21] D. A. Clare, J. Blum, and I. Fridovich, *J. Biol. Chem.* **259,** 5932 (1984).
[22] D. C. Speckhard, F. Y.-H. Wu, and C.-W. Wu, *Biochemistry* **16,** 5228 (1977).
[23] D. Chatterji and F. Y.-H. Wu, *Biochemistry* **21,** 4651 (1982).
[24] W. F. Bosron, F. S. Kennedy, and B. L. Vallee, *Biochemistry* **14,** 2275 (1975).

method for metal analysis, (2) the importance of accurately determining the concentration and the molecular weight of the enzyme, and (3) the critical steps of isolation and preparation of the enzyme. Traces of zinc introduced during the purification steps can lead to an accumulation of this metal ion in the protein. In addition, EDTA, used to remove the metal ions from the enzyme, binds strongly to the apoenzyme. Excessive zinc contents may result from this protein/EDTA complex binding adventitious zinc upon metal reconstitution.[25] On the other side, metal depletion of the protein can occur during dialysis. For instance, the applied purification procedure results in loss of magnesium. Thus, magnesium had earlier escaped detection, and its functional role was overlooked. Consequently, it is now recognized that two zinc ions and one magnesium ion are the intrinsic metal ions per subunit of the dimeric enzyme.[24,26]

Recently the high-resolution X-ray structure of the enzyme has confirmed these three metal-binding sites and established the organization of the three metal ions in metal-binding triads in each subunit.[27] A fourth cadmium-binding site remote from this triad has also been characterized.[28,29]

The similar chemical composition of the three metal-binding sites renders a specific labeling of each site difficult. Zn_4Mg_2-AP is completely demetalized by treatment with ammonium sulfate at pH 9. Two zinc ions per dimer restore the enzymatic activity, which is further regulated by binding of the second pair of zinc and magnesium.

The incorporation of cobalt(II) into apo-AP has been studied in the absence and presence of magnesium (Scheme IV).[30] In the absence of magnesium up to three pairs of cobaltous ions are bound. The correlation between enzymatic activity and increase in electronic absorption was

[25] Tight binding of a chelating agent to the protein is a frequent phenomenon. Therefore, care must be taken to completely remove the chelating agent before starting the reconstitution with metal ions.

[26] W. F. Bosron, R. A. Anderson, M. C. Falk, F. S. Kennedy, and B. L. Vallee, *Biochemistry* **16,** 610 (1977).

[27] J. M. Sowadski, M. D. Handschumacher, H. M. K. Murthy, B. A. Foster, and H. W. Wyckoff, *J. Mol. Biol.* **186,** 417 (1985).

[28] H. W. Wyckoff, M. D. Handschumacher, H. M. K. Murthy, and J. M. Sowadski, *Adv. Enzymol.* **55,** 453 (1983).

[29] Binding sites for adventitious metal ions on proteins are not uncommon. P. M. Harrison, G. C. Ford, D. W. Rice, J. M. A. Smith, A. Treffery, and J. L. White [*Rev. Port. Quim.* **27,** 119 (1985)] have described zinc binding sites on the ferritin molecule. In another study, J. M. Rifkind [*Biochemistry* **13,** 2475 (1974)] detected high-affinity sites for copper on hemoglobin. From these observations, it is becoming obvious that the excess of metal ions has to be removed after metal insertion, and that contamination with other types of metal ions should be carefully controlled.

[30] R. A. Anderson, F. S. Kennedy, and B. L. Vallee, *Biochemistry* **15,** 3710 (1976).

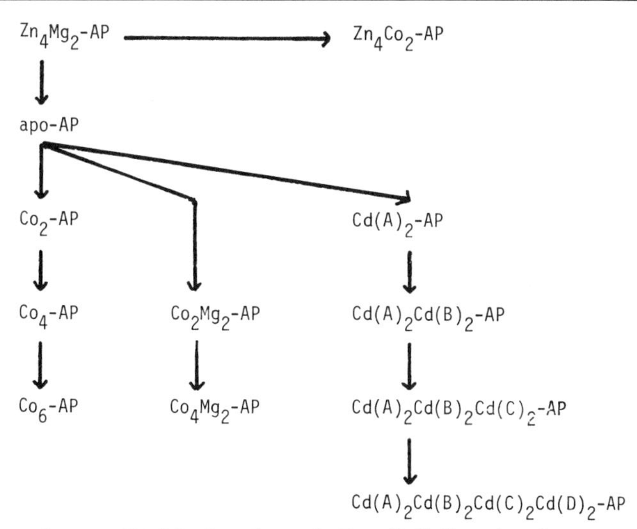

SCHEME IV. Metal exchange in *E. coli* alkaline phosphatase.

interpreted as arising from the first pair of cobalt(II) ions entering primarily the noncatalytic sites (Co_2-AP), the second pair entering the catalytic sites (Co_4-AP), and the third pair entering the magnesium sites (Co_6-AP). The magnesium site itself was probed with cobalt(II) in the species Zn_4Co_2-AP. In contrast, it was concluded that in the presence of magnesium cobalt(II) first enters the catalytic sites (Co_2Mg_2-AP). Only one more pair of cobalt(II) ions could be incorporated into this species forming Co_4Mg_2-AP. During these studies it had already been noted that two cobalt sites interact and that both phosphate and magnesium affect the electronic absorption spectra of the cobalt ion.

Much work has also been devoted to the preparation of cadmium-substituted derivatives of this enzyme for ^{113}Cd NMR studies.[31] A successive incorporation of cadmium into sites designated A, B, C, and D of the completely demetalized protein was noted resulting in Cd_2-, Cd_4-, Cd_6-, and Cd_8-AP (Scheme IV). With the available X-ray structure it is now possible to assign the spectroscopically determined sites: A is the catalytic zinc site, B the second zinc site, C the magnesium site, and D the metal-binding site remote from the metal triad. The following description of the stepwise incorporation of cadmium into the apoprotein is thought to illustrate the complexity of metal insertion into an apoprotein by taking into account the cooperativity of metal binding, the role of the metal ions in organizing the structure of the active site, and the migration of metal

[31] J. E. Coleman and P. Gettins, *Adv. Enzymol.* **55,** 381 (1983).

ions. Addition of 2 mol of cadmium per mol of protein at pH 6.5 creates Cd(A)$_2$-AP. Raising the pH to 7.5 makes possible the binding of another pair of cadmium at the B sites forming Cd(A)$_2$Cd(B)$_2$-AP. However, cadmium in this site is in fast exchange and the conformational flexibility in sites A and B is reduced only when the pH is further raised to pH 8.5. Therefore, the metal ions and pH influence the structure of the metal-binding sites. The fluxional behavior of the protein around the metal sites is completely frozen when the protein becomes phosphorylated. Phosphorylation leads to some migration of cadmium to the sites C. The third pair of cadmium ions enters these C sites [Cd(A)$_2$Cd(B)$_2$Cd(C)$_2$-AP]. The described cooperativity does not occur only within the metal-binding triad of one subunit, but it also extends to the triads of the two subunits. Under conditions for phosphorylation of Cd(A)$_2$-AP only in one subunit, cadmium migrates from the A site in one subunit to the empty B site in the other subunit to yield Cd(A)Cd(B)-AP* (the asterisk denotes the phosphorylated enzyme).

Blue Copper Oxidases and Homobinuclear Metal Sites

Laccases (EC 1.10.3.2) are monomeric glycoproteins of variable molecular weight and the best studied and least complicated of the blue copper oxidases. The four copper ions of laccase are classified as one Type 1 copper center, characterized by a distorted tetrahedral geometry with metal/thiolate coordination, one Type 2 copper center of tetragonal geometry with no sulfur-containing ligands, and one Type 3 copper center designating a binuclear copper site. The distribution of these types of copper centers among blue copper oxidases is shown below.

Enzyme	Type 1	Type 2	Type 3	Total metal content
Laccases	1	1	1	4
Ceruloplasmin	2	1	2	7
Ascorbate oxidase	3	1	4	12

Laccase can be completely demetalized with cyanide at pH 7 and reconstituted with cuprous ions. Exposure to air oxidizes the bound metal ions to the cupric state with concomitant restoration of enzymatic activity. To study the function of the copper ions in laccase one is interested in a separate manipulation of each type of copper center. To this end, it has been possible to remove specifically the Type 2 copper ion by treatment with the chelating agent bathocuproine disulfonate at pH 4.0 in the presence of ascorbate and guanidinium hydrochloride.[32] On the other hand, the Type 1 site was differentially labeled and thereby masked with mercury by taking advantage of sulfur in this site being a very strong ligand for heavy metal ions.[33] Metal exchange at the binuclear Type 3 site has not yet been reported for blue copper oxidases. However, procedures for metal exchange are known for the single Type 3 copper sites of hemocyanin and tyrosinase.[34,35] One copper ion in the binuclear site of a molluscan deoxyhemocyanin was specifically removed by treatment with cyanide.[36] The conditions for this procedure are very critical, since cyanide tends to remove all the copper ions from the protein. In addition, a selective oxidation of one of the two copper ions in deoxy hemocyanin was achieved by reduction with nitric oxide produced from nitrite. It was proposed that nitrite and added ascorbate protect the second copper ion from oxidation.[37] X-Ray crystallographic studies of an arthropodan hemocyanin, which has a completely different quaternary structure as compared to the molluscan hemocyanin, revealed that each copper ion is coordinated to three nitrogen atoms of histidines.[38] X-Ray crystallographic studies of these proteins at higher resolution will hopefully characterize the source of asymmetry in the site revealed by the preparation of these two derivatives, half-apohemocyanin and half-methemocyanin. The derivatives of the binuclear copper site in hemocyanin are listed in Scheme V.

Metal exchange studies are also reported for homobinuclear iron sites. While the two iron sites in hemerythrin exhibit nonidentical coordination environments,[39] the protein has fiercely resisted any metal exchange at its active site. In contrast, one iron ion has been selectively removed from the binuclear iron site of an acid phosphatase.[40] This was possible, because the removal of this iron ion at pH 4.9 in the presence of dithionite is

[32] R. Malkin, B. G. Malmström, and T. Vänngard, *Eur. J. Biochem.* **7,** 253 (1969).

[33] M. C. Morris, B. L. Hauenstein, Jr., and D. R. McMillin, *Biochim. Biophys. Acta* **743,** 389 (1983).

[34] E. I. Solomon, K. W. Penfield, and D. E. Wilcox, *Struct. Bonding (Berlin)* **53,** 1 (1983).

[35] C. Rüegg and K. Lerch, *Biochemistry* **20,** 1256 (1981).

[36] R. S. Himmelwright, N. C. Eickman, and E. I. Solomon, *J. Am. Chem. Soc.* **101,** 1576 (1979).

[37] A. J. M. Schoot Uiterkamp, *FEBS Lett.* **20,** 93 (1972).

[38] W. P. J. Gaykema, A. Volbeda, and W. G. J. Hol, *J. Mol. Biol.* **187,** 255 (1985).

[39] R. E. Stenkamp, L. C. Sieker, L. H. Jensen, J. D. McCallum, and J. Sanders-Loehr, *Proc. Natl. Acad. Sci. U.S.A.* **82,** 713 (1985).

[40] J. L. Beck, D. T. Keough, J. De Jersey, and B. Zerner, *Biochim. Biophys. Acta* **791,** 357 (1984).

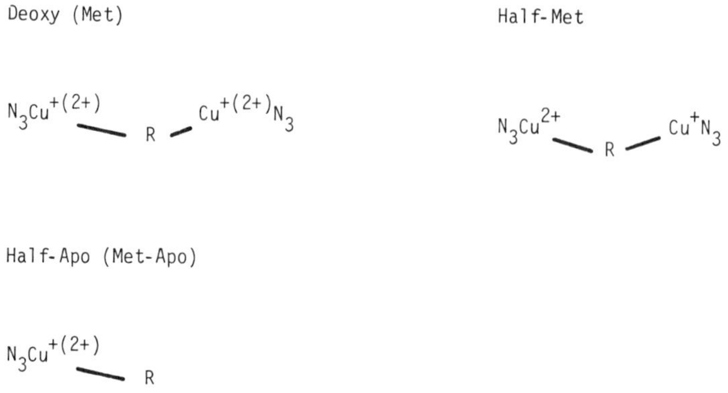

Deoxy (Met)

$N_3Cu^{+(2+)}$ — R — $Cu^{+(2+)}N_3$

Half-Met

N_3Cu^{2+} — R — $Cu^{+}N_3$

Half-Apo (Met-Apo)

$N_3Cu^{+(2+)}$ — R

SCHEME V. Modifications of the binuclear copper site in hemocyanin. In this representation R designates the endogenous bridging atom; it is not taken into account that the copper atoms may be bridged by exogenous ligands, e.g., dioxygen. Among those, some are able to break the endogenous bridge, resulting in derivatives called half-Met-2 or dimer. The latter is the EPR-detectable Met form.

much faster than the removal of the second iron ion. Conditions are given in the report to insert other metal ions into the vacant site, forming iron/metal-hybrid enzymes.

Thermolysin from Bacillus thermoproteolyticus (TL, EC 3.4.24.4)

The neutral protease thermolysin contains one zinc ion and four types of calcium ions. The latter are organized in a binuclear calcium site and two isolated calcium sites.[41] Selective removal of the catalytic zinc ion has been achieved with the chelating agent 1,10-phenanthroline.[42] Insertion of cobalt(II) into this derivative, Ca_4-TL, yielded $CoCa_4$-TL. The calcium ions in thermolysin can be displaced by lanthanide ions. In these lanthanide-substituted derivative only one lanthanide ion was found to bind to the binuclear calcium site, i.e., the resulting stoichiometry is $ZnLn_3$-TL.[41] A hybrid enzyme has been prepared with one bound Tb^{3+} ion and Co^{2+} in the active site. The pair Tb^{3+}/Co^{2+} in this derivative was used as a spectroscopic distance probe.[43] From the calculated distance of 13.7 Å between terbium and cobalt it was inferred that terbium occupies the binding place at the binuclear calcium site.

[41] B. W. Matthews and L. H. Weaver, *Biochemistry* **13,** 1719 (1974).

[42] B. Holmquist and B. L. Vallee, *J. Biol. Chem.* **249,** 4601 (1974).

[43] W. D. Horrocks, B. Holmquist, and B. L. Vallee, *Proc. Natl. Acad. Sci. U.S.A.* **72,** 4764 (1975).

[10] Introduction of Exchange-Inert Metal Ions into Enzymes

By HAROLD E. VAN WART

One of the most fruitful means of exploring the function of metal atoms in metalloenzymes has been through substitution of the native metal by others with different properties.[1] For example, the effects of the replacement on activity can reveal important facts about the role of the metal in catalysis. In addition, the preparation of new metallo species offers the opportunity to introduce into the enzyme metal atoms that are chromophoric or which have other spectral properties that can be used as probes of events occurring at specific binding sites. This has been a particularly fruitful approach for enzymes requiring zinc or magnesium, since these metals are spectroscopically silent.

Until relatively recently, metal substitutions in enzymes had been restricted to exchange-labile metal ions which could be easily incorporated by virtue of their rapid exchange rates. The slow rate of ligand substitution reactions of the inert metals has, in fact, been the primary reason for their lack of attention. However, once it was recognized that the incorporation of exchange-inert metals into enzymes had the potential to provide some unique information, means have been derived to prepare such metallo species. In this chapter, the principles that underlie the use of exchange-inert metals with enzymes are first reviewed and the types of information to be gained from such studies are briefly surveyed. Next, the procedures for the preparation of enzymes containing exchange-inert metals are described together with methods for the interpretations of the results.

Background and Underlying Principles

The variations in the rates at which ligands can be substituted or exchanged (these terms are used synonomously here) from the inner coordination sphere of metal ions is enormous. For example, the rates of water exchange for various metal aqua complexes span over 16 orders of magnitude.[2] Qualitatively, those complex ions for which the rates are very rapid have been termed "labile," while those for which the rates are

[1] B. L. Vallee and W. E. C. Wacker, *in* "The Proteins" (H. Neurath, ed.), Vol. 5. Academic Press, New York, 1970.

[2] J. E. Huheey, "Inorganic Chemistry," 3rd Ed., p. 548. Harper & Row, New York, 1983.

very slow are termed "inert." In reality, there is a continuous gradation in lability and Taube has suggested that a complex be considered labile if the exchange reaction in question is complete in 1 min under ordinary conditions.[3]

Many metals are exchange inert in at least one stable oxidation state, depending in part on the identity of the ligands.[3,4] For the biochemical uses to be described in the following sections, it is desirable to work with metal ion complexes that are considerably more inert than Taube's definition given above. This is because, after preparation of an enzyme containing an exchange-inert metal, the subsequent studies to be conducted can require hours or days to complete. Thus, for biochemical studies, attention has been centered on the most inert metal ions. In addition, the exchange-inert metal ions to be used with enzymes must be those that are stable under biological conditions. Thus, the biological applications of inert metals have been largely limited to Co(III) and Cr(III), since both are very inert and can be incorporated into enzymes via their divalent states (see below) under relatively mild conditions. Though other metals such as Ru(II) and Pt(II) could meet the stated qualifications under certain conditions, they have received little attention and will not be considered here.

With regard to the incorporation of exchange-inert metals into enzymes, there arises an immediate paradox. The very property of inertness that is to be exploited in the product would seem to make the preparation of the exchange-inert metalloenzyme difficult, since the ligand exchange reactions needed to introduce the metal into the protein are, by definition, slow. In principle, this problem can be overcome by two strategies. First, a metal ion that is labile with one set of ligands can be used to incorporate the metal into a protein site which, after substitution of the protein ligands, results in a metal complex that is exchange inert. For example, Werber and associates reported the use of [Co(III)(phen)(ATP)(O_2^-)] to incorporate Co(III) into myosin by exchange of O_2^- for a protein ligand.[5] Although it has recently been shown that the mechanism of incorporation of Co(III) into myosin in this reaction is *not* by Co(III) ligand exchange,[6] this type of approach is still theoretically possible.

A more generally applicable strategy of proven utility is to incorporate the metal ion into the protein in an exchange-labile oxidation state that is subsequently oxidized or reduced to an exchange-inert state. Two labile/inert redox pairs that fit this criterion are Co(II)/Co(III) and Cr(II)/Cr(III)

[3] H. Taube, *Chem. Rev.* **50,** 69 (1952).

[4] H. Taube, *Surv. Prog. Chem.* **6,** 1 (1970).

[5] M. M. Werber, A. Oplatka, and A. Danchin, *Biochemistry* **13,** 2683 (1974).

[6] J. A. Wells, M. M. Werber, J. I. Legg, and R. G. Yount, *Biochemistry* **18,** 4793 (1979).

and this accounts in large measure for the almost exclusive use of the inert Co(III) and Cr(III) ions with enzymes to date. Both intrinsic (tightly bound) and extrinsic (weakly bound) metal ions in metalloenzymes have been replaced by Co(III) and Cr(III).[7,8] If it is an intrinsic metal that is to be replaced by Co(III) or Cr(III), the enzyme containing the labile divalent ion is first prepared and the inert, trivalent ion is obtained by *in situ* oxidation. Extrinsic metal ions that bind weakly to the enzyme are more difficult to replace stoichiometrically by this procedure. However, if they bind tightly to a cofactor or coreactant, a Cr(III) or Co(III) complex of this species can be prepared for use with the enzyme.

It is appropriate before proceeding further to consider some of the basic properties of the Co(II)/Co(III) and Cr(II)/Cr(III) redox pairs. Co(II) complexes can adopt a number of different coordination geometries and numbers, including 4-coordinate tetrahedral or square planar, 5-coordinate bipyramidal or 6-coordinate octahedral complexes. Since Co (II) is a d^7 ion, all of its complexes are paramagnetic, whether they are high or low spin, and all are substitution labile. Co(III) is a d^6 ion and virtually all of its known complexes are octahedral.[9] Co(III) shows an affinity for nitrogen donor ligands, but will also form complexes with halides and oxygen ligands. The oxidation of Co(II) to Co(III) is favored in the presence of nitrogen donors and is facilitated at alkaline pH. Almost all octahedral Co(III) complexes are spin-paired ($e_g^6t_{2g}^0$) and diamagnetic. Only complexes such as $[Co(III)(H_2O)_3F_3]$ and $[Co(III)(F_6)]^{3-}$ are paramagnetic.[9] Two spin allowed (singlet–singlet) absorption bands are expected for octahedral Co(III) complexes corresponding to the $^1T_{1g} \leftarrow {}^1A_{1g}$ and $^1T_{2g} \leftarrow {}^1A_{1g}$ transitions. These transitions can be split by lowering the octahedral symmetry and are influenced by the steric arrangement of the ligands. The energy of the $^1T_{1g} \leftarrow {}^1A_{1g}$ transition, which is in the visible region, shows a dependence on the identity of the donor ligands, in accordance with the spectrochemical series. Thus, the wavelength of maximum absorbance decreases as the number of nitrogen donors is increased.

Cr(II) is a d^4 ion that forms mostly octahedral complexes. All are strong reductants and are stable only under anaerobic conditions. They are easily oxidized by molecular oxygen to give stable Cr(III) complexes. Most octahedral Cr(II) complexes are high spin ($t_{2g}^3e_g^1$) and paramagnetic, though $[Cr(II)(CN)_6]^{4-}$ is a low spin complex. Mononuclear high spin complexes have a single spin-allowed $^5E_g \leftarrow {}^5T_{2g}$ optical transition. This is the source of the sky-blue color of aquo-Cr(II). Cr(III) forms complexes

[7] J. I. Legg, *Coord. Chem. Rev.* **25**, 103 (1978).

[8] W. W. Cleland and A. S. Mildvan, *Adv. Inorg. Biochem.* **1**, 163 (1979).

[9] F. A. Cotton and G. Wilkinson, "Advanced Inorganic Chemistry," 4th Ed., Chap. 21. Wiley, New York, 1980.

with a wide variety of ligands and almost all are octahedral. All Cr(III) complexes are paramagnetic with three unpaired electrons and have three spin-allowed optical transitions.

Survey of Applications of the Inert Properties of Co(III) and Cr(III) to Enzymes

Although it is not the specific intention of this chapter to provide a review of the use of exchange-inert metals in enzymes, it is appropriate to delineate the types of applications in which the use of these metals has been most fruitful. The exchange-inert property of Co(III) and Cr(III) ions has been used in a number of different ways in different classes of metalloenzymes to obtain several types of important mechanistic information. Two very abundant classes of metalloenzymes are those in which Zn(II) is an intrinsic functional constituent and those that require Mg(II) as a cofactor for activity. Notable among the latter class are the enzymes that utilize some form of Mg(II) complex with a nucleotide or other phosphorylated species as a substrate or cofactor. Thus, it is not surprising that most of the applications of exchange-inert metals have been directed toward these two classes of enzymes. The major strategy for zinc metalloenzymes has been to substitute Co(II) for Zn(II) and oxidize it *in situ* to Co(III). The most widely used strategy for the Mg(II)-dependent enzymes has been to prepare either Cr(III) or Co(III) complexes of the phosphorylated coreactant for use with the enzyme. Regardless of which system is being studied, the types of information obtained can be divided into several categories, as exemplified below.

Role of Inner Sphere Coordination in Metal Ion Function

The substitution of an inert for a labile metal atom can provide a test of whether a ligand exchange reaction is involved in the function of the metal ion. Since ligand exchange reactions for an inert metal will be slow compared to catalysis, the inertness of the metal should inactivate the enzyme if such an exchange reaction is required for each round of catalysis. In zinc metalloenzymes such as carboxypeptidase A and carbonate dehydratase, some of the most frequently postulated mechanisms (e.g., the metal–carbonyl mechanism) require coordination of part of the substrate to the metal during each round of catalysis. For both enzymes, activity is retained on substitution of Co(II) for Zn(II), but activity is lost on oxidation of Co(II) to Co(III).[10–13] This has been interpreted to support the

[10] H. Shinar and G. Navon, *Biochim. Biophys. Acta* **334,** 471 (1974).
[11] H. Shinar and G. Navon, *Eur. J. Biochem.* **93,** 313 (1979).
[12] H. E. Van Wart and B. L. Vallee, *Biochem. Biophys. Res. Commun.* **75,** 732 (1977).
[13] H. E. Van Wart and B. L. Vallee, *Biochemistry* **17,** 3385 (1978).

importance of inner sphere coordination of the substrate to Zn(II) in the mechanism of these enzymes, although caution must be exercised in these types of studies, since the loss of activity can potentially be due to other factors, such as a change in coordination geometry or number (see below).

Many Cr(III) and Co(III) complexes of nucleotides have been shown to bind rapidly and firmly to their enzymes,[8] indicating that they are behaving like the respective Mg(II) complexes. This constitutes proof for these systems that the binding does not require the formation of an inner sphere bond from a residue on the enzyme to the metal and that the binding occurs via interaction with other portions of the complex. In contrast, Sperow and Butler have shown that, unlike the Mg(II) complex of pyrophosphate, the Cr(III) complex does not bind to yeast pyrophosphatase.[14] This indicates that an inner sphere bond to the metal is important for binding of substrate by this enzyme.

The role of inner sphere coordination to Mg(II) in the reaction catalyzed by the aspartate kinase–homoserine dehydrogenase complex from *E. coli.* has been studied by Takahashi and associates.[15] In this reaction, aspartate and ATP are substrates and threonine is a feedback inhibitor. Co(II) can substitute for Mg(II) as the extrinsic metal and *in situ* oxidation of reaction mixtures containing Co(II) and aspartate or threonine gave enzyme–Co(III)–aspartate and enzyme–Co(III)–threonine adducts, respectively. This provides evidence for the binding of aspartate and threonine in the first coordination sphere of the metal.

Differentiation of Metal Ions in Different Sites and between Those with Different Roles

Many metalloenzymes possess more than one metal binding site, making it difficult to study the structure and function of each separately. The selective incorporation of an inert metal into one or more of these sites allows the opportunity to study the other site(s) selectively or to differentiate between the structure and/or function of these sites. For example, *E. coli* alkaline phosphatase has three types (a catalytic, a structural, and a regulatory) of metal binding sites. The enzyme binds three atoms of Co(II) per subunit and Anderson and Vallee have shown that only one is susceptible to oxidation.[16,17] Experiments with various Zn(II), Co(II) hybrids suggest that it is the Co(II) atom in the catalytic site that is oxidized and that the incorporation of Co(III) into this site alone results in the loss of activity.

[14] J. W. Sperow and L. G. Butler, *J. Biol. Chem.* **251,** 2611 (1976).

[15] J. K. Wright, J. Feldman, and M. Takahashi, *Biochemistry* **15,** 3704 (1976).

[16] R. A. Anderson and B. L. Vallee, *Proc. Natl. Acad. Sci. U.S.A.* **72,** 394 (1975).

[17] R. A. Anderson and B. L. Vallee, *Biochemistry* **16,** 4388 (1977).

Many Mg(II)-dependent enzymes that utilize phosphorylated substrates or cofactors have more than one metal binding site. In many instances, one of the sites is that at which a Mg(II)–nucleotide complex binds and the other is a site at which a free metal ion binds. In studies using labile metal atoms, these two sites are difficult to study separately. However, the use of inert Co(III) or Cr(III) complexes of the phosphorylated substrate or cofactor allows saturation of the first site without allowing these ions access to the second site, which can then be saturated with other labile or inert metal ions.

For example, Mildvan and associates used the diamagnetic β,γ-bidentate [Co(III)$(NH_3)_4$ATP] complex to saturate the Mg^{II}ATP binding site of the catalytic subunit of cAMP-dependent protein kinase.[18] This allowed the paramagnetic Mn(II) ion to be placed selectively in the second site and permitted distances between the two sites to be estimated from the effects of Mn(II) on the nuclear relaxation rates of the [Co(III)$(NH_3)_4$ATP] complex. Balakrishnan and Villafranca[19] have used the opposite approach in their studies of glutamine synthase. They selectively incorporated Cr(III) and Co(III) into the nonnucleotide binding site to allow the selective binding of Mn(II)–nucleotide to the second site. The hybrid with diamagnetic Co(III) at the nonnucleotide site and Mn^{II}ADP or Mn^{II}ATP complexes in the second site enabled the EPR spectrum of Mn(II) to be examined cleanly without interference from the metal at the other site.

Use of Exchange-Inert Metal Nucleotide Complexes to Determine Enzyme Mechanisms

Cleland has used Cr(III) complexes of various nucleotides as "dead end" inhibitors in the investigation of the mechanism of various multisubstrate enzymes.[8] Such complexes can be used to predict inhibition patterns that can be used to differentiate between alternative kinetic mechanisms. For example, it has been shown that Cr^{III}ATP is a competitive inhibitor of hexokinase vs Mg^{II}ATP and a noncompetitive inhibitor vs glucose, enabling it to be shown that the mechanism is random and not ordered.[20] Lardy and associates have found that Cr^{III}ATP and Cr^{III}ADP inhibit ATP hydrolysis by adenosinetriphosphatase competitively, but inhibit ITP hydrolysis noncompetitively.[21] This has been interpreted in terms of the presence of two types (e.g., regulatory and catalytic) of nucleotide binding sites.

[18] J. Granot, H. Kondo, R. N. Armstrong, A. S. Mildvan, and E. T. Kaiser, *Biochemistry* **18,** 2339 (1979).

[19] M. S. Balakrishnan and J. J. Villafranca, *Biochemistry* **18,** 1546 (1979).

[20] K. D. Danenberg and W. W. Cleland, *Biochemistry* **14,** 28 (1975).

[21] S. M. Schuster, R. E. Ebel, and H. A. Lardy, *Arch. Biochem. Biophys.* **171,** 656 (1975).

A different type of use of inert metals has been proposed by Legg[22] for investigating the role of specific amino acid residues in catalytic mechanisms. He has used Co(III) as a site specific amino acid-modifying reagent. The rationale is that certain residues (e.g., tyrosine, histidine) can be chemically modified to enable them to chelate a labile metal ion such as Co(II). Subsequent oxidation to Co(III) freezes the ion on the residue, thus modifying it in the hope that the modification will affect activity and reveal the role, if any, of the residue in catalysis. Using this approach, it has been suggested that the tyrosine-248 residue of carboxypeptidase A is essential for peptidase but not esterase activity.[22]

Use of Exchange-Inert Metal Nucleotide Complexes to Determine Nucleotide Binding Geometry

Mg(II) can complex with nucleotides to produce a number of isomers that differ with regard to the number of phosphate oxygens that serve as ligands and their steric arrangement. It is unlikely that the enzymes that utilize such complexes as substrates bind the individual isomers with equal affinity. In order to study which isomers are preferentially utilized, complexes of the nucleotides with Cr(III) and Co(III) have been prepared.[8] Since these complexes are inert, they retain their configuration after isolation and cannot interconvert during their interaction with the enzyme. The Cr(III) complexes of ATP and other nucleotides are stable with either water or ammonia as the remaining ligands. However, only the triammine and tetraammine complexes of $Co^{III}ATP$ are redox stable.

$Cr^{III}ATP$ can exist as a number of different isomers. There are four tridentate forms in which one oxygen of the α-, β-, and γ-phosphate groups are each liganded to the metal, four distinct bidentate forms involving coordination of both the β- and γ-phosphates and a single monodentate form with a coordinated γ-phosphate.[8,23] The bi- and tridentate forms can have either the Λ left-hand or Δ right-hand screw sense. These complexes have been used in two ways to yield information about the active site conformation and mechanism of the respective enzymes. The first is related to the observation that either mixtures of the bidentate or tridentate isomers or the individual isomers themselves act as inhibitors with markedly different binding constants for different enzymes. For example, Cleland and associates have found that for a series of kinases, bidentate $Cr^{III}ATP$ was a much better inhibitor of hexokinase and glycerokinase than tridentate $Cr^{III}ATP$, the two forms were equally effective toward myokinase and arginine kinase, and tridentate $Cr^{III}ATP$ was a

[22] M. S. Urdea and J. I. Legg, *Biochemistry* **18,** 4984 (1979).

[23] D. Dunaway-Mariano and W. W. Cleland, *Biochemistry* **19,** 1496 (1980).

slightly better inhibitor of creatine, 3-phosphoglycerate, and acetate kinases.[8]

A second feature of these studies is that certain isomers of $Cr^{III}ATP$ can serve as substrates for the different enzymes. The kinases preferentially use bidentate isomers as substrates. It has been found that hexokinase, glycerokinase, arginine kinase, and creatine kinase utilize the Λ-β,γ-bidentate isomer, while pyruvate kinase, phosphofructokinase, and myokinase utilize the Δ isomer.[8] The rates for the first turnover and for product release vary for the different kinases, and allow mechanistic differences between the enzymes to be elucidated.

Use of Inert Metals to Provide Site Location Information

Another way to exploit the inert property of metals is to use them to provide some type of information about the location of the binding site on the enzyme. The most obvious example of this is the use of the metal as an affinity label that will not migrate after incorporation into the enzyme. Thus, it should be possible to proteolyze the exchange-inert metalloenzyme and isolate the peptide fragments bound to the metal, thus enabling the ligands to the metal to be identified. Takahashi and co-workers have pursued such an approach for aspartate kinase.[24,25] Since Co(II) substitutes for Mg(II) in the aspartate kinase reaction, *in situ* oxidation was used to prepare an affinity labeled Co(III)–enzyme–ATP complex. After proteolysis, a fragment containing this complex was isolated, providing partial localization of the binding site.

Wu and associates used Co(III) as an affinity label for the Zn(II) binding site in RNA polymerase.[26] The enzyme is composed of five subunits ($\alpha_2\beta\beta'\sigma$) which collectively contain 2 mol of tightly bound Zn(II) per mole of complex. During the isolation of the individual subunits, the Zn(II) is lost and it was difficult to determine which subunit(s) contained the metal-binding sites. Since Co(II) substitutes for Zn(II), oxidation was employed to immobilize the metal ion as Co(III). After isolation, it was found that at least one of the metal binding sites was on the β'-subunit.

Experimental Methods and Considerations

The strategies that have so far been used successfully to incorporate Co(III) or Cr(III) into enzymes can be divided into two classes. The first

[24] C. Ryzewski and M. T. Takahashi, *Biochemistry* **14,** 4482 (1975).

[25] J. K. Wright, J. Feldman, and M. Takahashi, *Biochem. Biophys. Res. Commun.* **72,** 1456 (1976).

[26] C.-W. Wu, F. Y.-H. Wu, and D. C. Speckhard, *Biochemistry* **16,** 5449 (1977).

involves introduction of the labile, divalent ion into a specific site followed by *in situ* oxidation. This is the best strategy for replacing an intrinsic metal ion by Co(III) or Cr(III) and will be the subject of this section. The second strategy involves incorporation by addition of a preformed Co(III) or Cr(III) complex of a cofactor or substrate to the enzyme. The preparation and purification of such complexes have been described in detail recently[8,23] and will not be repeated here. It is also possible, in principle, to incorporate an exchange-inert metal by preparing it with particularly labile ligands which, after exchange with ligands on the protein, becomes totally inert. However, this has not so far been achieved with Co(III) or Cr(III).

The first step in the preparation of a Co(III) or Cr(III) enzyme is the preparation of the respective Co(II) or Cr(II) species. Next, the oxidation to the trivalent state is carried out. The oxidation of Cr(II) to Cr(III) will occur spontaneously using dissolved oxygen as the oxidant. However, the ease of oxidation of Co(II) to Co(III) will vary markedly and will probably require the use of a weak oxidant such as hydrogen peroxide.

Procedures for the preparation of Co(II)–enzymes have been detailed in another chapter in this volume.[27] It is recommended that the use of buffers that can serve as nitrogen donors (e.g., Tris) and alkaline pH be avoided, since both favor Co(III) and can promote oxidation outside of the enzyme during preparation of the Co(II)–enzyme. For example, a solution of Co(II) ions in Tris buffer at pH 8 will be rapidly oxidized to a green complex by dissolved oxygen. The rate of oxidation of Co(II) to Co(III) in the enzyme will depend on several factors, including the identity of the ligands at the metal-binding site, the coordination geometry and number, and the accessibility of the site to oxidants. In general, only weak oxidants should be necessary and most studies have employed hydrogen peroxide at or near neutral pH at concentrations in the 0.1–20 m*M* range. The reaction times have varied from 10 min to over 12 hr. It is recommended that the precise conditions for each enzyme be established using past experience only as a guide.

It must be understood from the outset that the goal of the reaction is the *selective* oxidation of Co(II) to Co(III) without concomitant oxidation or modification of the protein. The first problem that arises is that of whether the oxidant used to convert Co(II) to Co(III) will simultaneously oxidize amino acid side chains, such as those of tyrosine, tryptophan, and cysteine. If so, the changes in the function of the enzyme that are observed cannot be attributed unambiguously to the incorporation of Co(III). A rational way of ascertaining whether oxidation of amino acid

[27] D. S. Auld, this volume [8].

residues is affecting the activity of the enzyme is to carry out control experiments in which another active metallo species of the same enzyme containing a metal atom that is not susceptible to oxidation is subjected to the same treatment. The lack of effect of the oxidant on such a species is a necessary, but not sufficient criterion for establishing that the oxidant has not harmed the enzyme. Amino acid analysis of the resultant Co(III)–enzyme should also aid in determining whether the oxidant has altered the protein.

A second point to be considered is that since hydrogen peroxide is a two-electron oxidant, while Co(II) is a one-electron reductant, a free radical will be a by-product of the reaction. Since this radical will be produced at or very close to the metal-binding site, the possibility must be considered that it will attack and modify a nearby residue or migrate to other sites where it can cause damage, including cross-linking of protein chains. This potential problem will not be revealed in a control experiment involving the reaction of the oxidant with the apoenzyme or with the enzyme containing a nonoxidizable metal, since the radical is only formed if the one-electron oxidation of the metal actually occurs. Thus, it is important to minimize the effects of radical damage whenever possible.

Both types of problems discussed above have been encountered in the oxidation of Co(II)–carboxypeptidase A with hydrogen peroxide.[13] Such treatment results in oxidation of amino acid residues and cross-linking of the protein, but in minimal production of Co(III)–carboxypeptidase. Thus, the oxidation reaction must be viewed as one in which the protein portion of the metalloenzyme is competing with the Co(II) atom for the oxidant. For some enzymes, such as alkaline phosphatase,[16,17] the oxidation of Co(II) can be carried out cleanly and selectively using hydrogen peroxide. However, for other enzymes, such as carboxypeptidase A, the oxidation of the protein is the dominant reaction.[12,13] For carbonate dehydratase reaction with hydrogen peroxide allows only partial conversion of Co(II) to Co(III) and the two species must be separated chromatographically.[10,11]

The problems with protein oxidation and radical damage discussed above in the preparation of Co(III)–carboxypeptidase have been overcome by changing the oxidant and reaction conditions. This case illustrates the options available if this problem were encountered in other systems. The problem of radical by-product formation has been overcome by carrying out the reaction in the presence of phenol, a radical scavenger. Alternatively, the use of a one-electron oxidant could have been explored. For example, it has been reported that $[Co(III)(phen)_2CO_3]^+$ is an effective oxidant for the conversion of Co(II)– to Co(III)–myosin.[6]

To circumvent the problem of protein oxidation, one may employ a different two-electron oxidant in the hope that it has greater selectivity for the target metal compared to the protein. For carboxypeptidase A, *m*-chloroperbenzoic acid was employed to selectively oxidize the Co(II) to Co(III) without oxidizing any residues on the protein. This reagent was chosen because it resembles certain competitive inhibitors of the enzyme and therefore might act as an active-site-directed oxidant. Small molar excesses of this reagent in the presence of equimolar phenol were found to rapidly oxidize Co(II)- to Co(III)-carboxypeptidase A without altering the enzyme. One concern with the use of such a reagent is that the reduced form (benzoic acid) or the phenol used as a scavenger could remain in the first coordination sphere of the Co(III), causing a steric alteration of the site. However, this can be easily investigated by separating these species from the enzyme by gel filtration and quantitating their recovery from their absorbances.

It is important to follow the time course of oxidation of the Co(II)-enzyme, which can be monitored by several means. It is certainly advisable to monitor the activity of the enzyme, since the effect of oxidation on activity will no doubt be of interest. Since all Co(II)-enzymes have optical bands due to the *d–d* transitions of the metal, the change in these bands on oxidation can be followed, provided that the enzyme is available in sufficient quantities and has a high enough solubility. If possible, it is strongly recommended that the EPR spectrum of the enzyme be examined during the course of the reaction. Since Co(II) is paramagnetic and Co(III) is almost always diamagnetic, the loss of the Co(II) EPR signal is consistent with (but not proof that) the oxidation of Co(II) to Co(III) has occurred. It should be noted that liquid helium temperatures are usually necessary for the observation of the Co(II) EPR spectrum. If the percentage loss in the EPR signal correlates well with the observed alterations in activity, this is strong evidence for a cause and effect relationship.

As an example of these procedures, data are shown in Figs. 1–3 for the oxidation of Co(II)- to Co(III)-carboxypeptidase A by *m*-chloroperbenzoic acid in the presence of phenol. The reaction is complete within 30 sec and Fig. 1 shows that small molar excesses of the oxidant simultaneously abolish both the peptidase and esterase activities of the Co(II)-enzyme. The oxidant has no effect on the Zn(II)-enzyme. The reaction also abolishes the EPR spectrum of the Co(II)-enzyme (Fig. 2) and the loss of the EPR signal correlates with the loss of enzymatic activities (Fig. 3). Thus, this represents the successful and selective oxidation of the active site Co(II) to Co(III).

An essential part of the characterization of the Co(III)-enzyme is the determination of its metal content. This can be achieved by analytical measurement by atomic absorption spectroscopy or other direct methods.

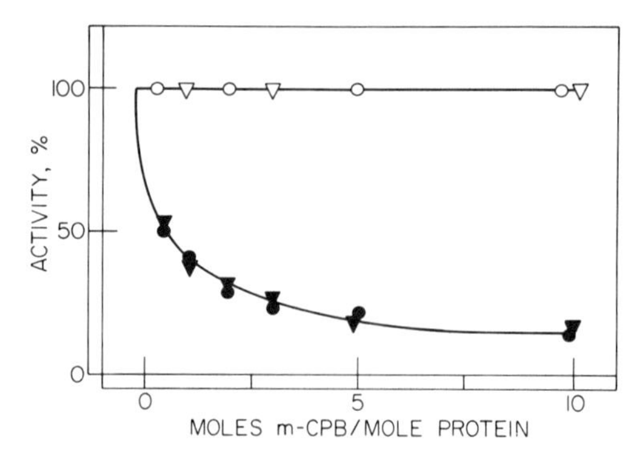

FIG. 1. Peptidase (○,●) and esterase (▽,▼) activities of Zn(II)–carboxypeptidase A (○,▽) and Co(II)–carboxypeptidase A (●,▼) on reaction with increasing concentrations of *m*-chloroperbenzoate (*m*-CPB). Activities are expressed as percentages of the unmodified controls. From Van Wart and Vallee.[12]

Alternatively, it may prove convenient to prepare the enzyme containing a stoichiometric amount of Co(II) with tracer amounts of ^{60}Co or ^{57}Co and to use the rate of γ-emission before and after oxidation to measure the Co content of the product. The inertness of the resultant Co(III)–enzyme

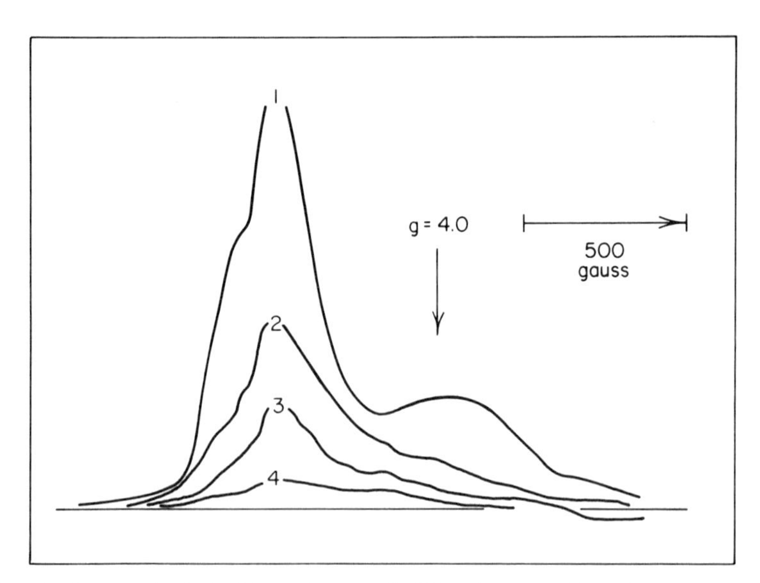

FIG. 2. Changes in the EPR spectrum of Co(II)–carboxypeptidase A on reaction with increasing concentrations of *m*-chloroperbenzoate (*m*-CPB). The ratios of moles *m*-CPB/mol protein are (1) 0.0, (2) 0.5, (3) 2.0, and (4) 10.0, i.e., those at which the activity measurements in Fig. 1 were performed. Microwave power, 20 mW; modulation frequency, 100 kHz; microwave frequency, 9.40 GHz; 5 K. From Van Wart and Vallee.[12]

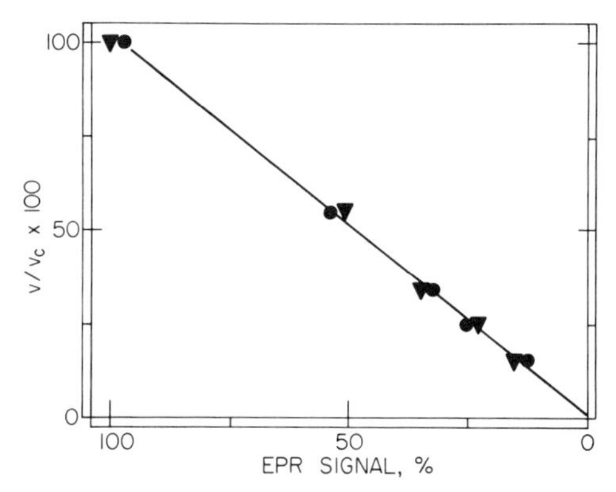

FIG. 3. Correlation of decreases in the peptidase (●), esterase (▼) activities and the EPR signal of Co(II)–carboxypeptidase after treatment with *m*-chloroperbenzoate. Activities are expressed as percentages of the unmodified controls and the EPR signal is the percentage of integrated area remaining after treatment. From Van Wart and Vallee.[13]

should not be assumed and the rate of loss of Co(III) from the enzyme should measured. The distorted environment of the metal in the protein can cause its ligand exchange reaction to be faster than in related model complexes and the results for enzymes are unpredictable. Thus, while Co(III)–carbonate dehydratase retains the Co(III) ion for long periods of time,[11] Co(III)–carboxypeptidase A loses the metal with a halflife of about 30 min.[13] As a test of inertness, it should be assessed whether the Co atom can be exchanged by other labile atoms, such as Zn(II). If it is truly Co(III), this exchange should not occur. It is recommended that the amino acid composition of the protein be determined to assess if damage to amino acid residues has occurred and that SDS–PAGE or another suitable technique be employed to investigate whether the reaction has altered the molecular weight or subunit structure of the protein.

The putative Co(III)–enzyme should have spectral properties consistent with those of Co(III). In addition to being diamagnetic, the enzyme should have a weak absorption band due to a ${}^1T_{1g} \leftarrow {}^1A_{1g}$ transition in the 400–600 nm region. The wavelength is shifted to the blue as the number of nitrogen donors increases. Octahedral Co(III) complexes are known to have weak circular dichroism and magnetic circular dichroism (MCD) bands. As an example, the visible absorption and MCD spectra of Co(II)- and Co(III)–carboxypeptidases are shown in Fig. 4. A very convincing means of establishing that the effects of oxidation on the activity or other properties of the enzyme are due to the formation of Co(III) is to demonstrate that reduction back to Co(II) restores all of the properties of the original Co(II)–enzyme. This has been achieved for Co(III)–carboxypeptidase A with dithionite,[13] for Co(III)–carbonate dehydratase with dithio-

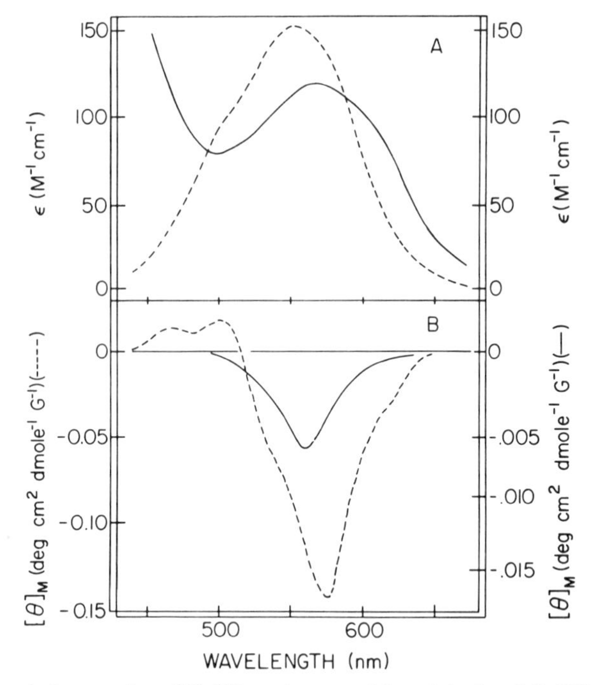

FIG. 4. Spectral properties of Co(II)–carboxypeptidase A (---) and Co(III)–carboxypeptidase A (——) at 4 °. (A) Visible absorption spectra with Zn(II)–carboxypeptidase A in the reference beam (B) magnetic circular dichroism spectra. From Van Wart and Vallee.[13]

nite and borohydride,[11] and partially for Co(III)–aspartokinase with dithiothreitol.[24]

The preparation of enzymes containing Cr(II) is best achieved by addition of Cr(II) ions to an anaerobic solution containing the apoenzyme. Subsequent exposure to atmospheric oxygen will rapidly oxidize the Cr(II) to Cr(III). Cr(III)–glutamine synthetase is the only example of a Cr(III)-enzyme in which the Cr(III) is bound entirely to protein ligands.[19] Balakrishnan and Villafranca prepared this species by first reducing Cr(III) chloride to Cr(II) in 0.1 m*M* HCl with zinc amalgam under a nitrogen atmosphere followed by the stoichiometric addition of Cr(II) to the apoenzyme under anaerobic conditions. Kowalsky has used Cr(II) perchlorate to reduce ferricytochrome *c* in acetate buffer at pH 4.78 and in cacodylate buffer at pH 7.0.[28] The Cr(II) solution can be prepared with tracer quantities of ^{51}Cr so that the presence of Cr in the product can be quantitated by measuring the rate of γ-emission. The conversion of

[28] A. Kowalsky, *J. Biol. Chem.* **244,** 6619 (1969).

Cr(III) to Cr(II) in the reduction with zinc amalgam and the extent of oxidation of Cr(II) to Cr(III) in the enzyme can be followed optically, since the green Cr(III) and blue Cr(II) complexes have distinct absorbance spectra. No reduction of a Cr(III)–enzyme back to the Cr(II) state has yet been reported.

Interpretation of Results

The first concern in interpreting the effects of incorporation of Co(III) or Cr(III) into an enzyme by *in situ* oxidation is to establish that it is the oxidation of the metal and not the protein that is responsible for the observed changes. This can be accomplished by the techniques described above. Presumably, the goal of preparing the Co(III) or Cr(III)–enzyme was to investigate the influence of having an exchange inert metal at that site on some property of the enzyme. Thus, a second concern is that the oxidation of Co(II) to Co(III) or Cr(II) to Cr(III) in the metal-binding site of the enzyme has changed more than the lability of the ion. It may also change the coordination number, coordination geometry, metal charge, pK of metal-bound water, or other properties of the site. Thus, a more difficult question is whether it is solely the change in lability that is responsible for the observed changes in the behavior of the enzyme.

In the case of the Co(II)/Co(III) pair, the Co(II) in the enzyme may exist in a number of coordination environments, including tetrahedral, five-coordinate, and octahedral states. Since it is highly likely that the resultant Co(III) species will be octahedral, the possibility that a change in coordination geometry or number is responsible for the changes in the properties of the Co(III)–enzyme must be considered. This question arose in the study of Co(III)–carboxypeptidase A.[12,13] Since it was known that metallocarboxypeptidases with a variety of coordination geometries and numbers are active enzymes, it was possible to argue in this case that it was the exchange-inert property of Co(III), rather than its coordination geometry, that was responsible for the loss in activity.[12,13] In systems where the oxidation of octahedral Co(II) to Co(III) has been carried out, this ambiguity is less of a problem.

In interpreting the properties of an enzyme containing an exchange-inert metal, one is cautioned not to assume without proof that the metal is behaving in an inert manner over the period of time during which the observations on the enzyme are being made. The environment of the metal in the enzyme may be quite distorted, leading to a labilization of one or more of the ligands. The spread of rate constants for "exchange-inert" metals is very large and it is possible that the metal can undergo exchange reactions in times shorter than those required for characterization. For example, Co(III)–carboxypeptidase A irreversibly releases the Co(III) with a half-life of 30 min,[13] which made its characterization and study difficult. Shinar and Navon[10] have found that the second-order rate constants for the binding of cyanide and azide to Co(III)-carbonate dehydratase are 0.06 and 0.07 M^{-1} sec^{-1}, respectively, which are considerably faster than the rates observed for Co(III) inorganic complexes. Thus, one must be cautious not to assume that the metal will exhibit classically inert behavior and remain unchanged at its site of oxidation for indefinite periods of time.

A last concern in the interpretation of the data is the assumption that the Co(III) or Cr(III) resides at the same site as the Co(II) or Cr(II). It is possible that, on oxidation, the metal migrates to a site with a more favorable geometry or to one with preferred ligands. This is a difficult, if not impossible, fear to eliminate. If it can be established that reduction back to the Co(II) or Cr(II) enzyme restores its original properties, one could effectively argue that any migration on oxidation must be local. However, the unequivocal proof that the metal retains the same ligands as before oxidation would take special efforts to establish.

[11] Use of Chelating Agents to Inhibit Enzymes

By DAVID S. AULD

In recent years, a large number of chelating agents have been synthesized, and studies of their capacity to complex metals have provided a vast body of knowledge for extensive experimentation in metalloenzymology. Table I[1-3] lists a number of the most frequently used nitrogen-, oxygen-, and sulfur-containing chelating agents and their stability constants toward Zn^{2+}, Cu^{2+}, Fe^{2+}, and Fe^{3+}. Nonchelating analogs of these agents [e.g., 1,7- or 4,7-phenanthroline for 1,10-phenanthrolines (OP)] are often used as controls.[4] If inhibition is solely due to a nonchelating property of the inhibitor (e.g., hydrophobicity in the case of the phenanthro-

[1] A. A. Schilt, "Analytical Applications of 1,10-Phenanthroline and Related Compounds." Pergamon, London, 1969.

[2] L. G. Sillen and A. E. Martell, *in* "Stability Constants of Metal-Ion Complexes," Special Publ. No. 17. The Chemical Society, London, 1964.

[3] L. G. Sillen and A. E. Martell, *in* "Stability Constants of Metal-Ion Complexes," Special Publ. No. 25. The Chemical Society, London, 1971.

[4] B. L. Vallee and W. E. C. Wacker, in "The Proteins" (H. Neurath, ed.), Vol. 5. Academic Press, New York, 1970.

lines) then the nonchelating analogs can be as effective an inhibitor of enzyme catalysis as the chelating agent.

Generally, some or all of these chelating agents—but not their nonchelating analogs—inhibit metalloenzymes, either by removing the metal or by forming a complex with it *in situ*. The effectiveness and mode of action of a chelating agent as an inhibitor of a metalloenzyme are usually judged by the kinetic parameters, K_i and $\bar{n}$ obtained from the plot[5]:

$$\log(V_c/V_i - 1) = -\log K_I + \bar{n} \log[I] \tag{1}$$

where V_i and V_c are the velocities of the enzyme in the presence and absence of the metal binding agent, I. K_i is the apparent inhibition constant and $\bar{n}$ is the average number of moles of chelating agent complexed per mole of metal.

If the value of $\bar{n}$ is greater than one, the chelating agent either is removing the metal from the enzyme or has multiple binding sites on the enzyme surface. In either case, the K_i and $\bar{n}$ values are apparent constants dependent on what part of the log–log plot is used in the calculation. This can be demonstrated by inspection of the case of metal removal as might be observed for an agent such as 1,10-phenanthroline. OP can form mono-, di-, and tridentate species, i.e.,

$$M^{n+} + L \underset{}{\overset{K_1}{\rightleftharpoons}} M(L)^{n+} \overset{K_2}{\rightleftharpoons} M(L)_2^{n+} \overset{K_3}{\rightleftharpoons} M(L)_3^{n+} \tag{2}$$

where

$$K_1 = \frac{[M(L)]^{n+}}{[M^{n+}]\,[L]}$$

The apparent stability constant for the di- and tridentate species are

$$\beta_2 = K_1K_2 = \frac{[M(L)_2^{n+}]}{[M^{n+}][L]^2} \tag{3}$$

$$\beta_3 = K_1K_2K_3 = \frac{[M(L)_3^{n+}]}{[M^{n+}][L]^3} \tag{4}$$

The values for β_2 and β_3 for a number of chelating agents interacting with Zn^{2+}, Cu^{2+}, Fe^{2+}, and Fe^{3+} are listed in Table I. The enzyme, on the other hand, can be considered a monodentate "ligand," i.e.,

$$M^{n+} + E \overset{K_E}{\rightleftharpoons} ME^{n+} \tag{5}$$

The association constant for the enzyme metal complex, K_E, will usually be greater than the value of K_1 for the chelating agent · metal complex but

[5] T. L. Coombs, J. P. Felber, and B. L. Vallee, *Biochemistry* **1,** 899 (1962).

TABLE I
METAL COMPLEXING AGENTS EMPLOYED TO INHIBIT METALLOENZYMES

Metal ion	Ligand	Log K_1	Log K_2	Log K_3	Log β_2	Log β_3	Reference
Zn^{2+}	1,10-Phenanthroline (phen)	6.6	5.8	5.2	12.4	17.6	1
	2,9-Dimethylphen	4.1	3.6		7.7		1
	4,7-Dimethylphen	6.9	6.2	6.0	13.1	19.1	1
	α,α'-Bipyridyl	5.3	4.5	3.8	9.8	13.6	1
	Pyridine-2,6-dicarboxylic acid (dipicolinic acid)	6.4	5.5		11.9		2
	Ethylenediamine tetra-acetic acid (EDTA)	16.4					2
	[(Ethylenedioxy)diethyl-enedinitrolo]tetraacetic acid (EGTA)	12.9					3
	8-Hydroxyquinoline (oxine)	8.5	7.3		15.8		3
	8-Hydroxyquinoline-5-sulfonic acid (HQSA)	8.0			15.0		3
	2,3-Dimercaptopropan-1-ol (BAL)	13.5			23.3		2
	Mercaptoacetic acid (thioglycolic acid)	7.9			15.0		2
Cu^{2+}	1,10-Phenanthroline (phen)	9.3	6.8	5.4	16.1	21.5	1
	2,9-Dimethylphen	5.2	5.8		11.0		1
	4,7-Dimethylphen	8.1	8.0	8.4	16.1	24.5	1
	4,7-Diphenylphen	5.7		3.8			3
	α,α'-Bipyridyl	8.2	5.5	3.3	13.7	17.0	3
	Dipicolinic acid	9.1	7.4		16.5		2
	EDTA	18.8					2
	EGTA	17.7					3
	8-Hydroxyquinoline	12.1	10.9		23.0		3
	HQSA	12.5	10.5		23.0		3
	Thioglycolic acid	8.6	7.2	.2	15.8		3
Fe^{2+}	1,10-Phenanthroline (phen)					21.3	1
	4,7-Dimethylphen	5.6				23.0	1
	4,7-Diphenylphen				21.8		2
	α,α'-Bipyridyl	4.2	3.7	9.6	7.9	17.5	3
	Dipicolinic acid	5.7	4.7		10.4		2
Fe^{3+}	Dipicolinic acid	10.9			17.1		2
Fe^{2+}	EDTA	14.3					2
Fe^{3+}	EDTA	25.1					2
Fe^{2+}	EGTA	11.9					3
Fe^{3+}	EGTA	20.5					3
Fe^{3+}	8-Hydroxyquinoline	13.7			26.3	36.9	3
Fe^{2+}	HQSA				15.7	21.8	2
Fe^{3+}	HQSA	11.6	11.2				2
Fe^{2+}	BAL				15.8		3

not greater than β_2 or β_3 (Table I). The concentration of chelating agent needed to effectively compete with the apoenzyme for the metal will have to be in the concentration range where the di- and tridentate species are favored [Eqs. (3) and (4)]. Thus for a metalloprotein ($M^{n+}E$) with a K_E value of 10^{10}, at 10^{-6} M OP the concentrations of $M^{n+}E$, mono-, di-, and tridentate OP complexes are 93, 4, 2, and 0.4% of the total while at 10^{-5} M OP they are 13, 5, 32, and 50%, respectively. A 10-fold change in chelator concentration changes the ratio of di- plus tri- to monodentate species by 184-fold. Under these conditions the slope of the inhibition curve [Eq. (1)] would have a value between 2 and 3.

If the value of $\bar{n}$ is one for di- or tridentate-forming chelators, it is likely that a ternary enzyme·metal·chelator (L) complex is formed. The K_i value obtained in this case would be the dissociation constant for (L). The existence of mixed complexes can be detected not only by their effect on catalysis, but also in some instances spectrally.[1,6] Thus, spectrophotometric studies indicate the formation of a stable 1 : 1 complex between 1,10-phenanthroline and alcohol dehydrogenase.[1,7]

Some chelators such as the phenanthrolines are particularly useful agents where activating ions such as Mg^{2+} or Mn^{2+} are needed for enzyme activity. Mg^{2+} forms only very weak complexes with OP having a K_1 values of 1 whereas β_3 is 17.6 for zinc (Table I). On the other hand, EDTA has K_1 stability constants of 9, 14, and 16 toward Mg^{2+}, Mn^{2+}, and Zn^{2+}, respectively[2,3] (Table I) and β_3 for Mn^{2+} is 7 toward OP.[2] For enzymes that are activated by Mg^{2+} or Mn^{2+} it is therefore more reasonable to choice Mg^{2+} as the activating cation and OP as the chelating agent to investigate the possible requirement of Zn^{2+} or another transition metal for the activity of the enzyme.[8]

Both time-dependent and instantaneous inhibition can be observed when enzymes are assayed in the presence of chelating agents. Preincubation of the enzyme and chelator for specified periods of time can lead to lose of enzymatic activity if (1) the metal is involved in structure stabilization of the protein or (2) rebinding of the metal is prevented by modification of a chemically labile ligand, such as a sulfhydryl group or (3) dissociation of the metal is a very slow process. If the latter mechanism is the cause of the time-dependent phenomenon, dilution of the enzyme · chelator complex to a concentration of chelator which is insufficient to bind the metal (usually below the value of K_1, Table I) should reverse the inhibition. If it does not, irreversible damage to the structure of the ligand site

[6] B. L. Vallee and T. L. Coombs, *J. Biol. Chem.* **234,** 2615 (1959).
[7] T. K. Li, D. D. Ulmer, and B. L. Vallee, *Biochemistry* **2,** 483 (1963).
[8] S. E. Pollack, T. Uchida, and D. S. Auld, *J. Protein Chem.* **2,** 1 (1983).

and/or enzyme has occurred. Examination for the presence or absence of this time-dependent inhibition in the absence of the substrate for the reaction is critical to evaluating the effect of the chelator in an assay.

Instantaneous inhibition should be examined by making the addition of the enzyme as the last component to the assay and keeping the assay to as short a time period as possible. The assay should be designed to give linear initial rate kinetics in the absence of the inhibitor. Inhibition by either forming a ternary enzyme · metal · chelator complex or removal of the metal can be "instantaneous" if the on and off rates for the chelator and metal are fast in comparison to the time for assaying the enzyme.

The kinetics for the removal of metal from carbonic anhydrase and carboxypeptidase by chelating agents show that the process occurs through two mechanisms: S_N1 and S_N2.[9,10] S_N1 involves the spontaneous dissociation of the zinc enzyme, followed by a rapid reaction of the free metal with chelating agent. In contrast, S_N2 reflects the rapid formation of a ternary complex, followed by a slow dissociation of the chelate-bound metal. Polydentate ligands, such as EDTA and nitrilotriacetic acid, follow an S_N1 pathway; certain bidentate ligands, such as 1,10-phenanthroline and pyridine carboxylic acids, follow an S_N2 pathway; other bidentate agents, such as α,α'-bipyridine, follow a mixed S_N1/S_N2 pathway.

Chelating agents are also used to remove the metal from zinc proteins to yield the respective apoproteins.[11] For different chelating agents, the effectiveness for this process varies with the metalloprotein concerned. As a practical consideration, EDTA should be avoided in preparing apoenzymes as it tends to bind nonspecifically to proteins,[12] thereby creating specious metal-binding sites, and leading to problems in reconstitution studies.

The inhibition of an enzyme by a chelating agent does not constitute conclusive evidence that it is a metalloenzyme; this can be demonstrated definitively only by metal analysis.[13] Similarly, the failure to observe inhibition with a particular chelating agent need not be an absolute gauge of the absence of zinc. Thus, EDTA does not inhibit yeast alcohol dehydrogenase in contrast to its effect on liver alcohol dehydrogenase but 1,10-phenanthroline inhibits both.[13]

[9] Y. Kidani and J. Hirose, *J. Biochem.* (*Tokyo*) **81,** 1383 (1977).

[10] E. J. Billo, *J. Inorg. Biochem.* **10,** 331 (1979).

[11] D. S. Auld, this volume [8].

[12] D. D. Ulmer and B. L. Vallee, *Adv. Chem. Ser.* **100,** 193 (1971).

[13] B. L. Vallee and A. Galdes, *Adv. Enzymol. Relat. Areas Mol. Biol.* **56,** 284 (1984).

Section II

Analytical Techniques

[12] Atomic Absorption Spectrometry

By WALTER SLAVIN

Introduction

Atomic absorption methods are the most widely used for the determination of metals in biological materials. This review will provide an overview of the field to help in the selection among the available techniques. The various elements will then be treated with recommendations for preferred methodology. The review is written at a point in time when most analysts still view the graphite furnace technique as an accessory to flame atomic absorption (AAS). Actually the two techniques are quite separate and independent. This review is thus written with the understanding that atomic absorption is two techniques: flame AAS and furnace AAS. This position will not be defended here but such a defense has been published recently.[1]

Flame AAS is generally the best analytical technique available for samples that are most easily collected as solutions, if the analyte concentration is above the mg/liter range in the solution. Precision of about 1% is routinely obtained and better precision is available if extra care is taken in the preparation of standards and if slightly more time-consuming methods are used. When the solution sample is suitably diluted, usually in simple media, the analyte may be quantitated in seconds. Automation is widely and relatively inexpensively available. Atomic spectroscopy is inherently very specific and interferences that will reduce the accuracy below the level of the precision are few. Those interferences that exist are well characterized and are generally controlled in a routine manner. The net result is that flame AAS should be used in every situation where it is applicable.

There are some exceptions to the generalizations above. The very refractory metals, e.g., W and Ta, are not reduced to an atomic vapor in the flame and are therefore not accessible to flame AAS. However the refractory metals are very rarely required in biological systems. If, as is so often the case, the analyte concentration approaches the μg/liter level or below, the precision becomes much poorer. In such cases other more sensitive methods are often preferable, for example furnace AAS. If many metals must be determined in each sample, inductively coupled plasma emission spectroscopy is often more rapid, though it is more expensive and it requires a higher level of operator skill than flame AAS.

[1] W. Slavin, *Anal. Chem.* **58,** 589A (1986).

If the sample is a solid, such as bone or tissue, dissolution is required prior to flame AAS. This slows the process and opens the opportunity for contamination during the various handling steps. It would be very convenient if there were a way to introduce solid samples directly into the flame. Papers have occasionally appeared in the literature but no solid sampling technique for flame AAS has been found to be generally acceptable for biological samples.

Flame AAS is well established and has been discussed exhaustively in many accessible books and reviews.[2,3] We have not felt the need to discuss here the theory or the general analytical technique.

In contrast to flame AAS, the furnace as we know it today, dates only from the early 1980s. There are no fully adequate books on this topic and the technique is often misunderstood. Most importantly, furnace AAS is usually the only available analytical technique for metals in biological samples when we go beyond the common metals: Na, K, Ca, and Mg. It is very important that the theoretical basis for furnace AAS be understood so that the technique can be applied correctly.

There are many reviews that compare furnace AAS with other analytical techniques for biochemical samples (e.g., see Refs. 2 and 3). Morrison[4] found that "furnace AAS offered the greatest opportunity" among the available methods, although "a variety of interferences plagued the technique." It is this specific deficiency of furnace AAS that the modern methods seek to remove. A compilation of papers on trace metals in medicine and biology was edited by Brätter and Schramel[5] including analytical procedures. Stoeppler[6,7] has reviewed furnace analytical methods in biological samples for many elements based on a great deal of experience.

The Furnace System

The concept of analytical furnace AAS was proposed by L'vov[8] just a few years after Walsh proposed flame AAS. The first paper appeared in 1959 and the technique became commercial in 1969. However, the com-

[2] B. Welz, "Atomic Absorption Spectrometry." VCH, Weinheim, Federal Republic of Germany, 1985.

[3] H. T. Delves, *Prog. Anal. At. Spectrosc.* **4,** 1 (1981).

[4] G. H. Morrison, *Crit. Rev. Anal. Chem.* **8,** 287 (1979).

[5] P. Brätter and P. Schramel, "Trace Element Analytical Chemistry in Medicine and Biology." de Gruyter, Berlin, 1980.

[6] M. Stoeppler, *Spectrochim. Acta* **38B,** 1559 (1983).

[7] M. Stoeppler, P. Valenta, and H. W. Nurnberg, *Z. Anal. Chem.* **297,** 22 (1979).

[8] B. V. L'vov, *Inzh.–Fiz. Zh.* **2,** 44 (1959); and, in English, *Spectrochim. Acta* **39B,** 159 (1984).

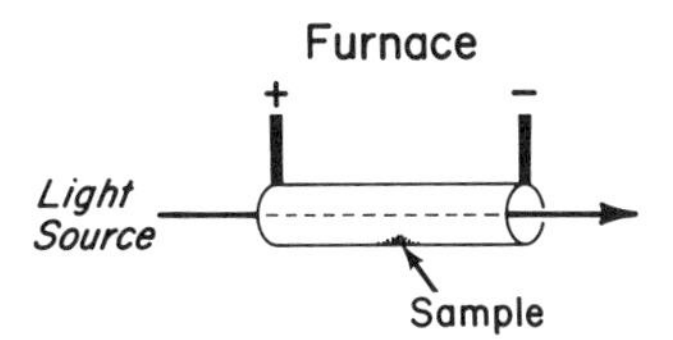

FIG. 1. L'vov concept.

mercial instruments took the easy route of adapting then available flame AAS instruments by replacing the flame with a graphite furnace. This was easy and convenient to do but it led to a decade of poor analytical performance, as we shall explain. In 1978 L'vov again took the lead[9] and showed why the commercial instruments were producing poor analytical data. He made some recommendations but a clearer understanding gradually evolved over the next few years and considerable improvement can still be expected in the future. The stabilized temperature platform furnace, STPF, is the name we have given to furnace AAS when it is used as closely as possible to the theoretical requirements. It is not a proprietary name or technique, but the analytical equipment must be thoughtfully designed for the requirements of furnace AAS. Essentially the same conditions are given the name "constant temperature furnace" by some workers.

L'vov's original concept was very simple and elegant. Take a small sample and quickly heat all of it to a particular high temperature that will convert all of it to an atomic vapor. This is illustrated in Fig. 1 by putting the sample within an electric furnace. All of the sample including the analyte will be converted to an atomic vapor and the analyte will absorb a portion of the light from a lamp containing a pure element. In this situation, the integrated absorbance, $A \cdot sec$, at the element resonance line will be proportional to the mass of element in the sample. Everything else that we do is just to achieve these simple conditions.

We can think of the furnace as a cell within which the sample vapor is partially confined. If we put the sample on the wall of the furnace, illustrated in Fig. 2, the various materials in the sample will vaporize as the wall heats up. The temperature and the rate at which the analyte will vaporize depend on the compounds in which the analyte is present. The L'vov theory requires that the gas-phase temperature while the analyte is an atomic vapor be the same for standards and samples. So we cannot tolerate the thermal ambiguity that occurs when we put the sample on the wall of the furnace. Therefore we add a small platform within the furnace

[9] B. V. L'vov, *Spectrochim. Acta* **33B**, 153 (1978).

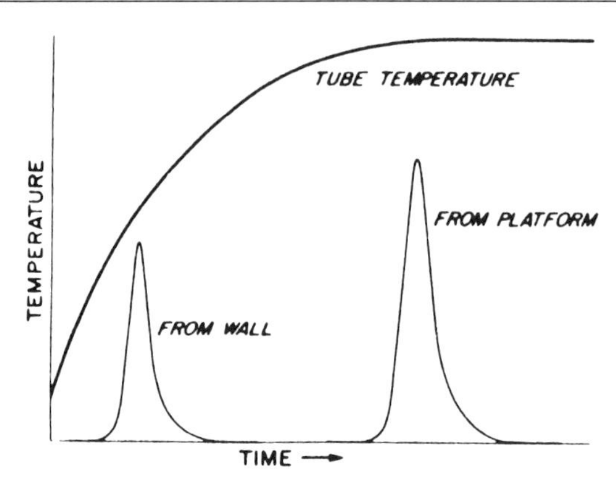

FIG. 2. Volatilization from the wall and the L'vov platform.

which is heated by radiation from the furnace walls. This delays vaporization until the walls and the gas within have settled to some quite stable temperature, illustrated also in Fig. 2.

Several other STPF conditions quickly become apparent from this model. We use rapid heating of the furnace because we want the steady conditions to be achieved before the analyte is vaporized. A furnace instrument must use fast signal processing because everything is happening quickly and the fast analyte signal must be accurately followed. Older flame AA instruments used slow analog circuits that, in some cases, may provide usable results. Unfortunately we cannot accurately predict when they will work or how badly they might fail.

We add a matrix modifier to stabilize the analyte to higher temperatures. There are several important advantages to this. We can then char at higher temperatures and reduce the magnitude of the background signal; but more importantly we can stabilize the analyte on the platform while the STPF conditions are coming to equilibrium. The matrix modifiers are sometimes a nuisance if they provide a large blank. But, for most metals, the technique does not work reliably without the modifier.

For the theory to be applicable, the atomic vapor must leave the furnace tube at a rate controlled only by gaseous diffusion. Therefore, the gas flow must be stopped in the tube during the atomization process.

The walls of the furnace tube are assumed to provide no chemical reactions with the sample. Ordinary graphite at high temperature is porous to many atomic vapors and some analyte is lost through the walls of the furnace if ordinary graphite tubes are used. A layer of dense pyrolytic graphite is deposited at high temperature on the graphite substrate. This pyrolytic graphite coating is not at all porous. Also it greatly reduces

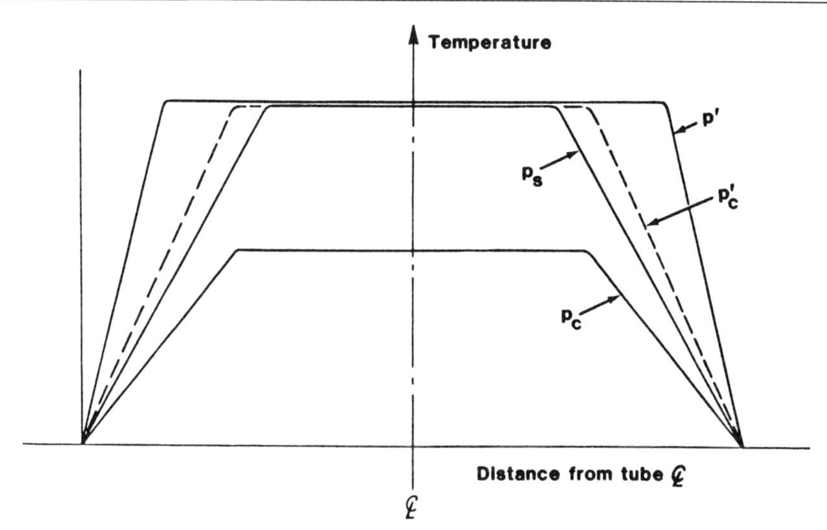

FIG. 3. Temperature profile along the length of the tube.

chemical effects between graphite carbon and the sample. Uncoated tubes can still be used, but the analytical performance is thereby degraded.

We have recently recommended what we call a "cool-down" step between the char and atomization steps. Its usefulness can be understood from Fig. 3 which shows a diagram of the temperature profile along the length of the tube. The curve marked p_s is the steady-state temperature after a few seconds. If the tube is heated from room temperature, the instantaneous heat distribution at the end of the max power step (between 1 and 2 sec) is shown by p'. This is because, with very rapid heating, the only effective method for heat dissipation is radiation, since both conductive and convective heat loss take finite time. The temperature profile after the char step is shown by p_c and, if the temperature is raised directly to the atomization temperature, the dotted p'_c curve results at the end of maximum power heating. After a few seconds both the p' and p'_c distribution will decay to the steady-state p_s curve mostly by conduction to the cold ends of the tube.

Thus, if we return the tube to room temperature after the char step, a longer isothermal zone, p', results at the instant the tube reaches atomization temperature. This reduces some of the disadvantageous results of the thermal gradient at the ends of the tube. This technique is particularly useful for the more refractory metals but we now use it routinely for all analyses.

Almost no real samples can be run without background correction. After an instrument has been in service for some time, the background

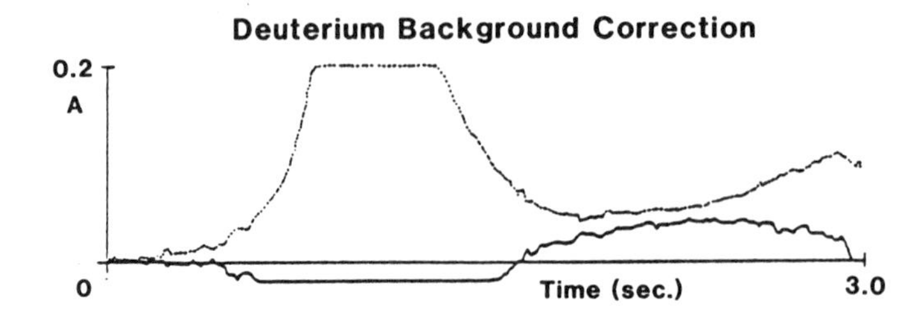

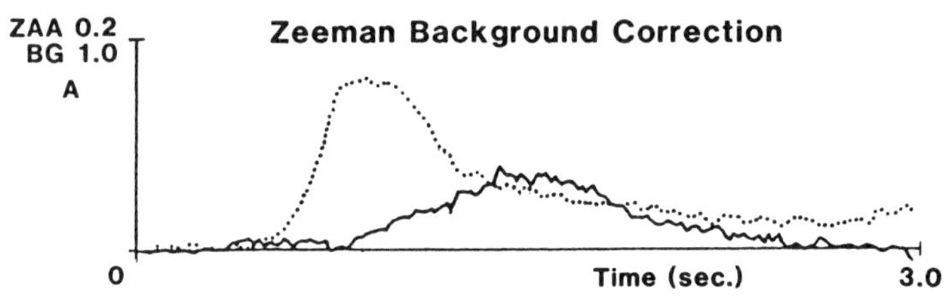

FIG. 4. Selenium in fish tissue.[2]

correction lamps do not always remain in adjustment and correction can introduce errors. This is, beyond doubt, the major advantage of Zeeman correction. Since the same source and optical system are used for both analysis and correction, nothing can go out of adjustment. For this reason Zeeman corrected systems can accomodate much higher backgrounds which produces greater accuracy, as well as lower detection limits, in real samples.

Continuum correction *often* introduces overcorrection errors for particular combinations of matrix and analyte. We have recently summarized dozens of these errors.[10] Figure 4 shows the problem that arises in tissue samples (fish in this case) when selenium is determined. There is a large negative signal caused by phosphate bands in the gaseous phase when a continuum corrector is used. The dotted line is background. There is no problem when Zeeman correction is used.

How does one decide if a Zeeman correction is necessary for a sample? If you are serious about furnace analyses and accurate results are required at low concentrations, a Zeeman corrected system is close to mandatory.

Method of Additions

We almost never use the method of additions. It requires that a preliminary measurement be made plus measurements on two or three addi-

[10] W. Slavin and G. R. Carnrick, *At. Spectrosc.* **7,** 9 (1986).

tions, each approximately equal to the unspiked absorbance. Thus it is always very time consuming. It is also always less precise than using a working curve prepared from standards. It is less precise because each result must be extrapolated back to the horizontal axis. If the additions are not chosen to be about equal to the initial signal, the precision quickly becomes even poorer, again because of the errors introduced by extrapolation. The usual excuse for using the method of additions is that, while slow and imprecise, it will at least correct for unknown errors, thus it is accurate. But that is not true. The most common errors are background correction errors and these are not corrected by the method of additions. There is no way to know whether the signal from the unspiked sample was caused by analyte or by background. Usually it is caused by an unknown mixture of the two.

We construct a working curve of standards plus matrix modifier and we expect to use that curve for all of the samples we must analyze. Sometimes we add approximately the concentration of the major components of the sample into the standard solutions, e.g., NaCl for seawater samples. Usually that is not necessary. In older papers on furnace AAS the method of additions was a last resort when all else failed, and it was often used with blind faith.

New Sample

The experimental conditions for the elements of interest in biological samples are summarized in Table I for graphite furnace AAS using Zeeman correction. These conditions are a little different from those that are optimum for instruments equipped with continuum background correctors. The char temperature is listed as "pretreatment" temperature in Table I. The characteristic mass, m_0, is a measure of analytical sensitivity that is defined below. The Zeeman furnace detection limits in real samples is usually within a factor of 2 of the listed m_0. The last column marked "rollover" is the absorbance to which the analytical curve is asymptotic using Zeeman correction. It is discussed further in texts on Zeeman correction.

When a new sample comes into the lab, standards are prepared to provide signals close to 0.15 and 0.3 *A* and the recommended matrix modifier is prepared as a diluent. We want to start with a simple situation so we dilute the sample until the inorganic solids are less than 1%. Later we determine if less dilution will also work. A firing using the instrument manufacturer's recommended conditions (e.g., Ref. 11) will provide a

[11] Perkin-Elmer Corp., "Techniques in Graphite Furnace AAS," Part No. 0993-8150. Perkin-Elmer, Ridgefield, Connecticut, 1985.

TABLE I
ZEEMAN BACKGROUND CORRECTION

Element	nm	Slit (nm)	Site[a]	Modifier	Pretreat (°C)	Atom (°C)	m_0[b]	Approx, rollover
Ag	328.1	0.7	P	0.005 mg PdCl	900	1800	1.3	1.8
Al	396.2	0.7	P	0.05 mg Mg $(NO_3)_2$	1700	2500	10	0.8
As	193.7	0.7	P	0.02 mg Ni	1300	2300	17	1.3
Au	242.8	0.7	P	0.05 mg Ni	1000	2200	13	1.6
Ba	553.6	0.2	W		1200	2550	6.5	1.8
Be	234.9	0.7	P	0.05 mg Mg $(NO_3)_2$	1500	2500	1.0	0.5
Bi	223.1	0.2	P	0.02 mg Ni	900	1900	24	1.8
Cd	228.8	0.7	P	0.2 mg PO_4 + 0.01 mg Mg $(NO_3)_2$	900	1600	0.35	0.8
Co	242.5	0.2	P	0.05 mg Mg $(NO_3)_2$	1400	2500	7	1.0
Cr	357.9	0.7	P	0.05 mg Mg $(NO_3)_2$	1650	2500	3.3	1.6
Cs	852.1	0.7	P	0.2% H_2SO_4	900	1900	5.3	1.8
Cu	324.8	0.7	P		1000	2300	8	0.7
Fe	248.3	0.2	P	0.05 mg Mg $(NO_3)_2$	1400	2400	5	0.8
Ge	265.1	0.2	P	0.05 mg Mg $(NO_3)_2$	950	2400	34	1.8
Hg	253.7	0.7	P	0.02 mg Pd	140	2000	85	0.3
Li	670.8	0.2	P		900	2600	1.4	1.8

Mn	279.5	0.2	P	0.05 mg Mg $(NO_3)_2$	1400	2200	2.2	1.1
Mo	313.3	0.7	W		1800	2650	9	1.8
Ni	232.0	0.2	P		1400	2500	13	0.7
P	213.6	0.7	P	0.05 mg La $(NO_3)_2$	1400	2650	3000	0.7
Pb	283.3	0.7	P	0.2 mg PO_4 + 0.01 mg Mg $(NO_3)_2$	850	1800	12	1.5
Pt	265.9	0.7	W		1300	2650	115	1.1
Rb	780.0	0.7	P		800	1900	2.4	1.8
Sb	217.6	0.7	P	0.02 mg Ni	1100	2400	38	1.4
Se	196.0	2.0	P	0.02 mg Ni + 0.025 mg Mg $(NO_3)_2$	900	2100	30	1.4
Si	251.6	0.2	P		1400	2650	40	—
Sn	286.3	0.7	P	0.2 mg PO_4 + 0.01 mg Mg $(NO_3)_2$	800	2100	23	1.8
Sr	460.7	0.7	W		1300	2600	1.4	1.8
Te	214.3	0.2	P	0.02 mg Ni	1000	2000	15	1.1
Ti	364.3	0.2	W		1400	2650	45	1.8
Tl	276.8	0.7	P	1% H_2SO_4	600	1400	17	0.6
V	318.4	0.7	W	0.05 mg Mg $(NO_3)_2$	1100	2650	40	1.3
Zn	213.9	0.7	P	0.006 mg Mg $(NO_3)_2$	700	1800	0.45	1.0

[a] P, platform; W, wall.

[b] Characteristic mass, m_0, in picograms for 0.0044 $A \cdot$ sec.

rough result. We then add a recovery aliquot to the sample containing about as much analyte as will produce an incremental signal between 0.2 and 0.3 $A \cdot$sec, unless that puts the total beyond the analytical range. If the recovery is complete, ±5–10%, the confidence can be high in achieving accurate results.

The most important diagnostic we have is the stable slope of the working curve. We have defined the characteristic mass, m_0, to represent analyte sensitivity. We measure m_0 in terms of pg that produce an integrated A signal equal to 0.0044 $A \cdot$sec. This sensitivity is analogous to the flame AAS term for sensitivity which is mg/liter for 1% absorption (0.0044 A). The m_0 values are specific for each analyte and independent of matrix. The values have been reported[12] and are summarized in Table I. Variations of about 15% may reflect differences between individual instruments and several papers are in preparation[13] on this topic. For an individual instrument, the day-to-day variation should be much less than 15%. This slope is matrix independent.

Probably the most common explanations of failure to achieve expected m_0 slopes are the usual contamination problems in a trace element laboratory. Some of these are slightly troublesome for flame AA or ICP but the problems are much more severe at the very low levels handled with the graphite furnace. Attempting to achieve a specific m_0 provides the chemist a very powerful means to quality control his laboratory operation.

The most common cause of failure to recover an added spike, outside of contamination, is loss of analyte at the chosen char temperature. A char study is done on the sample, not on a standard. The unspiked sample is used if the signal level is greater than 0.1 $A \cdot$sec, otherwise enough analyte is added to produce a signal between 0.2 and 0.4 $A \cdot$sec. Use the recommended char temperature and decrease the temperature in 200° steps from there. Remember that the sample matrix determines the satisfactory char temperature.

If the m_0 is correct and the recovery in the unknown matrix is satisfactory, the rest is detail. We prefer to use reference materials to confirm a method and to build confidence. The reference materials available for biological analyses were recently summarized.[14] We use NBS 1643 trace metals in water as a further test of our standards. Also remember that small volumes of solution left in a room change rapidly under ambient conditions because of solvent evaporation and contamination from settling dust.

[12] W. Slavin and G. R. Carnrick, *Spectrochim. Acta* **39B,** 271 (1984).
[13] Unpublished results from our laboratory.
[14] Y. Muramatsu and R. M. Parr, *I. A. E. A. RL/128* **December** (1985).

Specific Methods

Specific methods are suggested below for many trace metal determinations of biological interest. In the case of furnace methods we have in some cases recommended modern analytical conditions in preference to conditions used by earlier authors. In these cases we are very confident that these changes will not introduce unexpected errors but will improve the precision and accuracy of the results. We have avoided including methods which we believe require further study. For more suggestions regarding furnace AAS consult Ref. 15.

Sample Preparation and Contamination

For both flame and furnace AAS the analytical conditions are relatively independent of the specific element being determined once the sample is in solution in an appropriate medium for the technique. We will discuss sample preparation first and then the conditions for the individual elements.

A large proportion of the literature dealing with biological samples analyzed by furnace AAS is really devoted to sample preparation—and to the control of contamination since the two go together. We will discuss sample preparation in the material below devoted to individual elements, but treatment in full detail is considered beyond the scope of this chapter. The classic book on wet and dry ashing procedures by Gorsuch[16] and the long review by Delves[3] provide more detail. The book by Zief and Mitchell[17] on contamination in ultratrace analysis is also very valuable.

The special problems associated with the preparation of tissue samples for furnace AAS must be addressed. For trace metal analyses above the mg/liter range typical of flame AA and ICP, contamination plays a role for the common metals such as Na, K, Ca, Mg, and some others at lower levels. But for furnace analyses in the μg/liter range, many elements are sufficiently ubiquitous to pose a serious problem, e.g., Zn, Cd, Pb, Al, Fe, Mn, Cu. If samples are handled in stainless-steel syringes, then Co, Ni, Cr, and Mo become serious contaminants. Failure to address this problem adequately has produced many papers purporting to measure "normal" levels in nature that in fact are measuring only uncontrolled contamination. This is documented over and over again,[18] but the problem persists.

[15] W. Slavin, "Graphite Furnace AAS—A Source Book," Part 0993-8139. Perkin-Elmer, Ridgefield, Connecticut, 1984.

[16] T. T. Gorsuch, "Destruction of Organic Matter." Pergamon, Oxford, England, 1970.

[17] M. Zief and J. W. Mitchell, "Contamination Control in Trace Element Analysis." Wiley (Interscience), New York, 1976.

[18] J. Versieck and R. Cornelis, *Anal. Chim. Acta* **116,** 217 (1980).

While the various alternative sample preparation procedures for tissue have been addressed by many workers, a study of general utility was reported by Taylor[19] measuring Al in human tissue. Four common sample preparation methods for the analysis of human tissue were compared after the authors had reviewed the variety of methods that were proposed. The methods tested were (1) TMAH dissolution, (2) extraction with EDTA, (3) dry ashing in a muffle furnace, and (4) acid digestion. They preferred dry ashing for accuracy and because of the few steps necessary. EDTA extraction failed to remove Al from some tissues. Wet digestion and TMAH dissolution was tedious and prone to contamination. For their dry ash method they took 0.1–1 g of tissue into acid-washed silica crucibles and dried the sample to constant weight in a hot air oven at 130°. It was then transferred to a muffle oven and ashed overnight at 570°. After cooling, the ash was dissolved in 10 ml of 2% HNO_3.

Flame AAS is now used widely for Ca and Mg and AAS is sometimes used as an alternative to flame emission for Na and K. Fe, Zn, and Li in blood can be determined with the flame but the sample requirements are vastly smaller if the furnace is used. Metals in tissue have been determined in the flame after dissolution but the much greater sensitivity of the furnace suggests that most of these analyses should be done by furnace AAS. The individual elements are treated separately below.

Specific Determinations

Aluminum

The determination of Al in serum by graphite furnace AAS is certainly the most published furnace application because of the role of Al toxicity in dialysis therapy. A symposium on this topic was recently published[19] including discussion of methodology and clinical problems.

The causes of perceived analytical problems in some of the recent publications were assessed.[20] A general problem with the serum Al papers is that most authors have tried to address normal serum levels (typically less than 10 μg/liter by the same protocol they use for the Al levels that are found in dialysis patients, which are often 100 μg/liter or higher. Special care is required to obtain accurate normal levels. Some of these precautions are not necessary for serum Al levels higher than 50 μg/liter. Continuum correction can usually be used because the background is not

[19] A. Taylor, "Aluminum and Other Trace Elements in Renal Disease." Bailliere Tindall, England, 1986.

[20] W. Slavin, *J. Anal. At. Spectrosc.* **1,** 281 (1986).

large, but the precision is distinctly better when Zeeman correction is used.

Pyrolytically coated graphite tubes are important for the Al determination as is the use of a pyrolytic graphite platform. Argon should be used as the purge gas for the Al determination since nitrogen produces a smaller and more variable analytical signal. The Al line at 396.2 nm has recently been shown[21] to be preferable to the 309.3-nm line previously used, at least for Zeeman-corrected systems. The sensitivity of the two lines is similar but the 396-nm line provides a much wider range working curve.

Serum samples are diluted with an equal volume of an aqueous solution containing about 2 g/liter of $Mg(NO_3)_2$, selected to have as little Al contamination as possible. A 10-μl aliquot of this sample is delivered onto the platform. Standards containing 100, 200, and 300 μg/liter of Al are diluted with an equal volume of the same $Mg(NO_3)_2$ solution that was used for the sample. The Al blank for the $Mg(NO_3)_2$ solution is subtracted from each analytical result. This arrangement permits the working curve to be plotted in $A\cdot$sec signals versus μg/liter Al in the serum although, of course, the actual concentration of Al in the standards is half that plotted. We use the O_2-ash step at 600° but many workers now use air instead of O_2 for this ash step. The baseline offset is corrected automatically in modern instruments.

Contamination is the most important problem in the serum Al determination. Contamination can enter at many points: the metal needles used to collect blood, material leached from the serum collection tubes, impurities in the water, reagents and vessels used, and from air-borne dust. Al is usually present in room dust. Sample handling and separation of serum should be done in sealed containers to the extent possible because dust contamination in the sample preparation stage cannot be distinguished from endogenous Al. When the diluted samples are placed on the table of the autosampler, dust problems can usually be flagged by analyzing two separate serum aliquots from each sample in two separate autosampler cups. Failure to achieve agreement within twice the standard deviation of the method usually indicates a dust problem.

Errors arising from dust contamination, like background correction errors, are positive and additive. They are therefore less likely to be observed for high serum levels than for normal levels close to the detection limit.

Normal healthy serum Al levels are less than 10 μg/liter and may often be less than 1 μg/liter. It is probably necessary to use somewhat different analytical conditions to measure normal levels accurately as compared to

[21] D. C. Manning and W. Slavin, *At. Spectrosc.* **7,** 123 (1986).

levels of interest in dialysis patients. The analytical procedure should have a standard deviation approaching 0.2 μg/liter which, with an Al detection limit of 4 pg, requires that about 20 μg of serum be in the furnace. A good background correction system is mandatory.

Sample preparation for tissue Al was discussed in the general section above and Ref. 19 should be consulted. After dissolution, tissue samples can be treated just as described for serum Al. Urine can be analyzed for Al after a 1 + 4 dilution in the matrix modifier and 0.2% Triton X-100.

Antimony

Antimony can be determined in blood and urine using simple STPF conditions. The presence of large amounts of Fe produces an interference in the determination of Sb at 217.6 nm using continuum correction[22] and this will probably cause a small error for Sb in blood. The problem is avoided by using Zeeman background correction. Early furnace papers determining Sb in blood and urine found a variety of interferences that required separation of the matrix by extraction prior to deposition of the extract in the furnace. STPF conditions with Zeeman correction avoid these problems.

Arsenic

Using continuum background correction, phosphate will strongly interfere in the determination of As at the 193.7-nm line.[22] This interference is not present using Zeeman correction nor does it seem to be present using the less sensitive As line at 197.2 nm. This interference has caused a great deal of trouble in biological samples analyzed by furnace AAS. The presence of Al at concentrations greater than some 10 mg/liter will introduce a strong positive error in the As determination at 193.7 nm using continuum background correction.[23] This is not usually a problem in biological samples but it may be avoided using Zeeman correction or by accepting 2-fold poorer sensitivity and using the 197.2-nm line.

For toxicology, As has been determined[24] using STPF conditions and 20-μl aliquots of whole blood. The blood was diluted 1 + 1 in a solution containing 1% Triton X-100 and the matrix modifier.

There is not much information in the literature on the determination of As in tissue using STPF technology but there is no reason to expect any problem if Zeeman correction is used.

The semimetallic elements including As and Se react with H in an acid medium to form a volatile H_2M compound that can be removed in the gas

[22] F. J. Fernandez and R. Giddings, *At. Spectrosc.* **3,** 61 (1982).
[23] K. W. Riley, *At. Spectrosc.* **3,** 120 (1982).
[24] D. K. Eaton and J. R. McCutcheon, *J. Anal. Toxicol.* **9,** 213 (1985).

phase from the matrix solution. The hydride is typically conducted into a heated cell in the AAS optical system and the absorbance of the dissociated and atomized metal vapor is measured. Hydride techniques are more sensitive than flame AAS because a large sample volume, 10–50 ml, can be used. The concentrational sensitivity is similar to that of the graphite furnace but, of course, the hydride technique requires a large sample volume to achieve the sensitivity.

To use the technique for any of the semimetals, the sample must be pretreated to convert all species of the analyte to the same compound prior to analysis. The interferences reported deal with the difficulty of this conversion and with the process of dissociation of the hydride. Traditionally, the hydride is decomposed in a flame or electrically heated quartz cell which is limited in temperature by the melting point of quartz. Recently Willie *et al.*[25] showed that decomposition of the hydride in the furnace removed the decomposition interferences, permitting standardization against aqueous solutions of the analytes.

The equipment for hydride analysis is less expensive than the graphite furnace but it is much slower. Probably As can be determined with greater assurance of accuracy with the graphite furnace.

Barium

The determination of Ba is often required for forensic analyses where it is an indicator of gunshot residues. It is a sensitive determination in the furnace providing a detection limit near 1 μg/liter. But it is a test of instrumental performance because of the combination of the long resonance wavelength and the high temperature required.

Beryllium

Beryllium is sensitively determined in the furnace and is suitable for toxicology determinations. It has been determined in urine[26] using STPF conditions and Zeeman background correction with a detection limit of 0.05 μg/liter. Urine samples were diluted 1 + 3 in the matrix modifier and 20 μl of sample was added to the platform furnace.

Bismuth

Serum and urine samples have been analyzed for Bi after a 1 + 1 dilution in EDTA.[27] The results would probably have been improved us-

[25] S. N. Willie, R. E. Sturgeon, and S. S. Berman, *Anal. Chem.* **58,** 1140 (1986).
[26] D. C. Paschal and G. C. Bailey, *At. Spectrosc.* **7,** 1 (1986).
[27] R. L. Bertholf and B. W. Renoe, *Anal. Chim. Acta* **139,** 287 (1982).

ing the STPF conditions and the recommended Ni matrix modifier. Hydride methods can be used for Bi.

Cadmium

Since blood Cd provides an indication of recent exposure and urine Cd provides an indication of total body burden, both determinations are biologically important. Cadmium is one of the most widely determined metals using the graphite furnace. A direct STPF method for Cd in urine detected less than 0.04 μg/liter in the sample.[28] Zeeman correction was necessary because of the large background signals that accompanied the determination. The urine was diluted 1 + 4 in the matrix modifier.

Cadmium is equally well determined in blood and serum using the simple direct procedures mentioned above. A thorough STPF procedure has not yet been published although many successful furnace methods have been reported. Tissues of a wide variety have been readily analyzed for Cd with STPF methods, most especially with Zeeman correction. Backgrounds are usually very large when biological samples are analyzed for Cd and continuum correction methods have often been shown to provide errors. The World Health Organization sponsored a massive program on the measurement of Cd and Pb in kidney tissue and blood and their monograph[29] should be consulted. Bone samples have also been analyzed for Cd after appropriate dissolution. Several papers have described the determination of Cd in solid samples of hair, usually for toxicology purposes. When the STPF conditions were used the solid samples could be analyzed against solution standards.

Calcium (and Magnesium)

The preferred method for Ca and Mg in most biological materials is flame AAS. A reference method has been developed for Ca in serum that provides a coefficient of variation (CV) of about 0.2% of the amount present by very careful preparation of the standards. Similar results can be obtained for Mg. However, we will discuss simpler methods that provide an analytical rate greater than 100 samples per hour without requiring automation. The CV for Ca and Mg in serum at normal levels and urine at equivalent levels is better than 1%.

It is usually most convenient to determine Ca and Mg in an air-acetylene flame, although phosphate interferes for Ca by binding a portion of

[28] E. Pruszkowska, G. R. Carnrick, and W. Slavin, *Clin. Chem.* **29**, 477 (1983).

[29] M. Vahter, "Assessment of Human Exposure to Pb and Cd through Biological Monitoring." National Swedish Institute of Environmental Medicine and Karolinska Institute, Stockholm, Sweden, 1982.

the analyte in a refractory compound. The effect is suppressed by adding a swamping quantity of a metal that binds phosphate more efficiently. Typically La or Sr has been used for this purpose but other metals have been suggested. Alternatively, a hotter nitrous oxide–acetylene flame can be used without the suppressant but the hotter flame is not as convenient to use and it induces an ionization interference which must be suppressed by the addition of Cs or K.

A reliable and well-proven method[30] for both Ca and Mg in serum is to dilute the serum 1 + 49 in an aqueous solution containing 0.1% (w/v) La (as the chloride) and 0.5 *M* HCl for analysis in an air–C_2H_2 flame. Standards are prepared in the same diluent. If only Mg is required, the La is not necessary. Urine may be analyzed the same way although a greater dilution is often required. Tissue, feces, and diet can be analyzed for Ca and Mg after wet or dry ashing and suitable dilution, proceeding as for serum above.

It is rarely necessary to use the furnace for the determination of Ca or Mg in biological materials. Sometimes, however, ultramicro methods are called for and the furnace can be used for nl quantities of biological fluid and low μg amounts of tissue.

Magnesium and Ca are sensitive STPF determinations, with a characteristic mass, m_0, of 0.3 pg/0.0044 $A \cdot$sec for Mg and 0.8 pg for Ca. For Mg, this corresponds to 0.015 μg/liter in a 20-μl solution on the platform. Of course Mg is present at such high levels in biological materials that this extreme sensitivity is rarely useful. The ubiquitous presence of Ca and Mg makes the control of contamination the only major problem when the furnace is used for these determinations.

Chromium

The literature before 1980 reported Cr as a difficult determination in biological materials using the graphite furnace. This is chiefly because continuum correction with a deuterium arc provides very poor results at the long wavelength used for Cr (357.9 nm). By now many commercial instruments use continuum correction with a tungsten lamp which is an important help. Nevertheless, Cr is an element that is much better done with Zeeman correction, especially since so many biological samples require Cr measurement close to the detection limit. Pyrolytically coated graphite tubes are also particularly useful for Cr. Some workers use STPF conditions for Cr but deposit the sample on the tube wall. Our experience indicates that it is advantageous to use the platform, however. The char conditions are not greatly altered by the $Mg(NO_3)_2$ matrix modifier but

[30] D. L. Trudeau and E. F. Freier, *Clin. Chem.* **13**, 101 (1967).

small and variable amounts of Cr are lost during the char step if the matrix modifier is omitted.

Some workers have reported loss of volatile organic Cr compounds at low temperatures, especially in tissue samples and in plants. Recent papers seem not to be troubled by this problem, probably because of the improved experimental conditions. Some workers have used nitrogen as a purge gas instead of argon. It has been shown that narrow CN bands introduce a correction error at the 357.9-nm line in the presence of nitrogen. Argon should be used. Contamination of plasticware, pipets, etc. is a problem with Cr. Older Eppendorf pipets have stainless-steel springs which can yield a Cr signal.

Serum and blood Cr is an important determination now being done frequently with STPF conditions and Zeeman correction.[31] Serum is diluted 1 + 1 in the matrix modifier and 0.2% Triton X-100. Whole blood is diluted 1 + 3 in the same diluent. Air or oxygen are added during the char step to facilitate the ashing of the organic matrix. About 0.02 μg/liter Cr can be detected in the dilutions of either material using 40-μl aliquots in the furnace. Toxic levels of Cr have been found in the plasma of welders using these methods.[32] Urine has been analyzed for Cr by the same procedures.[33] Hair and other tissue can be analyzed for Cr using the same conditions suggested for blood and urine, after dissolution.[34]

There is increasing interest in measuring the speciation of Cr, particularly for toxicology, since Cr(VI) is known to be toxic while Cr(III) is relatively harmless. The hexavalent Cr can be precipitated with lead sulfate or absorbed onto Amberlite. Consult the literature for details.[15]

Cobalt

Cobalt is generally determined at the 242.5-nm line on Zeeman instruments and at the slightly more sensitive 240.7-nm line on continuum-corrected instruments. It is usually determined on the platform, but wall sampling works reasonably well because Co is a high temperature determination. Pyrolytically coated tubes are important for this determination.

Blood can be analyzed directly for Co in a 1 + 3 dilution of the blood in a mixture of the matrix modifier with Triton X-100. See Cr for more details. The detection limit is about 0.5 μg/liter in the dilution which is inadequate sensitivity to measure normal levels of Co. Increased sensitiv-

[31] A. J. Schermaier, L. H. O'Connor, and K. H. Pearson, *Clin. Chim. Acta* **157,** 123 (1985).

[32] B. W. Morris, C. A. Hardisty, J. F. McCann, G. J. Kemp, and T. W. May, *At. Spectrosc.* **6,** 149 (1985).

[33] W. Slavin, G. R. Carnrick, D. C. Manning, and E. Pruszkowska, *At. Spectrosc.* **4,** 69 (1983).

[34] G. Bagliano, F. Benischek, and I. Huber, *Anal. Chim. Acta* **123,** 45 (1981).

ity is obtained by ashing a larger aliquot of blood or plasma followed by extraction into an organic solvent which is loaded onto the platform. Several workers have erroneously reported "normal values" for Co in blood using direct methods or extractions from inadequate quantities. Normal values are not reliable by direct furnace methods if they are close to the detection limit.

Plant and animal tissue have been analyzed for Co after simple dissolution and dilution in the matrix modifier. However, the requirement for measurement at levels lower than can be achieved by direct methods usually calls for concentration by ion exchange or by organic extraction.

Copper

Serum is frequently analyzed for Cu in the flame.[2] The discussion under Zn, concerning the alternative in flame AAS between a 1 + 9 dilution of serum to remove viscosity effects and that of matching serum viscosity, applies equally to Cu. With the 1 + 9 dilution, Delves[3] reported a CV of 6% within batch and 8% day-to-day for a serum with 0.7 mg/liter of Cu. His detection limit was 0.03 mg/liter. Some workers add a commercial synthetic plasma expander to the standard solutions to match better the viscosity of the samples.[2]

To improve on this precision many papers recommend extraction of Cu into an organic solvent prior to flame AAS. With the improvements in modern furnace AAS it seems to me more precise, rapid, and potentially free of contamination to use the furnace in situations where direct flame AAS provides inadequate sensitivity for Cu in biological materials. Thus it is no surprise that more workers are turning to the furnace for this determination.

Copper is the only element commonly determined in the graphite furnace that suffers a significant loss of sensitivity (about 50%) when the transverse ac Zeeman corrector is used. Nevertheless, this small loss of sensitivity is more than compensated by the better background correction using the Zeeman effect. At the present time we have not found a matrix modifier that improves the determination of Cu.

Copper has been determined in plasma, red blood cells, ocular and cerebrospinal fluids[35] using simple 1 + 1 dilution in 1% Triton X-100 and standards prepared in water. The detection limit is about 0.1 μg/liter in the dilution. Many workers have determined Cu in urine with the modern furnace. Milk and infant formula have been analyzed for Cu using simple dilution methods and the STPF technique.[36] Delves[3] determined the Cu in

[35] M. C. McGahan and L. Z. Bito, *Anal. Biochem.* **135,** 186 (1983).

[36] D. J. Hutchinson, F. J. Disinski, and C. A. Nardelli, *J. AOAC* **69,** 60 (1986).

protein fractions that were separated from 2 μl of serum by cellulose acetate membrane electrophoresis.

Gold

The determination of Au in biological materials stems from its use in the therapy of rheumatoid arthritis. There are no papers using the new technology for serum, urine, or tissue but the opportunity should be attractive. A detection limit of about 1 μg/liter should be possible by direct procedures, once the sample was in solution. Older papers report the determination of Au in these materials plus tissue and feces.[37] The older methods report similar detection limits to that mentioned above but interferences were found that we do not expect with STPF conditions.

Iron

As reported for Zn below, serum Fe can be determined by flame AAS after direct dilution of serum variously proposed as 1 + 1[2] to 1 + 5.[3] Quantities of hemoglobin in serum too small to be visible as hemolysis can introduce large errors in serum Fe and protein precipitation is usually recommended prior to flame AAS. Total Fe-binding capacity (TIBC) is also determined by flame AAS. Tissues rich in Fe can be determined by direct flame AAS after digestion, as can diets and feces. Urine Fe requires solvent extraction for flame AAS and is probably better done by furnace AAS.

While flame AAS is adequate for routine determination of serum Fe, micro methods utilizing the furnace have been developed[38] for pediatrics, etc. Of course, deproteinization is still required, usually with trichloroacetic acid. Because of furnace sensitivity, the serum sample should be diluted 1 + 9 in a solution containing the matrix modifier and about 0.2% Triton X-100, and a 10-μl aliquot should be deposited on the platform. In the dilution, 100 μg/liter of Fe will provide a signal of about 0.4 $A \cdot$sec using STPF conditions. The same method will apply to Fe in tissue and other biological materials, once the sample is in solution.

Lead

While blood lead has been determined extensively by flame AAS, the method required large samples and troublesome sample pretreatment. In the early 1970s most workers used the Delves cup technique.[39] Now, however, the graphite furnace is increasingly the method of choice. A

[37] R. M. Turkall and J. R. Bianchino, *Analyst (London)* **106,** 1096 (1981).
[38] S. A. Lewis, T. C. O'Haver, and J. M. Harnly, *Anal. Chem.* **56,** 1651 (1984).
[39] H. T. Delves, *Analyst (London)* **95,** 431 (1970).

review of Pb poisoning[40] published in 1986 found that 25% of the 95 laboratories participating in a CDC proficiency test for blood Pb used the furnace and an equal number used the Delves cup. This ratio is shifting rapidly to the furnace.

For the STPF blood Pb method,[41] whole blood is diluted 1 + 9 in a solution containing 0.2% Triton X-100 and the matrix modifier. Alternatively, the matrix modifier may be added in a separate aliquot using an autosampler. Ten-microliter aliquots are deposited on the platform. The standards are made up with 0.2, 0.4, 0.6, and 0.8 ng of Pb in the same diluent that is used for the samples. Background correction is necessary for reliable blood Pb results but the backgrounds are not large and continuum correction provides good results. Zeeman correction provides better precision. A CV of about 2% was found with a Zeeman method between 200 and 1000 μg/liter.

There is controversy in the literature over the determination of serum Pb with the furnace because the serum levels are very low. We believe that STPF methods are mandatory for serum Pb and that Zeeman correction may be mandatory also. If, as has been reported, normal serum Pb is less than 1 μg/liter, it will not be possible to measure these levels with confidence with a direct method. A 20-μl aliquot of a 1 + 1 dilution of serum in the diluent mentioned in the previous paragraph will provide a detection limit a little lower than 1 μg/liter. For serum levels higher than this, the method can be used with confidence.

Urine Pb down to the 1 μg/liter level can be measured in 20 μl of a 1 + 1 dilution of urine in the matrix modifier. If a somewhat poorer detection limit is acceptable, a 1 + 3 dilution of urine is more reliably handled by the autosampler. Tissue Pb, particularly animal tissue, has been widely determined by this simple STPF method once the tissue is in solution. With many biological samples outside of blood, the major problem is caused by inadequate continuum background correction. Stimulated by reported problems in the determination of Pb in infant formula, a recent paper[42] used the simple STPF method stated above and Zeeman background correction. Reliable results were found with good precision. Zeeman correction is particularly valuable for the determination of Pb in complex inorganic matrices that produce large backgrounds.

Lithium

Lithium is administered for certain psychological disorders. It can easily be monitored in serum in a 1 + 9 dilution using flame AAS. For

[40] R. L. Boeckx, *Anal. Chem.* **58,** 275A (1986).

[41] E. Pruszkowska, G. R. Carnrick, and W. Slavin, *At. Spectrosc.* **4,** 59 (1983).

[42] J. R. Andersen, *Analyst(London)* **110,** 315 (1985).

direct urine Li measurements in the flame, somewhat greater dilution may be necessary.[2]

There are very few publications on Li in biological materials with the furnace and probably none using STPF conditions. Furnace sensitivity for Li is very great and a 20-μl aliquot on the platform will have a detection limit lower than 0.1 μg/liter. Before biological methods can be developed, more work is necessary on the optimum conditions in the STPF.

Magnesium

See section on calcium.

Manganese

Manganese has been widely determined in biological materials using flame AAS and many reference books will provide methods (e.g., Refs. 2 and 3). However Mn is an ideal STPF analyte, and the low levels almost always required in biological materials suggest that it should now be done in the furnace, usually by direct methods. The major problem with the determination of Mn is the control of contamination, which is discussed in an earlier section. Zeeman correction is particularly useful for Mn because, at the long wavelength used for Mn, 279.5 nm, the continuum correctors are not very effective.

Whole blood, serum, and blood cells have been analyzed for Mn by many workers using furnace methods, although contamination has probably produced large errors in many of these studies. The normal serum level is close to 1 μg/liter, probably lower.[18] Work in our laboratory found Mn results, in μg/liter, of 11.8 in normal whole blood, 1.7 in normal plasma, and 24.6 in packed red blood cells,[43] although contamination may have contributed to this work which was done before 1980. We now recommend that serum be diluted 1 + 2 in the matrix modifier and 0.2% Triton X-100 and 20-μl aliquots be deposited on the platform. This arrangement yields a detection limit of about 0.1 μg/liter. The same methods have been used for urine.[44]

Tissue samples were analyzed by the same technique after the samples were put into solution.[45] Hair samples were carefully cleaned and decomposed in a miniautoclave and a similar procedure was used on the solutions.[46]

[43] C. E. Pippenger, C. Garlock, F. Fernandez, W. Slavin, and J. Iannarone, *Adv. Epileptol.* **11** (1980).

[44] W. Frech, J. M. Ottaway, L. Bezur, and J. Marshall, *Can. J. Spectrosc.* **30,** 7 (1985).

[45] J. P. Dougherty, R. G. Michel, and W. Slavin, *Spectrosc. Lett.* **18,** 627 (1985).

[46] O. Guillard, J.-C. Brugier, A. Piriou, M. Menard, J. Gombert, and D. Reiss, *Clin. Chem.* **30,** 1642 (1984).

Mercury

Mercury is generally determined by the cold vapor technique using a mercury–hydride accessory for a flame AAS. See Welz[2] for an extensive discussion of the technique and biochemical applications. Mercury losses are known to occur if biological materials are dry ashed and wet ashing also poses a risk of loss. Low temperature and very fast ashing are used. The autoclave, or high pressure decomposition bomb, is particularly useful.

Mercury can be determined in the furnace although there is not much experience in the literature. Grobenski *et al.*[47] had difficulty retaining Hg in real samples using any of the matrix modifiers in the literature. They found success with several tissue samples by avoiding the char step altogether and atomizing after the sample was dried. Of course this produced large background signals but these were not troublesome using Zeeman correction. They deposited 20 μg of Pd on the platform before depositing the sample and heated the platform to 1000° for 10 sec. The tube was returned to room temperature, the sample deposited, and maximum power heating was used at 1000°.

Molybdenum

Molybdenum is a sensitive ($m_0 = 9$ pg) furnace determination although relatively few papers have been published on biological applications. It is one of the most refractory metals determined in the furnace and therefore the sample is deposited on the wall. Nevertheless the other STPF conditions are used, including the use of A·sec signals.

Plant tissue was analyzed for Mo[48] after wet ashing in HNO_3 and H_2O_2. The use of A·sec signals was important in obtaining accurate results because many matrix materials altered the absorbance profiles.

Nickel

Nickel is easily determined in the furnace from the wall or from the platform. We prefer the platform and matrix modifier for biological materials. Again, it is important to use A·sec signals.

The IUPAC reference method for Ni in urine and serum was prepared for the graphite furnace.[49] They extracted the Ni with APDC into MIBK and analyzed the organic phase in the furnace. Since 1 + 1 dilutions of urine or serum in 20-μl aliquots on the platform would provide a detection

[47] Z. Grobenski, R. Lehmann, B. Radziuk, and U. Völlkopf, *At. Spectrosc.* **7,** 61 (1986).

[48] M. Hoenig, Y. V. Elsen, and R. Van Cauter, *Anal. Chem.* **58,** 777 (1986).

[49] S. S. Brown, S. Nomoto, M. Stoeppler, and F. W. Sunderman, Jr., *Pure Appl. Chem.* **53,** 773 (1981).

limit in the sample of about 1 μg/liter and normal values are reported to be only slightly larger than that,[18] a direct method might not be sufficiently sensitive. Sunderman *et al.*[50] analyzed a protein-free filtrate of whole blood or serum for Ni. They found a detection limit of 0.1 μg/liter which may not be low enough for the 0.5 μg/liter normal values they found in serum.

Phosphorus

Phosphorus can be determined in the furnace although the STPF characteristic mass is only about 3 ng. This poor sensitivity comes about because the P resonance lines lie in the deep ultraviolet below the levels available to conventional furnace instrumentation. It is particularly important to use the platform, matrix modifier and A·sec signals.

Potassium

See section on sodium.

Selenium

Selenium is probably the most valuable determination on Zeeman corrected furnace instruments using STPF technology because competitive methods are slow and prone to manipulative errors at the low concentrations that are typically of interest in biological materials. Nevertheless, the volatility of many Se compounds, especially organoselenium compounds, produces troubles. Both Fe and P produce severe overcorrection errors when Se is determined with continuum correction, making Zeeman correction mandatory for Se in biological materials. There are many papers in the literature that have not used Zeeman correction for Se but they rely on delicate timing of the thermal program so that Se is not volatilized at the same time as the interferent. The review by Verlinden *et al.*[51] of the AAS determination of Se should be consulted.

Whole blood and serum may be analyzed for Se by a 1 + 2 dilution in the combined Ni and magnesium nitrate modifier which includes 1% Triton X-100. A 20-μl aliquot of this will provide a Se detection limit in the sample of about 3 μg/liter which is adequate for whole blood and plasma for which normal values range near 100 μg/liter.[52]

The determination of Se in urine has been reviewed.[53] High levels of

[50] F. W. Sunderman, M. C. Crisostomo, M. C. Reid, S. M. Hopfer, and S. Nomoto, *Ann. Clin. Lab. Sci.* **14,** 232 (1984).

[51] M. Verlinden, H. Deelstra, and E. Adriaenssens, *Talanta* **28,** 637 (1981).

[52] M. Verlinden, *Talanta* **29,** 875 (1982).

[53] H. J. Robberecht and H. A. Deelstra, *Talanta* **31,** 497 (1984), and *Clin. Chim. Acta* **136,** 107 (1984).

sulfur in urine were found to interfere in the Se determination unless the magnesium nitrate was present. But 1 + 4 dilutions of urine provided reliable results with detection limits below 10 μg/liter in the urine.[54]

Hydride methods are widely used for Se at concentrational detection limits similar to those of the furnace but requiring larger samples. For biological materials the modern furnace technique is generally preferable to hydride methods.

Silicon

There is some interest in the Si determination in biochemical samples. There is every reason to be optimistic about the potentiality of the furnace for this application but, at present, there are almost no practical publications. There are many physicochemical studies in the literature,[15] mostly with pre-STPF technology. We recommend that the platform be used (see Table 1), but we have not looked for an appropriate matrix modifier. The solution detection limit is probably very close to 1 μg/liter using a 20-μl sample. There have been scattered reports that some graphite furnace tubes provide a large Si blank because of contamination of the graphite materials.

Silver

Silver is sensitively detected in the graphite furnace and is relatively free of interferences when STPF conditions are used. There is very little literature on biochemical applications but there is a growing literature on environmental materials and, of course, metallurgical chemistry. The experimental conditions for Ag with STPF and Zeeman correction were recently studied[55] and Table I reflects the recommendation that Pd be used as a matrix modifier.

Sodium (and Potassium)

Sodium and K are usually determined in biological materials by flame emission spectroscopy and many convenient instruments are commercially available for this purpose. However, flame AAS is equally appropriate and a simple dilution, e.g., about 1 + 200 for serum, is usually the only sample preparation required.

The high concentrations of Na and K in biological materials cause very little demand for furnace analyses. While a few ultramicro applications have appeared, there is little useful experience using the furnace for these elements.

[54] G. R. Carnrick, D. C. Manning, and W. Slavin, *Analyst (London)* **108,** 1297 (1983).
[55] D. C. Manning and W. Slavin, *Spectrochim. Acta* **42B,** 755 (1987).

Tellurium

Tellurium is rarely sought in biological materials but it should pose no problems when determined with STPF conditions. Like As and Se, there are problems in the presence of phosphorus if continuum correction is used and Zeeman correction is recommended for biological materials.[22]

Thallium

Thallium is sensitively determined in the furnace which has been used frequently for toxicological problems. The determination is prone to interference in the presence of large amounts of chloride because vapor-phase chlorides of Tl are quite stable. In spite of the attention of several authors, the matrix modifier that we recommend is probably inadequate and a better one should be sought. The literature prior to the STPF technique showed considerable interferences which are not seen with modern technology.

Several workers report Tl in serum and in urine, some with aspects of the STPF technique but, so far, there are no definitive Tl methods in either matrix. If a detection limit of 3 μg/liter is satisfactory, dilution of whole blood, serum, or urine of 1 + 2 to 1 + 4 in the matrix modifier should work well with a 20-μl aliquot on the platform. For blood samples Triton X-100 should be added also.

Tin

In many general analytical studies Sn has proven troublesome in many matrices, particularly in the presence of sulfate. We have tested the reported problems and found no trouble with STPF conditions.[56] There seem to be no methodology papers on tin in specific biological fluids or tissues but the general procedures are very likely to work well. A dilution of blood or urine of 1 + 3 will provide a detection limit of about 4 μg/liter in the sample using a 20-μl aliquot on the platform.

Vanadium

Vanadium is a refractory element determined from the wall of the furnace but otherwise using STPF conditions.[57] Direct dilution furnace methods for V in serum or urine yield a detection limit less than 4 μg/liter

[56] E. Pruszkowska, D. C. Manning, G. R. Carnrick, and W. Slavin, *At. Spectrosc.* **4,** 87 (1983).

[57] D. C. Manning and W. Slavin, *Spectrochim. Acta* **40B,** 461 (1985).

in the sample. Normal levels are below this and workers have resorted to APDC extraction.[58]

Zinc

Zinc is determined in serum and urine by flame AAS. Simple dilution of the sample 1 + 4 in water permits the sample to be aspirated into the burner. At this dilution the burner and signal readout must be quite stable to provide precision approaching a CV of 1%. Some workers suggest a smaller dilution, e.g., 1 + 1,[2] but this risks difficulties in standardization because of the difference in viscosity between samples and standards. The viscosity problem is often (hopefully) controlled by preparing standards in a serum or urine matrix, subtracting a large blank. This is also not very satisfactory. The best compromise appears to be a 1 + 4 dilution and standardization against simple aqueous solutions. Delves[3] suggested 1 + 9 dilution in 6%, v/v, butanol of 200 μl of serum to control the viscosity effect completely for this determination, yielding a CV of about 4%, within batch, and 7% day-to-day. Modern instruments may provide somewhat better precision.

For urine, Delves[3] used a 1 + 5 dilution in butanol as above, yielding a CV of about 3% within batch. For whole blood he diluted 1 + 9 in 0.05% Triton X-100. In cerebrospinal fluid where the Zn content is one-tenth that in serum, a 1 + 1 dilution in 6% butanol was used. The analysis of feces and diet is usually done simply by flame AAS after dissolution of the sample.

For micro determination of Zn in serum or urine the furnace is used with STPF conditions. Because of the sensitivity of the furnace Zn method, the determination is more prone to contamination problems than Al, Pb, or Fe, approaching the problem of Ca, Na, or K. Work near the furnace detection limit probably requires clean room facilities. There is an isolated report that Fe causes an overcorrection error when continuum background correction is used,[59] making Zeeman correction advantageous for biological materials.

For pediatric or micro serum Zn, 10 μl of serum was diluted 100-fold and a 10-μl sample was deposited in the furnace.[60] The STPF technique was still too sensitive and the reference used a technique for reducing sensitivity that can introduce error. The Zn line at 307.6 nm appears to have about 1000-fold less sensitivity which makes it too insensitive. It would be preferable to dilute the sample 200-fold and use a 5-μl aliquot.

[58] L. Pyy, E. Hakala, and L. H. J. Lajunen, *Anal. Chim. Acta* **158,** 297 (1984).

[59] N. J. Miller-Ihli, T. C. O'Haver, and J. M. Harnly, *Anal. Chem.* **54,** 799 (1982).

[60] N. E. Vieira and J. W. Hansen, *Clin. Chem.* **27,** 73 (1981).

Because of the sensitivity, Zn has been determined in serum fractions separated by chromatography.[61] Ten-microliter urine samples were analyzed directly[60] in the furnace. Segments of rat brain tissue were analyzed for Zn using STPF conditions.[62]

Conclusion

While this review has attempted to be objective in its recommendations for various biochemical analyses, it is apparent that there is a bias for furnace AAS methods. Perhaps this is partly a result of experience, but I believe there is much more to it than personal judgment. Most analytical instruments are devices that compare a signal obtained on a sample with the signal obtained on an appropriate standard. Both Walsh and L'vov, inventors respectively of the flame AAS and furnace AAS techniques, looked forward in their initial papers to an analytical technique that was free of interference. This would permit the instrument to perform without standards, that is to provide absolute analytical results. So far neither flame nor furnace AAS has achieved this lofty goal. The limitation of flame AAS in this respect results from ambiguity as to how much analyte is transferred to the flame by the nebulizer and spray chamber as well as the complexity of flame chemistry. In contrast, all of the analyte is converted into an atomic vapor for most elements determined in the furnace,[63] and, for most elements, remain homogeneously distributed within the furnace tube. Thus there is a likelihood that absolute analysis may eventually be realized. Even lacking that desirable situation, the present STPF technology provides a high confidence that analytical results will be accurate. This coupled with the very great sensitivity of furnace AAS outweighs the major disadvantage, the slowness, since each determination requires about 2–4 min.

The furnace is particularly suited to the requirements of biochemical trace metal analysis since the technique provides its own in-built dry ashing procedure that can free the determination from the influence of the organic matrix. The residual inorganic ash is usually small enough so that the instrumental system is untroubled, especially if Zeeman correction is used. Additionally, biochemical analyses must often be made on small sample volumes and this is compatible with the requirements of furnace

[61] P. E. Gardiner, J. M. Ottaway, G. S. Fell, and R. R. Burns, *Anal. Chim. Acta* **124,** 281 (1981).

[62] N. W. Alcock, *Neurobiol. Zn,* Part A, p. 305. Liss, New York (1984).

[63] B. V. L'vov, V. G. Nikolaev, E. A. Norman, L. K. Polzik, and M. Mojica, *Spectrochim. Acta* **41B,** 1043 (1986).

AAS. Thus I believe that furnace AAS will gradually take over an increasing proportion of the biochemical trace metal determinations.

There are several problems that stand in the way of increased utilization of furnace AAS for trace metal analyses. The extensive history of furnace AAS interferences during the 1970s weighs against the credibility of the modern technology, but this is gradually passing. More importantly, many analysts fail to give adequate attention to the laboratory hygiene that is required at the very low levels accessible by furnace AAS. Several collaborative studies have shown that isotope dilution mass spectrometry, IDMS, and neutron activation analysis, NAA, provided more accurate data on round robin samples than furnace AAS. The most important reason for this observation is that IDMS and NAA equipment are expensive and require special skills. In that environment, the additional expense for adequate and clean laboratory facilities is easily justified. Given the same care and facilities, furnace AAS is at least equal to IDMS and NAA in accuracy for biochemical trace metal analysis.

[13] Multielement Atomic Absorption Methods of Analysis

By JAMES M. HARNLY and DONITA L. GARLAND

Introduction

Atomic absorption spectrometry (AAS) with carbon furnace atomization permits detection limits in the 0.1–10 ng ml^{-1} range (1–100 n*M* for most elements of biological interest) using sample volumes of 5–20 μl. These characteristics make carbon furnace AAS a very attractive tool for the determinations of metals in enzymes and biological samples. The biggest disadvantage of carbon furnace AAS is that it is a single element method. Separate experiments are required for the determination of each element.

Over the last 8 years, a prototype (not commercially available) multielement AAS has been developed and characterized by the Nutrient Composition Laboratory at USDA in collaboration with the University of Maryland.[1-5] This simultaneous multielement atomic absorption contin-

[1] J. M. Harnley, T. C. O'Haver, W. R. Wolf, and B. M. Golden, *Anal. Chem.* **51,** 2007 (1979).

[2] J. M. Harnly and T. C. O'Haver, *Anal. Chem.* **53,** 1291 (1981).

[3] J. M. Harnly, N. J. Miller-Ihli, and T. C. O'Haver, *J. Autom. Chem.* **4,** 54 (1982).

[4] J. M. Harnly and J. S. Kane, *Anal. Chem.* **56,** 48 (1984).

[5] J. M. Harnly, N. J. Miller-Ihli, and T. C. O'Haver, *Spectrochim. Acta* **39B,** 305 (1984).

TABLE I
SIMAAC COMPONENTS

Component	Supplier
300 W Cermax Xenon Arc Lamp (Model LX300UV) and Power Supply (Model PS300-1)	ILC Technology, Sunnyvale, CA
Echelle Polychromator (Spectraspan III)	Beckman Instrument Co., CA
Galvanometer (Model G325D) and Scanner Controller (Model CCX-650)	General Scanning Inc., Watertown, MA
11/34-VE Declab Minicomputer with Analog-to-Digital and Digital-to-Analog Converters, Real Time Programmable Clock, Digital I/O, Floating Point Processor, Video Terminal, and Line Printer	Digital Equipment Corp., Maynard, MA
Carbon Furnace and Programmer/Power Supply (Model HGA-500)	Perkin-Elmer Corporation, Ridgefield, CT

uum source spectrometer (SIMAAC) provides simultaneous detection of 16 elements, full background correction, calibration covering 5–6 orders of magnitude of concentration for each element, and compatibility with either flame or carbon furnace atomization.

Equipment

As shown in Table I and Fig. 1, SIMAAC is composed of commercially available components: a xenon arc lamp, an echelle spectrometer, and a dedicated minicomputer. A galvanometer with a quartz refractor plate has been mounted immediately behind the entrance slit of the echelle. Oscillation of the quartz plate back and forth produces repeated scans across a narrow wavelength region, or wavelength modulation.

The heart of the system is the dedicated minicomputer. Data are acquired for each element at a rate of 1120 intensity measurements per second. This is a total rate, for all 16 elements, of 17,920 measurements per second. Data can be acquired at this rate for up to 30 sec. During an atomization, the acquired data are stored on disk. After the atomization is complete, the data are recalled, absorbances are computed, and the peak height and peak area (for carbon furnace atomization) are computed.

Principles

A xenon arc lamp or white light source furnishes continuous intensity at all wavelengths. Atomic absorption appears as an inverted peak in the

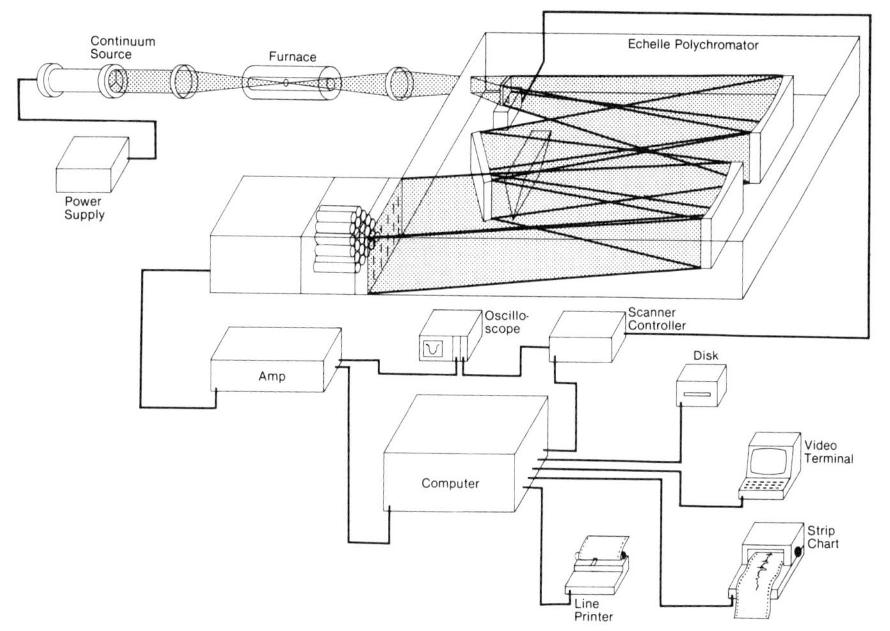

FIG. 1. Block diagram of SIMAAC.

continuum (Fig. 2). With wavelength modulation, rapid, repetitive scans are made across the absorption profile. Twenty discrete measurements are made during each scan. The 20 intensities are used to compute absorbances in several ways. The most sensitive absorbance (log $[I_0/I]$) employs the five measurements to either side of the profile (points 1–5 and 16–20 in

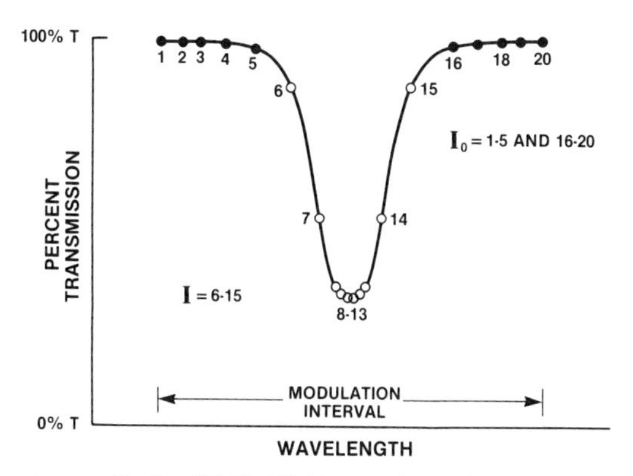

FIG. 2. Absorption profile for SIMAAC. Twenty intensity measurements are made during each pass across the profile.

TABLE II
DETECTION LIMITS FOR SIMAAC

Element[a]	Detection limit[b] (ng/ml)
Al	0.8
Ca	—
Co	2.0
Cr	0.4
Cu	0.4
Fe	1.0
K[c]	10.0
Mg	—
Mn	0.4
Mo	2.0
Na	—
Ni	2.0
V	3.0
Zn	0.9

[a] Used traditional AAS wavelength.
[b] The detection limit is defined as the concentration giving a signal three times the standard deviation of the baseline noise. Carbon furnace parameters shown in Table III.
[c] Less sensitive analytical line, 4044 nm.

Fig. 2) as the reference intensity (I_0) and the 10 measurements at the profile center (points 6–15 in Fig. 2) as the sample intensity (I). Less sensitive absorbances are computed using measurements 1 and 20 as I_0, and pairs of intensities in the wings (2 and 19, 3 and 18, 4 and 17, 5 and 16, and 6 and 15) as I. In this manner, a series of calibration standards produce a series of 6 calibration curves which can cover 5 to 6 orders of magnitude of concentration.

Wavelength modulation corrects for the intensity fluctuation of the xenon arc source and broad band spectral (background absorption) interferences. Modulating at 56 Hz effectively eliminates intensity fluctuations which occur at frequencies less than 30 Hz. In this manner, the instability of the xenon arc source is minimized and the SIMAAC detection limits for elements determined at wavelengths >280 nm are comparable to those for conventional AAS using hollow cathode lamps (HCLs). Below 280 nm the xenon arc is increasingly less intense and the detection limits become increasingly worse (Table II). Background absorption is simply a very low frequency (but sometime quite severe) component of intensity fluctuation. Consequently, all computed absorbances are inherently background corrected.

Use of a white light source instead of the narrow emission line of a HCL leads to less sensitivity and nonlinear calibration curves. To remedy this problem, a high resolution echelle spectrometer is used. The narrow spectral bandpass of the echelle substitutes for the narrow HCL line. Fortuitously, the echelle can also be used to detect as many as 20 elements simultaneously. Thus good sensitivity, linearity, and multielement detection are simultaneously achieved.

Compromise Parameters

Multielement carbon furnace AAS requires the selection of a single method for sample preparation and a single set of atomization parameters. Thus, multielement determinations require compromise parameters as compared to conventional single element AAS where all parameters are optimized for the determination of the element of interest. In addition, we have adopted the philosophy that the elemental determinations must be made by direct comparison of the sample measurements to calibration standards in 5% nitric acid. The sample preparation and the atomization parameters must be selected such that this method of calibration yields accurate results. Many single element methods employ the method of additions to minimize the sample preparation and correct for interferences. Multielement methods of additions are not possible, however, without violating the basic premise of the method, i.e., the matrix remains unchanged except for the addition of the element of interest.

Atomization Parameters

The data we have acquired in the past 8 years agree well with the analytical approach of the stabilized temperature platform furnace (STPF) advocated by the Perkin-Elmer Corporation.[6] This approach requires fast electronics ($<$10–20 msec per absorbance computation), accurate background correction, high-density pyrolytically coated tubes, atomization from a pyrolytic platform, rapid heating of the furnace, area measurements, argon as the purge gas, stopped gas flow during atomization, and matrix modification to stabilize the element during the ash cycle. The basic procedure is listed in Table III. The drying step (as noted) is a minimum of 60 sec. Longer times may be used to assure a slow, gentle drying of the samples and standards. The greatest source of error for

[6] W. Slavin, "Graphite Furnace AAS—A Source Book." Perkin-Elmer Ridgefield, Connecticut, 1984.

TABLE III
CARBON FURNACE ATOMIZATION CONDITIONS

Parameter	Measurement
Atomization mode	Platform (pyrolytically coated)
Signal measurement	Peak area
Sample and standard matrix	5% nitric acid
Drying	
Time (minimum)	30 sec ramp, 30 sec hold
Temperature	200–300°
Ashing	
Time (minimum)	30 sec ramp, 30 sec hold
Temperature	500°
Atomization	
Time	Maximum heating rate, 5–10 sec hold
Temperature	2700°
Purge	
Gas	Argon
Rate	0–20 ml min^{-1}

carbon furnace AAS arises from lack of reproducibility in the sample drying step.

The charring temperature is dictated by the most volatile element. We traditionally use 500° to permit the analysis of Zn, Pb, and Cd. This low charring temperature also helps prevent loss of the metals as volatile organic complexes. Single element methods employ much higher temperatures. This helps reduce background interferences but requires careful testing to ensure no prevolatilization of the metals. Undigested samples, even charred at 500°, should be checked for prevolatilization (i.e., is the analytical signal constant as a function of the charring temperature).

An atomization temperature at 2700° is used to allow detection of the widest range of elements. The data in Fig. 3 detail the effect of the atomization temperature on the peak areas. Atomization at temperatures of 2500° or higher permit Mo and V to be determined but are accompanied by a decrease in signal for the most volatile elements.

The argon purge flow can be a very useful tool for biological samples. A low flow (0–20 ml min^{-1}) is used to obtain maximum sensitivity. Elements which tend to occur at very high concentrations in biological systems (Ca, K, Mg, Na, and Zn) can be analyzed much more easily with high purge flows (100–300 ml min^{-1}). These high flows significantly reduce the sensitivity of these elements and eliminate the problem of residual material being carried over to the next atomization.

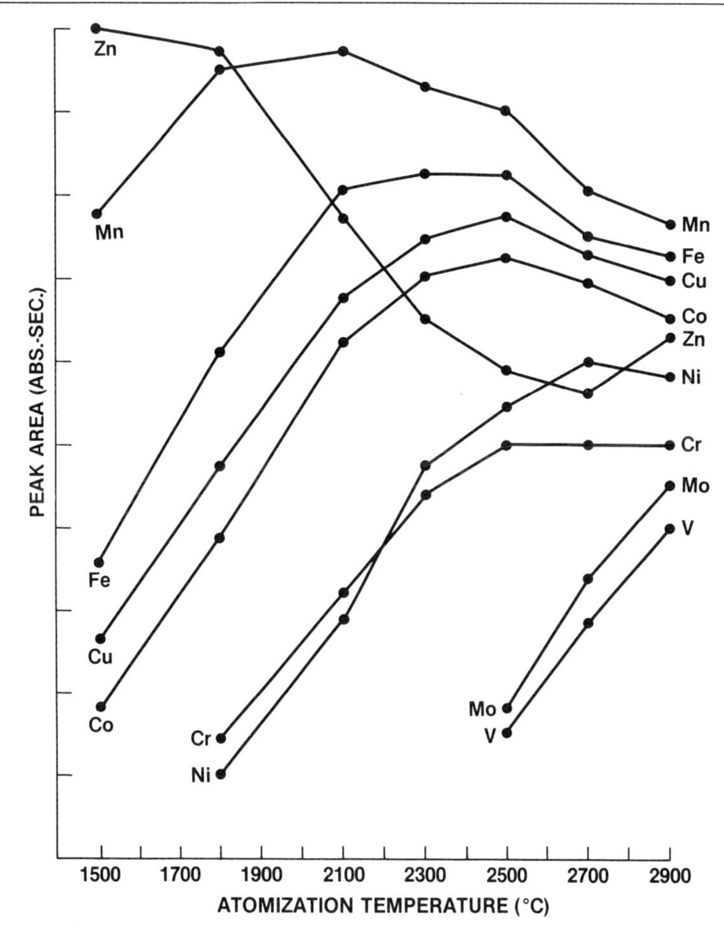

FIG. 3. Peak area of nine elements, atomized from a platform, as a function of the atomization temperature.

Sample Preparation

The goal of any sample preparation scheme is to place the sample into a form which can be accurately analyzed using the calibration standards. For metalloenzymes, this must be done without contaminating the sample (from glassware, plasticware, or reagents) or losing any of the sample (adsorption to container walls or volatilization). For multielement determination, the sample preparation method must be appropriate for all the elements.

The most desirable approach is no treatment at all. The feasibility of no preparative step is dependent on the nature of the sample and the accuracy desired. Highly purified enzymes can be analyzed directly and accurately depending upon the buffer in which they are dissolved. NaCl and KCl buffers (and halides in general) should be avoided. Chloride tends to form undissociated metal chlorides in the gas phase which suppress the analytical signal. HCl is highly volatile (compared to metal chloride salts) and is usually lost in the charring step. If a large excess of metal is present, however, stable chloride salts (such as NaCl, KCl, $MgCl_2$, $CaCl_2$) will remain in the furnace until the atomization step and will suppress the signals of less concentrated metals. Thus, a chloride-buffered matrix would be suitable for highly purified sample but not for a crude sample where high biological concentrations of Ca, K, Mg, and Na might be expected.

Two major digestive procedures which have been used extensively are (1) dry ashing, sometimes preceded by lyophilization, and (2) wet digestion with nitric acid and hydrogen peroxide. The lyophilization–dry ashing method is useful when sample volume is limited and has proven especially applicable to the determination of trace metals in blood serum and rat eye lens. During the lyophilization of blood serum (in a quartz test tube), the sample tends to draw away from the wall, forming a plug in the center of the tube. This reduces the chances of adsorption on the wall during the ashing step. The ashing oxidizes the organic component leaving only the inorganic constituents. Dissolution in nitric acid is preferred. Dissolution in HCl has been used for the determination of Cr in blood serum (by method of additions) but is inappropriate for multielement determinations because of the high concentration of Ca, K, Mg, and Na (forming stable chloride salts upon drying in the carbon furnace).

The nitric acid–hydrogen peroxide wet digestion method is more appropriate when larger sample volumes are available. The purity of the hydrogen peroxide has proven to be the biggest problem in the past.

Multielement standards are routinely made up in 5% HNO_3. Lower acid concentrations were tried, but the long-term stability of the standards was not satisfactory.

Data Handling

Although not at first obvious, data handling for a multielement and/or automated method is a compromise of the analyst's time. Because of the large volume of information acquired in a short time, it is almost impossible to examine the data as closely as desired. Reference standards and pool samples serve as valuable checks on the accuracy of the data.

Analytical Methods

Microsurvey Method

An approach to metalloenzyme determinations which is truely unique to SIMAAC is the microsurvey method. This method is used for multielement determinations in purified (and concentrated) biological samples. This method is designed to answer the question of whether trace elements are present as contaminants or in stoichiometric ratios. Results are based on two or three atomizations of a sample compared to standards in 5% HNO_3 and are assumed to have an accuracy of ±50%. This method is useful for acquiring data for extremely small sample volumes where there is insufficient sample to develop and validate a highly specific analytical method. The microsurvey method is very useful in obtaining information from limited samples or preliminary data for more thorough studies.

The assumptions of the microsurvey method are given in Table IV. The sample must be of high purity, the buffer must be well known, an appropriate blank must be available, and the furnace conditions listed in Table III must be used. Reported interferences for carbon furnace AAS are predominantly from the anion effects previously discussed. Organic materials may cause background absorption and prevolatilization of the metals. With an accurate background correction system and checks to ensure there is no prevolatilization, there should be few interferences arising from a purified sample. Samples in metal chloride buffers (i.e., NaCl) cannot be analyzed. Samples in solutions buffered with HCl can be analyzed if there are no other concentrated metals in the sample.

Recoveries of known additions of nine elements to eight common buffer solutions and deionized, distilled water are shown in Table V. Recoveries ranged from 46 to 123%. However, 68 of the 81 values fell between 80 and 120% recovery. Only 5 values fell below 75% and none was greater than 125%. The stated accuracy limits of ±50% are conservative. Yet, ±50% accuracy permits values differing by an order of magnitude or more to be detected. Thus the microsurvey method is well suited

TABLE IV
ASSUMPTIONS FOR THE MICROSURVEY METHOD

1. Purified sample
2. Known buffer (usually dialyzate)
3. Measure difference (sample—dialyzate)
4. Furnace conditions shown in Table III

TABLE V
ELEMENTAL RECOVERIES IN COMMON BUFFERS

Buffer	Al	Co	Cr	Cu	Fe	Mn	Mo	Ni	V
Deionized, distilled H_2O	109	105	97	102	101	99	105	96	104
0.1 *M* Tris, pH 8.4[a]	89	102	96	91	92	92	92	89	88
0.1 *M* MES, pH 6.1[b]	83	100	86	77	98	72	115	110	123
0.1 *M* imidazole, pH 7.0	75	95	105	94	101	93	85	100	85
Phosphate-buffered saline,[c] pH 7.2	53	88	98	93	98	58	108	102	110
0.1 *M* PIPES, pH 6.8[d]	94	91	97	76	94	82	104	96	105
0.1 *M* HEPES, pH 7.5[e]	87	93	100	66	88	82	92	68	96
0.1 *M* NaH_2PO_4	100	96	93	88	82	66	98	93	106
0.1 *M* Na_2HPO_4	88	96	96	46	76	62	94	71	96

[a] Tris(hydroxymethyl)aminomethane.
[b] 2-(*N*-Morpholino)ethanesulfonic acid.
[c] 0.005 *M* phosphate, 0.15 *M* NaCl.
[d] Piperazine-*N*,*N*-bis(2-ethanesulfonic acid).
[e] *N*-2-Hydroxyethylpiperazine-*N'*-2-ethanesulfonic acid.

for differentiating between contamination and metal concentrations high enough to have stoichiometric significance.

High Accuracy Methods

Highly accurate methods are traditionally highly specific methods. Development of a high accuracy method usually results in a thorough characterization of the sample or sample type requiring time and sample volume. Methods with accuracies of $\pm 5\%$ can be achieved for the carbon furnace, but are applicable only to the specific sample.

The impetus for developing high accuracy methods arises either from the importance of the results or from plans for a long-term project. The necessary time and energy for the development of a high accuracy method is not usually justified for "one of a kind" samples. A major requirement is sufficient sample volume to allow the method development, i.e., enough sample to permit recovery studies or determinations by another method to cross check the carbon furnace AAS results. Two examples of high accuracy methods are those developed for the determination of trace elements in blood serum[7,8] and rat eye lens.[9]

[7] S. A. Lewis, T. C. O'Haver, and J. M. Harnly, *Anal. Chem.* **56,** 1651 (1984).
[8] S. A. Lewis, T. C. O'Haver, and J. M. Harnly, *Anal. Chem.* **57,** 1 (1985).
[9] J. M. Harnly and D. L. Garland, unpublished results.

TABLE VI
COMPARISON OF VALUES (μg/liter) OBTAINED BY SIMAAC WITH OTHER METHODS FOR A BOVINE SERUM POOL

Method	Al	Co	Cr	Mn	Mo	Ni	V
SIMAAC, furnace[a]	12 ± 3	1.2 ± 0.4	0.34 ± 0.11	2.2 ± 0.4	12 ± 3	2.1 ± 0.4	1.4 ±0.7
ICP-AES[b]	12	—	—	2.5	7.6	<4.0	1.2
	—	—	—	—	20	—	—
ICP-AFS	—	—	—	2.4	—	—	—
CFAAS[c]	20	0.9	0.23	2.9	25	2.2	4.0
	9.9	—	0.2	2.5	17.6	—	—
	—	—	—	3.3	—	—	—
IDMS	—	—	0.27	—	—	—	—
NAA	—	1.3	0.33	2.5	15	—	—

[a] $n = 10$.
[b] Two independent investigators.
[c] Three independent investigators.

The blood serum method was developed for the purpose of characterizing a reference bovine serum standard. Approximately 1 year and 200 ml of serum were used in developing this method. The method used 2 ml of serum with the addition of 20 μl of 0.1 M $Mg(NO_3)_2$ as an ashing aid. The samples were frozen in silanized quartz test tubes, lyophilized, and then ashed at 480° for 18 hr. The ashes were then dissolved in 0.5 ml of 5% HNO_3 (a concentration factor of four). The samples were analyzed using calibration standards in 5% HNO_3.

The results for the seven trace elements in the bovine serum standard are shown in Table VI. Results were obtained from other laboratories and other methods. The agreement between methods and laboratories was excellent. It can be seen that all seven elements were close to the detection limits of the method as indicated by the standard deviation of the determinations.

Trace elements in rat eye lens were determined as part of a study of cataract mechanisms. The rat eye lenses used weighed between 25 and 40 mg. The concentration of many of the elements would be readily detected using flame AAS except the sample volume was too small. The samples were prepared by putting the preweighed lenses in quartz test tubes (one per tube). The lenses were ashed at 480° for 18 hr. If black specks were still visible in the white ash, 100 μl of 5% HNO_3 was added and the lenses were ashed another 18 hr at 480°. The ashes were then dissolved in 0.5 ml of 5% HNO_3. Trace metals were determined using standards in 5% HNO_3.

TABLE VII
ELEMENTAL CONTENT OF RAT EYE LENS

Element	Range (nmol/g)[a]
Zn	150–310
Fe	91–31
Mg	2200–2700
Ca	320–510
Na	3600–6700
K	200–790
Mn	0.5–1.0
Cu	3.3–38
Cr	0.6–22
Ni	<0.9[b]
V	<0.8[b]
Co	<0.4[b]
Mo	<0.8[b]

[a] Lens weights ranged from 27 to 39 mg.
[b] Assumes a lens weight of 40 mg.

Two sets of analytical conditions were used to determine the elements in the rat lens. For Ca, Fe, K, Mg, Na, and Zn, the conditions listed in Table III were used, but with a purge gas flow of 300 ml min^{-1} to reduce the analytical signal. The reduced signal and the extended calibration range of SIMAAC made it possible to determine these six metals at the very high concentrations. Seven other elements, Co, Cr, Cu, Mn, Mo, Ni, and V were determined using the conditions listed in Table III but with a purge gas flow of 0 ml min^{-1} to give maximum sensitivity.

A summary of the initial results for the rat eye lenses is shown in Table VII. The accuracy of the method for the six high elements was cross-checked using flame AAS and a pool sample composed of bovine lenses. Accuracies for the trace elements were also verified using recovery studies in the pool sample.

Acknowledgments

One of the authors (DLG) would like to acknowledge the National Heart, Lung, and Blood Institute for partial support of this research.

[14] Ion Microscopy in Biology and Medicine

By SUBHASH CHANDRA and GEORGE H. MORRISON

The increased sophistication of microscopic techniques allows studies of ultrastructural, cytochemical, and immunocytochemical aspects in biological systems. Intracellular structures and their ultrastructural abnormalities under pathologic conditions can now be studied at electron microscopic resolution. Recent advances in electron optical[1,2] and ion optical[3,4] techniques also provide elemental composition information in relation to cell morphology. Such techniques have become valuable tools for biological research due to the fact that elements are not only vital for life but also play a variety of important roles in intracellular regulatory events. Inorganic elements such as Mg, Ca, Zn, and Cu, have long been recognized as important cofactors for enzymatic reactions inside the cell. At the cost of enormous amounts of energy, the plasma membrane enzymes maintain a fine ionic balance (Na^+, K^+, Ca^{2+}, etc.) inside the cell. These physiologically important cations are intimately involved in bioelectric, contractile, and intracellular physiological events. Calcium alone has now become a major area of research, and subcellular fluxes of Ca^{2+} have been considered as "messengers" for physiological events.[5] An ionic imbalance or local redistribution inside the cell may reflect a symptom or causative factor for altered cell function. Ion microscopy, a technique that provides visual ion images[6] with cell morphology, is well suited for such studies. The ion microscope is capable of detecting all elements (isotopes) from hydrogen to uranium with sensitivities generally in the parts per million (ppm) range.

The technique of ion microscopy[3,7–11] is based on secondary ion mass

[1] A. P. Somlyo and H. Shuman, *Ultramicroscopy* **8,** 219 (1982).
[2] T. A. Hall and B. L. Gupta, *J. Microsc.* (*Oxford*) **163,** 193 (1984).
[3] R. Castaing and G. Slodzian, *J. Microsc.* (*Paris*) **1,** 395 (1962).
[4] R. Levi-Setti, G. Crow, and Y. L. Wang, *Scanning Electron Microsc.* **II,** 535 (1985).
[5] A. B. Borle, *Ergeb. Physiol., Biol. Chem. Exp. Pharmakol.* **90,** 3 (1981).
[6] Ion images represent the distribution of major isotopes of the particular element studied and do not discriminate between the ionized and bound states of that element. For example, a positive secondary ion image at mass 40 ($^{40}Ca^+$) represents the distribution of total Ca inside the cell rather than either free or bound Ca.
[7] G. H. Morrison and G. Slodzian, *Anal. Chem.* **47,** 932A (1975).
[8] M. S. Burns, *J. Microsc.* (*Oxford*) **127,** 237 (1982).
[9] P. Galle, J. P. Berry, and F. Escaig, *Scanning Electron Microsc.* **II,** 827 (1983).
[10] A. Lodding, *Scanning Electron Microsc.* **III,** 1229 (1983).
[11] R. W. Linton, S. R. Walker, C. R. Devries, P. Ingram, and J. D. Shelburne, *Scanning Electron Microsc.* **II,** 583 (1980).

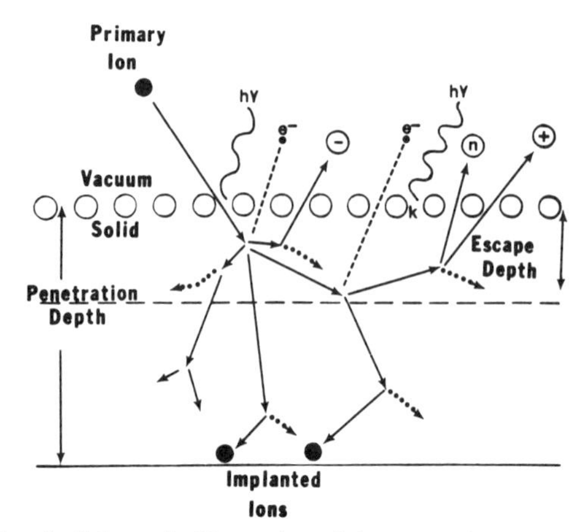

FIG. 1. Schematic illustration of the sputtering process.

spectrometry (SIMS) and depends on the sputtering phenomenon induced by bombardment (impact) of a solid sample surface with an energetic "primary" ion beam (Fig. 1). Upon impact, the primary ion beam energy dissipates into the sample surface in the form of a collision cascade, resulting in sputtering of a few top atomic layers of the sample surface. The sputtering products include photons, electrons, neutral atoms and molecules and positively or negatively charged ions. The sputtered ions, only a fraction of the sputtered particles, are referred to as secondary ions and represent the surface chemical composition of the specimen. The secondary ions thus produced are accelerated into a mass spectrometer for analysis and onto different display modes. The microscope mode of display produces visual ion images revealing elemental (isotopic) distribution in relation to cell morphology with a lateral resolution of $\sim$0.5 μm (Fig. 2). From Fig. 2 the distribution of K and Ca can be studied in the same normal rat kidney (NRK) cells. While K is more or less homogeneously distributed, Ca shows a remarkable distribution and reveals higher cytoplasmic intensities. A thorough understanding of the instrument and sample preparatory procedures is necessary before such studies can be undertaken with confidence. In this chapter, we briefly describe the CAMECA IMS-3f ion microscope, biological sample preparatory procedures, and the potential of ion microscopy in biology and medicine.

Instrumentation

A schematic of the CAMECA IMS-3f microanalyzer is shown in Fig. 3. The instrument is primarily composed of three sections: (1) primary ion optics, (2) secondary ion optics, and (3) detection/display system.

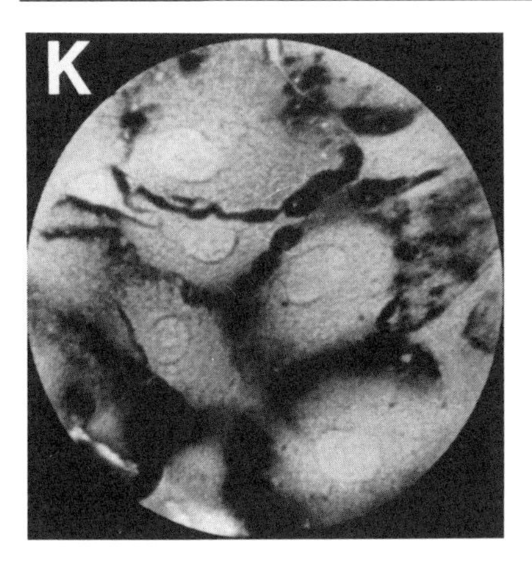

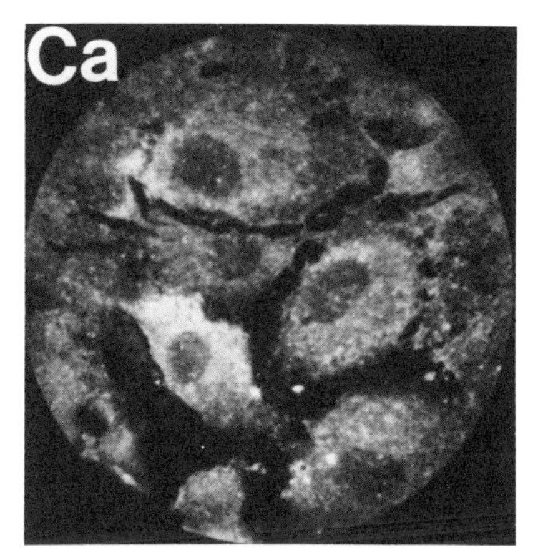

FIG. 2. Secondary ion images of $^{39}K^+$ and $^{40}Ca^+$ revealing subcellular distribution of K and Ca in cryofractured and freeze-dried NRK cells. Image exposure times: $^{39}K^+$ = 1/8 sec, $^{40}Ca^+$ = 100 sec.

Primary Ion Optics

The primary ion optics section consists of a primary ion source(s), a primary magnet, and an electrostatic lens system. The primary ion sources can produce either positive or negative ion beams. The duoplasmatron source is generally used to produce O_2^+, O^-, Ar^+, etc. beams. An oxygen beam is preferred for biological ion microanalysis due to its enhancing and stabilizing effect on positive secondary ion (Na^+, K^+, Ca^+, etc.) yields.[12,13] Similarly, a Cs^+ beam produced by the cesium liquid metal source enhances and stabilizes negative ion (Cl^-, P^-, S^-, etc.) yields.[14,15] Therefore, the selection of the primary beam depends on the nature of the element studied. The primary beam is accelerated to a desired energy (5–20 keV) and mass filtered by the primary magnet. The purified beam is then focused using the electrostatic lens system onto the specimen held at ±4.5 kV in the high vacuum (10^{-7}–10^{-9} Torr) sample chamber of the instrument. The sample potential is positive when positive secondary ions are of interest, and negative for negative secondaries. The primary beam current can be adjusted from less than 1 nA to about 8 μA while the spot size is controllable between approximately 1 and 500 μm in diameter.

[12] C. A. Anderson, *Int. J. Mass Spectrom. Ion Phys.* **2,** 61 (1969).

[13] C. A. Anderson, *Int. J. Mass Spectrom. Ion Phys.* **3,** 413 (1970).

[14] H. A. Storms, K. F. Brown, and J. D. Stein, *Anal. Chem.* **49,** 2023 (1977).

[15] P. Williams, R. K. Lewis, C. A. Evans, and P. R. Hanley, *Anal. Chem.* **49,** 1399 (1977).

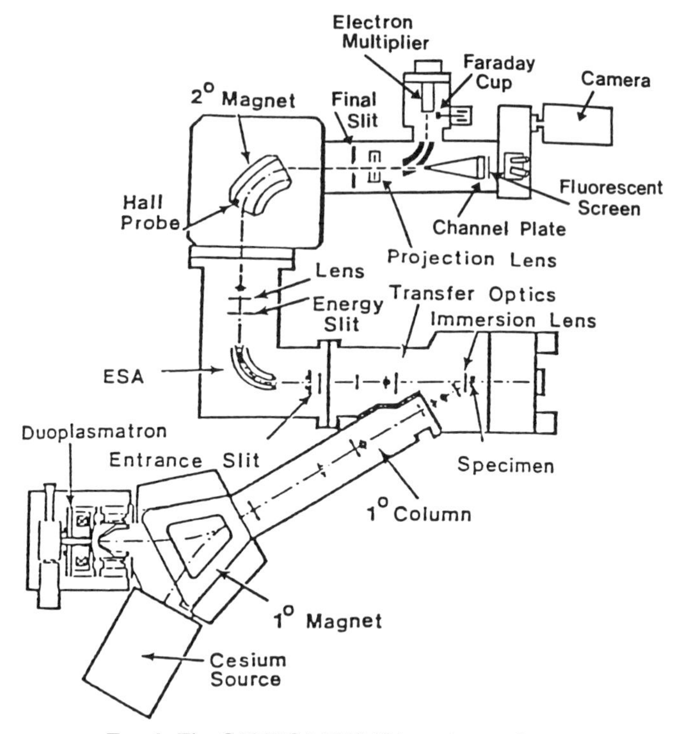

FIG. 3. The CAMECA IMS-3f ion microanalyzer.

Secondary Ion Optics

The secondary ion optics system is composed of three main components: (1) the immersion lens/transfer optics assembly, which extracts and focuses the secondary ions into the spectrometer section, (2) the mass spectrometer, which disperses the secondary ions according to their energy and momentum, respectively, and (3) the projection lens/detection assembly which either magnifies and focuses the mass filtered secondary ion beam for final imaging or directs it to the desired detection mode.

A schematic of the secondary ion optics is shown in Fig. 4. Since the sample is held at $\pm$ 4.5 kV, secondary ions of the same polarity are accelerated through an electrostatic immersion lens held at ground. The secondary ion beam then encounters a set of three electrostatic lenses (only one is energized at a time depending on the desired field of view: 25, 150, or 400 μm in diameter). The transfer optics then focus the image crossover on an aperture (the contrast diaphragm assembly which consists of four apertures: 20, 60, 150, and 400 μm in diameter in the Cornell

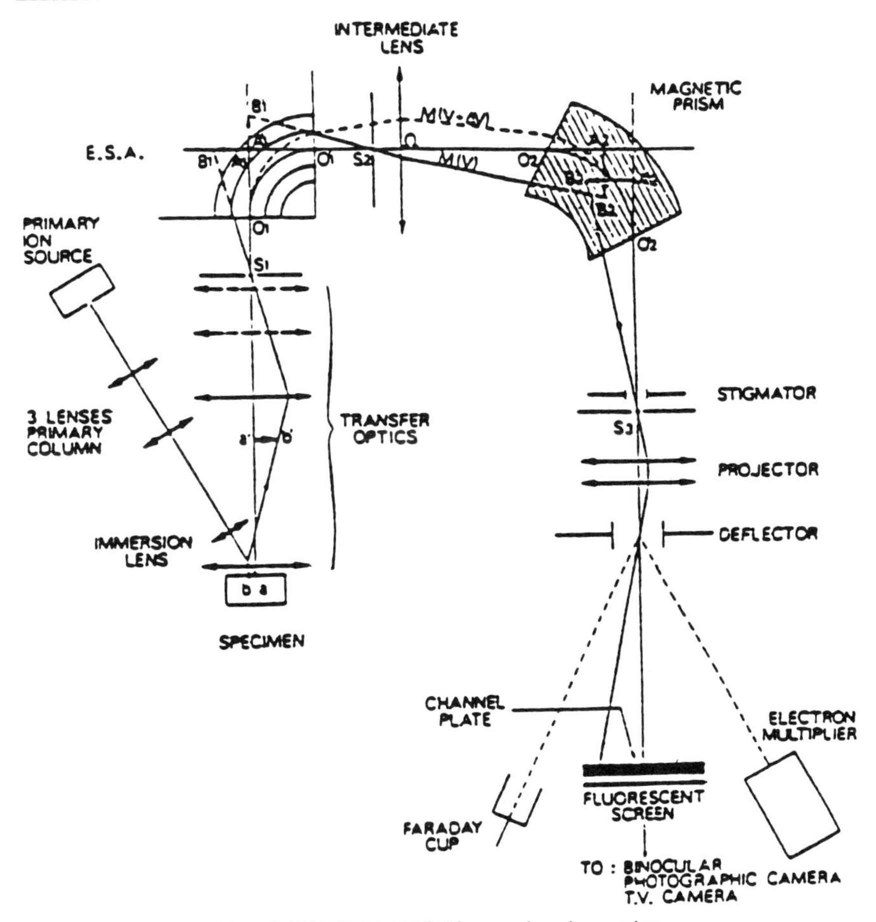

FIG. 4. CAMECA IMS-3f secondary ion optics.

instrument) located at the mass spectrometer entrance slit. The aperture size is inversely proportional to the spatial lateral resolution of the image and directly proportional to the secondary ion collection efficiency. After the contrast diaphragm, the secondary ion beam passes through a field aperture, which limits the analysis area to the central portion of the field of view. The field aperture is equipped with three interchangable apertures of 100, 750, and 1800 μm in diameter. The smaller field apertures allow higher mass resolution by reducing the acceptance angle of the mass spectrometer.

Next the secondary ion beam enters the mass spectrometer region. The CAMECA IMS-3f is equipped with a modified Nier–Johnson double focusing mass spectrometer which disperses the ions based on their energy and momentum using two different sections. The electrostatic analyzer section (ESA in Fig. 3 and Fig. 4) screens the secondary ions according to their energy, and the secondary magnet section allows the passage of one ionic species at a time based on the mass-to-charge ratio. Between the two sections of the mass spectrometer, an energy slit controls the energy bandpass between 0 and 130 eV and a mass spectrometer lens focuses the energy filtered ions into the correct plane of the secondary magnet for minimum image distortion and maximum transmission. A smaller energy bandpass reduces the secondary ion collection efficiency and increases the lateral and mass resolution of the secondary ion image. The 3f spectrometer is capable of a mass resolution of about 10,000. The energy and mass-filtered secondary ion beam then passes through the spectrometer exit slits (final slit in Fig. 3) and into the projector lens assembly. From here on, this beam can be subjected to desired detection/display modes.

Detection Modes

The mass-filtered secondary ion beam can be subjected to either the ion microprobe or the ion microscope mode of detection. The microprobe mode is used for mass spectra, depth profiles, line scans, and bulk analysis. The microscope mode allows imaging of elemental distribution with ~0.5 μm lateral resolution.

Figure 5 shows the secondary ion detection system of the CAMECA IMS-3f ion microanalyzer. In the microprobe mode, the secondary ion beam is deflected into either the electron multiplier or the Faraday cup. The electron multiplier is used for the signals levels up to 700,000 cps. To avoid electron multiplier degradation at higher counts, a Faraday cup is used. In this mode, a sample area of a few to several hundred square micrometers may be sampled for analysis. In the microprobe mode, mass spectral analysis may be performed where the secondary ion intensities are recorded versus mass-to-charge ratio for designated mass units anywhere from 1 to 250. Figure 6 shows a low mass resolution ($m/\Delta m$ = 300) mass spectrum from freeze-dried and cryofractured Chinese hamster ovary (CHO) cells grown in cultures on a silicon substrate. Peaks at masses 39, 23, 28, 24, and 40 are dominant and represent $^{39}K^+$, $^{23}Na^+$, $^{28}Si^+$, $^{24}Mg^+$, and $^{40}Ca^+$ secondary ion intensities, respectively. In biological ion microanalysis, a mass spectrum is obtained to get a general idea of the chemical composition of the specimen rather than absolute quantities. Furthermore, a low-resolution mass spectrum cannot predict mass interferences due to charged polyatomics, hydrocarbons, hydrides, oxides,

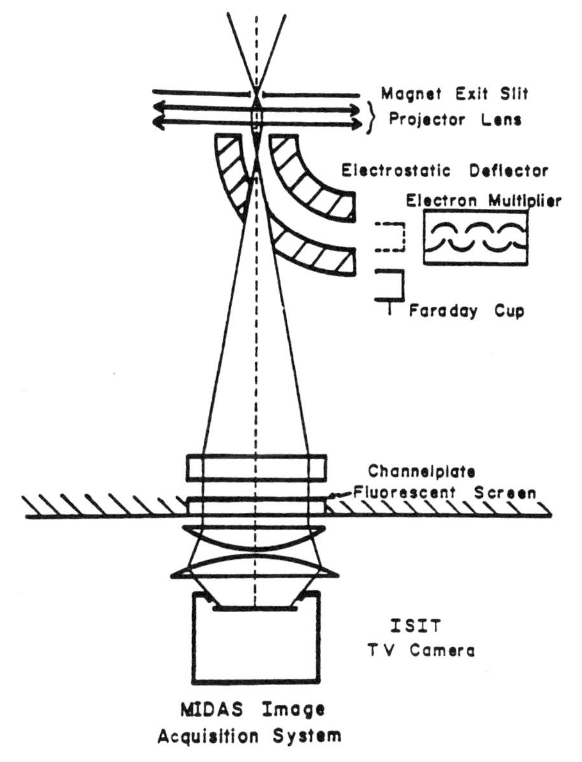

FIG. 5. CAMECA IMS-3f secondary ion detection system.

etc. For example, the mass 24 peak may represent a combination of Mg^+, NaH^+, and C_2^+ secondary ion intensities. A mass resolution of about 2000 is needed to separate these peaks. Figure 7 shows a high mass resolution spectrum at mass 24 indicating the dominant peak as Mg^+. In general, biological materials (not embedded in plastic) do not show significant mass interferences for physiologically important elements such as Na, K, Ca, and Mg. Elements of masses higher than about 50 should be checked carefully since they are more prone to polyatomic, hydrocarbon, oxide, etc. interferences.

In the depth profile microprobe mode of detection, the designated secondary ion intensities are recorded versus time of bombardment. This analysis provides the sample's composition in the z direction with a depth resolution of about 50 Å. This mode is the most important for semiconductor analysis; however, in biology it still remains to be explored thoroughly. Figure 8 shows a depth profile of $^{39}K^+$, $^{23}Na^+$, $^{24}Mg^+$, and $^{40}Ca^+$ secondary ion intensities from a freeze-dried cryosection of chick intes-

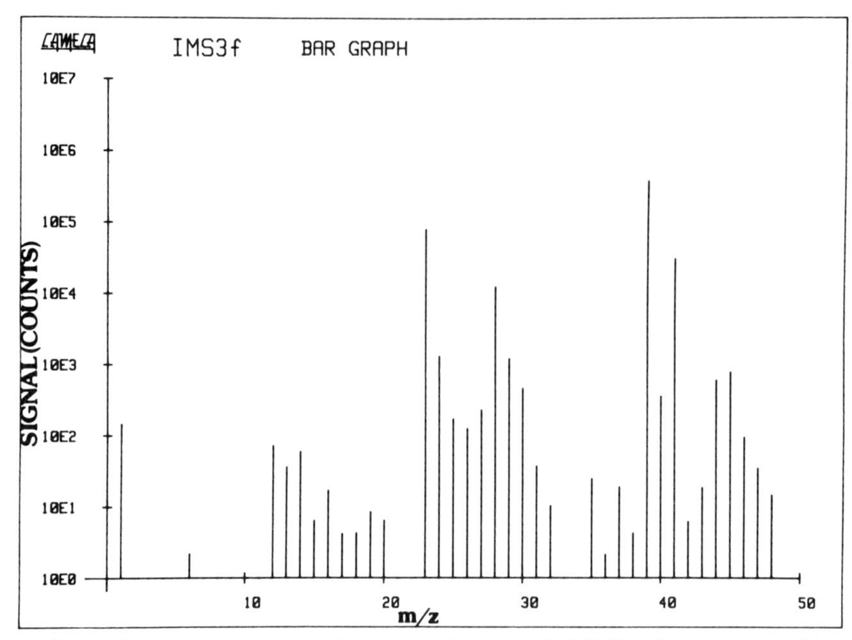

FIG. 6. A mass spectrum from fractured and freeze-dried CHO cells grown on a silicon substrate. The peak at mass 28 is due to the silicon substrate.

tine. A portion from a villus was centered for analysis. It is evident that K is the highest signal and Ca the lowest. The section remained intact during analysis for about 10 min and showed fairly constant secondary ion intensities with depth.

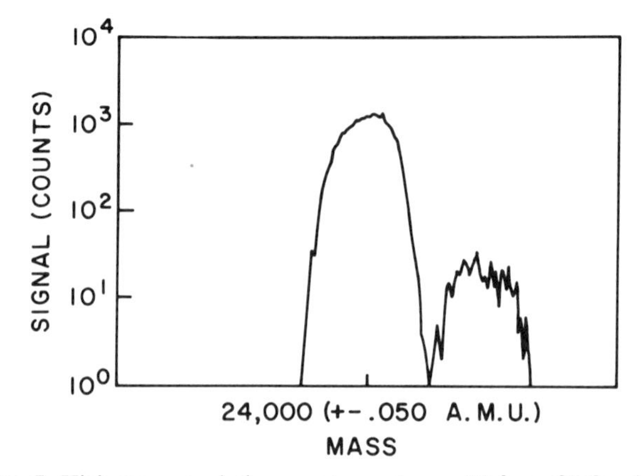

FIG. 7. High mass resolution spectrum at mass 24 from CHO cells.

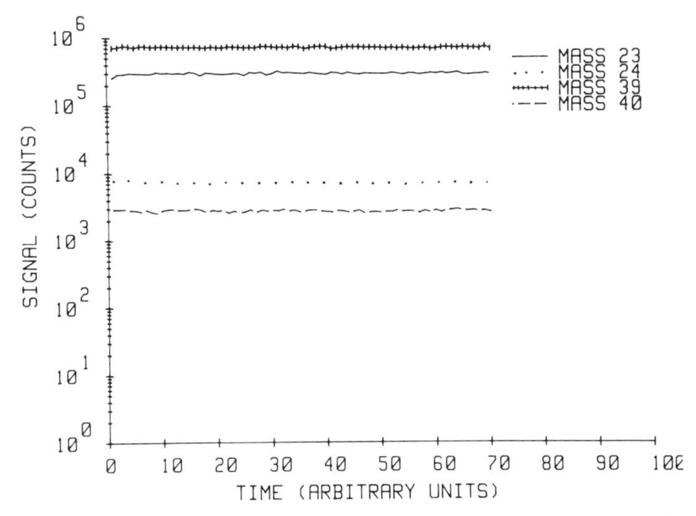

FIG. 8. Depth profile of $^{23}Na^+$, $^{24}Mg^+$, $^{39}K^+$, and $^{40}Ca^+$ in freeze-dried chick intestine cryosection.

Visual ion images provided by the microscope mode of detection is by far the most exciting mode for biological ion microanalysis. In this mode, the projector lens system focuses the mass resolved secondary ion image onto the microchannel plate while allowing variable image magnification (Fig. 5). The microchannel plate acts as an ion-to-electron converter/amplifier.[16] The electron cascades from the individual channels of the microchannel plate impact on a fluorescent screen, forming the desired visible ion image. The ion optics of the instrument, in this fashion, preserve the spatial distribution of the emitted secondary ions through the mass spectrometer so that a one-to-one correspondence is maintained between the position of a sputtered ion leaving the sample surface and its position in the final mass-resolved ion image. Selecting a different mass-to-charge ratio in the instrument takes less than a few seconds, so that multielement distributions can be studied from the same cells. Furthermore, since the elements are separated based on their mass-to-charge ratio, different isotopes of the same element can be studied. The visible ion images formed on the fluorescent screen may be recorded directly using a 35-mm camera (Fig. 2) or a low light level television camera for on-line digital image processing.

Image processing is necessary for quantification of ion images. For on-line image acquisition and processing, the Microscopic Image Digital Ac-

[16] M. Bernheim, *Radiat. Eff.* **18,** 231 (1973).

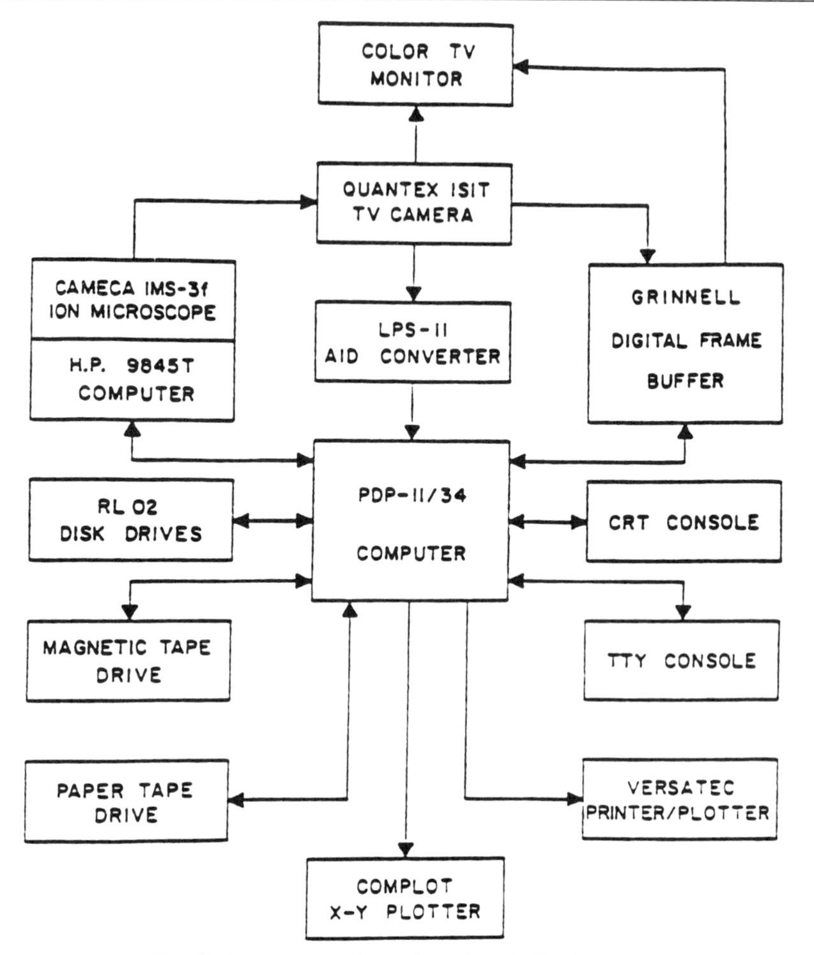

FIG. 9. Computer configuration of the MIDAS system.

quisition System[17] (MIDAS) has been used in our laboratory. A schematic of the MIDAS system is shown in Fig. 9. The ion image from the fluorescent screen of the ion microanalyzer is monitored by a Quantex ZX-26 Intensified Silicon Intensified Target (ISIT) low light level television camera.[18,19] The image is then digitized and stored in the Grinnell frame buffer. The Grinnell is capable of digitizing the image into 256 levels of gray or 4096 levels of color. The Grinnell is interfaced with a PDP 11/34A

[17] B. K. Furman and G. H. Morrison, *Anal. Chem.* **52,** 2305 (1980).
[18] Y. Talmi, *Anal. Chem.* **47,** 658A (1975).
[19] K. Kalata, this series, Vol. 114, p. 486.

minicomputer for control and a variety of ways for image processing. The computer is equippped with two removable RL02 disc storage units and a magnetic tape unit for data storage. This image acquisition system is capable of digitizing and storing an ion image in one-thirtieth of a second. Recently, an inclusion of a videocassette recorder with time-base correc- tor in the MIDAS system allows real time continuous image acquisition with an unlimited storage capacity.[20]

Sample Preparation for Biological Ion Microscopy

Studies using any sophisticated technique would not be meaningful unless the reliability of sample preparation is confirmed. An ideal sample preparation methodology should preserve the native elemental distribu- tion of the cell and at the same time satisfy the instrumental requirements. For ion microscopy, the sample should be compatible with the high vac- uum requirements and conducting since it is held at 4500 V. The sample surface should be planar so that no image distortion is produced due to surface irregularities (an absolute flatness is not a requirement, the sur- face irregularities below the instruments lateral resolution are generally tolerable). Since biological samples are nonconducting, they have to be sectioned (thin slices of tissue) and mounted on a conducting substrate. The substrate should not impart any contaminants at the same time. The high vacuum conditions of ion microscopy require sample fixation before analysis. This requirement introduces a classical problem in biological elemental microscopy. The hydrated matrix of biological tissues, the highly diffusible nature of physiologically important cations (Na^+, K^+, Ca^{2+}, etc.) and their asymmetric distribution inside and between inside and outside the cell requires a careful sample fixation. An ideal sample fixation should preserve the structural as well as chemical integrity of the living cell. Table I lists several methods of sample preparation for ion microscopy.

In the conventional method of sample fixation, routine electron micro- scopic techniques are used.[21,22] Biological tissues are cut into small pieces (~1 mm^3) and fixed using aqueous fixatives (e.g., glutaraldehyde and osmium tetroxide). The fixation is followed by dehydration using organic solvents (typically ethanol or acetone) and finally, embedment in plastic resins to produce a block of plastic-embedded tissue. The tissue can then be sectioned into 1- to 2-μm-thick sections typically used for ion micros- copy. The sections are then deposited on conducting substrates such as

[20] J. T. Brenna, M. G. Moran, and G. H. Morrison, *Anal. Chem.* **58,** 428 (1986).
[21] A. R. Spurr, *J. Ultrastruct. Res.* **26,** 31 (1969).
[22] R. Salema and I. J. Brando, *J. Submicrosc. Cytol.* **5,** 79 (1973).

TABLE I
SAMPLE PREPARATION METHODS

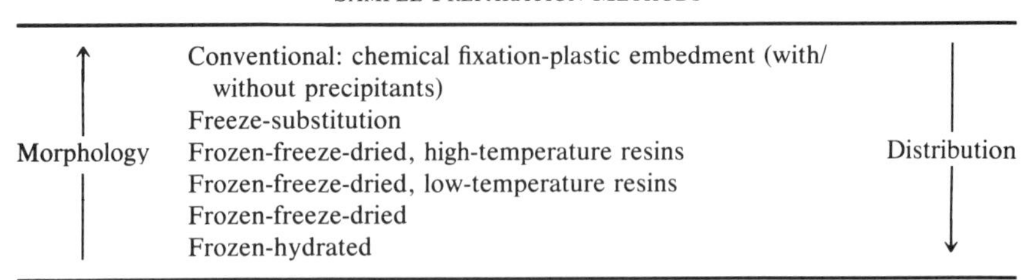

	Method	
↑	Conventional: chemical fixation-plastic embedment (with/without precipitants)	
	Freeze-substitution	
Morphology	Frozen-freeze-dried, high-temperature resins	Distribution
	Frozen-freeze-dried, low-temperature resins	
	Frozen-freeze-dried	
	Frozen-hydrated	↓

silicon, germanium, and tantalum, for ion microscopic analysis. Generally, the results obtained from this procedure reveal an excellent morphological preservation but suffer from gross loss/gain and translocation of diffusible elements.[23] We do not recommend this procedure for studying diffusible elements. This procedure may be used for highly immobile and tightly bound elements. Even then, the burden of proof remains on the experimenter. Within the conventional techniques (sometimes for Ca^{2+} localization), calcium-precipitating agents such as pyroantimonate and oxalate may be added in order to precipitate and immobilize Ca^{2+} in its native state. This method, however, suffers from similar artifacts as discussed above.

In freeze-substitution techniques, the tissue fixation is achieved by quick freezing the specimen either by submerging in cryogenic fluids such as Freon-22, propane, and liquid nitrogen slush, or slamming onto metal blocks supercooled by liquid nitrogen or liquid helium. The main criterion for selection of a cryogenic fluid is its capability to produce vitreous ice inside the specimen upon freezing.[2,24,25] Such a fixation preserves both the structural and chemical integrity. The next step in freeze-substitution is the removal of frozen water from the specimen at cold temperatures (at or below $-80°$) using various freeze-substitution media[26] (ether/acrolein in the ratio of 4 : 1 is popular). After dehydration, the specimen is impregnated with plastic. Sections from such tissues must be cut dry. Floating of freeze-substituted sections on water may result in loss and translocation of diffusible elements.[27] This methodology for ion microscopy is better than conventional methods since it shows better retention of diffusible

[23] K. M. Stika, K. L. Bielat, and G. H. Morrison, *J. Microsc.* (*Oxford*) **118,** 409 (1980).

[24] K. Zierold, *J. Microsc.* (*Oxford*) **125,** 149 (1982).

[25] G. J. Jones, *J. Microsc.* (*Oxford*) **136,** 349 (1984).

[26] M. A. Hayat, "Principles and Techniques of Electron Microscopy," Vol. 1, 2nd Ed. University Park, Baltimore, Maryland, 1980.

[27] R. L. Ornberg and T. S. Reese, *Fed. Proc., Fed. Am. Soc. Exp. Biol.* **39,** 2802 (1980).

elements.[28] However, the translocation at the subcellular level remains suspect and has to be confirmed using more rigorous methods such as frozen freeze-dried or frozen-hydrated.

The next two methods are improvements on freeze-substitution. After freezing the specimen, the frozen water is removed by freeze-drying at low temperatures (generally at or below $-80°$ since higher temperatures may be prone to increasing the size of ice crystals due to recrystallization). Plastic is then impregnated into the tissue at high[29] or low temperatures.[30] Once again, the sections from these specimens must be cut dry. In this methodology the impregnation of plastic into freeze-dried tissue remains the main suspect and generally the results are confirmed using more rigorous methods.

The last two methods of sample preparation are ideal for SIMS analysis for two reasons. First, they preserve the structural and chemical integrity of the specimen; second, these procedures do not require addition of any foreign compound (e.g., plastic) into the tissue before analysis. The latter is especially important since SIMS is sensitive to matrix (environment in which the element is present) effects. Indeed, upon plastic embedment, the plastic becomes the main matrix and, therefore, any qualitative as well as quantitative interpretations from such samples must be checked using nonembedding cryotechniques.

In the frozen freeze-dried procedure of sample fixation, small pieces (<1 mm^3) of tissue are fast frozen and cryosectioned at low temperatures.[23] Sections of 1 to 2 μm thickness are collected and pressed onto precooled silicon pieces. After freeze-drying, these sections can be analyzed by ion microscopy. Although this methodology has been perfected to a remarkable precision in the electron probe field,[31,32] it is yet to be perfected for ion microscopy. Pressing cryosections flat onto a silicon substrate is one of the most challenging problems in this methodology. This situation is further complicated by the curling of cryosections upon freeze-drying and even the dislodging of sections from the substrate upon primary ion beam impact. For these reasons in addition to the eroding (sample destroying) nature of SIMS analysis, thin sections ($\sim$100 nm as used for electron probe analysis) are vulnerable to quick destruction. The preferred sample thickness remains about 1–2 μm. Figure 10 shows K^+

[28] G. D. Ross, G. H. Morrison, R. F. Sacher, and R. C. Staples, *J. Microsc.* (*Oxford*) **129,** 221 (1983).

[29] F. D. Ingram and M. J. Ingram, *J. Microsc. Biol. Cell* **22,** 193 (1975).

[30] S. Chandra, G. H. Morrison, and R. Chiovetti, Jr., *Proc. Annu. Conf.—Microbeam Anal. Soc.* **20,** 125 (1985).

[31] R. D. Karp, J. C. Silcox, and A. V. Somlyo, *J. Microsc.* (*Oxford*) **125,** 157 (1982).

[32] A. P. Somlyo, M. Bond, and A. V. Somlyo, *Nature* (*London*) **314,** 622 (1985).

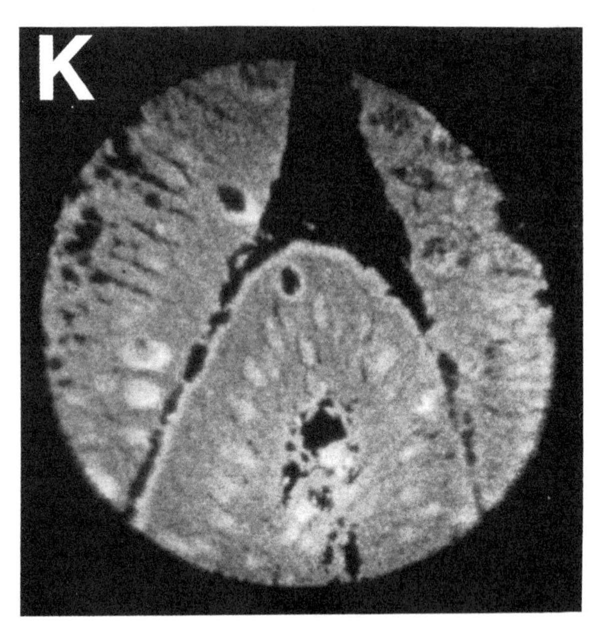

FIG. 10. Secondary ion image of $^{39}K^+$ from freeze-dried chick intestine cryosection. Image exposure time = 1/4 sec. Image field of view = 150 μm in diameter.

distribution in a freeze-dried cryosection of chick intestine. Individual villi are easily recognizable. Within a villus, goblet cells, epithelial cell layer, and individual nuclei are discernible.

A nonsectioning cryotechnique has been developed in our laboratory for cell culture systems.[33,34] The presence of nutrient media necessary for cell growth prevents the direct analysis of cultured cells for elemental studies. Any attempts to wash the nutrient medium results in exposure of cells to unphysiological reagents.[35,36] This challenge has been overcome by using a simple sandwich-fracture methodology. Cells can be cultured directly on a polished surface of sterilized silicon pieces (~1 cm^2) without showing any evidence of toxicity. Upon confluency, the cells are sandwiched using a clean smooth surface (pieces of Corning glass coverslip, silicon wafer, etc.). The sandwich is then fast frozen in cryogenic fluids and fractured by prying the two halves apart under liquid nitrogen. This

[33] S. Chandra, G. H. Morrison, C. W. Coulter, and S. E. Bloom, *J. Cell Biol.* **99,** 424a (1984).

[34] S. Chandra, G. H. Morrison, and C. C. Wolcott, *J. Microsc.* (*Oxford*) **144,** 15 (1986).

[35] M. R. James-Kracke, B. F. Sloane, H. Shuman, R. Karb, and A. P. Somlyo, *J. Cell. Physiol.* **103,** 313 (1980).

[36] J. Wroblewski and G. M. Roomans, *Scanning Electron Microsc.* **IV,** 1875 (1984).

procedure produces large areas containing hundreds of cells grouped together on the silicon substrate where the fracture plane has gone through the very top of the cell monolayer, removing the half-membranes and the extracellular nutrient media to the top surface of the sandwich. After freeze-drying at $-80°$ for about 24 hr, the fractured cells can be analyzed for subcellular elemental localization and ion transport studies. Images of K and Ca distribution shown in Fig. 2 were obtained from NRK cells prepared with this methodology.

In the frozen-hydrated procedure, the sample is fast frozen, cryosectioned, and analyzed frozen-hydrated. This approach eliminates the freeze-drying step. However, a cold sample stage in the instrument is necessary for analysis. A cold stage for the CAMECA IMS-3f ion microscope has been developed recently.[37] Preliminary studies have shown the ability of the ion microscope to analyze frozen-hydrated biological specimens.[38] These developments are recent, and a thorough evaluation of the frozen ice matrix in biological samples is continuing in our laboratory. Figure 11 shows the Ca distribution in PtK_2 rat kangaroo cells analyzed frozen-hydrated.

The direction of the arrows in Table I generally indicates the reliability of elemental distribution and recognizable morphological preservation in different methods of sample preparation for elemental microscopy.

Evaluation of Sample Preparation Based on Physiological Studies

Although preservation of the structural integrity in biological specimens can be checked precisely with electron microscopy, the chemical integrity at the subcellular level is difficult to evaluate. However, a few guidelines based on cell physiology may serve as rules of thumb in the evaluation of chemical integrity. It is generally known that plasma membrane enzymes maintain high K^+ and low Na^+ levels inside the cell, while the extracellular fluids have higher levels of Na^+ than K^+. Therefore, injured cells tend to accumulate Na^+ and lose K^+, and dead cells become high Na^+ cells. Calcium may also act as another indicator element. For example, in damaged cells, besides the loss of K^+ and gain of Na^+, a massive accumulation of cytoplasmic Ca especially in mitochondria has been characterized.[34,39]

Physiological experiments may also be conducted in order to confirm the validity of the sample preparation procedures and the instrument used for analysis. The use of ionophores to disturb the ionic balance is one

[37] M. T. Bernius, S. Chandra, and G. H. Morrison, *Rev. Sci. Instrum.* **56,** 1347 (1985).
[38] S. Chandra, M. T. Bernius, and G. H. Morrison, *Anal. Chem.* **58,** 493 (1986).
[39] A. P. Somlyo, A. V. Somlyo, and H. Shuman, *J. Cell Biol.* **81,** 316 (1979).

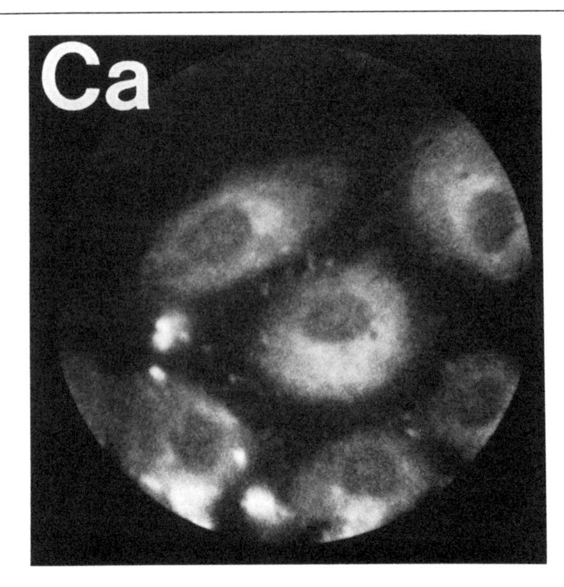

FIG. 11. $^{40}Ca^{+}$ secondary ion micrograph from cryofractured PtK_2 rat kangaroo cells analyzed using a cold sample stage in the ion microscope. Image exposure time = 120 sec. Image field of view = 150 μm in diameter.

example. NKR cells treated with 1 m*M* ouabain (a specific inhibitor of plasma membrane Na^{+},K^{+}-ATPase) for different time periods were sampled using the above mentioned sandwich-fracture methodology.[40] An ion microscopic analysis of these samples is shown in Fig. 12. At 0 min after treatment with ouabain, high K and low Na intensities are observed (see cell A for example). Cells sampled after 20 and 60 min of ouabain treatment clearly show an intracellular increase in Na^{+} and loss of K^{+} with time. A dead cell (arrow) with high Na^{+} and low K^{+} signals can be seen in the images made after 20 min of treatment with ouabain. Such a successful imaging of ion transport confirms the reliability of sample preparation and instrumentation used for analysis.

Ouabain is known to cause a dramatic alteration in K^{+}/Na^{+} ratios inside the cell. The high sensitivity of ion microscopy is capable of imaging rather small changes in physiologically important elements such as calcium. A 5-min exposure of NRK cells to 1 μM calcium ionphore A23187 resulted in a substantial loss of Ca gradient that existed between the nuclei and the cytoplasm of control cells. The first row of Fig. 13 shows the distribution of Ca and Mg in control NRK cells (0 min of

[40] S. Chandra and G. H. Morrison, *Science* **228,** 1543 (1985).

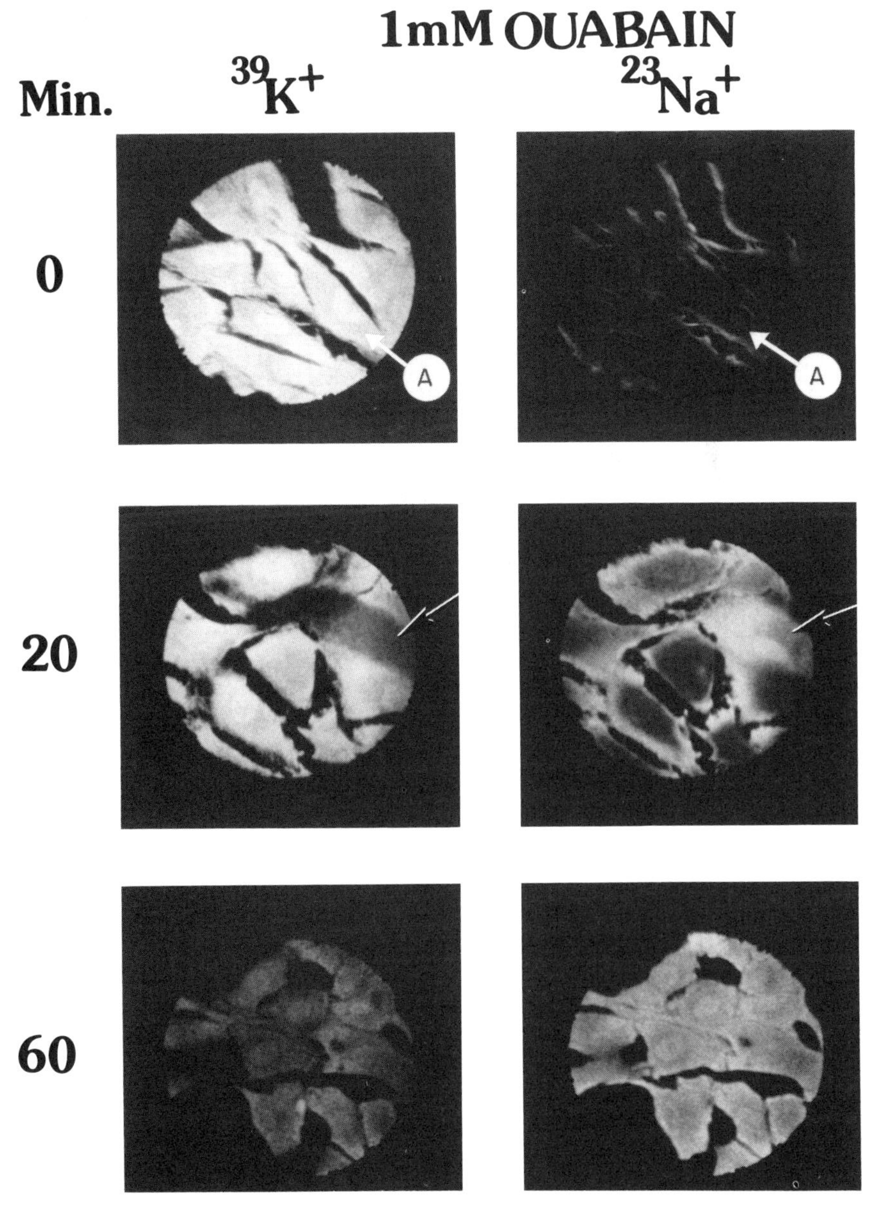

FIG. 12. Imaging Na^+–K^+ ion transport in NRK cells after inhibition of Na^+,K^+-ATPase with ouabain. Time in minutes indicates the exposure of cells to ouabain. The secondary ion images of potassium ($^{39}K^+$) are in the left column and the corresponding sodium ($^{23}Na^+$) images are on the right for each treatment. Brightness indicates relative ion intensity. Instrumental and photographic conditions were kept identical for these ion images. Bar = 40 μm. From Chandra and Morrison.[40] Copyright © 1985 by the American Association for the Advancement of Science.

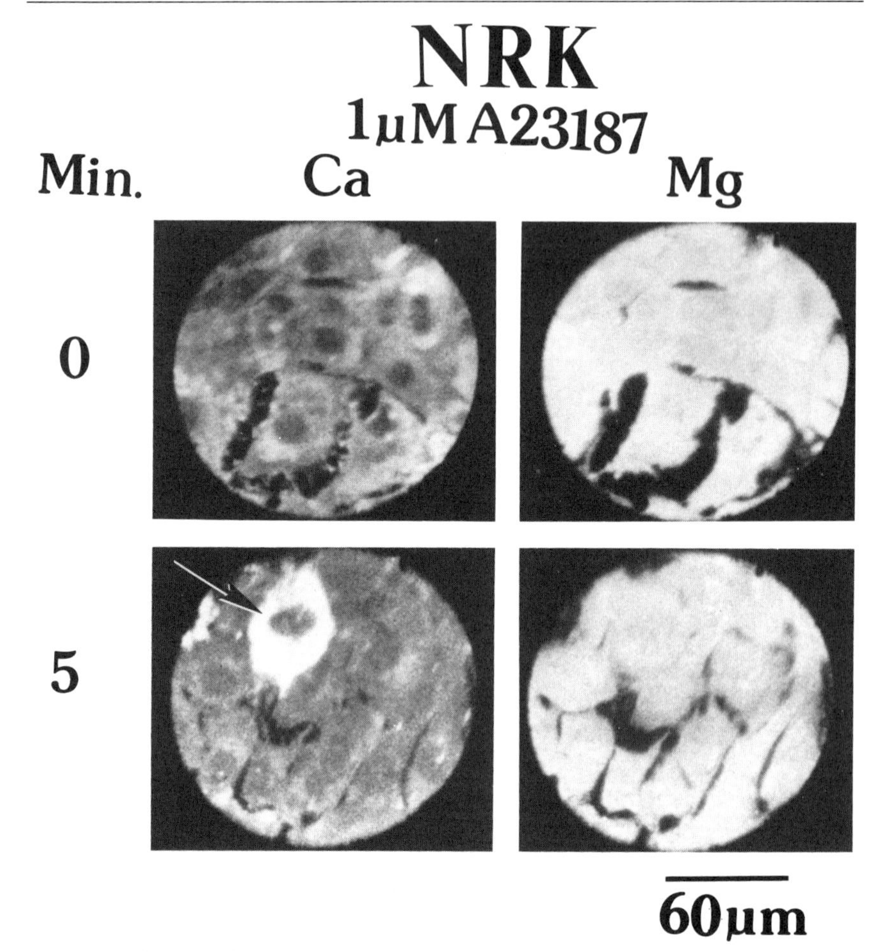

FIG. 13. $^{40}Ca^+$ and $^{24}Mg^+$ secondary ion images after the exposure of NRK cells to 1 μM calcium ionophone A23187. Time in minutes shows the exposure of cells to A23187. One injured cell (arrow) shows a massive cytoplasmic Ca accumulation. The same cell was also found to be high in Na (ion image not shown). Image exposure times $^{40}Ca^+$ = 100 sec; $^{24}Mg^+$ = 60 sec. Photographic processing times were optimized for quality of the individual ion image. The ion images of $^{40}Ca^+$ and $^{24}Mg^+$ from control cells were published by Chandra and Morrison.[40] Copyright © 1985 by the American Association for the Advancement of Science.

A23187 treatment). A large gradient of Ca is observed between the nuclei and the cytoplasm of cells. Even within the cytoplasm, higher Ca intensities are not homogeneously distributed. On the other hand, another physiologically important element, Mg, is found to be homogeneous throughout the cell at this resolution. Cells that were sampled after 5 min of

A23187 exposure (second row of Fig. 13) reveal that the Ca gradient between the nuclei and cytoplasm has been substantially reduced, reflecting a release of Ca from cytoplasmic sites to the nuclear region. One cell (arrow) shows a massive cytoplasmic Ca accumulation due to cell injury, and should be discarded from any physiological explanation.

Image Quantification

Quantification of secondary ion images is still a challenging problem in biological ion microscopy. The measured secondary ion intensity (I) for a particular element M has been defined[7] as follows:

$$I = \tau C_m S i_p a_0$$

where τ is the practical ion yield representing the ratio of the number of M^+ ions collected to the number of M atoms removed from the sample surface. C_m is the atomic concentration of the element M corrected for its isotopic abundance. S is the total sputtering yield (number of atoms of any kind removed per incoming primary ion). The primary beam current density is i_p, and a_0 is the area of the target surface. While i_p and a_0 are precisely controllable, τ and S depend on many factors. The practical ion yield of an element is inversely proportional to the exponential of the element's first ionization potential and depends on the chemical state of the element, the chemical and physical properties of the sample matrix, instrumental transmission, and sampling conditions. In biological cells, although the matrix is composed primarily of C, H, N, and O, areas such as vacuoles and extracellular fluids are prone to matrix effects (a term that accounts for both τ and S). In plant cells the situation is even more complex due to the presence of a cell wall, a structure totally different in chemical composition and density than either cytoplasm or nucleus. In plastic embedded tissue, sputter rates for the cell wall, nucleus, and cytoplasm of plant root cells were found to be different.[41] Such a differential sputtering of microstructures at the subcellular level can give rise to misleading secondary ion signals. The cryofractured and freeze-dried mammalian cell cultures, to a first approximation, do not suffer from this artifact.[42] So far very little is known about the severity of matrix effects in biological systems,[41,43,44] and these studies have utilized plastic-embedded tissues as model systems. It is clear that any evaluation of matrix effects should be made on either freeze-dried or frozen-hydrated matrices. The

[41] A. J. Patkin, S. Chandra, and G. H. Morrison, *Anal. Chem.* **54,** 2507 (1982).

[42] S. Chandra, W. A. Ausserer, and G. H. Morrison, *J. Microsc.* (*Oxford*) in press (1987).

[43] M. S. Burns, *Int. Conf. Secondary Ion Mass Spectrom.* **5,** 95 (1985).

[44] J. T. Brenna and G. H. Morrison, *Anal. Chem.* **58,** 1675 (1986).

frozen-hydrated matrix, being closest to real-life situation, should provide an ideal evaluation.

In order to quantitate secondary ion intensities at every discrete location of the ion image, standards of known elemental concentration and comparable τ and S at every discrete location have to be made. Preliminary studies for the quantification of ion images from tissue sections using ion implantation and digital image processing are promising.[44,45] For the cell culture system, cells with known concentration of electrolytes to produce a matching standard matrix that provides comparable τ and S are presently being investigated. These studies remain challenging and are an active area of research in our laboratory.

Combining Microscopic Techniques

The pooling of several microscopic techniques generally results in a better understanding of a problem. Ion microscopy coupled with electron microscopy provides a powerful combination for ultrastructural as well as elemental studies. This combination is especially important for ion microscopy since the ion microscope is not a morphological tool. In ion microscopy, the morphological recognition depends on the gradient of elements present within different cellular compartments. Ion images of $^{40}Ca^+$ and $^{24}Mg^+$ shown for control NRK cells in Fig. 13 are a good example of this effect. The cellular morphology cannot be recognized in the $^{24}Mg^+$ secondary ion image due to lack of a detectable Mg gradient between the nuclei and cytoplasm of cells. Such features, however, are clearly recognized in the $^{40}Ca^+$ ion image due to a detectable Ca gradient. Figure 14a shows a transmission electron micrograph revealing ultrastructural details of smooth muscle cells from rabbit renal artery. Figure 14b shows a $^{39}K^+$ ion image of this tissue obtained using the Cameca IMS-3f ion microscope. it is obvious that these techniques together enhance the microfeature recognition for elemental distribution, and provide a powerful combination to tackle biological problems.

A combination of ion and fluorescence microscopy recently provided a unique approach to study the subcellular distribution of total and free Ca in cultured cells.[46] The knowledge of stores and movements of Ca^{2+} should result in a better understanding of the role of this important ion under physiological and pathological conditions. Figure 15 shows total and free Ca distributions in NIH-3T3 cells measured by ion and fluores-

[45] W. C. Harris, Jr., S. Chandra, and G. H. Morrison, *Anal. Chem.* **55,** 1959 (1983).

[46] S. Chandra, D. Gross, W. W. Webb, and G. H. Morrison, manuscript in preparation.

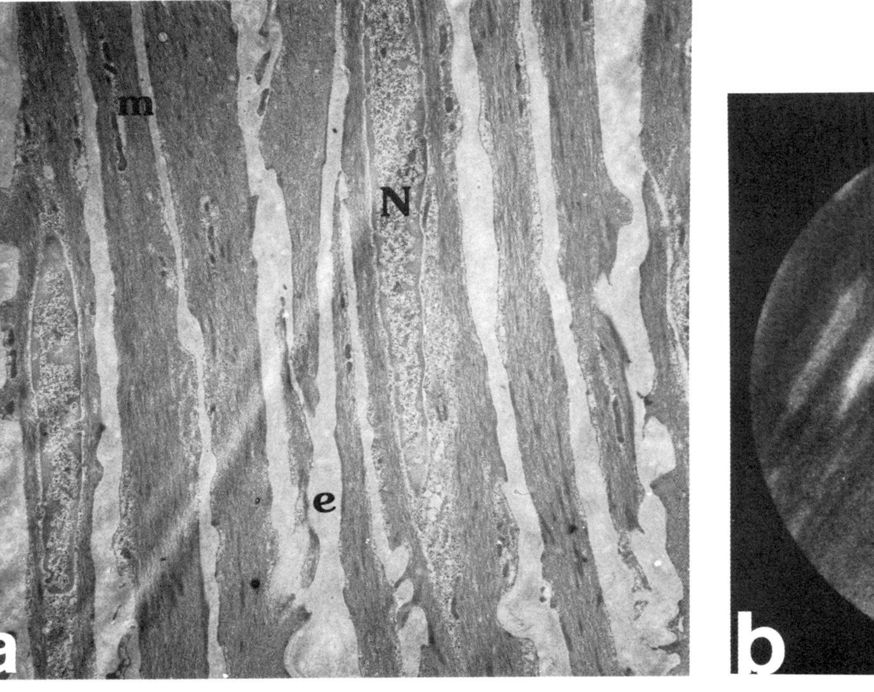

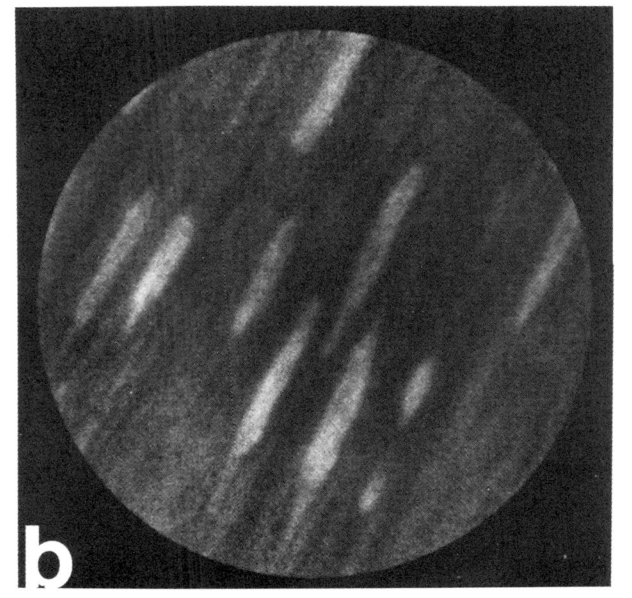

FIG. 14. Smooth muscle cells of rabbit renal artery after gluteraldehyde fixation and Lowicryl KAM embedment[30] (no claim is made for reliability of the K distribution). (a) A transmission electron micrograph shows nuclei (N), mitochondria (m), extracellular spaces (e), and contractile filament. Bar = 1 μm. (b) A $^{39}K^+$ secondary ion image obtained by the Cameca IMS-3f ion microscope. The nuclei, cell cytoplasm, and extracellular spaces are discernible. Circular field of view = 90 μm in diameter.

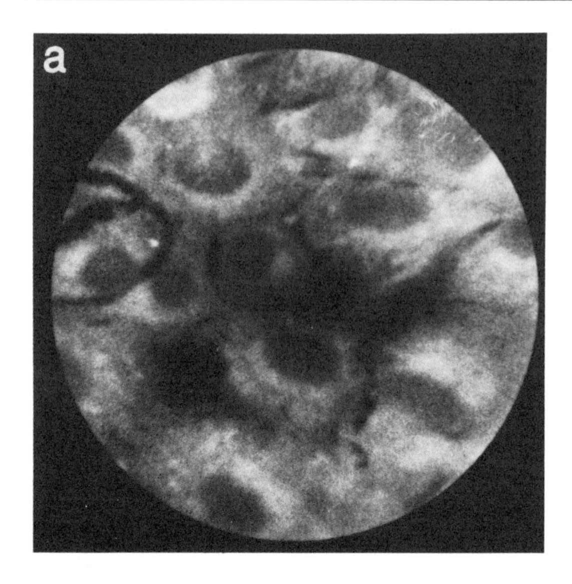

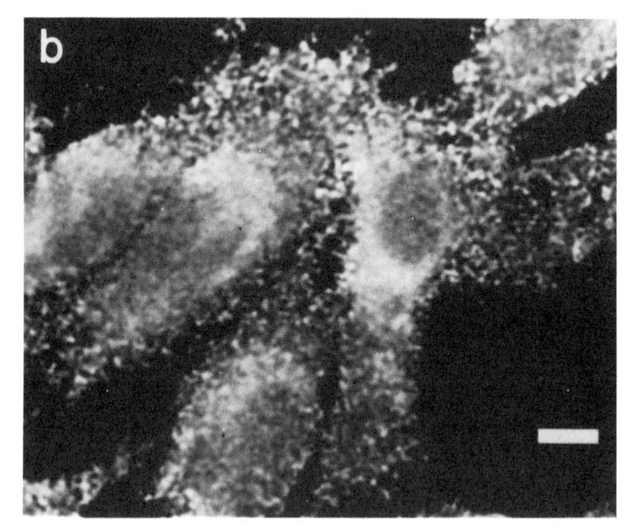

FIG. 15. The distribution of total Ca (a) as revealed by ion microscopy ($^{40}Ca^{+}$ ion image) in NIH-3T3 mouse fibroblast cells. Image exposure time = 100 sec. The circular field of view is 150 μm in diameter. Free Ca^{2+} (b) subcellular distribution in this cell line is also heterogeneous. (Courtesy of D. Gross and W. Webb, Cornell University, Ithaca, New York.) Bar = 10 μm.

cence microscopy (using Fura-2 Ca^{2+} indicator[47]), respectively. From the total Ca image, it is clear that stores of Ca concentrate more in the cell cytoplasm. Smaller cytoplasmic organelle endoplasmic reticulum has been shown as a Ca storage site in nonmuscle cells.[32] The free Ca distribution is not homogeneous throughout the cell. Slightly lower intensities from nuclear regions and heterogeneous cytoplasmic distribution are observed indicating a gradient of free Ca inside the cell.

Conclusion and Future Developments

The ion microscope provides a powerful technique for studying the subcellular distribution of elements in biological systems under physiological, pathological, and toxicological conditions. The direct ion imaging with cell morphology with a lateral resolution of $\sim 0.5\ \mu m$, an unambiguous mass spectrometric isotopic detection and high sensitivity of the technique are important features for biological research. The hydrated matrix of biological specimens requires careful sample fixation before analysis. Frozen-hydrated or frozen freeze-dried methods allow sample fixation that preserves both chemical and structural integrity of a living cell. With recent developments in high spatial resolution,[4,48] many smaller cytoplasmic structures fall within the range of ion microscopy. Ion image quantification remains an area for future research. The use of stable isotopes for quantification and prediction of ion transport also remains a challenging and underexplored area of research. The technique of ion microscopy has reached a stage where many important biomedical problems can presently benefit from its unique capabilities.

Acknowledgments

The authors acknowledge Walter Ausserer, Gregg Potter, and Howard E. Smith, Cornell University, for constructive criticism of the manuscript. This work was supported by National Institutes of Health.

[47] G. Grynkiewicz, M. Poenie, and R. Y. Tsein, *J. Biol. Chem.* **260,** 3440 (1985).
[48] M. T. Bernius, Y. C. Ling, and G. H. Morrison, *J. Appl. Phys.* **60,** 1904 (1986).

[15] Flame Atomic Emission Spectrometry

By TERENCE H. RISBY

Introduction

The intensities of the spectral deactivations of excited state atoms and their corresponding excitation energies form the basis for the identification and quantification of metals and nonmetals by atomic emission spectrometry. Therefore this method of analysis is dependent upon the ability to produce excited state atoms and to separate and quantify the spectral radiation associated with their deactivation. This chapter will discuss the theory associated with the production of excited state atoms but will limit discussion to instrumentation and applications associated with the use of flames as atom reservoirs (flame emission or flame photometry).

Theory

There are a number of processes involved in the conversion of the analyte solution to quantifiable excited state atoms.

$$[\text{Solution}] \xrightarrow[\text{nebulization}]{} [\text{Aerosol}] \xrightarrow[\text{desolvation}]{} [\text{Solid}] \xrightarrow[\text{volatilization}]{}$$

$$[\text{Vapor}] \xrightarrow[\text{dissociation}]{} [\text{Atoms}] \xrightarrow[\text{excitation}]{} [\text{Excited atoms}] \xrightarrow[\text{deactivation}]{} [\text{Atoms}]$$

There are many texts that discuss the theories of nebulization, desolvation, dissociation, excitation, and ionization.[1,2] Therefore, only a brief discussion of the theory of excition in relation to the production of the atomic emission signal will be presented. There are also a number of processes that can compete with the production of excited state atoms and thereby reduce the number of excited state atoms. For example, there may be insufficient energy to dissociate the vapor and excited molecular species are produced. These species deactivate by the emission of banded molecular spectra. Identification based on molecular spectra are seldom specific. Conversely, there may be sufficient energy available in the atom reservoir that the atoms are ionized. Both these processes will reduce the available concentration of atoms to be excited. Finally, there may be sufficient energy to atomize the vapor but insufficient energy to excite the

[1] J. A. Dean and T. C. Rains, "Flame Emission and Atomic Absorption Spectrometry," Vol. 1. Dekker, New York, 1969.

[2] R. Mavrodineau, "Analytical Flame Spectroscopy." Springer-Verlag, New York, 1970.

atoms. This problem is only observed with atom reservoirs of limited energy (e.g., flames).

Excitation

Excitation is the process whereby ground state atoms are converted to excited state atoms whose radiative deactivation can be quantified spectrometrically.

$$M_0 + \Delta E \rightarrow M_j \rightarrow M_0 + h\nu_{j,0} \tag{1}$$

If excitation occurs as a result of thermal collisions, then the equilibrium concentrations of ground and excited state atoms can be described by Boltzmann's distribution law:

$$N_j/N_0 = (g_j/g_0)e^{-E_{j,0}/KT} \tag{2}$$

where N_j and N_0 are the number densities of the excited state (M_j) and ground state (M_0) atoms, g_j and g_0 are the statistical weights of these states, $E_{j,0}$ is the energy difference between these states (excitation energy), K is Boltzmann's constant, and T is the temperature of the atom reservoir.

The intensity of the radiative deactivation of excited atoms is linearly related to the concentration of analyte atoms:

$$P = (h\lambda_{j,0}/4\pi)(g_j/g_0)A_{j,0}l[\mathrm{M}]e^{-E_{j,0}/KT} \tag{3}$$

where P is the radiant flux (erg/(cm^2 S sterad) in the direction of photodetection, $A_{j,0}$ is the transition probability (S^{-1}) for radiative deactivation, l the thickness (cm) of the atom reservoir along the axis of observation, and [M] is the concentration of the analyte atoms. Clearly, if excitation can be described by the Boltzmann's distribution law (thermal collisions), then, providing there are no spectral or chemical interferences, the sensitivity of the analytical method is dependent upon the temperature of the atom reservoir. Spectral interferences may be caused by excited molecular species or by self-absorption by ground state analyte atoms. These effects are demonstrated by deviations from linearity of the intensity–concentration curves. Chemical interferences are caused by the formation of stable molecular species that sublime through the atom reservoir without atomization.

Boltzmann's distribution law does not apply for nonflame atom reservoirs since excitation is produced by collisions with energetic electrons. Also, it does not apply if excitation occurs by suprathermal chemiluminescence flame reactions.

Suprathermal Chemiluminescent Excitation. Another mode of excitation of ground state atoms occurs when flame radicals recombine by three-body collisions in the presence of a stabilizing metal atom. The energy resulting from these collisions is transferred and excites the analyte atom.

$$M_0 + OH + H \rightarrow M_j + H_2O \quad (4)$$
$$M_0 + CH + O \rightarrow M_j + CHO \quad (5)$$
$$M_0 + H + H \rightarrow M_j + H_2 \quad (6)$$

Additionally, flame radicals can collide with analyte species and cause them to be dissociated and excited.

$$C + MO \rightarrow M_j + CO \quad (7)$$

Suprathermal chemiluminescent excitation is limited to those elements whose excitation energies are less than the energy of recombination.

Instrumentation

The instrumental components of an atomic emission spectrometer provide a means to convert the analyte solution to excited state atoms that can be identified and quantified. These components will be discussed for their relevance to the analysis of metals in enzymes and metalloproteins.

Nebulizer → Atom reservoir → Wavelength selector → Photon detector

Nebulizers

The function of the nebulizer is to transport quantitatively and reproducibly a representative quantity of the analyte solution to the atom reservoir. There are two types of nebulizers used in atomic emission spectrometers: direct (total consumption) nebulizers and indirect (pneumatic or ultrasonic) nebulizers. Direct nebulizers are used primarily for turbulen flames that have high burning velocities such as oxygen–acetylene, nitrous oxide–acetylene, and oxygen–hydrogen, whereas indirect nebulizers are used for laminar premixed flames such as air–acetylene, air–hydrogen, and nitrous oxide–acetylene and for inductively coupled plasmas. Indirect nebulizers produce analyte aerosols with uniform droplet size which increases the probability that analyte excited atoms are produced. However, as much as 90% of the sample is lost during transport to the atom reservoir. Direct nebulizers transport almost 100% of the analyte aerosol into the flame, but there is no control on the droplet size and large droplets pass through the flame without complete atomization.

TABLE I
FLAMES FOR ATOMIC EMISSION SPECTROMETRY

Fuel	Oxidizer	Flame temperature (°C)
Propane	Air	1925
Hydrogen	Air	2025
Hydrogen	Oxygen	2700
Acetylene	Air	2300
Acetylene	Nitrous oxide	2950
Acetylene	Oxygen	3100

The efficiencies of overall transport of the analyte solution and its subsequent atomization are dependent upon the solvent viscosity. Organic solvents can be used to increase the sensitivity of analysis by preconcentration extraction of the analyte atoms and by increasing nebulization efficiency. Aqueous solutions of enzymes or metalloproteins can be nebulized directly into flames providing that the sample matrix does not interfere with the analyte signal. Since most flame emission spectrometers use indirect nebulizers, relatively large amounts of analyte solutions (1 ml) are required to produce stable signals.

Atom Reservoirs

Atom reservoirs are sources of energy to convert analyte aerosols to excited state atoms. There are two general types of atom reservoirs: flames and non-flames.

Flames. The energy available in flames to produce excited state atoms is the result of thermal and suprathermal chemiluminescent collisions with analyte atoms. Various stable flames are suitable for atomic emission spectrometry and the selection is determined by the energy required to excite the analyte atom. Table I lists some of the commonly used flames and their temperatures.[1,2]

Clearly, more energetic flames, such as nitrous oxide-acetylene or oxygen–acetylene, are more likely to produce excited state atoms unless these flames have sufficient energy to ionize the atoms. In addition, if these flames are fuel-rich, they contain significant concentrations of species that can excite atoms through suprathermal chemiluminescent collisions.

Nonflame Atom Reservoirs. There are other types of atom reservoirs that have been used for atomic emission spectrometry such as electrical arcs, electrical sparks, dc or ac electrical discharges, and inductively coupled plasmas. These atom reservoirs excite atoms by electron impact, and by collision with energetic ions. These devices have sufficient energies to excite a greater number of elements and also improve the excitation efficiencies of others. However, with the exception of the inductively coupled plasma, their use is limited. The inductively coupled plasma is the most widely used nonflame atom reservoir since it is simple to use and the sample is introduced as a solution *via* an indirect ultrasonic nebulizer. This atom reservoir is discussed elsewhere in the text. Nonflame atom reservoirs have the advantage that they have sufficient energy to excite most elements to various excitation states and thereby offer the possibility of performing multielement analysis. Also, the availability of different excitation states allows analysis to be performed over large ranges of concentration by the selection of a particular radiative decay. Generally, the lowest excited state will provide the best sensitivity.

Wavelength Selectors

The requirements for the selective detection of the radiative deactivation of excited state atoms are severe since it is necessary to separate discrete wavelengths from the background radiation. Background emission originates from excited flame radicals or from other excited species within the sample matrix. Generally, high-resolution monochromators are required for wavelength selection to improve selectivity and hence, sensitivity. The typical line width at half-height for atomic emission produced in a flame is 0.001 nm, therefore monochromators with resolution of this order are necessary. This statement must be qualified, since it is possible to quantify certain alkali and alkaline earth elements with air–methane flames using interference filters.

Photon Detection

Generally, photomultipliers are used to convert the spectral radiation to an electrical current and, depending upon the instrument involved, phase sensitive lock-in amplifiers are used to amplify the electronically or mechanically chopped emission signal. The use of photographic film placed at the focal plane of the Rowland-circle spectrograph can be used to record the entire spectral emission from the atom reservoir, thus enabling semiquantitative multielemental analysis to be performed. More modern versions of this approach have replaced the photographic film with suitably placed masked photomultipliers (quantometers) or photodiode arrays, and coupled their output to multichannel analyzers or computers for rapid quantitative multielemental analysis. The limit of detection for elemental analysis by atomic emission spectrometry is defined as the concentration that produces a signal three times the background radiation at the same wavelength.

Applications

Sample Preparation

The determination of the metal or nonmetal content of solutions of enzymes or metalloproteins by flame atomic emission spectrometry can be performed directly without sample preparation, providing the analyte species is contained in a noninterfering matrix. However, most enzmyes and metalloproteins are contained in buffered solutions with defined ionic strengths in order to maintain their activities. These matrix species can interfere with the analytical determination by a variety of mechanisms.

Reagent Purity. Quantification and identification of metals and nonmetals in metalloproteins and enzymes by atomic emission spectrometry require that any reagents used in the separation of these biomaterials prior to analysis must have high purities. Clearly, all the reagents should be free of the analyte atoms, and reagent purities should assayed under similar conditions to the actual analysis. Separation of the enzymes or metalloproteins from biological materials often requires the copious use of water, and trace elemental contaminants in water could interfere with the subsequent analysis.

Apparatus. The separation of metalloproteins and enzymes from biological specimens generally involves the use of equipment and apparatus that is manufactured from metal, glass, and plastic which could contribute significantly to trace metal contamination. Obviously, the use of metal dissection scalpels and tissue homogenizers cannot be avoided if the source of the metalloprotein or enzyme is tissue. However, subsequent exposure of the specimen or its extract to glassware or plasticware can be an additional significant source of contamination. Glassware can be cleaned by immersion in a mixture of high-purity nitric and hydrochloric acids (aqua regia) and, after cleaning, the glassware should be stored filled with distilled water. The cleaning of plasticware with similar protocols will irreversibly modify its surface and therefore more gentle cleaning must be used. Boiling the plasticware in solutions of EDTA (1 *M*) will remove any extractable metals from the surface of plasticware without surface modification. The chemically cleaned plasticware should also be stored filled with distilled water until used. The preparation of standard solutions in this chemically cleaned glassware should be performed by first equilibrating the glassware to the concentration of element of interest for 24 hr and then discarding the solution and remaking the standard. If this approach is adopted, standard solutions should remain stable for days.

Matrix Destruction. Organic materials in the sample may decompose in the flame to excited molecular species that may interfere with the atomic emission signal. Also, it is possible that the element of interest bound in biomolecules is not in true solution, and therefore it is impossible to obtain a representative aliquot of the sample. Under these conditions it is desirable to destroy the sample and organic matrix prior to analysis by oxidation (for review see Ref. 3). The resulting inorganic ash is then dissolved in a suitable acid and analyzed. Oxidation can be produced directly by high temperature digestion with nitric, perchloric, or sulfuric acids. Care must be taken to avoid the formation of explosive metal perchlorates or volatile inorganic compounds. Also, the formation of insoluble sulfates should be avoided since they can occlude trace elements. Chemical digestion requires that acids of ultrahigh-purity be used. Alternatively, thermal oxidation can be performed, although volatile elements can be lost. Low-temperature ashers which use reduced pressure oxygen electrical discharges are more effective means for the destruction of organic materials, since they do not introduce elemental contaminants or lose elements by volatilization.

Solvent Extraction. If the sample has been oxidized and dissolved in acidic solvents, there is often a need to concentrate the element of interest by complexation with suitable ligands to provide neutral complexes that can be extracted into organic solvents. Organic solvents are useful since they also improve the sensitivity of the analysis by increased nebulization of the sample. Additionally, the decomposition of the organic solvents can increase the concentration of excited flame radicals that enhance the populations of excited state atoms by suprathermal chemiluminescent excitation.

Interferences

Analysis by atomic emission spectrometry can be hampered by various chemical or spectral interferences.

Chemical Interferences. Chemical interferences can be caused by the biological matrix or by contamination from the reagents. Trace elements can be removed by dialysis in which the enzyme or metalloprotein-bound element is unable to cross a semipermeable membrane. The only potential problem with this approach is that the element must be bound strongly by the biomolecule. Additional sources of chemical interference that may be

[3] T. T. Gorsuch, "The Destruction of Organic Matter." Pergamon, London, 1970.

encountered in atomic emission spectrometry are due to the reduction in the number of excited state atoms. If the element is easily ionized by the atom reservoir (e.g., Li, Na, K, Rb, and Cs), then ionization may reduce the concentration of available atoms. Ionization buffers, that are easily ionized, may be added to the sample to decrease ionization of the analyte atom. Alternatively atoms may form refractory oxides (e.g., AlO, BO, BeO, CeO, DyO, ErO, GdO, HfO, HoO, LaO, LuO, and NbO) or other stable species in the flame that sublime through the flame without atomization. The use of more energetic flames often reduces this problem. The nitrous oxide-acetylene flame also contains significant quantities of CN and NH radicals which have been proposed as a means to reduce stable analyte species.[4] Similarly, fuel-rich flames increase the concentration of flame radicals which can enhance excitation by suprathermal chemiluminescent excitation.

Spectral Interferences. There are a number of atomic and molecular species that can interfere with the atomic emission signal. The simplest way to establish the presence of spectral interferences is to record a spectrum of the wavelength range that includes the atomic emission line. The spectral resolution of the monochromator will determine the extent of spectral interferences and thereby define the limit of detection for the analyte element. If the origins of the spectral interferences are excited flame radicals (e.g., C_2, OH, or CH) that cannot be reduced by changing the fuel to oxidizer ratio, then alternative atomic radiative deactivations must be monitored. Stable molecular excited species (e.g., MgOH, CaOH, SrOH, CuH) derived from reactions between flame radicals and other elements in the sample matrix are another source of spectral interference. The production of these species can be avoided by prior separation of the analyte element from matrix elements. These interferences can also be reduced by increasing the concentrations of flame radicals that can reduce these stable analyte species by changing the fuel to oxidizer ratio or selecting alternate flame gases. Spectral interferences that result from atomic emissions from matrix elements can be removed by prior chemical separation. Self-absorption is another spectral interference that may be observed when analyte elements are present at high concentrations, and this interference can be reduced by sample dilution.

Standard Additions

Quantification of elements by atomic emission spectrometry is performed by comparison of the signals of analyte solution to a calibration curve. Clearly, there are potential problems with this approach, espe-

[4] G. F. Kirkbright, A. Semb, and T. S. West, *Spectrosc. Lett.* **1,** 7 (1968).

cially when the sample is contained in a interfering matrix. Therefore, quantification should be performed using the method of standard additions. The basis of this approach is to monitor the analyte signal as a function of the addition of known concentrations of analyte standards. The sample analyte concentration is found by extrapolation. If extensive sample pretreatment is required in the analytical protocol, then standard additions should be performed prior to sample pretreatment in order to establish the overall recovery of the analysis. Clearly, if the matrix is a constant for a given analytical determination, then calibration curves could be established using a set of standards prepared in the same matrix.

Metal Analysis

The only requirements for metal analysis by flame atomic emission spectrometry are that the radiative deactivation occur in a region of the spectrum that can be quantified in a flame and that significant numbers of excited state atoms are produced.

Nonmetal Analysis

Although the analysis of metals in enzymes and metalloproteins is the emphasis of this book, it must be mentioned that nonmetals can also be determined by atomic emission spectrometry. Generally, the radiative deactivation of excited nonmetal atoms occurs in below 200 nm which makes direct analysis using flames impossible. However, there are a number of nonmetal molecular species that emit intense radiation in certain flames (e.g., air–hydrogen and cooled hydrogen diffusion flames), and these emissions can be used for quantification (compare the flame photometric detector for gas chromatography). Alternatively, nonmetals can be determined indirectly using metal surrogates. For example, compounds of silicon (as silicomolybdic acid) and phosphorus (as phosphomolybdic acid) can be formed and quantified indirectly by the determination of molybdenum. Similarly, sulfur (as *barium* sulfate) and halides (as *silver* halides) can also be determined indirectly. Any nonmetal can be determined in this manner providing that a suitable metal complex is formed that can be selectively extracted from excess metal reagent.

Conclusions

The data contained in Table II show that flame atomic emission spectrometry can be used to identify and quantify a variety of elements that are important to studies on enzymes and metalloproteins. Clearly, these detection limits vary significantly and this method of analysis is limited in

TABLE II
DETECTION LIMITS FOR THE ANALYSIS OF SELECTED ELEMENTS BY FLAME ATOMIC EMISSION SPECTROMETRY[a]

Element	Wavelength (nm)	Flame	Detection limit	
			μg/ml	n*M*
Ag (Cl, Br, I)	328.1	N/A	0.3	2
Al	396.1	N/A	0.02	0.7
As	235.0	A/A	2.2	29
B	249.7	N/A	5	462
Ba(S)	553.6	N/A	0.004	0.003
Be	234.9	N/A	0.03	3
Ca	422.7	N/A	0.0009	0.02
Cd	326.1	N/A	4.2	37
Co	345.4	N/A	0.01	0.2
Cr	425.4	N/A	0.01	0.2
Cs	852.1	A/P	0.002	0.02
Cu	327.4	N/A	0.01	0.2
Fe	372.0	A/A	0.01	0.2
Hg	253.7	N/A	2.5	12
K	766.5	A/P	0.0002	0.005
Li	670.8	A/P	0.000003	0.0004
Mg	285.2	N/A	0.02	0.8
Mn	403.1	N/A	0.008	0.1
Mo (P, Si)	390.3	N/A	0.1	1
Na	589.0	A/A	0.0001	0.004
Ni	341.5	N/A	0.03	0.5
Pb	405.8	N/A	0.2	1
Sb	259.8	N/A	0.35	3
Sn	284.0	N/A	0.7	6
V	437.9	N/A	0.3	6
W	400.9	N/A	3	16
Zn	213.9	N/A	3	16

[a] N/A, nitrous oxide–acetylene; A/A, air–acetylene; A/P, air–propane. Elements in parentheses are determined indirectly. This information was obtained from primary references and from various reviews.

its application to the analysis of individual elements with low excitation energies. Detection limits are also expressed as nanomoles of element and, if it assumed that elements activate, block, or are required by enzymes or metalloproteins in stoichiometric amounts, then this table can be used to define the minimum amount of enzyme or metalloprotein required for analysis by flame atomic emission spectrometry. The major disadvantage of all atomic spectrometric methods of analysis is that they provide no information of the oxidation state of the element or its speciation. However, this disadvantage can be redressed by the use of selective reagents coupled with solvent extraction or by chromatographic separation. In the latter approach atomic spectrometry is used for selective detection of the chromatographic effluent. Flame atomic emission spectrometry generally requires at least 1 ml of analyte solution in order to provide a representative signal, which generally precludes this technique for sequential multielemental analysis. Multielemental interactions are becoming increasingly important in the field of enzymes and metalloproteins.[5]

[5] C. Veillon and B. L. Vallee, this series, Vol. 54, p. 446.

[16] Inductively Coupled Plasma-Emission Spectrometry

By KAREN A. WOLNIK

Since analysis by inductively coupled plasma-optical emission spectrometry (ICP-OES) was first described,[1,2] the technique has taken its place as a mainstay in the field of elemental analysis. The first commercial instruments were introduced in 1974, and in subsequent years there has been swift expansion in the number of commercial vendors and ICP-OES operating facilities. The ICP owes its popularity as an emission source to a number of attractive features but primarily to the fact that it enables simultaneous or rapid sequential multielement determination of major, minor, trace, and ultratrace elemental sample constituents without changing experimental parameters. This combination of speed, wide dynamic range, part per billion detection limits for many elements, and an unusual degree of freedom from matrix interference effects enables reliable analysis of a variety of samples with a minimum of sample pretreatment. A number of reviews have been previously published[3-7] and a comprehensive discussion of inductively coupled plasma spectroscopy can be found in a recent book[8] that includes the latest developments in the field as well as fundamental background.

[1] S. Greenfield, *Analyst* (*London*) **89,** 713 (1964).
[2] R. Wendt and V. A. Fassel, *Anal. Chem.* **37,** 920 (1965).
[3] V. A. Fassel and R. J. Kniseley, *Anal. Chem.* **46,** 110A (1974).
[4] V. A. Fassel and R. J. Kniseley, *Anal. Chem.* **46,** 1155A (1974).
[5] R. M. Barnes, *Crit. Rev. Anal. Chem.* **7,** 203 (1978).
[6] P. W. J. M. Boumans, *ICP Inf. Newsl.* **5,** 181 (1979).
[7] S. Greenfield, *Analyst* (*London*) **105,** 1032 (1980).
[8] A. Montasser and D. W. Golightly (eds.), "Inductively Coupled Plasmas in Analytical Atomic Spectrometry." VCH, New York, 1987.

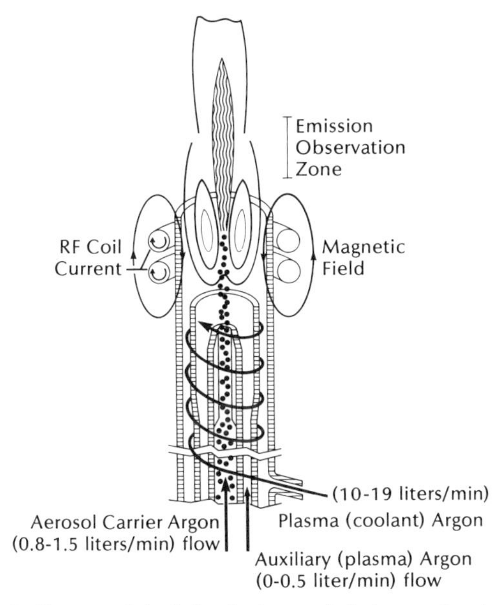

FIG. 1. Schematic diagram of the inductively coupled plasma showing a cross-section of the 3 concentric tube torch.

By definition, plasmas are ionized gases. The majority of ICPs used routinely today are composed solely of argon. The inductively coupled plasma (shown schematically in Fig. 1) is formed and sustained by the inductive coupling of time-varying magnetic fields with ionized gas. This produces a flow of electrons and ions in closed annular paths—the eddy current—which results in ohmic heating and further ionization. The plasma is initiated at the open end of a torch consisting of three concentric quartz tubes with flowing argon by providing seed electrons from a Tesla coil. The oscillating magnetic fields that sustain the plasma are created by high-frequency currents from a radio-frequency (RF) generator which flow through an induction coil, sometimes called the load coil. Plasmas formed in this manner reach temperatures of around 10,000 K in the region of maximum eddy current. A flow of argon introduced tangentially through the outer tube of the torch actually forms and centers the plasma in addition to preventing the torch from melting.

At a frequency of around 30 MHz, the plasma develops an annular or torroidal shape so that at the proper flow velocity of carrier gas, sample aerosol can be effectively injected into the "hole" in the doughnut shaped plasma. The analyte experiences a temperature of about 7000 K in the axial channel with residence times of around 2.5 msec prior to reaching the observation zone. The combination of high temperatures and long particle–plasma interaction times in the chemically inert argon environment leads to complete solute vaporization and a high degree of atomization with minimized depopulation processes such as metal oxide formation. Emission from monopositive ions as well as from neutral atoms is observed in the ICP.

RF Generator

The ICP radio frequency generators commercially available in the United States are predominantly the constant frequency type with crystal-controlled oscillators as opposed to the free-running, variable frequency generators popular in Europe. The two frequencies that have been approved by the Federal Communications Commission (FCC) and are used with ICPs are 27.12 and 40.68 MHz. The 27.12 MHz generator with 2.5–3 kW output is the most widely used. A matching or tuning circuit is used in conjunction with the fixed frequency RF generators to match the impedance of the plasma and load coil to the generator output. Proper tuning maximizes forward power (i.e., power coupled to the plasma) and minimizes reflected power. Many RF generators are equipped to automatically adjust power output to compensate for small changes in plasma impedance that occur when sample is introduced. The 2.5 kW generator contains a power amplifier tube that has a finite lifetime (on the order of 3000 hr of operation) and therefore requires periodic replacement.

Torch

Nearly all ICP torches are constructed of quartz in a 3 tube configuration. Occasionally the sample injection tube may be made of some other material, for example aluminum oxide, for special applications such as the determination of trace levels of silicon or when solutions containing hydrofluoric acid are being analyzed. The most critical torch configuration parameters are the concentricity of the tubes and the position and inner diameter of the injector orifice. The torch must be centered in the induction coil and the position of the injector tip with respect to the coil affects the ease of initiation as well as the analytical performance of the plasma.

There are several commercially available, so-called "standard," ICP torches with slight variations in design that provide comparable performance. These torches require flows of 10–20 liters/min of coolant argon in the outer tube. Many researchers have described modified torch designs that reduce argon consumption and thereby reduce ICP operating costs.[9]

[9] G. M. Hieftje, *Spectrochim. Acta* **38B,** 1465 (1983).

A number of these new designs, referred to as "low-flow" or "mini" torches, are now also commercially available. An additional advantage to the reduction of argon flow is an accompanying reduction in required operating power. This makes possible ICP systems which use less costly, lower power solid-state RF generators.

Sample Introduction

Samples may be introduced into the ICP as solids, liquids, or gases. Solids are usually vaporized by some type of thermal process prior to entering the plasma. Gaseous introduction is used for the determination of elemental constituents of flowing gas streams and gas chromatographic effluent as well as for determination of specific elements that can be converted chemically to volatile species such as the hydrides of As and Se. The most frequent application of ICP-OES, however, is the analysis of samples in solution. The discussion in this chapter will be confined to liquid sample introduction of aqueous solutions although organic solutions can also be analyzed successfully by ICP. The sample introduction process is critical to the accuracy and reliability of IC-OES determinations and the importance of this component of the ICP system cannot be overemphasized.

Pneumatic nebulizers, similar in principle to those used in atomic absorption spectrometry, are commonly used to convert liquid sample into an aerosol which is transported by the carrier gas through a glass spray chamber into the plasma in a continuous or steady-state mode. The spray chamber facilitates the removal of large droplets and design of the chamber affects transport and analytical stability. As a rule of thumb, anywhere from <1 to 5% of the sample solution being nebulized actually reaches the plasma.

Pneumatic nebulizers operate at carrier gas flows from 0.5 to 3 liters/min with solution aspiration rates ranging from <1 to 3 ml/min maximum. These nebulizers are capable of providing $<1\%$ relative standard deviation for repeated determinations of standard solutions. In the concentric glass nebulizer, liquid flows through a central orifice and carrier gas through a narrow (10–35 μm) annulus surrounding that orifice. Cross-flow nebulizers employ a right-angle arrangement of gas and sample flow inlets. The concentric glass nebulizer is somewhat fragile and less tolerant of solutions with a high salt content, whereas with cross-flow nebulizers, alignment of the gas and sample inlets is critical. User adjustable cross-flow nebulizers are available, however proper adjustment is tedious and may be inadvertently changed by temperature extremes or jarring of the nebulizer. So-called "fixed" cross-flow nebulizers, in which the sample and gas inlets are permanently aligned by the manufacturer, are much more durable. The author has found a "fixed" cross-flow nebulizer with a polypropylene body, saphire gas orifice, and platinum sample needle to be a rugged, corrosion-resistant, reliable general purpose nebulizer with good tolerance to salt solutions (1% or less) when humidified argon is used as the carrier.

Nebulizer clogging is a problem caused by sample solutions containing particulates or high salt content. Salts from solutions with a salt content greater than 1–2% may gradually deposit on the gas orifice. This causes a drift in emission intensity and a resultant drift in calibration. The carrier argon can be humidified by flowing through a frit in a water reservoir prior to the nebulizer. This helps increase the tolerance of the nebulizer to salt solutions and reduces drift. A wash solution interspersed between samples also improves nebulizer performance. The wash solution should correspond to the matrix of the sample solution as closely as possible, i.e., if the sample is in a 5% nitric acid solution, the wash solution should be 5% nitric acid. Care must be taken when switching from one solution to another to minimize the amount of air that enters the plasma since this produces instability in the ICP and in the worst case may extinguish the plasma.

Although pneumatic nebulizers aspirate solution, the rate of delivery can be affected by viscosity differences between sample and standard or by the depth of solution in the sampling container; therefore, it is recommended that solution delivery be controlled by a pump. Relatively inexpensive peristaltic pumps are commonly used for this purpose. It is important to be sure that the pump tubing is compatible with the sample matrix and does not introduce contamination. Tubing composed of Viton is acid resistant and retains its resiliency over several weeks of operation. Whenever pump speed requires frequent readjustment to maintain the same delivery rate or severe pulsations in aerosol production are observed, the tubing should be replaced. The author's experience has been that controlled washout and sample equilibration times provide the most reproducible results and minimize calibration drift. This controlled timing can readily be accomplished via an autosampler although simply watching a clock or setting a timer suffices. Flow injection analysis (FIA) techniques for ICP,[10] in which the sample solution is automatically injected into a stream of flowing liquid, are increasing in popularity. FIA techniques reduce sample consumption, and nebulizer blockage due to samples with a high salt content is reduced or eliminated since the nebulizer

[10] S. Greenfield, *Spectrochim. Acta* **38B,** 95 (1983).

and spray chamber are continually being washed, i.e., no air reaches the plasma.

A third type of pneumatic nebulizer that has a more limited popularity is the high-solids nebulizer. High-solids nebulizers[11] are based on the Babington design in which liquid (pumped) flows over the gas orifice. These nebulizers can handle solutions containing up to 50% solids, such as orange juice, without clogging; however, precision is often degraded. Since larger amounts of solid material enter the plasma, interference effects are increased and, indeed, deposits may form on the tip of the sample tube which causes plasma instability. Still, the use of high-solids nebulizers reduces sample pretreatment steps for some analyses and, therefore, if interferences can be overcome, poorer precision may be tolerated.

Ultrasonic nebulizers[12] convert liquid to aerosol via an ultrasonic transducer which requires a power source. Although they are more expensive than pneumatic nebulizers, ultrasonic nebulizers provide at least an order of magnitude improvement in detection limits for most elements when the aerosol is desolvated prior to reaching the plasma. With ultrasonic nebulizers 10 times more sample is delivered to the plasma and without desolvation interferences are severe; however the desolvation process itself suffers from transport interference effects due to differences in the sample and standard matrices. Relatively few laboratories routinely use ultrasonic nebulizers.

There are other specialized methods of liquid sample introduction, such as electrothermal vaporization of microliter sample volumes, which have been omitted from this chapter for the sake of brevity. Complete information on sample introduction techniques for ICP-OES can be found in Ref. 8.

An additional aspect of sample introduction that deserves mention here is the interface of high-performance liquid chromatography (HPLC) (reversed-phase, ion-pair, or ion-exchange) to the ICP for speciation studies. In the past this has been accomplished by direct, low dead volume coupling of HPLC columns to conventional pneumatic nebulizers. Normal HPLC flow rates and nebulizer sample delivery rates are compatible and band spreading is minimal since the aerosol approximates a gas and the flow rate through the spray chamber and torch is high. However, because of restrictions on HPLC injection volumes and because conventional nebulizers operate with <5% efficiency, sensitivity is frequently

[11] J. F. Wolcott and C. B. Sobel, *Appl. Spectrosc.* **32,** 591 (1978).

[12] K. W. Olson, W. J. Haas, Jr., and V. A. Fassel, *Anal. Chem.* **49,** 632 (1977).

insufficient to enable analysis at the trace levels often encountered in actual samples. Current research efforts center on developing a sample introduction technique for HPLC-ICP that provides improved sensitivity.

Spectrometer

In order to take full advantage of the multielement capabilities of the ICP, inductively coupled plasmas are usually interfaced to either scanning monochromators or polychromators (fixed array multichannel spectrometers referred to by some as "direct readers"). Most commercial systems rely on photomultiplier tube detectors although photodiode arrays are also used as detection devices. A polychromator has fixed multiple exit slits with multiple photomultiplier tube detectors and provides simultaneous multielement analysis. With scanning monochromators, elements are determined sequentially with a single detector; however, with computer-controlled rapid scanning instruments, several elements (i.e., different wavelengths) may be determined in a relatively short time. With a polychromator some type of computer or processor is necessary for retrieval and storage of the output from the numerous photomultipliers. A computer is also useful for processing the large volumes of data produced by a polychromator. The main advantage of a polychromator is that a large number of elements (50 or more) can be determined simultaneously whereas a scanning instrument is somewhat slower (30 elements in 10 min is typical) but more versatile since it provides total flexibility in wavelength selection.

Commercial spectrometers, both polychromators and scanning monochromators, vary with respect to focal length, transfer optics, and dispersion optics (i.e., ruled, holographic, or eschelle grating, etc.). These options in turn affect light throughput, wavelength range, resolution, and straylight rejection. In order to detect emission wavelengths below 200 nm a vacuum or purged spectrometer is recommended to eliminate absorption by oxygen. Nearly all present day instruments are equipped with some type of automatic background correction via measurement of off-peak emission. With commercial scanning monochromators, there are a number of methods used for scanning, rotating the grating being the most common, and a variety of computer-controlled peak finding routines. A means of displaying wavelength scans is standard for scanning instruments. It is also quite helpful with ICP polychromators if the region around each analytical line can be scanned and displayed; however, this feature is not available on older models. The software and hardware used for controlling these instruments and for data acquisition vary considerably with respect to sophistication, ease of modification, and flexibility.

All of the preceding factors and options must be considered carefully when purchasing an ICP system since the spectrometer and controller are the core of the system and its most expensive components.

Interferences

Interferences that may occur during analysis by ICP-OES can be divided into two main categories, spectral interferences and matrix interference effects. Spectral interferences stem from spectra emitted by components of the sample solution other than the analyte or from shifts in (argon) background continuum caused by the introduction of sample aerosol. Spectral interferences lead to errors in the measurement of analyte emission unless proper corrections are made. Background correction schemes which employ some type of off-peak background measurement at the same time the sample solution is being introduced into the plasma compensate for spectral interferences arising from increases or decreases in the intensity of the plasma continuum. Sometimes judicious selection of background correction wavelengths also enables correction for partially resolved spectral line overlaps and wing overlaps. Nonresolved spectral overlaps can be avoided by selection of an alternate line for the analyte if this is possible and the alternate line provides sufficient sensitivity. When direct spectral overlaps cannot be avoided, empirical corrections can be made for the spectral effect of one element as observed at the wavelength of another. This requires knowledge of the identity of the interferent, quantitation of its effect on the wavelength of interest and determination of its concentration in the sample being analyzed. Several wavelength tables[13–16] have been published for the ICP and these references are useful when evaluating spectra.

Matrix interference effects are caused by components of the sample solution other than the analyte and result in increased or decreased analyte emission. Although matrix effects are less severe in the ICP than in other emission sources or in atomic absorption and when present are

[13] G. R. Harrison, "Massachusetts Institute of Technology Wavelength Tables." MIT Press, Cambridge, Massachusetts, 1969.

[14] R. K. Winge, V. J. Peterson, and V. A. Fassel, "Inductively Coupled Plasma-Atomic Emission Spectroscopy: Prominent Lines." US Environmental Protection Agency, Athens, Georgia, 1979. Available to the public through the National Technical Information Service, Springfield, Virginia 22161.

[15] M. L. Parsons, A. Forster, and D. Anderson, "An Atlas of Spectral Interferences in ICP Spectroscopy." Plenum, New York, 1980.

[16] R. K. Winge, V. A. Fassel, V. J. Peterson, and M. A. Floyd, "Inductively Coupled Plasma-Atomic Emission Spectroscopy: An Atlas of Spectral Information." Elsevier, Amsterdam, 1985.

easily overcome, they can lead to erroneous analytical results if not taken into consideration. These interferences may arise during any of the various steps involved in the complex process of transforming sample solution into optical emission measured by a spectrometer.

Physical interferences due to viscosity, density, or surface tension differences between sample and reference solutions can occur during aerosol formation and transport. The degree and magnitude of physical interference effects are dependent on the sample introduction system and can be reduced or eliminated by matching the composition of the sample and reference solutions so that their physical properties are similar.

Chemical interferences such as the formation of refractory oxides are virtually nonexistent in the plasma due to the high excitation temperatures and inert argon environment of the ICP; however, some chemical effects caused by sample concomitants can occur during nebulization[17] and desolvation. Often nebulizer operating parameters can be selected which reduce these effects.

Interferences that take place during the volatilization–atomization process affect the emission of both atoms and ions in the plasma and result from differences in energy transfer caused by sample concomitants. Volatilization–atomization interferences can be diminished by an increase in plasma power and by using a reference standard with a composition similar to that of the sample.

Ionization interferences are caused by a change in the ionization equilibrium of the analyte due to the presence of an easily ionized element in the plasma. There are conflicting reports concerning the observation of this type of matrix effect in the ICP. The nebulizer–torch combination used and selected operating parameters may have a major influence on the observation of ionization interferences. In any case, ionization interferences have rarely been reported for relative levels of easily ionized element to analyte of less than 10,000 to 1.

Optimization

When determining a single element, instrument optimization ordinarily involves investigating and selecting a set of operating parameters which provide the best performance for determination of that particular element. Since no one set of parameters is ideal for all elements, with multielement determinations optimization results in a set of compromise operating conditions and evaluation of what constitutes best performance

[17] R. F. Browner, M. S. Black, and A. W. Boorn, *in* "Developments in Atomic Plasma Spectrochemical Analysis" (R. M. Barnes, ed.), p. 242. Heyden, London, 1981.

becomes more arbitrary. Practically, the operating variables which can be readily controlled by the analyst are forward power, plasma observation region, argon flow to the torch, argon flow and pressure to the nebulizer, and sample delivery rate (pumped) to the nebulizer. For scanning instruments, analytical wavelength is also selectable. The task of finding optimum parameters is further complicated by the fact that the aforementioned parameters are interdependent. Simplex or simplex-like algorithms are favored by many for multivariant optimizations and several newer scanning instruments employ this technique to automate the optimization process.[18,19]

A number of criteria can be used to evaluate "best" performance. These are net signal-to-background ratio, signal-to-noise ratio, analyte peak precision, and freedom from interferences. Note that when determining net signal-to-background ratios, background correction should not be used for on line background measurements since ideally background corrected continuum emission (i.e., for unstructred background) results in zero signal. Often the optimum argon flow (coolant) for a particular torch is recommended by the manufacturer leaving power, viewing height (plasma observation zone), nebulizer argon flow, and sample delivery rate the parameters which need optimization. Signal-to-background ratio is the performance criteria most commonly used for optimization.

Selection of a test solution for the optimization process depends on the requirements of the planned analytical procedure. For example, if a multielement determination with emphasis on trace levels of Se were desired, a solution containing approximately 10 times the normal Se detection level might be used to evaluate operating parameters. Sensitivity (signal-to-background ratio) would be the primary criteria evaluated with perhaps a second solution containing traces of Se in combination with high salts used to evaluate interferences. The excitation energy required to achieve emission for different elements at different wavelengths varies and this should be taken into consideration when choosing an element or elements for investigating operating parameters. Solution matrix also affects performance and selection of test solutions for optimization should be considered carefully.

To empirically arrive at compromise operating parameters for a general multielement procedure, this author selects four elements that are anticipated at low levels in most samples and cover a range of excitation potentials, e.g., Cu, atom line at 324.7 nm; Cd, atom line at 228.8 nm; Mn,

[18] J. J. Leary, A. E. Brookes, A. F. Dorrzapf, Jr., and D. W. Golightly, *Appl. Spectrosc.* **36,** 37 (1982).

[19] P. R. Demko, *Am. Lab.* **17,** 97 (1985).

TABLE I
TYPICAL ICP OPERATING PARAMETERS FOR MULTIELEMENT ANALYSIS

Parameter	Characterization
Incident power	1.0–1.6 kW
Reflected power	<5 W
Observation height	15–18 mm above load coil
Coolant gas	8–18 liters/min argon
Plasma (auxiliary) gas	<1 liters/min argon
Nebulizer	Concentric, cross-flow, or ultrasonic
Solution nebulization rate	0.6–3 ml/min

ion line at 257.6 nm; and Mo, ion line at 202.0 nm. Using the recommended coolant flow, parameters are investigated in order of the magnitude of their effect on signal-to-background ratio: power, viewing height, nebulizer gas flow, and sample delivery rate. Thus, after the optimum power is determined and set, viewing height is studied, the optimum determined and set and so on. If auxiliary argon is desired (mostly to safeguard the torch, particularly when organic solutions are being analyzed) this is set prior to optimization and rarely exceeds 1 liter/min. Since these parameters interact rechecking is helpful. Typical operating parameters are listed in Table I. Whenever a component of the analytical system, such as the torch, nebulizer, or solution matrix, is changed, operating parameters should be reevaluated.

System Performance

There are several figures of merit used to measure and compare ICP-OES system performance. Commercial vendors place a good deal of emphasis on detection limits; however, for pneumatic nebulization of water, most published detection limits are within a factor of 2 or 3 to one another and are therefore comparable. A listing made in 1978 of the best reported detection limits for both pneumatic and ultrasonic nebulizers (Table II) remains representative of the present day state-of-the-art. Detection limits are reported in the literature as the concentration equivalent to 2 or 3 times the standard deviation of numerous (10 or more) blank determinations. The International Union of Pure and Applied Chemistry (IUPAC) has adopted 3 as the preferred factor. Detection limits calculated in this manner are sometimes referred to as instrument detection limits since they may differ from the lowest detectable quantity in actual samples.

One of the main attributes of the ICP-OES is the excellent precision

TABLE II
BEST REPORTED DETECTION LIMITS[a]

Element	Line[b] (nm)	Detection limits (ng/ml)	
		Ultrasonic nebulizer with desolvation	Pneumatic nebulizer
Ag	I (328.07)	—	2
Ag	I (396.15)	0.2	1
As	I (193.76)	2	25
Au	I (267.59)	—	0.9
B	I (249.77)	0.1	0.2
Ba	I (455.40)	0.01	0.05
Be	I (243.86)	0.003	0.03
Bi	I (298.80)	10	50
C	II (103.40)	—	100
Ca	II (393.37)	0.0001	0.0005
Cd	I (228.80)	0.2	0.3
Cd	II (226.50)	0.07	0.4
Ce	II (418.66)	0.4	2
Co	II (238.89)	0.1	0.4
Cr	II (267.72)	0.08	0.5
Cr	I (357.87)	0.1	1
Cu	I (327.40)	0.06	0.3
Dy	II (353.1)	—	4
Er	I (400.8)	—	1
Eu	II (381.97)	—	0.06
Fe	II (259.94)	0.09	0.2
Ga	I (417.21)	0.6	3
Gd	II (342.25)	0.4	2
Ge	I (265.12)	0.5	2
Hf	II (339.98)	—	10
Hg	I (253.65)	1	50
Ho	II (345.7)	—	3
I	I (206.16)	—	10
In	I (451.1)	—	30
Ir	I (322.1)	—	70
K	I (766.5)	—	30
La	II (408.67)	0.1	0.4
Li	I (670.78)	0.02	0.3
Lu	I (451.9)	—	8
Mg	I (279.55)	0.003	0.01
Mn	II (257.61)	0.01	0.06
Mo	I (386.41)	0.3	0.5
N (NH)	(336.0)	—	100
Na	I (588.99)	0.2	0.1
Nb	II (309.43)	0.2	1
Nd	II (401.22)	0.3	1.5
Ni	I (352.45)	0.2	2

(continued)

TABLE II (*continued*)

Element	Line[b] (nm)	Detection limits (ng/ml)	
		Ultrasonic nebulizer with desolvation	Pneumatic nebulizer
Os	I (290.9)	—	6
P	I (253.56)	15	30
Pb	II (220.35)	1	15
Pb	I (283.31)	2	10
Pd	I (360.95)	2	6
Pd	II (248.89)	2	6
Pr	II (422.5)	—	10
Pt	I (265.95)	0.9	2
Re	II (209.2)	—	25
Rh		—	3
Ru	I (379.8)	—	60
S	I (182.03)	—	30
Sb	I (217.5)	—	15
Sc	II (361.3)	—	0.4
Se	I (196.03)	1	15
Si	I (251.6)	—	2
Sm	II (359.26)	—	0.5
Sn	I (190.0)	—	6
Sr	II (407.77)	0.003	0.02
Ta	II (296.51)	5	70
Tb	II (350.92)	0.1	0.5
Te	I (238.58)	—	15
Th	II (401.91)	—	3
Ti	II (334.94)	0.03	0.2
Tl	I (377.6)	—	75
Tm	II (346.22)	—	0.15
U	I (385.96)	1.5	8
V	II (309.31)	0.06	0.2
W	II (276.43)	0.8	5
Y	II (371.03)	0.04	0.08
Yb	II (369.42)	0.02	0.1
Zn	I (213.86)	0.1	0.3
Zr	II (343.82)	0.06	0.3

[a] Taken from a list compiled by P. W. J. M. Boumans and R. M. Barnes, *ICP Inf. Newsl.* **3,** 445 (1978).
[b] I indicates an atom line; II indicates an ion line.

that can be achieved with the technique. Short-term precision or stability is usually measured as the relative standard deviation of 10 consecutive determinations of a standard solution with element concentrations at least 100 times their detection limit. Relative standard deviations of 1% or less are expected. Precision for samples containing high salt levels may be somewhat poorer and as concentrations approach the detection limit, precision is degraded.

Long-term precision or stability is measured in the same way as short-term precision, however, the determinations are evenly spaced over some specified time span, usually several hours. Long-term stability is obviously affected by calibration drift and this figure of merit provides the user with some indication of how frequently recalibration will be necessary. It is not unusual to carry out analyses of 3–4 hr with less than 3% change in calibration but for quality control purposes, calibration should be checked after every 5–10 samples by analyzing a standard solution of known concentration. Although somewhat expensive, a mass flow controller that maintains a constant and accurate flow of argon to the nebulizer, greatly reducing drift and improving long-term stability, can be purchased for roughly $1000 and easily installed in the nebulizer gas line.

For scanning instruments wavelength reproducibility, i.e., the ability of the spectrometer system to locate the desired wavelength, is another measure of instrument performance. Wavelength reproducibility affects both short-term and long-term precision and is reflected in those figures of merit for ICP scanning spectrometers.

Provided the system is equipped with adequate electronics, the linear range for most elements in the ICP extends at least 4 or 5 orders of magnitude beyond the detection limit. In addition, programs are available that enable accurate quantitation using the nonlinear portion of the calibration curve which extends the working range another order of magnitude. Although tables which list the linear range for different elements at different wavelengths are useful guidelines, the observed linear range for an ICP system varies with the nebulizer, torch, and operating conditions used due to variations in aerosol droplet size, flow velocity, etc. Therefore, the linear range of the system for a particular element should be verified prior to sample analysis if high concentrations of that element must be determined.

The most important measure of instrument performance and one that is dependent on all aspects of the ICP system as well as the analytical procedure being used is accuracy. Analysis of certified standard reference materials whose composition approximates the sample matrix of interest is frequently used to assess accuracy. A variety of certified reference materials, including a variety of freeze-dried biological materials, can be obtained from the National Bureau of Standards (NBS) in Gaithersburg, Maryland or the International Atomic Energy Agency (IAEA) in Vienna, Austria. Aqueous elemental reference standards can be requested from the U.S. Environmental Protection Agency (EPA), Environmental Monitoring and Support Laboratory, Cincinnati, Ohio.

If appropriate reference materials are unavailable, a fortified (spiked) sample may be used to calculate the method recovery of the elements being studied. The assumption is made that the added elements are in the same form in which they exist in the sample. Replicate samples should be analyzed to determine the concentrations present initially. The difference between the natural levels and the quantities found in the spiked sample should correspond to the amounts added to the sample and are calculated as percentage recovered.

Because ICP-OES can be such a precise technique, the analyst must be cautious about equating reproducible results with accuracy. The accuracy of the ICP determination step can be verified by diluting the sample and reanalyzing if concentrations are sufficiently high or by quantitating by the method of standard additions. Both of these procedures alert the analyst to the presence of sample matrix effects that have not been properly rectified by the calibration scheme. Indeed for a few analyses of a material that will not be routinely analyzed, standard additions is a quick way to achieve accurate results without lengthy method development provided the standards are added at an appropriate level and that linear range is sufficient. Standard addition quantities from 1/3 to 5 times the amount present in the sample are generally acceptable.

Calibration

Due to the long linear range of the ICP, calibration can be accomplished using a simple 2 point standardization. The low standard is commonly a standard blank. The high standard should be at a concentration at least 100 times the detection limit.

For a multielement standard that approximates a particular sample matrix, it is necessary to assure that the standards used for elements at high levels are not contaminated with elements to be determined at lower levels and that spectral interferences are absent. Rather than risk the possible errors caused by combining standards at different concentration levels, some researchers use separate solutions of elements at high and low levels and rely on selected operating conditions to reduce matrix effects instead of using a matrix matched standard. Fortunately, for most biological analyses, the elements desired at higher levels in the standard (Ca, Mg, Na, and K) are rarely contaminated at significant levels with elements other than alkali and alkaline earths. Purchased solutions of phosphorus must be checked for contamination on an individual basis. Even when the same concentration levels are used for multielement standards, possible interactions must be considered when combining elements. Some combinations of elements or their associated anions can lead to a change in standard concentration when combined because of formation of insoluble salts, etc. Several elements also form oxyanions which may cause problems when mixed with other elements.

Because of errors resulting from contamination, ICP standards must be extremely pure, 99.999% purity is recommended. It is convenient to purchase pure, fairly concentrated (0.1–1%) individual standard solutions and dilute and combine as desired. As purchased, these solutions (usually in dilute nitric or hydrochloric acid) are stable for a number of years. For most elements in proper combinations, acidified standard solutions at concentrations of 1 part per million or more have a shelf life in excess of 1 year when stored in tightly closed Teflon or linear polyethylene bottles. Note that for matrix matched standards, the standard solution need only approximate the relative levels found in the sample matrix, e.g., 100 part per million Ca, Mg, Na, K, and P with 1 part per million Fe, Mn, Cu, etc. for plant analysis. A detailed description of a typical ICP method for the analysis of rat tissues, including calibration information and examples illustrating the precision and accuracy of the technique, is provided by Rader *et al.*[20]

Internal standards are not widely used for biological applications of the ICP since the elements of interest have varying excitation potentials and therefore exhibit different responses to interferences or altered conditions in the ICP. For multielement determinations it would be necessary to add several different internal standards to approximate and ultimately compensate for calibration errors resulting from pronounced differences between the sample and standard matrices or changes (drift) in ICP operating parameters.

[20] J. I. Rader, K. A. Wolnik, C. M. Gaston, E. M. Celesk, J. T. Peeler, M. R. Spivey Fox, and F. L. Fricke, *J. Nutr.* **114,** 1946 (1984).

[17] Inductively Coupled Plasma-Mass Spectrometry

By José A. Olivares

Inductively coupled plasma-mass spectrometry (ICP-MS) is an instrumental technique for elemental and isotopic analysis developed in 1980.[1] Since then, commercial instruments have been introduced by two instru-

[1] R. S. Houk, V. A. Fassel, G. D. Flesch, H. J. Svec, A. L. Gray, and C. E. Taylor, *Anal. Chem.* **52,** 2283 (1980).

ment manufacturers.[2,3] Reports from users of commercial ICP-mass spectrometers show that ICP-MS can be widely applied for the elemental and isotopic analysis of a wide variety of substances, including biological fluids. The technique is not totally free of interferences and requires some amount of sample preparation and careful quality control in order to obtain useful information from the analysis. Nevertheless, ICP-MS is capable of delivering detection limits between 0.01 and 0.1 μg/liter in dilute aqueous–acid solutions, often 10–100 times superior to its counterpart, inductively coupled plasma-atomic emission spectroscopy (ICP-AES). Because ICP-MS relies on a mass spectrometer for separation and detection of the elements, reliable isotopic information can be gathered with precision and accuracy often between 0.1 and 1%.

Several reviews have appeared on ICP-MS.[4-6] The overall scope of this chapter is to provide a description of the technique and its potential pitfalls as it applies to elemental and isotopic analysis of elements generally of interest to the biochemist. A few examples of methods employed in current ICP-MS analysis are included.

Instrumentation

Figure 1 is a schematic diagram of the basic components in commercially available ICP-mass spectrometers. Samples are usually introduced into the ICP in the form of an aerosol generated by nebulizing an aqueous solution of the sample. Solution uptake of 1 ml/min, with nebulization efficiencies of 1–3% for pneumatic nebulizers,[7] and 10% for ultrasonic nebulizers,[8] is common. The aerosol is entrained in a flow of argon gas (~1 liter/min) and carried to the central channel of the ICP. In the atmospheric pressure ICP, the aerosol is subjected to a reaction zone having a gas temperature estimated at about 5000 K, where the aerosol is atomized and ionized. A portion of the central gas channel of the plasma, containing the ionized sample, is allowed to flow into the mass spectrometer high-vacuum chamber via a sampling interface composed of a water-cooled nickel cone with a 0.5- to 1-mm-diameter orifice. The sampled gas is allowed to expand as a supersonic jet in the first vacuum region of the mass spectrometer chamber where the pressure is maintained at ~1 Torr.

[2] VG Plasma Quad, Vg Instruments, Inc., Inorganic Division, 300 Broad Street, Stamford, Connecticut 06901.

[3] ELAN, Sciex, 55 Glencameron Road, No. 202, Thornhill, Ontario, Canada L3T 1P2.

[4] A. R. Date, *Trends Anal. Chem.* **2,** 225 (1983).

[5] D. J. Douglas and R. S. Houk, *Prog. Anal. At. Spectrosc.* **8,** 1 (1985).

[6] R. S. Houk, *Anal. Chem.* **58,** 97A (1986).

[7] B. R. Bear, *M.S. dissertation.* Iowa State University, Ames, Iowa (1983), and V. A. Fassel and B. R. Bear, *Spectrochim. Acta* **41B,** 1089 (1986).

[8] K. W. Olson, W. J. Haas, Jr., and V. A. Fassel, *Anal. Chem.* **49,** 632 (1977).

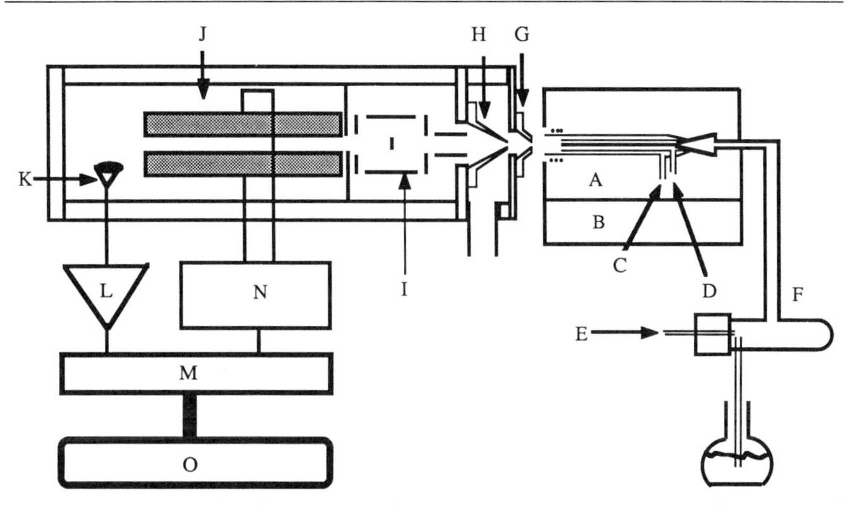

FIG. 1. Schematic of main components in an ICP-mass spectrometer: (A) ICP, torch, and impedance matching network, (B) rf generator, (C) inlet for argon outer gas, (D) inlet for argon auxiliary gas, (E) inlet for argon aerosol gas, (F) pneumatic nebulizer, (G) nickel sampling cone, (H) stainless-steel skimmer, (I) ion focusing lenses, (J) quadrupole mass analyzer, (K) electron multiplier detector, (L) pulse counting equipment, (M) computer/instrument interface, (N) analyzer control electronics, (O) computer.

A portion of the supersonic jet is skimmed by a properly positioned[9] stainless-steel skimmer and is introduced into a second high-vacuum region maintained at or below 5×10^{-4} Torr, depending on the vacuum pump arrangement used by the manufacturer. In this region the ions are collected and focused into a quadrupole mass spectrometer, which may be located in yet another differentially pumped vacuum chamber in order to maintain the pressure at or below 1×10^{-5} Torr for mass analysis and detection.

Signal Considerations

Ionization Efficiency

The degree of ionization of an element introduced into the ICP can be estimated from the Saha equation, if local thermodynamic equilibrium is assumed to exist in the plasma[10]:

$$\frac{M^+}{M^0} = \frac{1(2\pi m_e k T_{ion})^{3/2}}{n_e h^3} \frac{Q^+}{Q^0} e^{-IE/kT} \qquad (1)$$

[9] J. A. Olivares and R. S. Houk, *Anal. Chem.* **57**, 2674 (1985).

[10] P. W. J. M. Boumans, "Theory of Spectrochemical Excitation." Hilger and Watts, London, 1966.

where n_e is the electron density, m_e is the mass of the electron, k is Boltzmann's constant, T_{ion} is the effective temperature of the ion, h is Planck's constant, Q^+ and Q^0 are the partition coefficients of the ion and the neutral species, and IE is the ionization energy of the element. The degree of ionization for most elements is given in Fig. 2, as predicted by Eq. (1). Generally all of the elements with ionization energies below that of Ar (15.755 eV), the main gaseous component of the plasma, exhibit some degree of ionization in the plasma and are observed as M^+ ions. In fact, only He, F, and Ne have a lower degree of ionization than Ar and are not normally detected as ions. However, there is considerable evidence that the plasma is not at local thermodynamic equilibrium,[11] because the degree of ionization experimentally observed in the plasma is generally higher than predicted by the above equation. Similarly, elements with a second ionization energy below 15.755 eV can be observed as M^{2+} ions, but to a much lower degree. To the operator, Fig. 2 should serve as a reference point for initial calculations of relative signals to be obtained in an analysis. Where more accurate values of the ionization efficiency for a particular element are needed, signal comparison with the ion signal of an easily ionized element (~100%) of a nearby mass (or after correction for mass discrimination caused by the instrument) is customary.

Signal Range

Present ICP-MS instruments operate in the ion-counting mode due to the low ion currents (10^4 to 10^5 ions/sec per mg/liter of analyte) observed at the normally low sample operating concentrations (<20 mg/liter). Although commercial counting electronics are available for counting rates up to 100 MHz, electron multiplers used as ion detectors cannot produce discrete pulses at ion currents above 10^7 ions/sec (when used in the ion counting or saturated mode),[9] and are often nonlinear in the 10^6–10^7 ions/sec range. These signals, although of low intensity, are adequate for achieving high sensitivities for trace and ultratrace analysis, and are representative of a very small fraction of the ions created in the ICP which enter the sampling orifice and survive the extraction process.

The lower limit of the signal-operating range in ICP-MS is set by the observed background level. This background, mostly due to photons from excited argon atoms and from intense UV light emission from the ICP, has a minimal component from the electron multiplier dark current (<1 count/sec). The background level in ICP-MS instruments can generally be kept at 10–20 counts/sec with the use of ion lenses, which block the light

[11] R. S. Houk, H. J. Svec, and V. A. Fassel, *Appl. Spectrosc.* **35,** 380 (1981).

$$\left(\frac{M^{+}}{M + M^{+}}\right) \times 100\%$$

$$\left(\frac{M^{+2}}{M + M^{+} + M^{+2}}\right) \times 100\%$$

H 0.1																	He
Li 100	Be 75											B 58	C 5	N 0.1	O 0.1	F 10^{-3}	Ne 10^{-5}
Na 100	Mg 98											Al 98	Si 85	P 33	S 14	Cl 0.9	Ar 0.04
K 100	Ca 1 99	Sc 100	Ti 99	V 99	Cr 98	Mn 95	Fe 96	Co 93	Ni 91	Cu 90	Zn 75	Ga 98	Ge 90	As 52	Se 33	Br 5	Kr 0.6
Rb 100	Sr 4 96	Y 98	Zr 99	Nb 98	Mo 98	Tc	Ru 96	Rh 96	Pd 93	Ag 93	Cd 85	In 99	Sn 96	Sb 78	Te 66	I 29	Xe 8.5
Cs 100	Ba 9 91	La 10 90	Hf 98	Ta 95	W 94	Re 93	Os 78	Ir	Pt 62	Au 51	Hg 38	Tl 100	Pb 0.01 97	Bi 92	Po	At	Rn
Fr	Ra	Ac															

Ce 2 98	Pr 10 90	Nd * 99	Pm	Sm 3 97	Eu * 100	Gd 7 93	Tb * 99	Dy * 100	Ho	Er * 99	Tm 9 91	Yb 8 92	Lu
Th 100	Pa	U	Np	Pu	Am	Cm	Bk	Cf	Es	Fm	Md	No	Lw

FIG. 2. Calculated values for degree of ionization for M^{+} and M^{2+} at T_{ion} = 7500 K, $n_e = 1 \times 10^{15}$ cm^{-3}. Elements marked by asterisks yield significant M^{2+}, but partition functions were not available. Reprinted with permission from Houk.[6] Copyright © 1986 American Chemical Society.

path, and off-axis electron multiplier detectors. Furthermore, counting statistics predicts that the noise level is limited to the square root of the count,[12] but it is generally observed to be slightly larger than this. Thus, depending on the degree of ionization of the element, the abundance of the isotope being observed, and the background noise, the linear signal range in ICP-MS is usually between three and five orders of magnitude.[5]

Precision

As noted above, for a given sample the observed count can be statistically reproduced within the square root of the total count. The precision should, therefore, increase with larger counting times. The relative signal precision observed in ICP-MS is only partially limited by counting statistics. Other signal instabilities can occur from fluctuations in gas flow rates, nebulization efficiencies, plasma tuning, the physical characteristics of the sampling interface, the tuning electronics of the mass analyzer. Because each of these noise sources probably occurs at different frequencies, it is advisable to select the experimental observation times to be used in the analysis only after a careful study of the precision at various observation time periods. Thus, the instrumental complexity in ICP-MS limits the minimum precision. In fact, no one set of conditions has been found by this author, using a prototype ICP-MS instrument, for which day-to-day reproducibility remained constant without the need for recalibration. Internal standardization is, therefore, recommended for determinations requiring high precision for long periods of time. Nevertheless precisions on the order of 1% relative can be achieved with good experimental procedures.

Mass Discrimination

There are several potential sources of mass discrimination in ICP-MS: the ion sampling interface, the electrostatic ion lenses, the quadrupole mass spectrometer, and the detector. Therefore the optimal conditions for determination of a single element are not usually the best conditions for determination of a group of elements, especially if the elements have large differences in their masses. Thus compromise conditions for a group of elements must usually be found with standards for multielemental analysis. In isotope ratio determinations optimal signal and noise conditions can be manipulated to give the best precision, for both samples and stan-

[12] F. M. Harris, G. W. Trott, T. G. Morgan, A. G. Brenton, E. E. Knigston, and J. H. Beynon, *Mass Spectrom. Rev.* **3**, 209 (1984).

TABLE I
DETECTION LIMITS OF SELECTED ELEMENTS

Element	Detection limit (μg/liter)	Ref.[a]
Mg	0.1	1
Al	0.05	2
Ca	1.0	2
V	0.16	2
Mn	0.1	3
Fe	0.7	2
Co	0.05	3
Ni	0.03	2
Cu	0.03	2
Zn	0.2	3
As	0.04	3
Se	0.75	4
Mo	0.04	3
Cd	0.06	3
Hg	0.02	3

[a] References: (1) A. W. Boorn, E. S. K. Quan, and D. J. Douglas, *Annu. Meet. Fed. Anal. Chem. Spectrosc. Soc.* **11,** paper 493 (1984); (2) VG Isotopes, Elemental and Isotopic Analysis Applications Note 02-663, September 1985; (3) A. L. Gray, *Spectrochim. Acta* **41B,** 151 (1986); (4) A. R. Date and A. L. Gray, *Spectrochim. Acta* **40B,** 115 (1985).

dards, in order to correct for mass discrimination and to achieve accuracy in the ratio.

Detection Limits

A compilation of the best detection limits observed for selected elements is given in Table I. Along with its dependency on the degree of ionization of the element, the detection limit is also dependent on the relative abundance of the isotope used for the determination and on the mass of the isotope if expressed as mass per volume. Generally, the best detection limits are observed when the instrument is optimized for an individual element and the analytical solution is a dilute acid, e.g., 1% HNO_3. In multielement analysis an overall set of instrument conditions is used to obtain the best detection limits across the entire spectrum. Therefore, the detection limits are not as good.

Spectral Interferences

Chemical Background Interferences

The chemical background signal observed in ICP-MS is limited to the mass region below 81 Da. In this region elemental and molecular ions are observed due to the materials introduced into the ICP via the aerosol gas line. During normal operation, ions are not observed from erosion of the sampling cone which otherwise indicates imminent interface failure. Thus when nebulizing water, the background is due to Ar, H_2O, and simple molecular species from their reactions in the plasma or the extraction region leading to the mass spectrometer. An extensive list of background ions observed with a number of commonly used solvents is given in Table II, as well as the elemental species with which they potentially interfere. A single-source comprehensive listing of the background observed with 5% concentrations of HNO_3, HCl, and H_2SO_4 has been compiled by Tan and Horlick[13] using the Sciex ELAN instrument. Although the ion background list is extensive, many of the interferences are low in intensity and can be manipulated to some extent by choosing the instrument parameters and solvent or an interference-free isotope can be used for the analysis. Calcium is mot susceptible to interference, but judicious choice of isotope, solvent, and operating parameters can provide detection limits of 1 μg/liter, as seen in Table I.

Isobaric Interferences

Current ICP-MS instruments use quadrupoles as mass spectrometers and are inherently limited in resolution to approximately 1 amu. Therefore, there are a finite number of isobaric interferences, due to the isotopic composition of the elements, which affect analysis. These can be deduced from an isotope table of the elements. For example, a mass spectral resolution of $m/\Delta m = 28{,}500$ is required for the separation of ^{58}Fe from ^{58}Ni at 10% peak valley definition,[14] a task that cannot be performed with current quadrupole technology. Software-implemented corrections for known isobaric interferences can be performed with the commercial instruments, but should be scrutinized carefully. Whenever possible these interferences should be avoided by choosing an interference-free isotope or by extracting the interfering element from the analytical solution.

[13] S. H. Tan and G. Horlick, *Appl. Spectrosc.* **40,** 445 (1986).

[14] J. Eagles, S. J. Fairwaeather-Tait, and R. Self, *Anal. Chem.* **57,** 469 (1985).

TABLE II
BACKGROUND ICP-MS SPECTRA WITH DIFFERENT SOLVENTS

Solvent	Ionic species	Nominal mass	Potential interference with	Ref.[a]
$H_2O + HNO_3$	^{12}C	12	C	1
	^{13}C, ^{12}CH	13	C	1
	^{14}N	14	N	1, 2
	^{16}O	16	O	1, 2
	^{16}OH	17	O	1, 2
	$H_2{}^{16}O$	18	O	1, 2
	$H_3{}^{16}O$	19	F	1
	$H_2{}^{18}O$, H_2DO	20	Ne	1
	$H_3{}^{18}O$	21	Ne	1
	N_2, CO	28	Si	1, 2
	N_2H(?)	29	Si	1
	NO	30	Si	1, 2
	NOH(?)	31	P	1
	O_2	32	S	1, 2
	O_2H	33	S	1
	$^{18}O^{16}O$	34	S	1, 2
	$^{18}O^{16}OH$	35	Cl	1
	^{36}Ar	36	Ar, S	1
	^{36}ArH	37	Cl	1
	^{38}Ar	38	Ar	1
	^{38}ArH	39	K	1, 3
	^{40}Ar	40	Ar, K, Ca	1–3
	^{40}ArH	41	K	1
	^{40}ArD	42	Ca	1
	?	43	Ca	1
	CO_2	44	Ca	1
	$^{36}Ar^{16}O$	52	Cr	2
	$^{40}Ar^{14}N$, $^{38}Ar^{16}O$	54	Cr, Fe	1–3
	$^{40}Ar^{16}O$	56	Fe	1–3
	$^{36}Ar^{36}Ar$	72	Ge	2
	$^{38}Ar^{36}Ar$	74	Ge, Se	2
	$^{36}Ar^{40}Ar$	76	Ge, Se	1–3
	$^{38}Ar^{40}Ar$	78	Se, Kr	1–3
	$^{40}Ar^{40}Ar$	80	Se, Kr	1–3
	$^{40}Ar^{40}ArH$	81	Br	1
H_2O + HCl or H_2O + $HClO_4$	Cl	35,37	Cl	2
	$H^{35}Cl$	36	Ar	2
	$H^{37}Cl$	38	Ar	2
	$^{35}Cl^{16}O$	51	V	2, 3
	$^{35}Cl^{16}OH$	52	Cr	3

(continued)

TABLE II (*continued*)

Solvent	Ionic species	Nominal mass	Potential interference with	Ref.[a]
	$^{37}Cl^{16}O$	53	Cr	2, 3
	$^{37}Cl^{16}OH$	54	Cr, Fe	3
	$^{35}Cl^{16}O_2$	67	Zn	3
	$^{36}Cl^{16}O_2H$	68	Zn	3
	$^{37}Cl^{16}O_2$	69	Ga	3
	$^{37}Cl^{16}O_2H$	70	Zn, Ge	3
	$^{40}Ar^{35}Cl$	75	As	2, 3
	$^{40}Ar^{37}Cl$	77	Se	2, 3
$H_2O + H_2SO_4$	S	32,33,34	S	2
	$^{32}S^{16}O$	48	Ca, Ti	2, 3
	$^{33}S^{16}O$, $^{32}S^{16}OH$	49	Ti	2, 3
	$^{34}S^{16}O$	50	Ti, V, Cr	2, 3
	$^{34}S^{16}OH$	51	V	3
	$^{32}S^{16}O_2$	64	Ni, Zn	2, 3
	$^{32}S^{16}O_2H$, $^{33}S^{16}O_2$	65	Cu	2, 3
	$^{34}S^{16}O_2$	66	Zn	2, 3
	$^{34}S^{16}O_2H$	67	Zn	3
Organic	$^{12}C_2$	24	Mg	4
	$^{13}C_2$	25	Mg	4
	^{12}CO	28	Si	4
	^{13}CO	29	Si	4
	$^{12}CO_2$	44	Ca	4
	$^{13}CO_2$	45	Sc	4
	$^{40}Ar^{12}C$	52	Cr	4
	$^{40}Ar^{13}C$	53	Cr	4

[a] References: (1) D. J. Douglas and R. S. Houk, *Prog. Anal. At. Spectrosc.* **8,** 1 (1985); (2) R. S. Houk, *Anal. Chem.* **58,** 97A (1986); (3) J. W. McLaren, D. Beauchemin, A. P. Mykytiuk, and S. S. Berman, "Applications of Inductively Coupled Plasma/Mass Spectrometry." The Spectroscopy Society of Canada Workshop, Toronto, October 1985; (4) VG Isotopes, Elemental and Isotopic Analysis Applications Note 02-663, September 1985.

Molecular Interferences

Molecular interferences due to the matrix composition introduced into the plasma (discussed above), the formation of metal oxides (MO^+), and metal hydroxides (MOH^+) are also observed in ICP-MS. Elements with M–O dissociation energies of 4–9 eV are believed to survive the ICP to some extent.[5] These molecular species may also be formed in the extraction process preceding mass spectrometry. The intensity of MO^+ and MOH^+ species, relative to the M^+ signal, can be altered by careful choice of plasma operating conditions, e.g., aerosol gas flow rate and plasma forward power.[9,15] Since such parameters drastically affect the formation of these molecular species, it is quite conceivable that their formation is greatly dependent on instrument design. It is therefore appropriate for the analyst to identify possible sources of oxide interferences prior to the analysis, to remove the elements in question, or to minimize their oxide contribution to the analytical signal. A comprehensive study of the formation of MO^+ and MOH^+ species has been carried out by Vaughan and Horlick.[16] Table III lists possible oxide and hdyroxide interferences for selected elements.

Ionization Interferences

As discussed above, the formation of doubly charged ions is expected and observed in ICP-MS. Because of their double charge, M^{2+} ions are observed at half the mass of their M^+ ion counterparts. Like molecular interferences, the intensity of M^{2+} relative to M^+ ions is dependent on plasma forward power and aerosol gas flow rate, and should be minimized when a potential interference exists.[9,15] Vaughan and Horlick[16] have compiled a listing of elements for which M^{2+} species are observed (see Table III).

Nonspectral Interferences

Matrix Effects

ICP-MS signals have been found to be affected severely by high concentrations of dissolved solids, at least to a greater extent than its counterpart ICP-AES. It is generally agreed that the total solid content of the analytical solutions should be kept below 0.2% for continuous nebulization.[6] High levels of concomitant salts in the analyte solution can affect the total electron density in the ICP, especially when elements of low ionization energy are involved,[17] although instrumental conditions may determine the severity of this effect.[18] The analytical signal can be decreased below that observed under conditions of low concomitant salt

[15] G. Horlick, S. H. Tan, M. A. Vaughan, and C. A. Rose, *Spectrochim. Acta* **40B,** 1555 (1985).
[16] M. A. Vaughan and G. Horlick, *Appl. Spectroc.* **40,** 434 (1986).
[17] J. A. Olivares and R. S. Houk, *Anal. Chem.* **58,** 20 (1986).
[18] D. C. Gregoire, *Geol. Surv. Can.,* **86-1B,** 39 (1986).

TABLE III
OXIDE, HYDROXIDE, AND DOUBLY CHARGED INTERFERENCES FOR SELECTED ELEMENTS[a]

Mass	Elements	Interferences
24	Mg	$^{48}Ca^{+}$
40	Ca	$^{24}Mg^{16}O$
42	Ca	$^{26}Mg^{16}O$, $^{84}Sr^{2+}$
43	Ca	$^{86}Sr^{2+}$, $^{87}Sr^{2+}$
44	Ca	$^{87}Sr^{2+}$, $^{88}Sr^{2+}$
56	Fe	$^{40}Ca^{16}O$
58	Ni	$^{42}Ca^{16}O$
59	Co	$^{43}Ca^{16}O$
60	Ni	$^{44}Ca^{16}O$
61	Ni	$^{45}Sc^{16}O$
62	Ni	$^{46}Ca^{16}O$, $^{46}Ti^{16}O$
63	Cu	$^{47}Ti^{16}O$
64	Zn, Ni	$^{48}Ca^{16}O$, $^{48}Ti^{16}O$
65	Cu	$^{49}Ti^{16}O$, $^{130}Ba^{2+}$
66	Zn	$^{50}Ti^{16}O$, $^{50}V^{16}O$, $^{50}Cr^{16}O$, $^{132}Ba^{2+}$
67	Zn	$^{51}V^{16}O$, $^{134}Ba^{2+}$, $^{135}Ba^{2+}$
68	Zn	$^{52}Cr^{16}O$, $^{135}Ba^{2+}$, $^{136}Ba^{2+}$ $^{137}Ba^{2+}$, $^{136}Ce^{2+}$
75	As	$^{149}Sm^{2+}$, $^{150}Sm^{2+}$, $^{150}Nd^{2+}$, $^{151}Eu^{2+}$
76	Se	$^{151}Eu^{2+}$, $^{152}Sm^{2+}$, $^{152}Gd^{2+}$, $^{153}Eu^{2+}$
77	Se	$^{153}Eu^{2+}$, $^{154}Sm^{2+}$, $^{154}Gd^{2+}$, $^{155}Gd^{2+}$
78	Se	$^{155}Gd^{2+}$, $^{156}Gd^{2+}$, $^{156}Dy^{2+}$, $^{157}Gd^{2+}$
80	Se	$^{159}Tb^{2+}$, $^{160}Gd^{2+}$, $^{160}Dy^{2+}$, $^{161}Dy^{2+}$
110	Cd	$^{94}Zr^{16}O$, $^{96}Mo^{16}O$
111	Cd	$^{95}Mo^{16}O$
112	Cd	$^{96}Zr^{16}O$, $^{96}Mo^{16}O$
113	Cd	$^{97}Mo^{16}O$
114	Cd	$^{98}Mo^{16}O$
198	Hg	$^{182}W^{16}O$
199	Hg	$^{183}W^{16}O$
200	Hg	$^{184}W^{16}O$
202	Hg	$^{186}W^{16}O$

[a] From Vaughan and Horlick.[16]

concentration, although a slight increase in the boron analyte signal with increase in concomitant phosphorus has been reported.[18] Possible analyte signal suppression can also be confirmed by a decrease in the intensity of the observed argon ion signal.[17] Matrix matching seems to correct some of these problems. By resorting to internal standardization, keeping the matrix interfering element at low concentrations (~50 mg/liter), and proper adjusting of the aerosol gas flow rate (even to the extreme of using an ultrasonic nebulizer, rather than the more common pneumatic type), successful analyses can be carried out at the 10 μg/liter level for trace rare earth elements in a uranium matrix.[19]

The matrix may also play a critical role in instrument precision. High salt concentrations introduced into an ICP have caused fluctuations in the plasma tuning circuitry, suggesting physical changes in the electrical composition of the plasma.[17] Undoubtedly, if these plasma changes cannot be compensated for properly by the tuning circuitry within the time frame of data acquisition, the result will be an additional source of noise imposed on the measured signal.

Methods

Isotope Dilution vs Standard Addition Determinations

Because of the capability of providing reliable elemental and isotopic information, determinations can be carried out by both isotope dilution and standard additions. Both methods have been used successfully for a number of determinations with ICP-MS, and have recently been compared quantitatively by McLaren *et al.*[20]

The use of isotope dilution requires at least two stable isotopes of the element to be determined, along with the commercial availability of one of the isotopes in a highly enriched form. A spike of the enriched isotope is added to an aliquot of the sample in a concentration approximately equal to that of the analyte in the sample, and allowed to equilibrate. The enriched sample is introduced into the ICP-mass spectrometer after treatment to remove any unwanted interference or to preconcentrate the sample. A ratio of the reference isotope to the enriched isotope is taken from the net signals (after background subtraction with an appropriate blank) and the concentration of the analyte determined from the following formula:

$$C = \frac{M_s K(A_s - B_s R)}{V(BR - A)} \tag{2}$$

where C is the analyte concentration in the sample in micrograms per liter, M_s is the mass of the stable isotope in micrograms, V is the volume of the sample in liters, A is the natural abundance of the reference isotope,

[19] M. D. Palmieri, J. S. Fritz, J. J. Thompson, and R. S. Houk, *Anal. Chim. Acta* **184,** 187 (1986).

[20] J. W. McLaren, A. P. Mykytiuk, S. N. Willie, and S. S. Berman, *Anal. Chem.* **57,** 2907 (1985).

B is the natural abundance of the spike isotope, A_s is the abundance of the reference isotope in the spike, B_s is the abundance of the spike isotope in the spike, K is the ratio of the atomic weight of the element to the spike isotope mass, and R is the measured isotope ratio after spike addition.

In the standard addition method three aliquots of the sample are prepared: an unspiked sample and two spiked samples with approximately two and four times the concentrations of the elements to be determined. After any required sample preparation (e.g., extraction), the signals at the most abundant and most interference-free isotope of each of the elements to be determined are recorded and an appropriate blank subtracted. The analyte concentration is determined by first-order linear regression.

It is clear that the latter procedure is the most time consuming, but is the procedure of choice if monoisotopic elements are to be determined or if there are severe interferences which cannot be appropriately corrected. Mclaren *et al.*[20] showed excellent agreement in results obtained, with both methods, with accepted values in the preconcentration of Mn, Co, Ni, Cu, Zn, Cd, Pd, and Cr (0.04–2 μg/liter range, with 1–10% relative standard deviations) in seawater on silica-immobilized 8-hydroxyquinoline. Mn and Co were determined by standard additions alone. Total analysis time (excluding sample preparation) for the isotope dilution method was approximately 5 min for 10 repetitions, thus demonstrating high sample throughput. Higher precision in the determinations can, of course, be achieved at the expense of analysis time.

Isotope Ratios

Bioavailability studies on human subjects should undergo much improvement with the application of ICP-MS, as stable isotopes are substituted for the more hazardous radioisotopes used in earlier studies. Furthermore, the convenient form of sample introduction into the ICP-mass spectrometer makes alternative methods based on fast atom bombardment-mass spectrometry[14] and thermal ionization-mass spectrometry[21] less desirable unless any possible increase in precision is justifiable.

Serfass *et al.*[22] have demonstrated the use of ICP-MS to determine iron enrichment in erythrocytes of infants. ^{57}Fe and ^{58}Fe make good tracers for incorporation studies because of their low natural abundance (2.19 and 0.33 atom%, respectively). Furthermore, ^{57}Fe has no natural isobaric interferences and is available in enriched form at a much lower price than the ^{58}Fe isotope. ^{58}Fe does not have these qualities, but is well suited for

[21] J. R. Turnland, M. C. Michel, W. R. Keyes, J. C. King, and S. Margen, *Am. J. Clin. Nutr.* **35,** 1033 (1982).

[22] R. E. Serfass, J. A. Olivares, R. S. Houk, L. S. Ostedgaard, and S. J. Fomon, *Am. J. Clin. Nutr.* **41,** 829 (1985).

TABLE IV
$^{57}Fe/^{54}Fe$ RATIOS[a] IN BLOOD OF INFANTS FED ^{57}Fe-ENRICHED MILK FORMULA[b]

Infant number	Age (days) 84	112	140	168	196
1	0.376	0.375	0.423	0.433	0.428
2	0.377	0.380	0.461	0.446	0.450

[a] Natural abundance ratio = 0.3763.
[b] SRM 1577, 0.3753 (expected: 0.3763); SRM (spiked), 0.4918 (expected: 0.4932).

these studies because of its lower abundance. Iron isotope ratios were taken by ICP-MS after the red blood cell samples were digested with hot nitric and perchloric acids and brought up to volume with 0.1% nitric acid. The blood samples were taken at 28-day intervals from babies who were fed iron-fortified whey-adjusted formula from 84 to 111 days of age, low-iron whey-adjusted formula from 112 to 125 days of age, tracer administration on day 126, and iron-fortified whey-adjusted formula until an age of 196 days. The tracer was administered to two babies using 12 mg of ^{57}Fe, as $FeSO_4 \cdot 7H_2O$, per quart of formula on days 126 and 127; 1.44 mg of ^{58}Fe, as $FeSO_4 \cdot 7H_2O$, in 5 ml aqueous solution containing 75 mg of ascorbic acid between feedings on day 126; and as in the second case above, but doses on day 126 and 127. The results are shown on Table IV for $^{57}Fe/^{54}Fe$ ratios and on Table V for $^{58}Fe/^{57}Fe$ ratios. The ratios were

TABLE V
$^{58}Fe/^{57}Fe$[a] RATIOS IN BLOOD OF INFANTS FED ^{58}Fe-ENRICHED MILK FORMULA[b]

Infant number	Age (days) 84	112	140	168	196
One dose					
3	0.153	0.148	0.191	0.192	0.183
4	0.156	0.145	0.201	0.196	0.194
Two doses					
5	0.151	0.150	0.294	0.292	0.295
6	0.154	0.149	0.273	0.264	0.256

[a] Natural abundance ratio = 0.1507.
[b] SRM 1577, 0.145 (expected: 0.1507).

taken after monitoring ~15 mg/liter solutions for 100 msec on each isotope for 50 sequences, with appropriate blank subtraction. Only 5–10 μg of total iron was consumed in each determination. The relative standard deviations for the $^{57}Fe/^{54}Fe$ data range from 0.4 to 1.8% and 4 to 5% for the preenriched $^{58}Fe/^{57}Fe$ samples. The latter precisions are not as good due to the lower abundance of the ^{58}Fe isotope, corresponding to ~50 μg ^{58}Fe/liter in these analyses. It is clear from both sets of data that a plateau was reached and maintained between days 140 and 196 of life. These precisions are adequate for iron enrichment measurements simply by using ^{57}Fe as the tracer; in special cases the use of ^{58}Fe may be justified. These $^{58}Fe/^{57}Fe$ data have been independently corroborated.[23] Ting and Janghorbani have also described a procedure for isotopic analysis of iron in fecal matter employing enriched ^{57}Fe with isotope dilution analysis.[24]

Zinc bioavailability has also been explored by Serfass *et al.*[25] using ICP-MS. The ^{70}Zn isotopic enrichment of cows' blood plasma and milk was measured after extraction of the Zn with 0.04% diethylammonium diethyl dithiocarbamate in carbon tetrachloride from dissolved samples of ashed lyophilized solids in dilute HCl, pH 2.5–3.0, followed by reextraction into 1.2 *N* HCl. The results are shown in Table VI for the bovine plasma zinc enrichment, where intravenous injections of ^{70}Zn were made after 0, 4, and 8 hr. No corrections for mass discrimination were necessary since the natural zinc standard and the spiked standard are very close to the expected values. The zinc enrichment after each injection, along with its decline, can be followed by monitoring the $^{70}Zn/^{67}Zn$ ratio. In the milk samples collected, the zinc decline was much more prolonged after peaking around 12 hr after the first isotopically enriched zinc injection.[26] Similar results have been observed using ^{67}Zn as the enrichment isotope referenced to ^{64}Zn or to ^{68}Zn.

For human bioavilability studies the above zinc extraction procedures were used successfully (~98% extraction) with human fecal and urinary samples, as well as infant formulas. Serfass *et al.*[27] prepared infant formulas with enriched ^{67}Zn skim milk solids from a cow and added ^{70}Zn as the chloride to the formula. Isotope ratios of infant stools were determined

[23] M. Janghorbani, B. T. G. Ting, and S. J. Fomon, *Am. J. Hematol.* **21,** 277 (1986).

[24] B. T. G. Ting and M. Janghorbani, *Anal. Chem.* **58,** 1334 (1986).

[25] R. E. Serfass, G. L. Lindberg, and J. A. Olivares, *Fed. Proc., Fed. Am. Soc. Exp. Biol.* **44,** 933 (1985).

[26] R. E. Serfass, G. L. Lindberg, J. A. Olivares, and R. S. Hank, *Proc. Soc. Exp. Biol. Med.* **186,** 113 (1987).

[27] R. E. Serfass, J. J. Thompson, and E. E. Ziegler, *Fed. Proc., Fed. Am. Soc. Exp. Biol.* **45,** 588 (1986).

TABLE VI

$^{70}Zn/^{67}Zn$ RATIOS IN BOVINE PLASMA AFTER INTRAVENOUS ADMINISTRATION OF ENRICHED ^{70}Zn[a]

Time (hr)	Plasma Zn (μg/dl)	$^{70}Zn/^{67}Zn$ ratio	RSD (%)
Standard	—	0.175	1.5
Preinjection	91	0.1743	0.2
0.25	114	2.74	1.5
0.50	111	1.79	0.7
1.0	111	1.142	0.4
2.0	104	0.814	0.8
4.0	104	0.584	0.7
5.0	119	1.54	1.3
6.0	116	1.09	0.9
8.0	109	0.861	0.7
9.0	129	2.09	2.6
10.0	121	1.52	2.2
12.0	116	1.18	1.2
24.0	106	0.574	1.0
Enriched standard[b]		0.281	0.8

[a] 5 mg ^{70}Zn at time = 0, 4.02, and 8.02 into the jugular vein.
[b] Enriched standard calculated value = 0.287.

for the intrinsic label (^{67}Zn) and for the extrinsic label (^{70}Zn) referenced to ^{68}Zn. The average precisions for the $^{67}Zn/^{68}Zn$ and $^{70}Zn/^{68}Zn$ ratios were 1.5 and 1.9% RSD, respectively, using a Sciex ELAN ICP-MS instrument. The results clearly showed that the extrinsic and intrinsic labels follow similar appearance patterns in the infant stools.

Conclusion

The use of ICP-MS in biological problems is still in its infancy. But methodologies commonly used in elemental and isotopic analysis of inorganic problems with ICP-MS can be used to solve similar biological problems. The versatility offered by direct solution introduction into the ICP-mass spectrometer, the large sample throughput, and the high precision of the method provide the analyst with the necessary tools to gather data that were prohibitively difficult to acquire previously by other means. The pitfalls that can be encountered in the use of ICP-MS for elemental and isotopic analysis are slowly being recorded in the literature and should provide a well founded base for the development of sound methods of analysis.

Acknowledgments

Grateful appreciation is extended to the many ICP-MS laboratories that sent their latest work to me, much of which did not get included in this chapter because of the biochemical nature and methodological approach of the volume. I am indebted to Dr. Robert S. Serfass for releasing much of the information in the last section for publication here, and for his review, and to Dr. Dean Matson for proofreading the manuscript. I also thank Dr. Richard D. Smith and Battelle Northwest Laboratories for supporting this work.

[18] Atomic Fluorescence Spectrometry

By ROBERT G. MICHEL

The basic experimental arrangement for atomic fluorescence spectrometry (AFS) is shown in Fig. 1a. A light source that emits the characteristic spectral line radiation of the element of interest irradiates an atom cell which can be any one of a number of devices including either a laboratory flame or a plasma. The function of the atom cell is to completely break down any sample into its constituent atoms. The atoms of interest absorb radiation from the light source and then radiation is reemitted as fluorescence. The fluorescence can be detected, at right angles to the irradiation direction of the light source (Fig. 1a), by use of a conventional photomultiplier tube and a low-resolution monochromator (dispersive AFS) or the monochromator can be omitted altogether (nondispersive AFS) in which case the fluorescence would normally pass through a filter onto the photomultiplier tube. The omission of the monochromator allows more fluorescence to be gathered and sensitivity is improved. Figure 1b shows two typical ways in which an atom can fluoresce. It can be seen that it is possible to observe fluorescence at a variety of wavelengths depending on the transitions involved. The most sensitive transition is usually chosen.

The primary advantages of atomic fluorescence are its high sensitivity for the determination of metals in samples, the capability of measuring the concentration of up to 12 elements simultaneously (in one commercial instrument) with relative freedom from spectral interference effects, and the long linear range of the calibration curves compared to atomic absorption. The multielement capability of atomic fluorescence is possible because several light sources can be arranged around a flame or plasma and hence the fluorescence of several metals can be excited and detected virtually simultaneously. High sensitivity for AFS can be realized by use of a high-intensity light source because the fluorescence signal is propor-

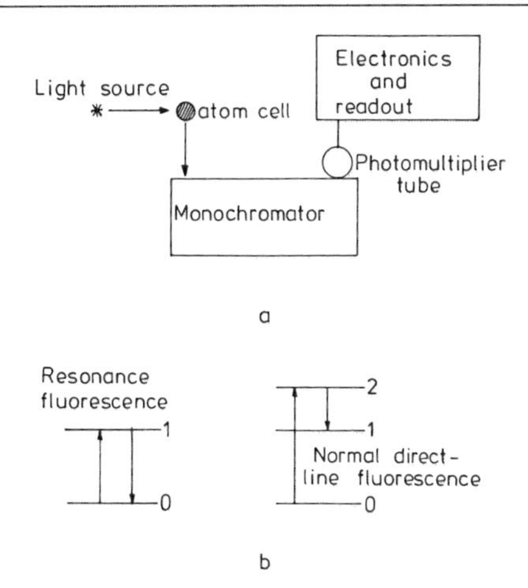

FIG. 1. (a) Experimental set-up for atomic fluorescence spectrometry. (b) Fluorescence of an atom.

tional to the intensity of the light source. Unfortunately, this is currently a disadvantage of the technique in that high-intensity conventional light sources are not easily available for a wide range of metals. Therefore, it is not possible in practice to realize AFS detection limits, for all metals, that are significantly better than competitive techniques such as flame atomic absorption (flame AA) or inductively coupled plasma (ICP)-optical emission spectrometry. Very high sensitivity can only be realized by laser-excited AFS in flames and graphite furnaces, but these instruments are expensive and not yet sufficiently well developed to be used for routine determinations of the concentrations of metals in samples. Good sensitivity in flames is available for a number of elements by use of high-intensity hollow cathode lamps in conjunction with nondispersive detection of the atomic fluorescence. Unfortunately, both nondispersive and dispersive AFS in flames are complicated by the necessity for background correction instrumentation which is difficult to set up and, as yet, none of the various correction methods has been demonstrated to be both reliable and applicable to a wide range of analyses.

Two practical approaches to atomic fluorescence are described here. The first concerns hollow cathode lamp excited AFS in the ICP. The second concerns the determination of mercury by cold vapor atomic fluorescence spectrometry. Both approaches possess some advantages over competing techniques, particularly ICP-optical emission spectrometry (ICP-OES) and cold vapor AA, respectively.

Hollow cathode lamp (HCL) excited AFS in an ICP in its commercial form is a nondispersive instrument that allows the determination of up to 12 elements simultaneously in a sample. It is affected by far fewer spectral interferences than ICP-OES but some physicochemical interferences are possible. For some groups of elements, sensitivity is comparable to ICP-OES. For the refractory elements, the sensitivity is worse than ICP-OES. There is not alot of information in the literature on this technique even though a commercial instrument is available. Hence, the range of its applicability has not been demonstrated.

The determination of mercury in many samples by use of cold vapor AFS is usually more sensitive than AA. The method is practical to set up because it is possible to use a conventional atomic absorption instrument, and the available mercury light sources are intense enough to realize significant gains in sensitivity over the atomic absorption method. In addition, the cold vapor technique is not affected by background problems. This discussion is limited to the cold vapor determination of mercury because the determination of other metals in samples is almost always limited by a combination of both light source and background correction problems and the lack of availability of commercial instrumentation. The determination of mercury by atomic fluorescence has been shown to be simple to use and has been used widely enough to justify its inclusion here as an accepted method.

Theory

The fluorescence signal size (B_F) is given by the following expression:

$$B_F = KbAn_0YI_0S$$

where K is a constant. The ratio of the amount of light absorbed to that emitted is called the quantum efficiency (Y) which is ideally equal to one. For analytically useful transitions in flames and plasma Y is usually at least 0.1. A is the Einstein transition probability, I_0 is the intensity of the light source integrated over its whole linewidth of emission, and S is a geometrical factor which accounts for the amount of light that is collected from the atom cell and detected by the detection system. It is not possible to collect all the fluorescence because it is emitted isotropically. The path length of the atom cell is b, and n_0 is the number of atoms in the ground state which is proportional to the concentration of atoms being measured.

The main feature of the above equation is that the signal size is directly proportional to both the light source intensity, I_0, and the atom concentration. Calibration curves for AFS with HCL excitation are linear with a relative slope of 1 (on a logarithmic plot) at low concentrations and bend back over toward the concentration axis with a limiting slope of -0.5 at

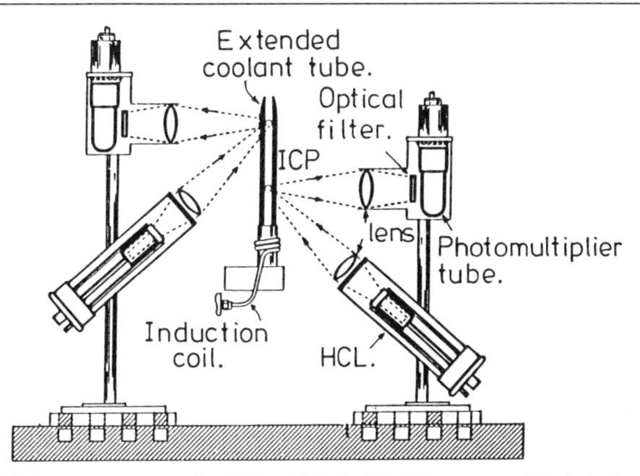

FIG. 2. Multielement system for HCL-ICP-AFS. (Courtesy of Baird Corporation.)

high concentrations. The curvature at high concentration is caused by self-absorption in the atom cell. The range of concentrations over which the calibration curve is linear can be about four orders of magnitude. This is about one to two orders greater than atomic absorption and it makes dilution steps a little less critical during method development.

Hollow Cathode Lamp Excited-Inductively Coupled Plasma-Atomic Fluorescence (HCL-ICP-AFS)

Instrumentation

HCL-ICP-AFS instrumentation is commercially available (Baird Corporation, Bedford, MA, is the only vendor). The main features of the instrument are highlighted here, together with their implications for sensitive and selective determination of the elements in samples.

Light Source. The multielement capability of AFS is realized in the Baird instrument by use of several light sources, as illustrated for two hollow cathode lamps (HCLs) in Fig. 2 where each HCL is shown pointed upward at an angle into the ICP. The HCLs used for AFS are special high intensity versions of the ones that are usually used for atomic absorption and up to 12 of them can be arranged in a circle around the atom cell.

Detection System. Each HCL has associated with it a photomultiplier tube detector which is shown placed vertically above each HCL in Fig. 2. In front of each photomultiplier tube is a filter. The filter allows a range of wavelengths to pass through it that includes the chosen atomic fluorescence wavelength of the element excited by the HCL. Spectral resolution is provided by the specificity of the spectral lines in the light source for the element of interest. A low-resolution monochromator can and often is used for AFS but filters pass more total light onto the photomultiplier tube and still provide sufficient resolution (in the range 2–10 nm) to exclude much of the background from the plasma. The wavelength that is used for the AFS measurement is chosen by the manufacturer to give the best sensitivity and is not normally variable by the operator.

Inductively Coupled Plasma. In ICP-OES the emission from most metals is detected from a region low down in the plasma just above the induction coil. This is a high background region but the optimum sensitivity is achieved there because the signals are so large. Many of the most sensitive transitions for OES are from the ions of the elements because of the high degree of ionization in this plasma. These same transitions cannot be used for HCL-ICP-AFS because HCLs predominantly emit atom lines. However, atoms can be found higher in the plasma where there is not as much energy for ionization. AFS measurements are therefore made there. The actual region used varies in detail from element to element depending upon which region gives the best sensitivity (signal-to-noise ratio). Two different regions of irradiation are indicated for the two light sources illustrated in Fig. 2. The use of these higher regions of the plasma causes some of the elements that form refractory oxides to give relatively poor sensitivity because of the lower temperature (3200–3800 K) high up in the ICP. This problem can be alleviated in various ways as discussed later, but there is an intrinsic problem[1] with the sensitivity for the refractory elements because, for many elements, it is not possible to use an observation height which provides both enough energy to dissociate refractory compounds and to simultaneously avoid the loss of atoms through ionization. The greatest problem occurs for those elements which form both refractory compounds and have low ionization potentials such as barium and strontium.

The ICP that is used for AFS is different from the OES-ICP. It has one of its silica tubes (the outer coolant tube of Fig. 2) extended upward, by about 60 mm, around the plasma in order to retard air entrainment and confine the atoms in the plasma. This results in a taller plasma discharge throughout which is distributed the atoms of interest for AFS measurements. In addition, it has recently been shown[2] that if the central tube of the ICP is made of graphite then a relatively low radio-frequency (rf) power is required to dissociate the oxides of some of the refractory oxide elements.

[1] R. J. Krupa, G. L. Long and J. D. Winefordner, *Spectrochim. Acta* **40B,** 1485 (1985).

[2] D. R. Demers, *Spectrochim. Acta* **40B,** 93 (1985).

TABLE I
HCL-ICP-AFS INSTRUMENT OPERATING CONDITIONS[a]

Parameter	Working range	Nominal value
rf power	400–800 W	600 W
Observation height	80–140 mm	125 mm (easily dissociated elements); 90 mm (refractory elements)
Gas flow rates		
Coolant (argon)	8–12 liters min^{-1}	10 liters min^{-1}
Carrier (argon)	1.2–2.5 liters min^{-1}	2 liters min^{-1}
Propane	0–60 ml min^{-1}	None (easily dissociated elements); 10 ml min^{-1} (alkaline earths), Al, Cr; 50 ml min^{-1} (refractory elements)
Sample uptake rate	0.7–2 ml min^{-1}	1 ml min^{-1}
Average (dc) HCL current	—	80–100% of rated maximum

[a] From Jansen and Demers.[3]

The plasma gas flows are, in general, lower than those used in the ICP-OES technique as are the rf powers (Table I[2,3]). The use of the lower power and flow reduces the temperature of the plasma, but this is to some extent compensated for by the extended coolant tube. The extension causes a taller ICP discharge, due to less quenching by air entrainment, and the plasma temperature is then maintained in higher parts of the plasma.

Detection Limits

Table II gives a comparison of the detection limits for HCL-ICP-AFS with detection limits for the same elements by flame Aa and ICP-OES. The detection limits for the various metals can be divided into several groups.[3]

Easily Dissociated, Weakly Ionized Elements. This category of elements includes copper, cadmium, lead, nickel, selenium, and silver. A significant atom population is available for excitation because these elements do not ionize strongly in the AFS region of the plasma and their compounds are easily dissociated into atoms. Excitation is favorable because hollow cathode lamps primarily emit characteristic radiation of atoms rather than ions.[1] It turns out that detection limits for this group of

[3] E. B. M. Jansen and D. R. Demers, *Analyst (London)* **110,** 541 (1985).

TABLE II
APPROXIMATE DETECTION LIMITS FOR VARIOUS ATOMIC SPECTROMETRIC TECHNIQUES[a]

Element	Detection limits (ng/ml of solution)		
	Flame atomic absorption	Inductively coupled plasma optical emission	Inductively coupled plasma atomic fluorescence
Aluminum	30	20	20[b]
Antimony	30	30	30
Arsenic	100	50	400
Barium	8	0.5	500
Bismuth	20	2	20
Boron	700	4	500[b]
Cadmium	0.5	4	0.05
Calcium	1	0.1	0.4
Chromium	2	5	6[b]
Cobalt	6	6	2
Copper	1	3	0.5
Gold	6	10	3
Iron	3	3	3
Lead	10	30	30
Lithium	0.5	3	1
Magnesium	0.01	0.1	0.5
Manganese	1	1	1
Mercury	200	20	30
Molybdenum	30	8	30
Nickel	4	10	2
Platinum	40	—	30
Potassium	2	60	2
Selenium	70	50	40
Silicon	60	10	300
Silver	0.9	3	1
Sodium	0.2	20	1
Sulfur	—	50	1000
Tantalum	1000	20	2000[b]
Tellurium	20	40	20
Thallium	9	40	20
Tin	100	30	200[b]
Titanium	50	2	300[b]
Tungsten	1000	40	300[b]
Vanadium	40	5	200[b]
Zinc	0.8	2	0.06

[a] Detection limits were taken from a Perkin-Elmer (Norwalk, CT) manual on analytical methods for AAS (1982) except for the AFS detection limits which were taken, with permission, from various unpublished information on HCL/ICP/AFS from Baird Corporation (Bedford, MA) and from Demers.[2]

[b] Propane added to the ICP to enhance the sensitivity of these elements.

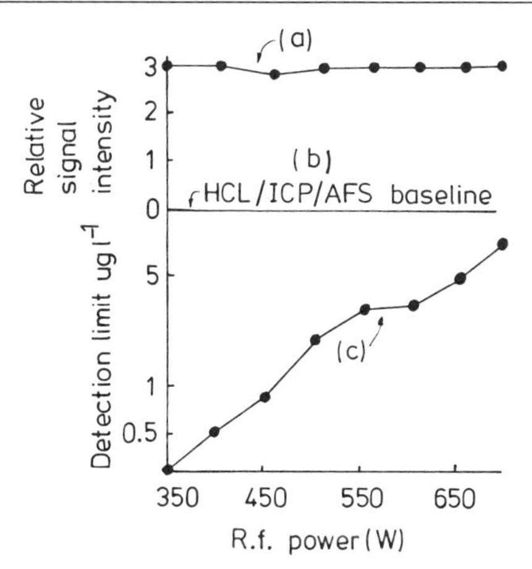

FIG. 3. Variation[3] of copper signal (a) and base line (b) and detection limit (c) with rf power for HCL/ICP/AFS. Observation height 110 mm above the load coil.

elements are comparable to the other techniques listed in Table I, and their signal sizes and detection limits do not change drastically with observation height in the plasma (110–140 mm above the load coil) due to the easily achieved population of atoms. The detection limit is optimum when a relatively low radio-frequency (rf) power of about 400 W is applied to the ICP because the background increases with power[3] (Fig. 3 illustrates this for copper).

Easily Dissociated, Strongly Ionized Elements. The alkali metals fall into this category. Although they are easily dissociated from their compounds they are also easily ionized. Ion lines are not strong in hollow cathode lamps hence the ions cannot be excited and used for analysis. An increase in rf power promotes ionization and hence the detection limits for the alkali metals are best at low rf powers. The change in detection limit with rf power[3] is much larger than for the previous group of elements (Fig. 4). The observation height is not critical although slightly better signals are obtained in the 110–140 mm region above the load coil. The detection limits for these elements are comparable to the other two techniques listed in Table II.

Refractory Elements. The refractory atoms are difficult to dissociate from their oxides because of the relatively low temperature of the AFS plasma region. Hence, one of the main disadvantages of HCL-ICP AFS is the poor sensitivity for the refractory oxide forming elements (e.g., molybdenum, boron, and aluminum) relative to other techniques.

The sensitivity of refractory metal determination can be improved somewhat (Table II) by the addition of a carbon-containing gas, such as propane, into the premix chamber of the sample introduction system.[2,3]

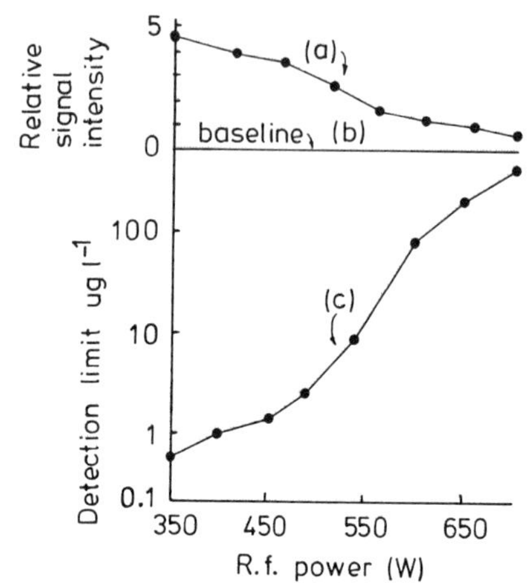

FIG. 4. Variation[3] of potassium signal (a) and base line (b) and detection limit (c) with rf power for HCL/ICP/AFS. Observation height 110 mm above the load coil.

This gas is carried into the ICP along with the nebulized droplets and the argon plasma support gas, and the signals are increased from elements that tend to form refractory compounds. Molecular spectra from carbon-containing species have been observed in the presence of propane which indicates that propane combustion creates carbon species that may prevent the formation of refractory oxides by some form of reduction reaction.

The range of observation height over which sensitive AFS signals for these metals can be observed is a little lower in the plasma (80–100 mm above the load coil) than for the other metals, because a higher temperature is required to dissociate the compounds even when propane is present.

The signals from some (boron, aluminum, beryllium, silicon) but not all (molybdenum, and most others) of the refractory metals are strongly dependent on the rf power to the plasma. The higher the power the more energy there is to break down the oxides and hence detection limits are optimum at high plasma powers (700–850 W). Recent data show[2] that lower rf powers can be used for Al, B, Be, and Si if the central tube of the ICP is made of graphite but no information is available for other refractory oxide-forming elements.

Propane also helps to break down some moderately hard to dissociate elements such as chromium,[2] which can be detected without propane but its detection limit can be improved with propane. Figure 5 shows the variation in signal and detection limit of chromium with and without propane. It can be seen that propane allows lower rf powers to be used for such elements without sacrificing sensitivity.

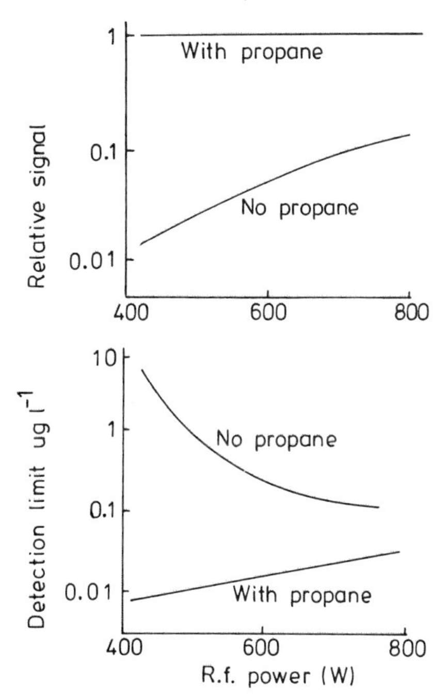

FIG. 5. Variation[2] of chromium signal (top) and detection limit (bottom) with rf power, with and without propane (propane flow 10 ml min^{-1}, fluorescence detected at 357.9 nm).

The optimum range for the propane flow rate is between 5 and 40 ml min^{-1}. Elements which form moderately strong oxides (e.g., Al, Ba, Be, Ca, Cr, Sn, Sr, and Yb) require a propane flow rate at the lower end of the range. Elements which form stronger oxides (B, Si, Ti, V, W) require a flow rate at the higher end of the range (Table I).

The presence of propane causes the ICP to have a triangular green area (C_2, Swan band emission) which extends in size in proportion to the amount of propane.[2] The strong oxide-forming elements are detected within the green region whereas the moderately strong oxide-forming elements are detected above the green region.

The three sets of optimum observation heights, rf powers, and propane flows for the determination of the three groups of elements are summarized in Table I.

Spectral Interferences

Spectral interferences, compared to OES in the ICP, are absent, as would be expected of an AFS or AAS instrument based on hollow cathode lamps, because only the characteristic radiation of the element of interest is available for absorption by the element of interest. In the case of AFS it is the exceptions that prove the rule and only five spectral interferences can be found[4] for direct spectral overlap of the excitation wavelength of one element with that of another. For example, the platinum wavelength at 271.903 nm from the platinum HCL can be absorbed by iron at 271.904 nm. This is a problem because iron fluoresces at so many wavelengths that it is not possible to find an alternative platinum fluorescence wavelength that does not have an iron line within the relatively wide spectral bandwidth of the interference filter. This spectral interference can be removed by use of a small monochromator[4] that has a narrow spectral bandwidth at the primary platinum wavelength of 265.9 nm.

In many types of AFS instrument a spectral interference can be caused by scatter of the incident source radiation off large droplets and undissociated particles in the atom cell which, without background correction, could be mistaken for fluorescence. However, the ICP is so efficient at breaking down droplets and particles, that there have been relatively few scatter signals found in HCL-ICP-AFS.[4,5] Scatter is observed when samples are analyzed which are high in aluminum, calcium, lanthanum, and silicon (at approximately the 1% level in solution), but no scatter has been seen for samples high in sodium, iron, and nickel. The current approach to compensate for such scatter signals is to add to all the standards equivalent concentrations of the species that are causing the scatter signals. This approach is called matrix matching. Silicon can be volatilized as silicon fluoride in order to remove its scatter interference. More sophisticated methods of background correction have not yet been applied to the one commercial version of this instrument.

Physicochemical Interferences

This type of depressive, or solute vaporization, interference has only been seen for the alkaline earth elements in the presence of aluminum, phosphorus, and silicon. The temperature of the AFS region of the ICP is relatively low compared to the OES region, but it is still high enough to remove the depressive interference of phosphate on the determination of calcium if a high enough rf power is used. This is illustrated in Table III where it can be seen that the signal from calcium in a phosphorus-containing wine sample increases with increase in rf power.[3] The calcium concentration agrees with the value obtained by flame atomic absorption only at

[4] D. R. Demers, *Symp. Instrum. Multi-Element Anal.* (1984). Julich, FRG.

[5] D. R. Demers, "The Spectral Line and Background Interference Maze in ICP–AES: A Way out via a New Approach." Paper 123 at the 1981 Pittsburgh Conference, Atlantic City, New Jersey.

TABLE III
INFLUENCE OF rf POWER IN HCL-ICP-AFS ON THE DETERMINATION OF CALCIUM IN A WINE SAMPLE[a,b]

Nominal rf power (W)	Ca (mg/liter)
500	0.45
600	1.09
700	2.02
800	2.05

[a] From Jansen and Demers.[3]
[b] At 2.5 liters min^{-1} carrier gas flow rate and after 5-fold dilution. The concentration in the wine sample as determined by atomic absorption analysis was 1.94 mg/liter.

power levels above 700 W. A decrease in carrier gas flow rate will also achieve the same effect but it results in a decrease in the stability of the plasma. The observation height is not critical for minimization of the phosphate on calcium interference.

Silicon, if it causes a physicochemical interference, can be volatilized as a silicon fluoride in order to remove its effect.

Introduction of Organic Solvents into the ICP

In the ICP, organic solvents burn to produce a green plume which is the same as that formed when propane is used to help in the dissociation of refractory oxides. Unfortunately, organic solvents also form incandescent carbon particles, due to incomplete combustion, which can be seen as a bright yellow luminosity in the ICP. This causes a high background which unacceptably degrades the signal-to-noise ratio of all measurements. The addition of oxygen (at about 300 ml min^{-1}) removes both the yellow and green regions simultaneously but this is disadvantageous because the refractory oxide-forming elements cannot be determined at all due to the lack of reducing conditions. The nonrefractory oxide-forming elements can still be determined with high sensitivity in the presence of oxygen.

It has been found that dioxane forms a green plume in the ICP without the yellow incandescence. This indicates that dioxane is burned more efficiently than other organic solvents which may well be related to the oxygen in the dioxane molecule. No other organic solvents have been found to have this effect, but it allows the refractory oxide-forming elements to be determined in either dioxane, or organic solvents that are diluted with two to three parts of dioxane. In addition, a small amount of oxygen can be bled into the sample introduction system in order to prevent carbon build up at the tips of the various tubes that support the ICP.

Simultaneous Multielement Analyses

In order to carry out a determination of several elements simultaneously it is necessary to ensure three things: first, that the signals from each element are big enough to allow the required sensitivity; second, that all interferences can be compensated for without changing the operating conditions. If these are not possible, then several elements cannot be determined simultaneously. Often operating conditions are used that allow some compromise between the sensitivity of all the elements to be analyzed and the extent of interferences. Third, it is necessary to have a long linear dynamic range of the calibration curves so that elements of fairly widely differing concentrations can be analyzed in one solution, rather than having to dilute the solution for some of the elements. AFS does have calibration curves with about 4 orders of magnitude of linearity so this does not pose a problem, and is similar to ICP-OES in this respect.

The differences between the effects of rf power for the three groups of elements make it a little difficult to find compromise conditions for simultaneous multielement analyses by HCL-ICP-AFS. The main problem is that high rf powers are required for the determination of the refractory elements. The degradation in the detection limits for the alkali metals at such high rf powers could be a factor of 30 (potassium[3]). This is less serious for elements like copper where the degradation in detection limit with increased rf power is only about a factor of four. The detection limits for the alkali metals are relatively good to start with, so this may not be a problem. The compromises necessary for multielement analyses are alleviated by the addition of propane because this allows lower powers to be used for some elements without too much loss in sensitivity. It has been shown[2] that an ICP with a central tube made of graphite reduces the rf power necessary to dissociate the refractory metal oxides, and allows compromise multielement conditions to be obtained more easily.

Neither the extent of physicochemical interferences nor the applicability of ICP-AFS to biological samples have been demonstrated by reports in the literature. Some of the considerations of introducing biological samples into an ICP for OES analyses are likely to be similar for HCL-ICP-AFS except that spectral interferences will not be present for AFS and some less than predictable physicochemical interferences may cause a problem in AFS. Considerations of sample viscosity, amount of sample required etc., which are associated with the very similar sample introduction systems of both AFS and OES instruments, will be virtually identical.

Determination of Mercury by Cold Vapor Atomic Fluorescence Spectrometry

The determination of mercury in samples by atomic absorption and atomic fluorescence spectrometries was reviewed in detail in 1975 by Ure.[6] There are two aspects of the determination of mercury which are of importance. First, the sample is treated in some way in order to convert the mercury into the correct form for introduction into the spectrometer. This form is usually the cold mercury vapor in order to take advantage of the volatility of mercury in its atomic state. The second aspect concerns the measurement itself which is discussed here. The sample treatment aspect is discussed elsewhere in this book and is not repeated here but, in summary, sample pretreatment depends on the original form of the sample. For example, it is common to digest solid samples by use of normal wet digestion procedures, with proper attention to the possible loss of mercury by volatilization. Also, various pyrolysis procedures can be used. Mercury ions in solution can be reduced by use of tin chloride to release the mercury vapor which is separated from the solution by simple bubbling with an inert gas, or by use of another more sophisticated gas separation apparatus. This gas can then be swept directly into the spectrometer for measurement of the mercury signal. Also, the mercury can be collected and concentrated on silver or gold amalgams or wools. These can be subsquently heated to release the mercury vapor so that it can be swept into the spectrometer by an inert gas such as argon. Mercury in gaseous samples can be collected and concentrated on gold-coated filters or wools and then vaporized into the spectrometer. All these procedures apply equally well to both atomic absorption and atomic fluorescence. Only the atomic fluorescence measurement itself is discussed in detail here. In the following it is assumed that the mercury is in the gaseous state and has been entrained into a flowing argon stream for introduction into the atomic fluorescence spectrometer.

Advantages of the Atomic Fluorescence Measurement

Selectivity. West[7] has compared atomic absorption and atomic fluorescence with respect to the relative effect of background molecular absorp-

[6] A. M. Ure, *Anal. Chim. Acta* **76,** 1 (1975).
[7] C. D. West, *Anal. Chem.* **46,** 797 (1974).

tion on the two techniques and demonstrated that the errors in atomic fluorescence are likely to be significantly less. This theoretical analysis did not account for the relative effects of scatter of light source radiation on the two techniques which is an important cause of errors in some forms of atomic fluorescence spectrometry. However, the release of mercury vapor from samples by the techniques summarized above does not introduce significant amounts of particulate matter that can cause scatter in the cold vapor atomic fluorescence atom cell. Accordingly, it is routine to be able to achieve very low background signals in AFS systems that are designed for cold vapor mercury determination without incurring errors caused by scatter or molecular absorption. This selectivity stems primarily from the ease with which mercury can be separated from the sample matrix. Other metals require high temperature atom cells that contain many species, related to both the sample and the atom cell, which can cause high background signals.

Sensitivity. The 253.7 nm mercury transition that is normally used is an efficient fluorescence transition and many different types of mercury light sources can be used to give intense fluorescence signals. This, together with the inherently low background of the cold vapor mercury generation approach, results in sensitivity for mercury which is excellent for many applications.

A primary limitation of all low-level mercury determinations is the size of the blank signal which is set by contamination levels in the sample collection and handling procedures. It is more likely that sensitivity will be improved by careful attention to contamination control than by improvements in instrumentation. Contamination control for trace metal analyses is discussed elsewhere in this book.

Other Advantages. The linear dynamic range of the AFS calibration curves are superior to those of AAS. In addition, the windowless AFS cell described here does not fog due to moisture as does the enclosed silica cell that is often used in AAS.

Determination of Mercury by Use of a Conventional Atomic Absorption Spectrometer for Atomic Fluorescence Spectrometry

The following discussion looks at a method that was published recently by Ebdon and Wilkinson[8] which uses a standard atomic absorption spectrometer and is therefore adaptable for use in most laboratories. This is in preference to custom built instruments such as those of Hutton and Preston[9] or Cavalli and Rossi,[10] although from an analytical point of view

[8] L. Ebdon and J. R. Wilkinson, *Anal. Chim. Acta* **128,** 45 (1981).
[9] R. C. Hutton and B. Preston, *Analyst (London)* **105,** 981 (1980).
[10] P. Cavalli and G. Rossi, *Analyst (London)* **101,** 272 (1976).

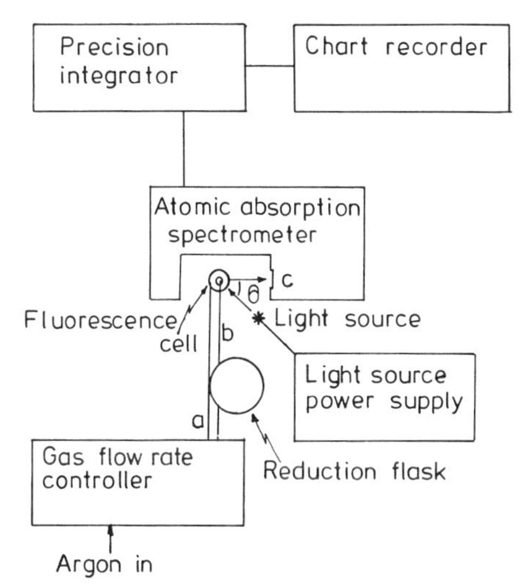

FIG. 6. Block diagram[8] of instrumentation for the determination of mercury by cold vapor atomic fluorescence spectrometry. a and b are gas lines for the argon sheath gas, and the argon for flushing the mercury into the atom cell, respectively. The angle θ is the angle subtended by source, atom cell and slit and can be 45° or 90° (see text).

any one of these instruments is appropriate and similar, and will lead to high sensitivity for the determination of mercury. The system of Hutton and Preston has potential for greater sensitivity because it is nondispersive. However, they did not achieve greater sensitivity than the dispersive systems of the other authors. The reduction unit that was used by Ebdon and Wilkinson[8] was a flask which contained a tin chloride solution through which was bubbled argon after addition of a sample solution. It is not described here as this is only one of several approaches that can be used to generate mercury vapor all of which are described elsewhere in this book.

Atomic Fluorescence Cell and Spectrometer. A block diagram of the instrument of Ebdon and Wilkinson[8] is shown in Fig. 6 and Table IV gives a list of required equipment and operating conditions, together with some suggested alternative equipment where appropriate. The instrument consisted of a flame atomic absorption spectrometer operated in a flame emission mode. The burner/premix chamber apparatus was replaced with the fluorescence cell which is shown, together with the reduction vessel, in Fig. 7. The mercury vapor generated in the reduction vessel was swept by argon through a 6-mm-i.d. Pyrex glass tube and the fluorescence was excited by irradiating the mercury vapor just above the top of the tube as indicated in Fig. 7. Approximately 150 capillary (melting point) tubes were glued around the Pyrex tube and held in place with a shortened Hirsch filter tube of 22 mm i.d. These provided a means of directing a

TABLE IV
EQUIPMENT REQUIRED FOR DETERMINATION OF MERCURY BY COLD VAPOR ATOMIC FLUORESCENCE SPECTROMETRY AND A CONVENTIONAL ATOMIC ABSORPTION SPECTROMETER[a]

Component	Model	Example suppliers	Operating conditions
Spectrometer	Any atomic absorption instrument	Any	Flame emission mode Wavelength 253.7 Bandpass 2.0 nm
Excitation source	Mercury microwave excited electrodeless discharge lamp in a	EDT Research Ltd., London, NW10, England	36 W incident power cooled by an air-flow of about 3 liters min^{-1}
	Broida 3/4 wave cavity, powered by a	From either supplier of generators below	
	Microtron 200 microwave generator or/	Electromedical Supplies, Wantage, Oxfordshire, England	
	Raytheon PGM10 microwave generator	Raytheon microwave div. Waltham, MA	
Alternative sources	Mercury radiofrequency excited electrodeless discharge lamp	Perkin-Elmer Corp., or/ Westinghouse Electric Corp., Horseheads, N.Y.	5 W incident power
	Low pressure mercury discharge lamp	Philips	4–90 W
Gas flow rate controller	Panchromatogram	Pye-Unicam, Cambridge, England	Reference outlet for flushing flask and column 1 outlet for sheath gas. Column 2 outlet closed.
Chart recorder	Any	—	Typically 5 min/in and 10 mV f.s.d.
G.C. integrator	TP503	Honeywell	Typically 1% threshold level, 1 sec response time

[a] Adapted from Ebdon and Wilkinson.[8]

second argon flow to provide a sheath around the observation area in a stiff laminar fashion (Fig. 7). This sheath was similar to an idea investigated by Cavalli and Rossi[10] and mitigated oxygen quenching of the fluorescence caused by air entrainment into the sample vapor stream. (Argon has a small cross-section for quenching of fluorescence compared to air and nitrogen.) It also prevented the loss of sample due to draughts and lateral diffusion of the vapor out of the observation region.

The light source in the paper of Ebdon and Wilkinson was an air-cooled mercury electrodeless discharge lamp powered by a microwave

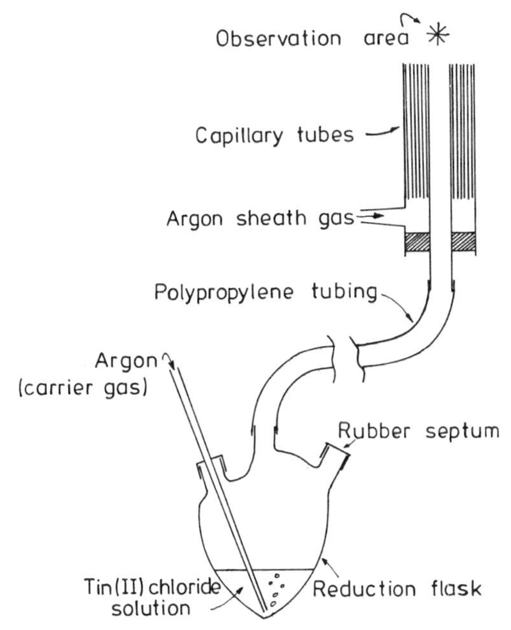

FIG. 7. Fluorescence cell with reduction vessel.[8]

generator because these were available in their laboratory. Alternative light sources could be a Philips low-pressure mercury vapor discharge lamp of low power (Hutton and Preston[9] used a 4-W lamp and Cavelli and Rossi[10] used a 90-W lamp) or a Perkin-Elmer or Westinghouse radio-frequency excited electrodeless discharge lamp. In all cases the manufacturers operating instructions should be followed. For either electrodeless discharge lamp the incident operating power should be optimized by successive measurements at different powers. At each new operating power the lamp should be allowed to stabilize for at least half an hour before a measurement of AFS signal intensity and stabilty is made.

The spectral bandwidth of the spectrometer is not critical. A bandwidth of 2 nm is a suggested starting point. The main approach is to open the spectrometer slit width so that a large signal size is obtained which gives a good signal-to-noise ratio at the concentrations to be measured.

If the fluorescence cell is placed in the position which exactly replaces the flame in the instrument, then the optical system of the spectrometer should have no problem seeing the fluorescence. It will probably be necessary to optimize the signal size through lateral and vertical movements of the fluorescence cell. This should ensure that the spectrometer detection system is observing as many of the fluorescing atoms as possible within the limitations of the spectrometer's particular design characteristics, such as the light-gathering power of its monochromator. During the optimization, the light source should always be at the same height as the entrance optics of the spectrometer. It is usual to allow the light source to irradiate the atoms at a 90° angle to the line of sight of the spectrometer entrance optics, but Ebdon and Wilkinson[8] found that an angle of 45° was optimum. This angle gave the minimum background from stray light detected by the spectrometer. It cannot be said whether or not this 45° angle is either possible or will result in an improved background level in any particular atomic absorption spectrometer that may be used by the reader. Experimentation will be necessary to find the optimum detection geometry.

The transient signals should be integrated rather than their peak height measured. A separate precision integrator is useful, such as the type provided as an accessory for some chromatographic instruments. In some cases, a modern atomic absorption spectrometer may have a method of measuring peak area which is normally for use in conjunction with a hydride generation apparatus. It is also possible that the facilities that are available on AA instruments for integrating furnace AA peaks will be usable for the present application. It may be necessary to amplify the signal from the atomic absorption spectrometer in order to ensure that the integrator is given the proper working voltage range. The specifications of the integrator should be studied to find out whether or not the appropriate signal conditioning is part of the integrator.

Method for Determination of Mercury

Switch on both argon gas flows. Inject 0.5-ml aliquots of either standard or sample solutions through the rubber septum into the tin(II) chloride solution (3 ml of a solution containing 40 g of chloride in 100 ml of 1 *M* HCl) in the reduction flask. Sweep the evolved mercury into the fluorescence cell and record the peak area of the fluorescence signal. Inject each solution at least five times and calculate the mean peak area. Elution time varies as a function of the mass of mercury injected but is typically complete in 30–100 sec.

Gas Flow Rates. Optimization of the flow rates of the two argon flows should result in a plot similar to that of Fig. 8. Ebdon and Wilkinson found an optimum sheath flow rate of 5.0 l min^{-1} and a carrier argon flow of 0.4 a min^{-1}. A lower carrier gas flow rate was not used because the precision and shape of the peaks were degraded even though larger peak areas were obtained at 0.3 l min^{-1}. At the optimized flow rates the signal size was about a factor of 10 greater with the sheath gas flowing than without it flowing.

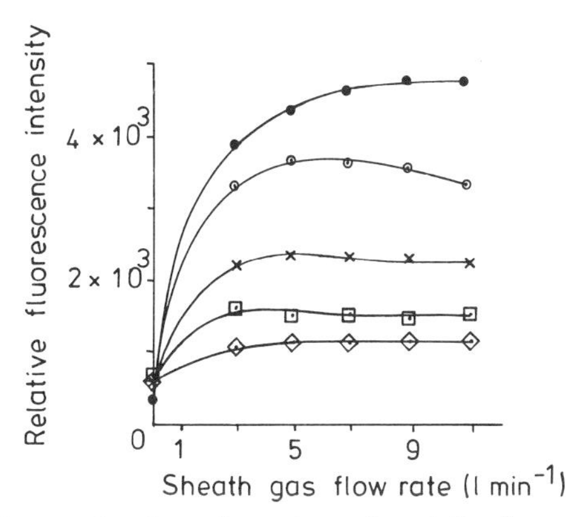

FIG. 8. Effect[8] of argon sheath gas flow rate on the relative fluorescence intensity (peak area) at various carrier gas flow rates: (●) 0.3; (⊙) 0.4; (×) 0.5; (⊡) 0.7; (◇) 0.9 l min^{-1}.

Limit of Detection and Linear Dynamic Range. The limits of detection of the most sensitive of the mercury analyses published in the literature for tin chloride reduction/aeration systems are usually in the 10–100 pg range. Ebdon and Wilkinson obtained a detection limit of 10 pg for 0.5 ml injections of the sample, which is a concentrational detection limit of 20 ng/liter. Precision was about 1–3%, at all concentrations above the 3 μg/liter level. A typical working calibration curve which illustrates the linear dynamic range is shown in Fig. 9. The HCL-ICP-AFS system of Lancione and Drew[11] can also be used in a continuous flow, tin chloride/cold vapor mode and yields a detection limit in the same range as other AFS systems. All these detection limits will be improved if some sort of concentration of the mercury is carried out using silver or gold wool systems.

Performance of the Method. Ebdon and Wilkinson[8] evaluated the performance of their system by the determination of mercury in barley seeds and NBS Orchard leaves. Samples (0.5 g) were allowed to stand overnight in concentrated nitric acid (20 ml for orchard leaves, and 10 ml for barley seeds) in a stoppered flask. They were then refluxed at 80° on a water bath for 30 min. The resulting solutions were diluted to 50 ml for the seeds, or 100 ml for orchard leaves, by use of a potassium dichromate solution to ensure the stability of the mercury (0.3 g of dichromate in 1 liter of 1.6 *M* HNO_3). The results of the analyses demonstrated very good accuracy and precision (Table V). The accuracy of the Orchard leaves analysis was tested by comparison with the NBS standard value, and that of the barley seed was tested by standard addition and recovery calculations (Table

[11] R. L. Lancione and D. M. Drew, *Spectrochim. Acta* **40B,** 107 (1985).

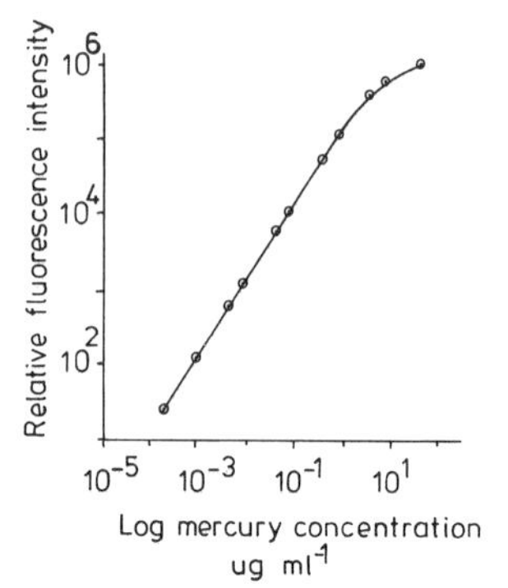

FIG. 9. Working curve[8] for mercury at 253.7 nm logarithm of the peak area (relative fluorescence intensity) against logarithm of mercury concentration.

TABLE V
MERCURY CONTENT OF ORCHARD LEAVES AND BARLEY SEED[a]

Sample	Mean Hg found[b] (ng g^{-1})	Range (ng g^{-1})	SD of mean (ng g^{-1})	RSD (%)
SRM 1571	157[c]	153–159	1.0	0.6
Barley seed	10.74	10.55–10.98	0.07	0.6

[a] Taken from Ebdon and Wilkinson.[8]
[b] The values quoted are the means of 7 separate determinations: for each determination, 5 measurements were made on the same digest.
[c] Certificate value, 155 + 15 ng g^{-1}.

TABLE VI
RECOVERIES OF PHENYLMERCURY(II) ACETATE[a] (2.5 ng Hg g^{-1}) ADDED TO BARLEY SEEDS[b,c]

Replicate number[d]	Hg found	Recovery (%)
1	13.22	99.2
2	13.31	102.8
3	13.23	99.6
4	13.24	100.0
5	13.18	97.6
6	13.23	99.6

[a] 2.5 ng Hg/g.
[b] Containing 10.74 ng Hg/g.
[c] Taken from Ebdon and Wilkinson.[8]
[d] Each reported figure is the mean of 5 measurements of the same digest.

VI). These data demonstrated that no mercury was lost during the analysis of the barley seed samples.

Conclusion

Atomic fluorescence spectrometry remains a technique with a great deal of potential for exceptional sensitivity and selectivity. However, there is only one commercial instrument now and this does not give higher sensitivity than competing techniques. Therefore, the relative advantages of the various techniques that are currently available must be compared in detail in the context of the analyses to be carried out before a choice of technique can be made.

The mercury analysis described here will realize about a factor of 10 better sensitivity over atomic absorption measurements on the same instrument and it is relatively easy to implement.

Acknowledgments

This work was supported in part by the National Institutes of Health under grant GM 32002 and the donors of the Petroleum Research Fund administered by the American Chemical Society, Research Corporation, and the University of Connecticut Research Foundation. The author was supported by a Research Career Development Award from the National Institutes of Environmental Health Sciences under grant ES 00130.

[19] Electrochemical Methods of Analysis

By JANET OSTERYOUNG

This chapter is restricted to electrochemical techniques involving net current flow due to redox reactions at electrode surfaces, and in particular to voltammetric techniques, those in which current is measured at various potentials. Other techniques, such as conductance, potentiometry, and coulometry, are not discussed. The accessibility of voltammetric techniques to the average user depends on availability of reliable and easily operated commercial equipment. More widespread use of voltammetry in recent years has resulted directly from new instruments for its implementation. Correspondingly the following discussion focuses on just a few techniques which can be readily implemented and have a basis in theory and published applications which guide interpretation of results. The reader should note, however, that commercial instrumentation is, at the time of this writing, changing rapidly. The optimist would predict considerable improvement in the sophistication and flexibility of these instruments even within the next few years.

Voltammetry has the following general features which make it a useful analytical tool. (1) The maximum current is directly proportional to concentration. This may be contrasted with the logarithmic relation of potentiometry or the expontential relation of spectrophotometry. Thus aspects of calibration, sensitivity, detection limits, and so on, common to all chemical analysis, are simplified. (2) The proportionality constant between current and concentration (i.e., the sensitivity) contains the value of the Faraday constant, 96,484 C $equiv^{-1}$ (coulombs/equivalent), and is therefore very large, or in other words, the current signal is very sensitive. (3) The quantitative relations between current and concentration are generally very well known. Especially when diffusion controls the current, the sensitivity can be calculated simply and accurately from first principles, and is generally not affected seriously by minor changes in experimental conditions. (4) The sensitivity depends on the nature of the analyte only through its diffusion coefficient in the solution, D, and the stoichiometry of its electrochemical reaction, expressed as the transfer of n electrons per molecule of analyte. These numbers are easily estimated from other data and do not have wide ranges. Thus sensitivities are nearly constant for a given experiment. This allows one to estimate concentrations of unidentified substances. (5) The amount of charge passed, and hence the amount of material converted, in a typical voltammetric measurement is so small that the technique can be considered nondestructive. (6) The response is generally not affected by turbidity. (7) By measuring current as a function of potential one obtains qualitative information about the analyte and about other components of the solution. It should be emphasized that voltammetry has inherently very poor resolution. However, it is also inherently a multielement technique which can be used simply to establish, for example, that a solution has no metals present above some threshold concentration.

The voltammetric behavior of metallic elements depends heavily on the precise nature of the bonding between the metal atom and its ligands. The central chemical problem of developing voltammetric methods of analysis for metals is to establish the proper complex form of the metal to optimize the voltammetric response with respect to the usual analytical criteria of sensitivity, resolution, detection limit, accuracy, and so on. This situation makes it obvious that voltamemtric techniques are ideally suited to characterizing specific metal species in solution. They have been used widely for that purpose, primarily in studies of natural water systems.[1] However, the subject of metal speciation is beyond the scope of this chapter.

Voltammetric Techniques

Voltammetric instruments change the potential of the electrode of interest (the indicator electrode) with respect to a reference electrode according to some preset program using parameters selected by the experimenter. They also measure the current flowing through the indicator electrode (and through a counter electrode which completes the circuit). The output is values of potential and current which can be displayed during the experiment. The plot of current versus potential is called a voltammogram. Here we discuss rotating disk voltammetry, normal and differential pulse voltammetry, staircase and square wave voltammetry, and linear scan voltammetry. Potential programs for these techniques are shown in Fig. 1 together with typical voltammetric response.

Rotating Disk Voltammetry

Several views of a rotating disk indicator electrode are shown in Fig. 2. It consists of a conducting disk sealed into an insulator, the whole assembly having the form of a cylinder. When this is rotated about the cylindrical axis in solution, steady-state convection to the disk is established, which in turn gives rise to a steady-state current. The resulting voltammogram for an uncomplicated reversible reaction is S-shaped, described exactly by the equation

$$i/i_L = 1/(1 + \xi\theta) \qquad (1)$$

where i is the current, i_L the limiting (or plateau) current, ξ the quantity $(D_O/D_R)^{2/3}$ where D_O and D_R are the diffusion coefficients of the oxidized and reduced forms, respectively, and

$$\theta = e^{nF(E-E^{\circ\prime})/RT} \qquad (2)$$

where F is the value of the Faraday, R the gas constant, T the absolute temperature, $E^{\circ\prime}$ the formal potential, and n the stoichiometric number of electrons for the reaction

$$O + ne^- = R \qquad (3)$$

The value of F/RT at 25° is 38.92 V^{-1}. Consider the material initially in solution to be the oxidized form, O; when the potential, E, is sufficiently

[1] H. W. Nuernberg, *Sci. Total Environ.* **12,** 35 (1979).

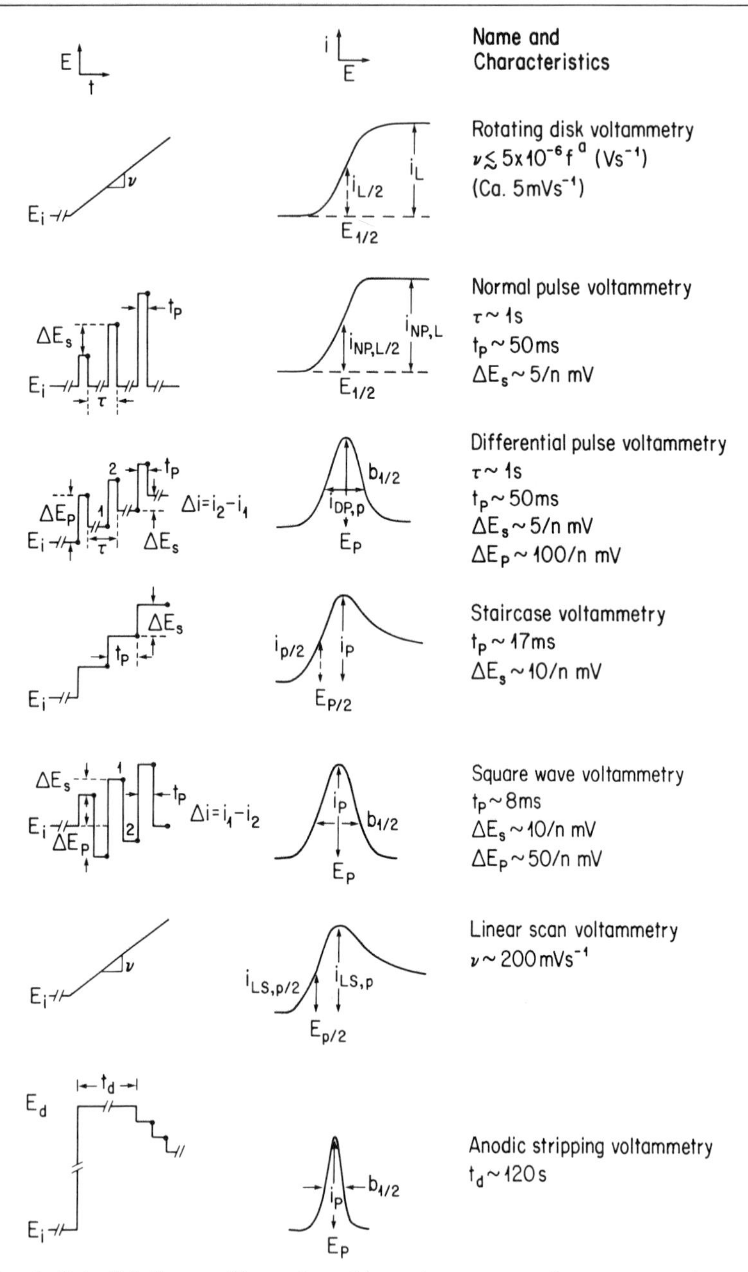

FIG. 1. Potential–time profiles and resulting voltammograms for common voltammetric techniques. For each voltammogram characteristic currents and potentials, and in some cases shape (half-width, $b_{1/2}$), are indicated. (a) Rotation rate in rpm. Anodic stripping is illustrated with staircase stripping; any of the above but rotating disk could be used as well.

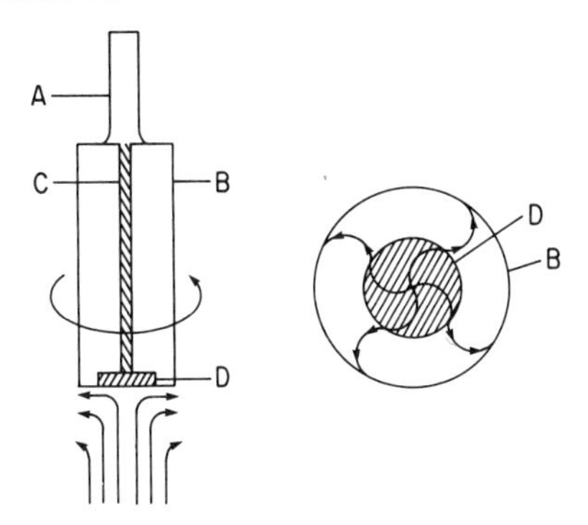

FIG. 2. Rotating disk electrode (a) side and (b) bottom views. A, brass spindle; B, Teflon insulator; C, electrical contact; D, electrode material. Arrows show direction of rotation of electrode and of fluid flow about the rotating electrode.

positive of $E^{\circ\prime}$, θ is large and i is zero. As the potential is made more negative, the driving force for the reaction increases, θ decreases, and i approaches i_L, the limiting value determined by the rate of transport to the electrode.

The limiting current, i_L, which is directly proportional to the bulk concentration of O, C_O, is given by

$$i_L = 0.620nFAD_O^{2/3}C_O\omega^{1/2}/\nu^{1/6} \tag{4}$$

where A is the electrode area (cm^2), ω the rotation rate (rad sec^{-1}), and ν the kinematic viscosity (cm^2 sec^{-1}) of the solution. Using typical values (A = 0.75 cm^2, $D = 8 \times 10^{-6}$ cm^2 sec^{-1}, ν = 0.01 cm^2 sec^{-1}) gives $i_L = 40nC_O\omega^{1/2}$ or $i_L = 12nC_Of^{1/2}$, where f is the rotation rate in revolutions per minute (rpm). Thus for $n = 2$ and $f = 600$ rpm the current is $i_L = 600C_O$ or the sensitivity is 600 μA/mM.

The sensitivity of the current measurement depends on the rotation rate of the electrode, a parameter of the experiment which usually can be varied arbitrarily. At higher rotation rates, material is swept toward the electrode surface at a higher rate, which in turn yields higher currents. Since i_L depends on the square root of ω and ω has units of rad sec^{-1}, i_L depends inversely on the square root of time, a relation characteristic of voltammetric currents controlled by diffusion.

Rotating disk voltammetry gives exceptionally accurate and precise quantitative results. The limiting current relationship is well-obeyed and robust. As indicated in Fig. 1, it is generally carried out with a ramp potential for experimental convenience. However, the rate of potential change must be negligible with respect to the rotation rate for the quantitative relations to be obeyed. This limits one to scan rates less than about the quantity 0.05ω expressed in units of mV $\sec^{-1}$. The technique is practical over the range $9 < \omega < 900$ rad $\sec^{-1}$. At the lower end of the range natural convection begins to perturb the current, while the higher end is limited by the onset of turbulence. Rotating disk measurements are not customarily used for voltammetric analysis,[2] but they should be much more widely used, for rotating disk voltammetry is the most accurate and reliable technique in situations requiring a solid electrode where neither sensitivity nor resolution is a problem.

Pulse Voltammetry

Several other voltammetric techniques can be grouped together under the general heading of pulse voltammetry. In each case the potential program consists of a sequence of constant values through which the electrode is moved by instantaneous or stepwise changes. The measured current is sampled in a specific pattern synchronized with the potential program. The potential programs, current-sampling schemes, and typical voltammograms are shown in Fig. 1. For the simple reversible reaction scheme of Eq. (3) the response for all of these cases can be expressed as

$$i = [nFAD_O^{1/2}C_O/(\pi t_p)^{1/2}]\psi \qquad (5)$$

where the dimensionless current function, π, depends on the potential program and describes the shape of the voltammogram.

Normal Pulse Voltammetry. This technique is simply an automated way of carrying out a sequence of individual experiments. That is, by some means (renewing the electrode surface, stirring the solution, etc.) the same initial conditions ($i = 0$) are established during each period ($\tau - t_p$) (see Fig. 1) before the potential is stepped from E_i. The resulting voltammogram is S-shaped and is given by Eq. (1) with $\xi = \sqrt{D_O/D_R}$. Thus the limiting current, $i_{NP,L}$, is given by

$$i_{NP,L} = nFAD_O^{1/2}C_O/(\pi t_p)^{1/2} \qquad (6)$$

and $\psi = 1/(1 + \xi\theta)$. The magnitude of $i_{NP,L}$, and hence the sensitivity, depends on the pulse width, t_p, which is the time parameter of the experiment. Since $i_{NP,L}$ is proportional to $t_p^{-1/2}$, the smaller t_p is the larger is the current. However, t_p cannot be made arbitrarily short because of problems of response time which depend on both the instrument and the configuration of the electrochemical cell. Times longer than 10 msec are unlikely to present experimental problems. For instrumental convenience $\tau/t_p \geq 10$, so $\tau \geq 0.1$ sec. For typical values ($A = 0.03$ cm^2, $D = 9 \times 10^{-6}$

[2] S. Bruckenstein and P. R. Gifford, *Anal. Chem.* **51,** 250 (1979).

$cm^2\ sec^{-1}$), $i_{NP,L} = 5nC_O/t_p^{1/2}$. Thus for $n = 2$, $t_p = 10$ msec, $i_{NP,L} = 98C_O$, or the sensitivity is 98 $\mu A/mM$.

Of the pulse techniques, normal pulse voltammetry is the simplest both in theory and implementation. It is especially attractive for analytical problems in which it is useful also to elucidate mechanisms of reaction.[3] Although humans appear to prefer a peak-shaped response (e.g., differential pulse, see below) to the limiting current of normal pulse (cf. Fig. 1), the latter is more robust and better defined both chemically and mathematically, and therefore provides a more reliable and accurate analytical signal.

Differential Pulse Voltammetry. In the differential pulse mode (Fig. 1) the potential is stepped through the range of interest, and the current flowing at each value of potential depends on the previous history of potential and time. As the output is a difference current, one might expect the response to resemble the derivative of the normal pulse voltammogram. In fact the current, i_{DP}, is very nearly that which would be obtained from differencing the normal pulse current, i_{NP}, i.e., $i_{DP}(E) \cong i_{NP}(E + \Delta E_S) - i_{NP}(E)$.[4] Thus, by simple calculus the current function at the peak has the value $(1 - \sigma)/(1 + \sigma)$, where $\sigma = \exp(nF\Delta E_p/2RT)$, and the peak current is given by

$$i_{DP,P} = i_{NP,L}(1 - \sigma)/(1 + \sigma) \tag{7}$$

The quantity $(1 - \sigma)/(1 + \sigma)$ is determined only by the choice of pulse height, ΔE_p, and is always less than unity. For $n\Delta E_p = 50$, 100, and 200 mV the value of $(1 - \sigma)/(1 + \sigma)$ is 0.45, 0.75, and 0.96, respectively. The sensitivity of differential pulse is less than that for normal pulse (ψ is less), but detection is generally better, that is, signal-to-noise ratios are better, because the difference subtracts proportionately more background than signal. The differential signal is also more sensitive than the limiting current of normal pulse to kinetic complications.

Staircase Voltammetry. This pulse technique has a potential program resembling a staircase (Fig. 1) and an unsymmetrical peak-shaped response with a very complex mathematical description, [cf. ψ, Eq. (5)]. Voltammograms resemble those of linear scan voltammetry (described below) and for $n\Delta E_s \leq 8$ mV are the same, independent of t_p, provided the current is sampled at time $t_p/4$ on each step, rather than at time t_p.[5] Under those conditions the theory for linear scan can be applied to staircase with scan rate $\Delta E_s/t_p$. The advantage of staircase with respect to linear scan

[3] C.-L. Ni, J. Osteryoung, and F. C. Anson, *J. Electroanal. Chem.* **202,** 109 (1986).

[4] M. Lovric, J. J. O'Dea, and J. Osteryoung, *Anal. Chem.* **55,** 704 (1983).

[5] R. Bilewicz, R. A. Osteryoung, and J. Osteryoung, *Anal. Chem.* **58,** 2761 (1986).

voltammetry is that the signal contains less unwanted background. It is very important practically, because digitally controlled instruments usually employ staircase voltammetry as a surrogate for linear scan voltammetry.

Square Wave Voltammetry.[6] Square wave voltammetry (Fig. 1) also has a complicated mathematical description, but the response is nicely symmetrical, centered on the reversible half-wave potential for reversible systems, and affords excellent rejection of background currents. For $nE_{sw} = 50$ mV and $n\Delta E_s = 10$ mV, $\psi = 0.93$, so the sensitivity is better than that for differential pulse, nearly as good as that for normal pulse [cf. Eqs. (5) and (6), $\psi = 1$]. The combination of low background currents and excellent sensitivity gives it the best, i.e., the lowest, detection limits of the direct voltammetric techniques. It also has the great advantage of speed: typical times required to complete an entire scan are less than 1 sec.

Linear Scan Voltammetry

Linear scan voltammetry has been the most widely used voltammetry technique. The potential program is simple (Fig. 1) and current is measured continuously, so the equipment is simple and inexpensive, but the response is ill-shaped (Fig. 1) and has a complicated mathematical description. The current is given by

$$i_{LS} = nFAD_O^{1/2}C_O(nFv/\pi RT)^{1/2}\psi \tag{8}$$

where v is the rate of potential change with time. Noting that nFv/RT has dimensions of reciprocal time, in fact Eq. (8) is exactly analogous to Eq. (5). The peak value of i_{LS} is found from the maximum value of ψ (= 0.252) and is given by

$$i_{LS,P} = 2.69 \times 10^5 n^{3/2} A D_O^{1/2} v^{1/2} C_O \tag{9}$$

at 25°, where the units are i (μA), A (cm^2), D_O (cm^2 sec^{-1}), v (V sec^{-1}), C_O (mM). For $n = 2$, $A = 0.03$ cm^2, $D_O = 9 \times 10^{-6}$ cm^2 sec^{-1}, $v = 0.2$ V sec^{-1}, $i_{LS,P} = 31C_O$ or the sensitivity is about 30 μA/mM.

The main difficulties with this technique, apart from its complex mathematical description, are that the unsymmetrical shape can make it difficult to measure the current accurately, and unwanted background currents can be large and difficult to subtract. It is especially hard to measure the current for a peak occurring on the long sloping tail of a preceding peak.

[6] J. Osteryoung and J. J. O'Dea, *Electroanal. Chem.* **14,** 209 (1986).

Stripping Voltammetry[7]

Stripping voltammetry is a technique which involves concentrating a sought-for species by an electrochemical reaction into, or onto, the surface of an electrode.

Anodic Stripping Voltammetry. In this case a metal is deposited into a drop or thin film of mercury of volume small compared to that of the analyte solution. Deposition is carried out by applying a constant potential (E_d, Fig. 1) sufficient to reduce the metallic species in stirred solution. The stirring must be controlled, to ensure reproducible mass transport of the species being reduced to the electrode surface; failing that the solution must be exhaustively electrolyzed. Following deposition, the solution is allowed to become still and then the potential is scanned to more positive values to oxidize (strip) the metal from the amalgam (Fig. 1).

If the deposition potential is such that the deposition occurs at its limiting current, and the fraction of analyte removed from solution is insignificant, then $i_L = kAC_O$, where k is the mass transport coefficient, and the amount of material deposited (mol) is $i_L t_d/nF$, where t_d is the deposition time (Fig. 1). The concentration of metal in the amalgam is $C_{M(Hg)} = kAC_O t_d/nFV_{Hg}$, where V_{Hg} is the volume of mercury. It is this concentration which is determined in the stripping step. Equation (5) applies to the stripping voltammogram and thus the stripping peak current is proportional to the initial concentration of the analyte in solution (C_O). For times t_d of a few minutes, concentration factors, $C_{M(Hg)}/C_O$, of the order of 1000 are readily obtained. Thus ASV is used for the determination of metal ions in the nanomolar concentration range since voltammetric techniques (Fig. 1) can be used to determine metals in the micromolar range. When thin mercury films are employed, the volume of mercury is smaller and the concentration factors can be 10^5–10^6. When the film is thin with respect to the characteristic time of the voltammetric stripping technique (i.e., $l^2 \ll Dt_p$, where l is the film thickness), the stripping peak current depends only on $t_d i_L$ and not on the volume of mercury in the film. In this case Eq. (5) no longer applies, but i_p is still proportional to C_O.

Thin mercury films are usually deposited on a rotating glassy carbon substrate from a solution of Hg(II); film thicknesses in the range 0.1–1 μm are suitable.[8] Films can last for a day of repetitive use. Mercury is the solvent but may also form compounds with the deposited metal. This can affect the oxidation peaks just as reduction peaks are affected by complex formation in the analyte solution.

[7] F. Vydra, K. Štulík, and E. Juláková, "Electrochemical Stripping Analysis." Ellis Horwood, Chichester, England, 1976.

[8] T. R. Copeland and R. K. Skogerboe, *Anal. Chem.* **46,** 1257A (1974).

A large number of metals, most commonly Pb, Cd, Cu, Bi, and Zn, can be determined by ASV. Care must be taken in obtaining pure reagents, water, and clean glassware, particularly when determinations are made at very low concentration levels. This is particularly true when determining lead, which is widely dispersed as a contaminant. Detection limits are roughly comparable to those of flameless atomic absorption spectroscopy and in routine practice are generally limited by control of contamination rather than by the inherant sensitivity of the technique. Experts working in clean rooms report determination of Pb at 1 pg/liter.[9]

Cathodic Stripping Voltammetry. In this case an insoluble film is deposited onto the surface of an electrode by oxidation and stripped away cathodically. For example a silver electrode, in the presence of sulfide, will readily form an adherent layer of $Ag_2S(s)$.[10] The potential at which the silver sulfide forms is more negative than that required to form silver ion in solution. The potential can be adjusted, as described above, so that the reaction is diffusion controlled and limited by the concentration sulfide in solution. The silver sulfide may then be reduced by a scan to negative potential which strips the material from the surface.

Cathodic stripping may also be applied to mercury electrodes; in recent years a variety of materials of biological interest have been concentrated by adsorption or specific reaction at the surface of a mercury electrode and then determined by cathodic stripping.[11]

Apparatus

Instruments. The techniques of Fig. 1 can be carried out with instruments of a wide range of sophistication. The basic requirements are shown in the block diagram of Fig. 3. Rotating disk voltammetry is a steady-state technique; that is, the current is independent of time. Thus in this simple case the experimenter is an adequate logic and timing module as well as potential programmer. The complex waveforms, e.g., square wave voltammetry, in contrast nearly require and certainly are made more powerful by control through digital computer. Then the potential program, logic and timing, and current output modules are comprized of digital-to-analog and analog-to-digital converters together with central processor, programmable clock, memory, and appropriate programs. This in addition provides the benefit of data processing, including routines for determining peak position and height, smoothing, subtraction of background currents, averaging, and calibration in units of concentration.

[9] L. Mart (Kernforschungsanlage, Juelich, Federal Republic of Germany), personal communication.

[10] K. Shimizu and R. A. Osteryoung, *Anal. Chem.* **53,** 584 (1981).

[11] M. Iwamoto and R. A. Osteryoung, *J. Electroanal. Chem.* **169,** 181 (1984).

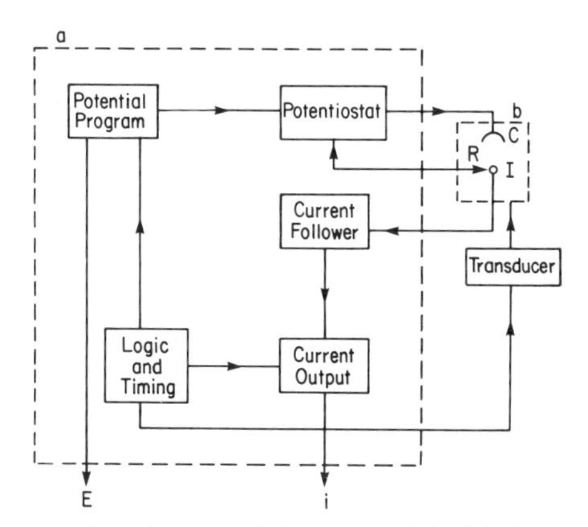

FIG. 3. Block diagram of a voltammetric instrument (a) with electrochemical cell (b). The indicator, counter, and reference electrodes are designated by I, C, and R, respectively.

The transducer is a device synchronized with the potential program to control the electrode mechanically or by some other means. It could, for example, be a variable-speed motor which could rotate the indicator electrode at a selectable frequency during the deposition time, t_d (Fig. 1), in anodic stripping voltammetry.

Voltammetric procedures are properly tied to specific instruments, and manufacturers are typically relied upon to provide "Application Notes" which translate published methods into specific procedures. Short and generally readable discussions of principles of analog instrumentation important to voltammetry can be found in recent books.[12] The literature of manufacturers regarding specific instruments should be consulted carefully.

Cells. Electrochemical cells should be easily cleanable, capable of being sealed, and provide fixed, reproducible orientation of the cell components. Common cell designs employ a cell top rigidly mounted to a support. The experiment is prepared for by appropriate assembly of electrodes, sparging tubes, and so on, with the top. The top incorporates a flanged gasket with a simple clip which makes it possible to attach a cell bottom, containing the solution, in a few seconds. Thus it can be convenient to work with several cell bottoms as one examines the properties of a variety of samples.

Typical applications involving determination of metals in aqueous solution require operating at potentials where oxygen is reduced. To avoid interference, oxygen is removed by sparging the solution with compressed, purified nitrogen or argon gas. Oxygen of the atmosphere readily diffuses back into the cell after sparging, even if a continuous flow of the

[12] A. J. Bard and L. R. Faulkner, "Electrochemical Methods." Wiley, New York, 1980.

inert gas is maintained over the surface of the solution. It is useful to operate the cell at the slight positive pressure provided by an outlet trap with a few centimeters of water. Under these conditions the oxygen concentration of the analyte solution can be reduced to $\sim 1\ \mu M$. However this level can fluctuate considerably during the course of a working day. This is the main limitation on constancy of background currents when determining metals in the submicromolar range.

Counter Electrodes. In the typical three-electrode configuration, the counter electrode simply completes the current circuit, and in consequence whatever reaction required to do this takes place at that electrode. In buffered aqueous solution using a platinum counter electrode these reactions would be usually the oxidation or the reduction of water (reduction of H^+ in acid solution). These reactions cause little change in the solution. In some cases, however, the counter electrode should be isolated from the analyte solution by using a tube with glass frit or "thirsty glass" plug. In the latter case the vendor's instructions for maintenance should be followed carefully.

Reference Electrodes. Because of the overwhelming preponderance of methods involving aqueous solutions in this field, we restrict this brief discussion to reference electrodes suitable for aqueous solutions. Problems with voltammetric experiments often can be traced to a problem with the reference electrode. Many of these problems could have been avoided by simple attention to the instructions for maintenance and use provided by the vendor.

The purpose of the reference electrode is to provide a working laboratory standard of potential for the experiment. To be useful the reference electrode must maintain a constant potential, that potential must have a known value with respect to the conventional thermodynamic scale of potential, and the resistance of the electrode must be acceptably low. In voltammetric measurements potentials are generally known with a precision of a few millivolts. Comparable uncertainty in the value of the reference potential is tolerable.

Reference electrodes in common routine use include calomel [$Hg/Hg_2Cl_2(s), Cl^-(C)$], silver–silver chloride [$Ag/AgCl(s), Cl^-(C)$], and to a lesser extent mercury–mercurous sulfate [$Hg/Hg_2SO_4(s), SO_4^{2-}(C)$]. The saturated calomel electrode (SCE) is saturated with potassium chloride. Commercial reference electrodes generally incorporate a junction which isolates the reference electrode but provides electrical contact between it and the anlyte solution. To protect this junction and prevent cross-contamination the entire reference assembly is then placed in a "salt-bridge" as described above for the counter electrode. For proper operation there must be flow of solution ($\sim 1\ \mu l/hr$) through the junction from the reference compartment into the bridge compartment. Reference electrodes which display high resistance (with respect to vendor's specifications) are best discarded and replaced.

Reference electrodes react adversely to mechanical or thermal shock, shaking, and temperature cycling. Because they are inherently open to the atmosphere (although superficially they may appear sealed), the liquid phase must be cared for according to instructions to maintain constant composition. Calmoel electrodes are especially prone to long-term drift in potential due to diffusion-controlled approach to equilibrium. It is prudent to intercompare laboratory reference electrodes regularly.

Indicator Electrodes. Choice of the material of the indicator electrode depends on the potential range one desires to examine and specific electrocatalytic properties of the material in the intended applications. In general the electrode material does not act simply as an inert conductor of electrons, so the proper electrode specified by the method being followed must be used.

The range of potential accessible with a given electrode depends on anodic and cathodic reactions of the solvent, supporting electrolyte, or the electrode material itself. In buffered aqueous solutions the limiting cathodic reaction is usually reduction of H_3O^+ or H_2O and the range is extended to more negative values by about 60 mV for every unit increase in pH. On platinum or carbon electrodes this limit is about -0.2 V (vs SCE) and on mercury electrodes about -1.1 V in *M* strong acid. The positive limit on platinum or carbon electrodes is oxidation either of the electrode material or water and depends on pH in the same way. In *M* strong acid this limit is about $+1.3$ V at platinum and $+1.6$ V at carbon. Mercury is oxidized in noncomplexing media at about $+0.4$ V, but is oxidized more easily (at less positive potential) in solutions containing ions such as chloride which form stable complexes or insoluble compounds with mercury. It reacts slowly but quantitatively with oxygen in solution, and therefore it is important to deaerate the analyte solution before bringing it into contact with a mercury electrode.

Mercury is the best electrode material for examining redox processes in the negative potential range (< 0.4 V vs SCE). Because it is a liquid, each bit of newly formed surface has the same properties, in contrast with solid electrode materials, which are heterogeneous with respect to microstructure, even if nominally "pure." Because of its very large surface tension, mercury is difficult to handle mechanically. In common practice it is used for voltammetry in two ways, as a dropping mercury electrode (dme) or as a hanging mercury drop electrode (hmde).[13] A third choice is

[13] W. Kemula and Z. Kublick, *Adv. Anal. Chem. Instr.* **2,** 123 (1963).

the static mercury drop electrode (smde),[14] which is essentially an automated hmde.

The dme is conveniently constructed from a 8–20 cm length of glass capillary tubing having internal diameter 50–70 μm. A head of mercury of about 30–80 cm provides dropping rates in the range 3–8 sec and drop areas at maximum size (i.e., when the drop falls) of $\sim$0.01 cm^2. Of course the current flowing through the drop as it grows is proportional to the area of the drop [Eq. (5)], and so increases as the area increases. When this type of electrode is used with a pulse technique, the drop period can be mechanically synchronized with the period of the experiment (cf. Fig. 3), so that the current is sampled at the same drop area in each cycle. Thus fluctuations in current due to changing area are not apparent in the voltammogram. This strategy is generally employed with normal and differential pulse voltammetry. Voltammograms obtained with the dme are, for historical reasons, called polarograms. Normal and differential pulse polarography are simply these voltammetric techniques employed with the dme as the indicator electrode.

Staircase voltammetry can be employed at the dme either with one period (one staircase tread) per drop or in a rapid-scanning mode in which the entire potential range is covered during the period of the drop. The resulting polarogram in the former case resembles the normal pulse polarogram, whereas the latter case gives the staircase response (Fig. 1).

The dme properly cared for provides an inexpensive, trouble-free, and reliable indicator electrode. The important elements in proper care are designed to prevent contamination of mercury in the reservoir providing the head and to prevent the rather fine capillary from becoming clogged. The mercury should be in contact only with inert materials such as glass and hard plastics (not Tygon) and should be protected from gross contact with ambient air using a moisture-trapping filter. The capillary tip should be washed thoroughly with distilled water after use and allowed to dry completely before the flow of mercury stops.

The hmde and smde should be employed in their commercial versions according to manufacturer's instructions. The hmde is far less expensive, but requires considerable technical expertise in use to give reliable results. The smde has a history of mechanical problems but at this writing it is a mature design with steadily improving records of reliable and trouble-free routine performance.

Any user of metallic mercury needs to keep in mind that its vapor in equilibrium at room temperature and pressure is lethal.[15] Normal good

[14] W. M. Peterson, *Int. Lab.* **10,** 51 (1980).

[15] NIOSH, "Registry of Toxic Effects of Chemical Substances." U.S. Government Printing Office.

laboratory housekeeping and ventilation completely protect the worker from exposure to mercury vapor. However prudence requires routine checks of the vapor concentration and prompt, thorough attention to spills. Contaminated mercury can be returned to the supplier for purification.

Solid Electrodes. A wide variety of conducting and semiconducting solids are also used to make indicator electrodes. Of these carbon is the most useful in negative potential ranges. The exact behavior of carbon electrodes depends on how the carbon is prepared. The most common form used for routine voltammetry is glassy (or vitreous) carbon, a hard, dense, isotropic material. The electrode is prepared from the material by sealing a billet into an insulating cylinder, making electrical contact at the back through the axis of the cylinder, and polishing the front face normal to the axis to a mirror finish. The seal bewteen the carbon disk and the insulating material must be smooth and leakproof for reliable performance. The technology for achieving a perfect seal is not straightforward, but commercial products generally are reliable. Vendors of rotors for rotating disk electrodes all sell electrodes especially designed for use with their equipment.

Temperature-cycling causes differential expansion and contraction of the conductor and insulator and eventually ruins the seal. This process is gradual rather than catastrophic, and causes increasing background currents with age. Particular attention must be given to storing electrodes in cold climates where aggressive energy conservation may lead to very low ambient temperatures on holidays and weekends.

In order for any solid electrode to work well it must be polished before use and in addition put through some electrochemical pretreatment procedure. Any reliable analytical procedure should specify these treatments. Polishing is notoriously an art, difficult to control or to describe clearly in writing.[16] This appears to be an especially important part of the overall procedure for glassy carbon electrodes. Individual procedures must be sought for detail; however, a few generalizations can be made. Electrodes which are very well polished (that is, smooth on a submicron scale) have low background currents but often little catalytic behavior. Normally fast processes become slow which leads to lower currents than anticipated. Thus it is possible to polish an electrode "too well." Polishing is done with slurry of abrasive; removing all traces of this slurry by vigorous and thorough washing is essential for obtaining reproducible results.[17]

[16] M. Schreiner, J. J. O'Dea, N. Sleszynski, and J. Osteryoung, *Anal. Chem.* **56,** 116 (1984).
[17] S. Dong and T. Kuwana, *J. Electrochem. Soc.* **131,** 813 (1984).

Analytical Procedures

One normally wishes to find a published procedure for the exact sample type and analyte one is concerned with and to follow that procedure exactly. One usually finds several procedures, each on related but not identical sample types, perhaps specifying equipment different from that which is available and employing materials or processes not readily usable in the setting for the application. It is therefore important to have a feeling for what features of a procedure are essential, what may be changed with caution, and what modifications are major enough to require some investigation of their effects.

A reputable procedure contains the following information. (1) The electrochemical instruments, electrodes, and any accessories are specified. Any nonstandard (noncommercial) equipment is described carefully. The electrode area is given. The name of the technique used is given as specified by the instruction manual of the instrument. In addition the potential program and current-sampling scheme should be given explicitly. (2) The electrochemical process on which the procedure is based is described as fully as it is known. In particular, the value of n on a specified time scale and D for the reactant are reported. Some investigation should have been made of the kinetics of the process. In many cases it is found to be diffusion-controlled over some range of conditions which should be specified. (3) The optimum chemical conditions are specified. Behavior over the range of conditions investigated should be summarized briefly. (4) Sensitivity, linear range, and detection limit (clearly defined) are given explicitly. The sensitivity should be compared quantitatively with that expected for the technique employed.

Changes in Time Scale. Often one must employ different time parameters than those specified by the procedure. When these changes are minor they can be expected to have only minor effects. For example, instruments used in the United States tend to employ times synchronized to the 60 Hz line frequency and therefore multiples of 16.7 msec, whereas in Europe, which generally has 50 Hz power mains, the characteristic timing is a multiple of 20 msec. An instrument providing an effective pulse time in the normal pulse mode of 48 msec would be expected to yield a sensitivity which is $(30/48)^{1/2} = 0.79$ times that of one with effective pulse time of 30 msec [cf. Eq. (6)].

Analog instruments usually operate at fixed times of current sampling. The new generation of digitally based instruments appearing at the time of writing generally allows wide variation of this parameter. This makes it much easier to study electrochemical processes and provides an important parameter to be optimized in developing new methods. However it

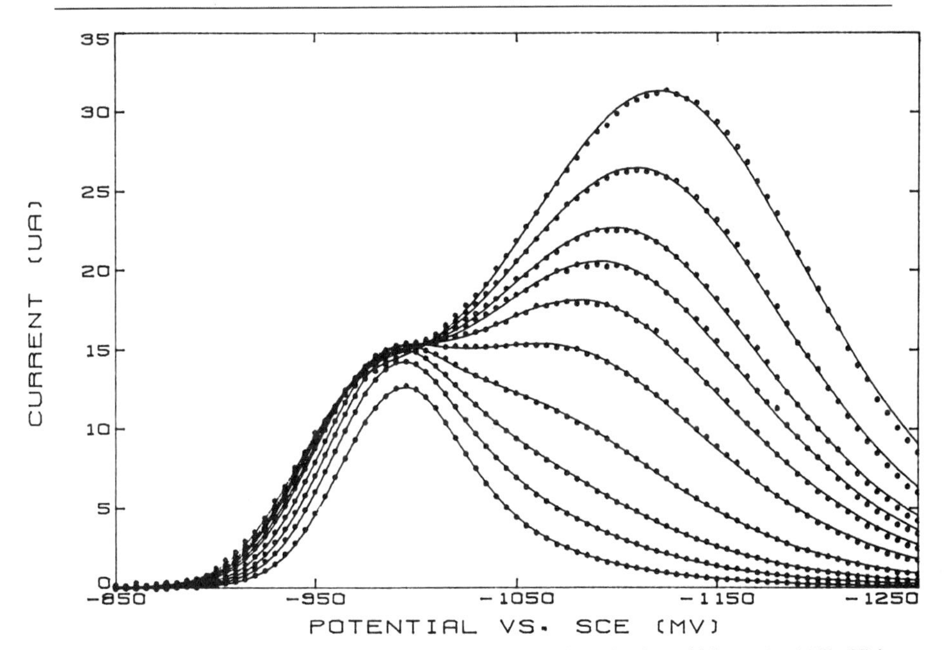

FIG. 4. Square wave voltammetric current as a function of pulse width, t_p: 1 mM Zn(II) in 1.05 M $NaNO_3$; EG&G Model 303 smde, electrode area 0.0265 cm^2; ΔE_s = 5 mV, ΔE_p = 25 mV. Experimental (●) and theoretical currents (—) with times (msec) in ascending order of curves (at 1100 mV): 50, 20, 10, 5, 2.5, 1.667, 1.25, 1, 0.714, 0.5. Reproduced with permission from J. J. O'Dea, R. A. Osteryoung, and J. Osteryoung, *J. Phys. Chem.* **87,** 3911 (1983); copyright © 1983 American Chemical Society.

places an added responsibility on the routine user to use acceptable time parameters.

Differences in time scale cause unexpected changes in the response when they correspond to differences in the mechanism controlling current flow. For example, at longer times the process controlling the current may be diffusion, whereas at shorter times some chemical rate may control the current. Thus the time scale affects the voltammogram not only through the time dependence displayed in Eq. (5), but through changes in ψ, which determines not only the magnitude but also the shape of the response. An example is shown in Fig. 4 which displays voltammograms for the reduction of Zn(II).

Changes in time scale also affect mechanism by changing the amount of material converted. Consider the current $i_{NP,L}$ of Eq. (5). The number of moles of product formed per unit electrode area is

$$\int i dt/nFA = 2D_O^{1/2} C_O t_p^{1/2}/\pi^{1/2} \tag{10}$$

which is about $3 \times 10^{-3}\, C_O t_p^{1/2}$ mol cm^{-2}. A monolayer of product is in the range of $\sim 5 \times 10^{-10}$ mol cm^{-2}. Thus voltammetry of micromolar solutions on time scales of a few tens of milliseconds produces a monolayer of product. If this product remains on the surface instead of diffusing away into solution it can affect dramatically the voltammetric currents. The normal pulse response at short times can be dramatically better than that at longer times because of the smaller amount of product formed.[18]

Changes in Concentration. Voltammetric techniques typically yield linear calibration curves over a thousandfold change in concentration and often, by changing time parameters, can be used over a factor of 10^4 or 10^5 change in concentration. The lower end of the range is determined by sensitivity and by the reproducibility of the background signal and can be thought of as a function of s_b/m, where s_b is the standard deviation of the background signal (or the analyte signal at very low levels) and m is the slope of the calibration curve (i.e., sensitivity). Background currents are quite large and can be affected by surfactants present in the sample. There also can be instrumental problems associated with measuring currents with a large background component. Measurement of ~100 pA is straightforward, which for the sensitivity of normal pulse voltammetry [Eq. (6) and following text] corresponds to 1 n*M* concentration. But detection limits ($\sim 3s_b/m$) for normal pulse are typically ~1 μM, because background currents are so large they have uncertainties on the order of 30 nA. Differential techniques subtract a greater proportion of the background than the signal and therefore yield better detection limits ($\sim 10^{-7}$ *M* for differential pulse and 10^{-8} *M* for square wave).

The high end of the concentration range can be determined by surface reactions [cf. Eq. (10) and following text]. In this case concentration and the characteristic time act in the same way. A reaction well-behaved at 10^{-4} *M* and 3 sec (e.g., staircase polarography) but ill-behaved at 10^{-3} *M* can be restored to good performance by decreasing the time to 30 msec (e.g., normal pulse polarography). On the other hand, most analytical instruments cannot handle the very large currents associated with measurements in concentrated solutions at short times. The current $i_{NP,L}$ at 30 msec in a 1 m*M* solution is 30 μA, which is unacceptably large.

Stripping voltammetry represents a special case in which sensitivity should never be a problem, but the upper limit of concentration is quite low. For anodic stripping voltammetry at the thin film mercury electrode, detection limits are almost always determined by factors of sampling and contamination rather than by the signal-to-background ratio of the measurement. Detection limits in the range of 1–10 p*M* are common.[7] Inter-

[18] W. O'Deen and R. A. Osteryoung, *Anal. Chem.* **43,** 1879 (1971).

metallic compounds which form in the amalgam at higher concentrations distort the stripping voltammograms and make it impossible to achieve quantitative results. Practical upper limits are ~0.1–10 μM; stripping voltammetry should not be used when the detection limits of direct voltammetry are adequate.

Changes in Technique. Procedures can often be adapted readily to a different technique. For example, procedures which specify differential pulse voltammetry often work as well or better if square wave voltammetry is used instead. But sometimes attributes of the process on which the analysis is based and which have not been well characterized will change performance dramatically when the technique is changed. This often arises from a change in time scale. For example, the time scale of staircase polarography is the drop time (or equivalently the period of the staircase, t_p) which is a few seconds. On the other hand, the time scale of the same experiment in the rapid-scanning mode is $(RTt_p/\Delta E_s nF)$, where ΔE_s is the step height. For $n\Delta E_s$ = 10 mV, t_p = 10 msec, this is 0.026 sec, or the experimental time scale is about 100 times shorter. The comments about time scale above apply here.

In some cases change in technique is associated with marked change in potential program. For example, normal pulse is very different from all the other pulse techniques in the form of the potential–time sequence. Indeed, while somewhat less powerful (having higher detection limits and poorer resolution) than the differential techniques, it provides finer control of the reaction by providing the smallest conversion of material and control of the initial potential from which the pulse is applied, and therefore often gives the best analytical results.

The question of substituting techniques becomes especially important in the context of the literature on polarography of metal ions. There is a large body of methods which employ so-called dc polarography (equivalent to what is described here as staircase polarography) which have unacceptable criteria of linear range and detection limit by today's standards.[19] However they are based on reliable experimental work on well-defined chemical systems. It is legitimate and useful to adapt these procedures simply substituting modern techniques such as square wave voltammetry. In some cases the change in technique will produce unanticipated changes in the response, and the method will have to be modified further.

Chemical Conditions. The reaction on which an analytical method is based is usually complicated, consisting of several linked chemical and

[19] I. M. Kolthoff and J. J. Lingane, "Polarography," 2nd Ed., Wiley (Interscience), New York, 1952.

electrochemical steps. An objective in developing an analytical method is to make these complexities irrelevant by establishing chemical conditions which control rates of reaction and the position of equilibria to advantage.

It is desirable to have a large (~0.1 *M*) concentration of electrolyte in the analyte solution. This serves several purposes. It provides good conductivity, which avoids experimental problems associated with a highly resistive solution. It ensures that the ion fraction of the analyte is negligible, which in turn ensures that the fraction of current carried by the analyte in the bulk of solution is small. Consequently the simple theories of diffusion can be used to predict the current accurately. Finally, high electrolyte concentrations compress the ionic distribution in the interfacial region between electrode and solution, which eliminates some undesirable kinetic effects.

Electron-transfer reactions of metal ions in aqueous solution invariably depend on hydrogen ion activity, and more generally on the acidic and basic properties of the solution. In common practice the pH dependence of a reaction is studied by using Britton–Robinson buffer (a phosphate–acetate–borate mixture giving adequate buffer capacity over the pH range ~2–10). The components of the buffer act also as ligands and as adsorbates at the surface of the electrode, so a single acid–base pair is a better choice for a buffer at a specific pH. Even when the electrochemical reaction does not have an explicit pH dependence it is wise to employ a well-buffered solution to avoid changes in pH near the electrode. Reduction of traces of oxygen consumes protons and can make the pH at the electrode surface much higher than in the bulk solution if the solution is unbuffered. Reduction of 2 μM oxygen would correspond to a pH at the electrode surface of 8.9 in unbuffered neutral solution.

Reductions generally consume protons and electrochemical reductions generally are faster and more well-behaved in acidic than in basic solution. Also, since the basic components of buffer systems available for basic solutions are inherantly better ligands than those for acidic solutions, it is more difficult to manipulate the pH and the complex form of the analyte independently in basic solution. Above pH 8.3 solutions scavenge carbon dioxide, which makes it difficult to maintain fixed composition. At higher pH values glassware is attacked by hydroxide. And finally, the performance of the indicator electrode is often better in acidic than in basic solutions. Thus for reductions of metal ions in aqueous solution, low pH values usually provide the best conditions.

In ideal situations, the triple objective of controlling pH, complex ion distribution, and ionic strength is met by a solution having only one or two components added to the water. For example, 0.1 *M* HCl is an excellent electrolyte solution for electrochemical studies, providing good conductivity, buffer capacity, and metal-complexing ability, and readily available in pure form. At the next level of complexity, acetate buffers made from acetic acid and sodium carbonate have the same properties. When more exotic solutes are required, special care must often be used for purification.

Selected Methods

The methods of this section are selected to provide specific examples of some of the general points made above.

Determination of Cd(II).[20] The procedure assumes that the sample has been prepared appropriately to yield the aquo ion Cd^{2+} with acceptable levels of impurities. The method employs anodic stripping voltammetry with a square wave voltammetric stripping step. The speed of the square wave technique makes it possible to carry out the experiment in the presence of oxygen, thus avoiding deaeration of the solution, and in the presence of convection, thus avoiding long equilibration times.

An EG&G PARC Model 384B polarographic analyzer is used with a Model 303 static mercury drop electrode and a Model 305 stirrer. The large drop size is used and found by weight to be 4.47 mg which, assuming spherical geometry, corresponds to an area of 0.023 cm^2. Potentials are measured with respect to a saturated calomel reference electrode (saturated with NaCl), and a Pt wire is used as counter electrode. The sodium rather than potassium salt is used in the reference electrode to avoid precipitation of $KClO_4$ at the junction, as the analyte solution contains perchlorate.

The applied waveform is shown in Fig. 5. The delay period following the extrusion of the drop consists of a conditioning time, a deposition time (t_d), and an equilibration time (t_e). During the conditioning period a potential is applied which strips any metallic impurities from the mercury drop. During the deposition period, the solution is stirred with a 10 × 3 mm magnetic stir bar with the stirrer on "autofast" mode. The equilibration period allows the concentration of metal in the drop to become uniform and the solution to become quiet. The net current signal is not changed significantly by residual convection. The square wave parameters are f = 100 Hz, ΔE_s = 2 mV, E_{sw} = 25 mV. The 15-point symmetrical sliding array of the instrument is chosen as the data-smoothing routine.

All chemicals are reagent grade and used without further purification. Water is purified by passing distilled water through a four-cartridge Milli-Q purification system. Stock solutions of 0.05 *M* Cd(II) are prepared from

[20] M. Wojchiechowski, W. Go, and J. Osteryoung, *Anal. Chem.* **57,** 155 (1985).

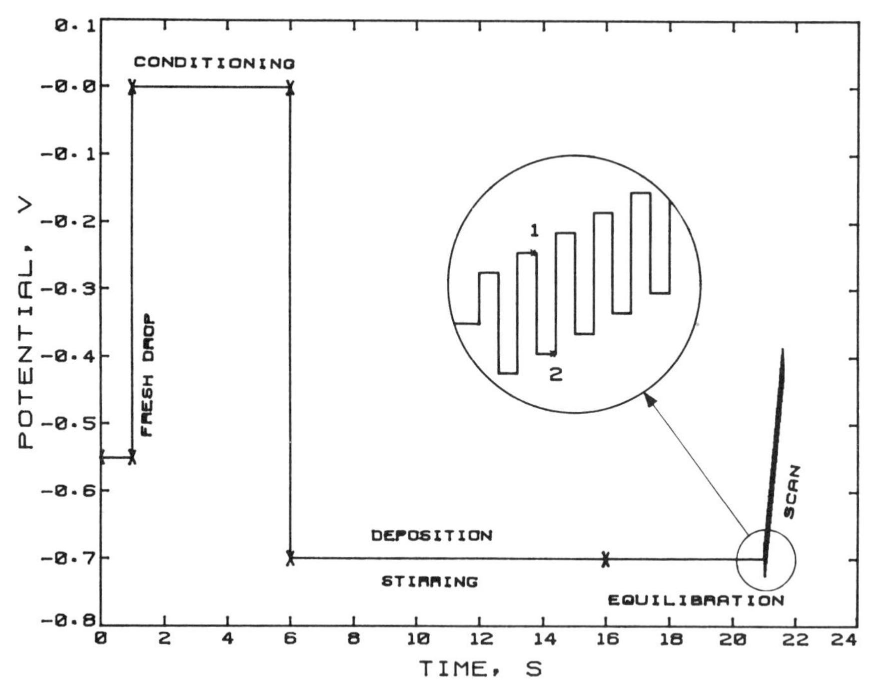

Fig. 5. Wave form for square wave anodic stripping voltammetry. See text. Reproduced with permission from Wojciechowski *et al.*[20]; copyright © 1985 American Chemical Society.

Fisher Scientific $Cd(NO_3)_2{\cdot}4H_2O$ and standardized by potentiometric titration with EDTA. Results in the presence of oxygen (undeaerated solution) are referenced to those after deaeration for 15 min with argon passed through a copper catalyst (Chemical Dynamics Corp. BASF R3-11) and a sparging bottle of the supporting electrolyte, 0.1 *M* $HClO_4$.

Peak currents are linear in deposition time over the range 5–120 sec with slope 0.343 μA/min. The sensitivity at t_d = 120 sec is 8.85 $\mu A/\mu M$. Detection limit is obtained as $3s_b/m$ where s_b is the pooled standard deviation of the calibration curve in the range 0.1–1 μM Cd(II) and m is its slope. Under the conditions employed detection limits are independent of deposition time in this range and equal 0.05 μM. As the main purpose of this study was to establish the utility of the square-wave technique for determinations in the presence of oxygen, no special precautions were taken for control of contamination, and no attempt was made to optimize detection limits. Typical voltammograms are shown in Fig. 6.

Determination of As(III).[21] As(III) is determined by differential pulse

[21] D. J. Myers and J. Osteryoung, *Anal. Chem.* **45,** 256 (1973).

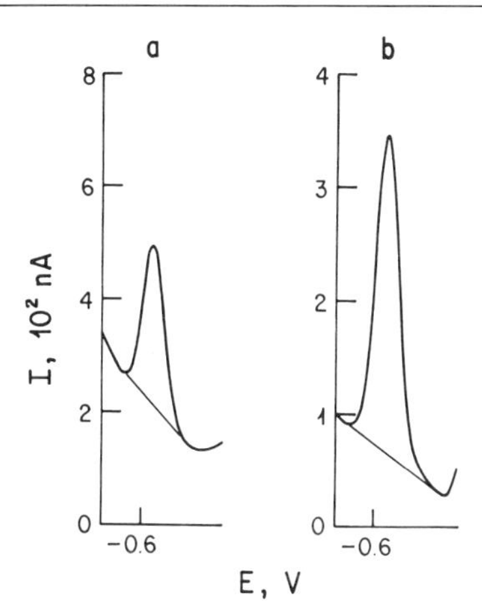

FIG. 6. Square wave anodic stripping voltammograms of (a) nondeaerated and (b) deaerated solutions of 0.1 μM Cd(II) in 0.1 M $HClO_4$. t_d = 30 sec, t_e = 5 sec. Reproduced with permission from Wojciechowski *et al.*[20]; copyright © 1985 American Chemical Society.

polarography in 1 M HCl solutions. As(V) is not electroactive under these conditions and hence the method is specific for As(III).

The analysis employs an EG&G PARC Model 174 polarographic analyzer with Model 174/170 drop timer, together with Houston Instruments Omnigraphic 2000 *x-y* recorder. The pulse width in the differential pulse mode is 56.67 msec and the current output is an integral over the last 16.67 msec of the pulse, giving rise to an effective pulse time of 48.3 msec.

The dropping mercury electrode is made from Sargent capillary tubing and has a flow rate of 2.5 mg sec^{-1} and natural drop time of 5.73 sec in 1 M HCl at open circuit with a mercury head of 48.0 cm. The counter electrode is a 1/4-in. carbon rod placed directly in the cell, and the reference electrode is a Beckman Model 391-70 saturated calomel electrode. Solutions are deaerated with prepurified nitrogen passed through a chromous scrubber and then distilled water before entering the cell. A sparging tube is used in the cell and a two-way stopcock used to direct the gas flow over the surface of the solution during voltammetric measurements.

A stock solution of 2.156×10^{-2} F As(III) was prepared by dissolving primary standard Baker Analyzed As_2O_3 in distilled water with the required amount of NaOH. All other chemicals were reagent grade and used without further purification. The reduction was examined in HCl, H_2SO_4, HNO_3, and a variety of carboxylic acid buffer solutions in the concentration range 0.1–1 M. The best medium was found to be 1 M HCl. In other electrolytes, most notably 1 M H_2SO_4, the medium of choice for dc polarography, the rate-determining step is slower and hence peak cur-

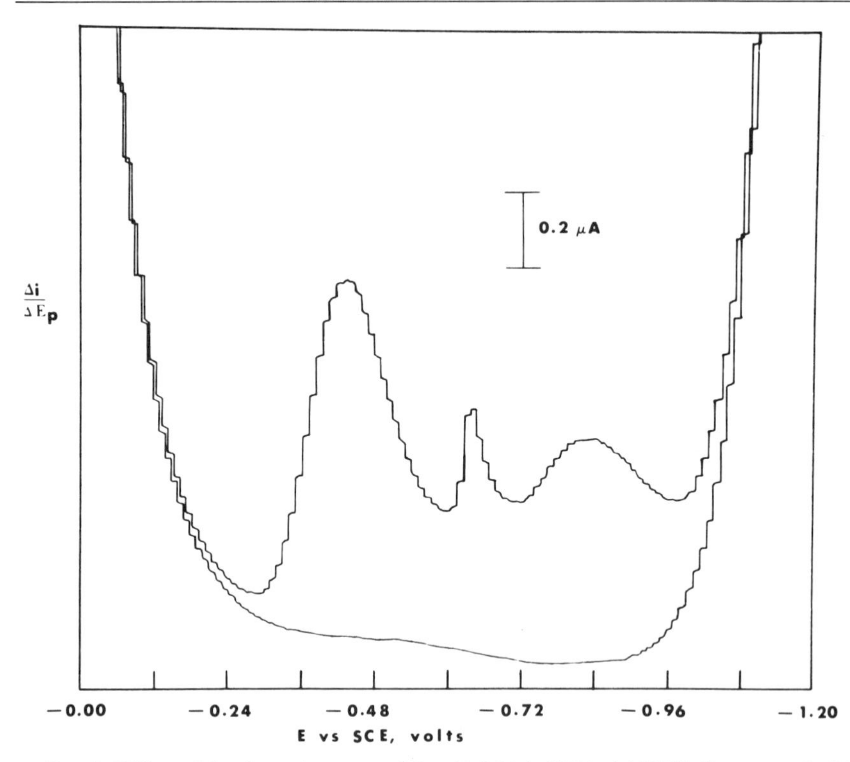

FIG. 7. Differential pulse polarogram of 8×10^{-6} M As(III) in 1 M HCl. Scan rate, 5 mV sec^{-1}; drop time, 2 sec; pulse amplitude, 50 mV. Reproduced with permission from Myers and Osteryoung[21]; copyright © 1973 American Chemical Society.

rents are less. A typical voltammogram is shown in Fig. 7. The peak at −0.43 V is used for quantitative determination.

Pulse amplitude is 100 mV. The peak is distorted by slow instrumental response at nonzero scan rates, and sensitivity decreases with scan rate. Performance at a given scan rate is improved by decreasing drop time, with corresponding decrease in signal. The sensitivity over the range 3 μM–30 nM is 342 μA/mM with scan rate 2 mV sec^{-1}, drop time 2 sec, and pulse amplitude 100 mV. Satisfactory calibration curves are obtained up to 3×10^{-4} M As(III), though there is some decrease in slope. The detection limit calculated as ts_b/m, where m is the slope of the calibration curve over the range 0.6–6 μg liter As(III), s_b its pooled standard deviation, and t the Student's t statistic, was found to be 0.2 μg/liter at 95% confidence.

Pb(II), Sn(II), Sn(IV), Tl(I), and Tl(III) interfere. In the presence of these species, record the polarogram, then add an mount of Ce(IV) sufficient to oxidize As(III) to inactive As(V). Excess Ce(IV) is reduced to Ce(III) by the Hg(0) present in the cell. Record as second polarogram and determine As(III) by difference. This procedure works well at concentrations of As(III) of 20 μg/liter or higher.

As(V) can be determined by this method if it is first reduced chemically to As(III). The most effective way of doing this is to reduce As(V) to $AsCl_3$ with Cu(I) in concentrated HCl, followed by extraction of $AsCl_3$ into benzene and back-extraction into water.

Conclusions

The above discussions provide a basic description of the analytical attributes of voltammetry together with some practical details of its implementation. Commercial instruments for carrying out these techniques have not reached the level of sophistication taken for granted with most forms of spectroscopy.[22] But recent improvements have finally made these techniques usable by nonexperts for both routine analytical applications and physicochemical investigations. By virtue of speed, sensitivity, robustness, and species specificity they should receive steadily broadening use for determining and characterizing trace metals in biological systems, especially at trace levels.

Acknowledgments

This chapter was made possible by financial support from the Guggenheim Foundation and from the National Science Foundation under Grant CHE 8305748. The author wishes to thank R. A. Osteryoung for helpful comments.

[22] J. Osteryoung, *Science* **218,** 261 (1982).

[20] Neutron Activation Analysis

By JACQUES VERSIECK

Introduction

Since its original description in the 1930s by G. Hevesy and H. Levi, neutron activation analysis (NAA) has developed into one of the most powerful and versatile techniques for elemental analysis in general and trace elemental analysis in particular. Its high cost did not prove to be a deterrent to its extensive application: because of its excellent sensitivity and remarkable specificity it has been used by researchers in the biological and medical field all over the world.

Over the course of the years, the fundamental principles on which the technique is based remained unchanged, but the remarkable developments in detection instrumentation and electronic equipment that took place, especially in the 1970s, gave impetus to considerable refinement. Thus, neutron activation analysis started as a single-element technique when only Geiger–Müller counters and NaI(Tl) detectors were available; with the development of high resolution Ge(Li) detectors—or, more recently, high-purity (HP) Ge detectors—and associated multichannel analyzers, it became a really multielement technique.

In radiochemical neutron activation analysis (RNAA), separations (for example, by adsorption on inorganic ion exchangers and ion-exchange resins, distillation, precipitation, or solvent extraction) are applied to eliminate matrix activities or to remove interfering activities. In recent years, purely instrumental neutron activation analysis (INAA) came into vogue. Instrumental neutron activation analysis does not involve any chemistry: samples are irradiated and counted over a period of time to obtain the desired information about the elements of interest. Spectral interference-free γ-ray energies are available for most elements and where there are two isotopic species at the same energy, their contributions can be computed. On the other hand, all biological matrices contain large amounts of sodium, potassium, and chlorine, which become highly radioactive upon irradiation so that their γ-ray photopeaks swamp the energies of other isotopes of interest. Because ^{24}Na (half-life or $t_{1/2}$ = 15.02 hr), ^{42}K ($t_{1/2}$ = 12.36 hr), and ^{38}Cl ($t_{1/2}$ = 37.2 min) are not removed but simply allowed to decay, most short-lived species in trace amounts cannot be accurately determined. Similarly, ^{32}P ($t_{1/2}$ = 14.28 days) produces an intense Bremsstrahlung which prevents the determination of several long-lived but low energy radioisotopes, such as ^{51}Cr ($t_{1/2}$ = 27.71 days, γ-line of 320.1 keV). Under such conditions, radiochemical separations have to be performed.

Available neutron irradiation facilities provide a wide range of neutron fluxes and energies. The three main sources of neutrons for irradiation are nuclear reactors, radioactive neutron sources, and electron and ion accelerators which produce high-energy neutrons. Nuclear reactors with their high neutron fluxes give the most intense irradiation and, consequently, permit the highest sensitivities for the detection and quantitative measurement of various elements. Many operate routinely at fluxes of 5×10^{12}–10^{14} n cm^{-2} sec^{-1} (neutrons per cm^2 per second) and even higher so that elements can be analyzed within the range of 10^{-9}–10^{-12} g and lower. As most neutron activation analysis uses research reactors as sources of neutrons, other sources will not be considered further. Similarly, *in vivo* techniques will not be treated in this chapter.

Nuclear reactors consist of uranium or uranium oxide, more or less enriched in ^{235}U, dispersed in a moderator of graphite, water, or deuterium oxide which slows down the fission or fast neutrons to thermal and epithermal energies. Figure 1 shows the reactor core and the irradiation positions in the 250 kW swimming pool nuclear reactor Thetis in the University Institute for Nuclear Sciences in Ghent.

Theory and Mathematical Formulas[1-3]

Neutron activation analysis relies on the fact that elements can be made radioactive by neutron bombardment. Two physical processes, one prompt and one delayed, are associated with this activation. For example, in the reaction between thermal neutrons and manganese, $^{1}n + ^{55}Mn \rightarrow ^{56}Mn + \gamma$, the prompt γ-ray is emitted, as usually, within 10^{-12} sec of the absorption of the neutron. Some analytical procedures utilize this prompt radiation, but they will not be discussed here. The usual form of neutron activation analysis concerns the delayed event resulting from the decay of the activated nuclide: ^{56}Mn decays with a $t_{1/2}$ of 2.582 hr, giving off characteristic γ (846.6, 1811.2, and 2112 keV) and β^- radiation (2850 keV max) and forming stable ^{56}Fe.

In the case of copper, there are two naturally occurring stable isotopes: ^{63}Cu and ^{65}Cu. During irradiation, a small fraction of the first will capture a thermal neutron to form ^{64}Cu and a different small fraction of the second to form ^{66}Cu nuclei. These radioisotopes decay, respectively, with a $t_{1/2}$ of 12.74 hr (main γ-lines: 511 and 1345.8 keV) and 5.10 min (main γ-lines: 833.6 and 1039 keV).

If a sample contains N nuclei of a particular stable isotope, the steady rate of formation of its (n,γ) product nuclei is $N\Phi\sigma$, in which Φ is the thermal neutron flux density in n cm^{-2} sec^{-1} and σ the (n,γ) cross-section of the target nuclei—the effective cross-sectional area per nucleus toward the capture of thermal neutrons (generally expressed in barns: one barn equalling 10^{-24} cm^2). If the (n,γ) product is radioactive, the nuclei will decay at a rate of λN^*, λ being the radioactive-decay, first-order rate constant of that species of radioisotope ($\ln 2/t_{1/2}$ or $0.6931/t_{1/2}$) and N^* the number of the radioactive nuclei present. Thus, at any given time during a

[1] F. Adams, R. Gijbels, J. P. Op de Beeck, D. De Soete, P. Van den Winkel, and J. Hoste, *CRC Crit. Rev. Anal. Chem.* **1,** 455 (1971).

[2] H. J. M. Bowen, *CRC Crit. Rev. Anal. Chem.* **10,** 127 (1981).

[3] V. P. Guinn and J. Hoste, *in* "Elemental Analysis of Biological Materials," p. 105. International Atomic Energy Agency, Vienna, Austria, 1980.

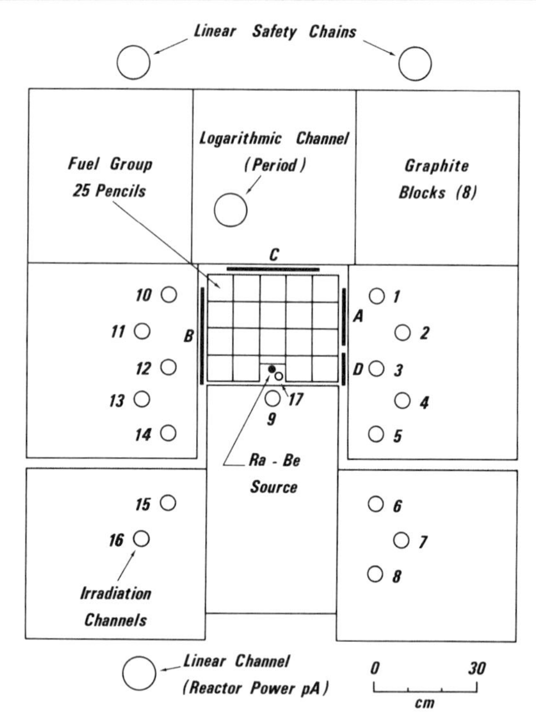

FIG. 1. Schematic representation of the reactor core and the irradiation channels (rabbit conducts) (1–17) of the 250 kW "swimming pool" research reactor Thetis (5% ^{235}U enrichment) in the Institute for Nuclear Sciences at the University of Ghent. A and B, safety rods; C, shim rod; and D, control rod.

steady irradiation, the net rate of formation of radioactive product nuclei is equal to $N\Phi\sigma - \lambda N^*$. After integration, the basic NAA equation is written as follows[3]:

$$A_0 = N\Phi\sigma(1 - e^{-\lambda t_i}) \quad (1)$$

A_0 being the activity (disintegration rate per second, in becquerel, Bq) exactly at the end of the irradiation, thus at zero decay time, and t_i being the duration of the irradiation period.

This basic neutron activation analysis equation applies to each radioisotope produced in the sample by (n,γ) reaction during the irradiation. The total disintegration rate of the activated sample is thus the sum of all the different A_0 values.

The expression in parentheses in the basic equation, $(1 - e^{-\lambda t_i})$, is called the saturation factor, or S. It varies from 0 to 1: zero for $t_i = 0$ or an extremely short irradiation relative to the half-life, $t_{1/2}$, of the radioiso-

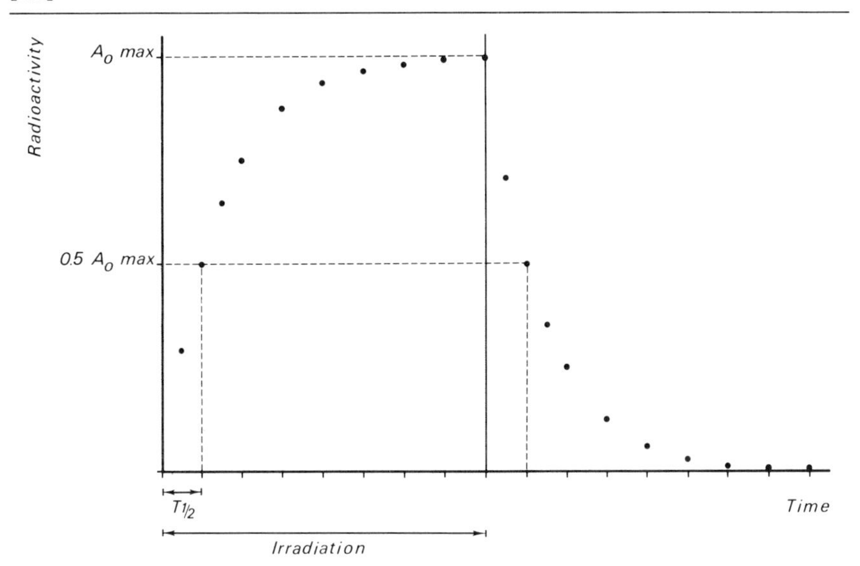

FIG. 2. Growth and decay of radioactivity during and after irradiation of an element in a nuclear reactor (e.g., ${}^{1}n + {}^{55}Mn \rightarrow {}^{56}Mn + \gamma$). The figure shows that the activity tends asymptotically to a maximum as the period of activation, relative to the half-life of the radioisotope ($t_{1/2}$), increases (according to the formula for the saturation factor or S—see text). So, little is to be gained by prolonging the irradiation time (t_i) beyond 6 or 8 half-lives (for $t_i = 6t_{1/2}$, $S = 0.9844$; for $t_i = 8t_{1/2}$, $S = 0.9961$).

tope; one for $t_i \gg t_{1/2}$ or an infinitely long irradiation. It is a dimensionless quantity, rapidly and asymptotically approaching the limiting value of 1 (at saturation) with increasing irradiation time relative to the half-life of the activated radioisotope species (see Fig. 2). At saturation, when $S = 1$, A_0 is simply equal to $N\Phi\sigma$, i.e., to the steady state of the formation of the radioisotope (rate of decay equals rate of formation).

After irradiation, the induced radioactivity decreases exponentially with the $t_{1/2}$ of the radioisotope species: the decay factor, or D, is equal to $e^{-\lambda t_d}$ or $e^{-0.6931 t_d / t_{1/2}}$, t_d being the time between the end of the irradiation and the moment of the measurement.

The term N in the basic neutron activation analysis Eq. (1) is equal to $w\Theta N_A/AW$, in which w is the weight of the element, Θ the fractional (or percent) isotopic abundance of the target nuclide, N_A Avogadro's number (6.022×10^{23} atoms/gram-atom), and AW the atomic weight of the element. Thus, the basic equation can be rewritten as follows:

$$A_0 = \frac{w\Theta N_A}{AW}\Phi\sigma(1 - e^{-\lambda t_i}) \tag{2}$$

In practice, because of experimental uncertainties regarding Φ, σ, and detection efficiency of measurement systems, Eqs. (1) or (2) are rarely employed. If a relative rather than absolute method is used, it is not necessary to know the numerical values of the above mentioned parameters. When a sample and a standard are irradiated in identical conditions and the activated sample and standard are counted on the same radiation detector, the much simpler equation may be applied:

$$\frac{A'_0 \text{ (sample)}}{A'_0 \text{ (standard)}} = \frac{w \text{ (sample)}}{w \text{ (standard)}} \tag{3}$$

where A'_0 (sample) and A'_0 (standard) are the counting rates, in counts per second, of sample and standard, and are equal to εA_0 (sample) and εA_0 (standard), where ε is the detection efficiency of the radiation detector. Since w (standard)—the mass of the element in the standard—is known, w (sample)—the mass of the element in the sample—can easily be calculated.

In actual practice, it is not necessary to use A'_0 (sample) and A'_0 (standard). Since the decay of a radioisotope follows the expression $A_t = A_0 e^{-\lambda t_d}$, the comparation equation can also be written for any stated decay time, t_d, since the end of the irradiation:

$$\frac{A'_{td} \text{ (sample)}}{A'_{td} \text{ (standard)}} = \frac{w \text{ (sample)}}{w \text{ (standard)}} \tag{4}$$

Since sample and standard cannot be counted on the same detector at the same time, one of the two measured counting rates must be calculated to what it would have been at the moment the other was counted. To make the adjustment, equation

$$A'_{t'd} = A'_{td} e^{-\lambda(t'_d - t_d)}$$

is used.[3]

Instead of coirradiating a standard for each element to be determined, preference may be given to the application of a single or multielement comparator technique.[4]

Equation (1) applies to all the numerous nuclear reactions giving rise to radionuclides and almost all the elements present in the sample become radioactive.

During irradiation, samples are exposed to thermal (energy or $E < 0.5$ eV), epithermal (1 MeV $> E >$ 0.5 eV), and fast or fission neutrons ($E >$ 1 MeV). The distribution at the site of a sample depends on the type of the moderator and on the distance traveled by the neutrons.

[4] L. Moens, F. De Corte, A. De Wispelaere, J. Hoste, A. Simonits, A. Elek, and Z. Szabo, *J. Radioanal. Nucl. Chem.* **82,** 385 (1984).

The type of the nuclear reaction and its cross section varies with the energy of the neutrons. The absorption of thermal and epithermal neutrons leads to (n,γ) reactions. At low neutron energies, the cross sections of many nuclides obey the $1/v$ law (v = velocity), which means that their cross sections are inversely proportional to the neutron velocity. Cross sections for monoenergetic neutrons at a velocity of 2200 m sec^{-1}, which is the most probable velocity in a Maxwellian distribution for 20° and corresponds to an energy of 0.0253 eV, are generally represented by σ_o. Effective cross sections or "average thermal cross sections," measured in a well-thermalized neutron source whose conventional flux is known by determination with an $1/v$ detector, are currently indicated by σ_{th}. The deviations from the $1/v$ law for most reactions up to about 0.5 eV are so small that the differences are within the accuracy of the measurements and σ_0 and σ_{th} can be substituted for one another. Above 0.5 eV, reaction cross sections no longer follow the $1/v$ law. The wide energy region between 0.5 eV and 1 MeV is referred to as the resonance region because many isotopes show high cross sections (resonance peaks) occurring at discrete neutron energies in this region. To calculate reaction rates, an effective cross section I_0—usually called resonance integral—has been defined. The energy of 0.5 eV is called the cadmium threshold because a cadmium foil of 0.5–1.0 mm thickness acts as a filter which is passed by neutrons only at energies above this level. The ratio of the thermal to resonance activation is measured conveniently by the cadmium ratio, the reaction rate of thermal and resonance neutrons to the reaction rate of resonance neutrons or

$$R_{Cd} = \frac{\sigma_{th}\Phi_{th} + I_0\Phi_{epi}}{I_0\Phi_{epi}} \tag{5}$$

The element most frequently used to measure epithermal neutron fluxes (Φ_{epi}) and to calibrate other resonance integrals is gold. Indeed, in the reaction $^{197}Au(n,\gamma)^{198}Au$ there is one resonance peak which is well established to be 1551 ± 20 barn. To measure Φ_{epi}, an "infinitely thin" gold foil is irradiated once with and once without cadmium envelopment.

The presence of fast or fission neutrons gives rise to threshold reactions of the type (n,p), (n,α), and $(n,2n)$. The first and the second entail the production of, respectively, $Z - 1$ and $Z - 2$ nuclides (Z = atomic number). Examples are $^{56}Fe(n,p)^{56}Mn$, $^{31}P(n,\alpha)^{28}Al$, and $^{54}Fe(n,\alpha)^{51}Cr$. Although the cross sections for these reactions (average fast neutron cross sections or σ_f) are generally low compared to those for thermal neutron activation, serious interferences may occur. For example, when serum is irradiated in a neutron flux with a thermal-to-fast ratio (Φ_{th}/Φ_f) of 10, phosphorus (roughly 130 μg/ml) produces an "apparent" aluminum concentration of approximately 94 ng/ml (whereas the true aluminum level is now generally believed to be less than 5 or 10 ng/ml)! A possible solution, namely to remove the phosphorus before the irradiation, would deprive the procedure of its most important advantage—its relative lack of susceptibility to contamination errors (vide infra).

Radioactive Decay[3,5]

Radioisotopes formed by (n,γ) reaction during irradiation in a nuclear reactor decay by negatron (β^-) emission, isomeric transition (IT), positron (β^+) emission, or electron capture (EC).

Beta transitions show a broad electron spectrum, from zero to a finite maximum energy, which is characteristic of the nuclide. Because of their continuous nature, beta spectra of different isotopes are difficult to individualize and therefore very rarely used for quantitative measurements in neutron activation analysis. Fortunately, almost all β^- emitters decay in some or all of their disintegration to an excited state of the product nucleus, which then promptly drops to its ground state under the emission of γ-rays with one or more characteristic, sharply defined energies.

Decay by IT is seen in cases where the (n,γ) product of the irradiation is a metastable nuclear isomer. In most of such cases, the decay to the ground state of the product nucleus occurs under the emission of γ-rays with one or a few characteristic energies. Examples of such metastable isomer (n,γ) products are ^{60m}Co, ^{69m}Zn, ^{77m}Se, ^{86m}Rb, and several others.

A number of (n,γ) products decay in some of their disintegrations by β^+ emission or EC. β^+ particles rapidly slow down in matter and then undergo an annihilation reaction with a negative electron to produce two 511 keV γ-rays emitted in opposite directions. In EC decay, characteristic X-rays are emitted; in many cases, it is also associated by γ-ray emission.

Most (n,γ) product radioisotopes decay to a stable isotope. There are, however, a number of instances in which the decay leads to another radioisotope. In some cases, these radioactive daughter substances are the most suitable nuclides for counting. Well-known examples are ^{99}Mo ($t_{1/2}$ = 66.02 hr) which decays to ^{99m}Tc ($t_{1/2}$ = 6.02 hr, γ-line, 140.5 keV), ^{115}Cd ($t_{1/2}$ = 53.5 hr) which decays to ^{115m}In ($t_{1/2}$ = 4.5 hr, γ-line, 336.3 keV), and ^{113}Sn ($t_{1/2}$ = 115.1 days) which decays to ^{113m}In ($t_{1/2}$ = 1.658 hr, γ-line: 391.7 keV).

Detection of Emitted Radiation

When an element has been separated in a radiochemically pure state, it can be counted using a simple, cheap beta or gamma counter. When

[5] C. E. Crouthamel, F. Adams, and R. Dams, "Applied Gamma-Ray Spectrometry." Pergamon Press, Oxford, England, 1970.

radiochemically impure fractions or intact activated samples are used for analysis, more sophisticated measurement systems are needed.

For simple γ-counting, thallium-activated sodium iodide scintillation detectors—NaI(Tl) detectors—are standard items. For more than 15 years, they were the reference detectors in neutron activation analysis laboratories throughout the world. The counting efficiency depends on their size and shape and the energy of the γ-photons. Well-type detectors have a higher efficiency but a poorer resolution than cylindrical detectors.

The routinely available resolution of 7.5–8.0% at the 661.6 keV ^{137}Cs photopeak of NaI(Tl) detectors is sufficient to discriminate a few photopeaks in a spectrum under proper conditions. So, they have been used in connection with multichannel analyzers of 256–400 channels, e.g., for the determination of manganese, copper, and zinc in serum and packed blood cells by the author in the early 1970s.[6]

For multielement determinations, increased selectivity is required; it is found in semiconductor detectors among which the lithium-drifted germanium [Ge(Li)] or high-purity germanium [HP-Ge] types are generally the choice for instrumental neutron activation analysis.[1,7] The mean property is an energy resolution down to approximately 1.8 keV (full width at half maximum or FWHM) for the ^{60}Co radiation at 1332.5 keV; thus, expensive multichannel analyzers with some 4000 channels are needed to make efficient use of their high resolution. A typical detection system consists of a shielded, encased, liquid nitrogen-cooled crystal, coupled to a high-voltage supply, a preamplifier, a linear amplifier, and a multichannel pulse–height analyzer. The spectrometer usually includes electronic components such as a pile-up rejector (pile-up peaks are peaks appearing from the coincident counting of two γ-rays or of an X-ray and a γ-ray), a percentage dead-time meter, a baseline restorer, a preset timer (for various selected counting clocktimes or livetimes), and an oscilloscope display. The large amount of data generated makes it very useful to have a computer to handle the information. When a pulse–height spectrum is read out, the energies and net photopeak counts (with their standard deviation) can be rapidly calculated and printed. The equipment can be programmed to compare the areas of specified peaks in the spectra of samples and standards and to compute the composition of the sample. To explain the rudiments of the analyses of a complex spectrum it is best to start with the pulse–height spectrum generated by a single radioisotope emitting γ-rays of only one energy. The features of interest are the charac-

[6] J. Versieck, A. Speecke, J. Hoste, and F. Barbier, *Z. Klin. Chem. Klin. Biochem.* **11,** 193 (1973).

[7] R. Dams, F. De Corte, J. Hertogen, J. Hoste, W. Maenhaut, and F. Adams, *in* "Physical Chemistry Series 2, Vol. 12: Analytical Chemistry. Part 1" (T. S. West, ed.), p. 1. Butterworths, London, 1976.

teristic, narrow, almost Gaussian photopeak (total absorption peak) and the Compton scattering continuum (partial absorption with escape of a secondary gamma) which extends as a plateau from the beginning of the spectrum (low energy region) to the "Compton edge," where it decreases rapidly. For a radioisotope emitting γ-rays of more than one energy for a mixture of radioisotopes, the pulse–height spectrum is the sum of the individual spectra, and hence exhibits a number of photopeaks, Compton continua, and Compton edges.

There are several useful catalogs of γ-ray energies.[5,8–10] They show that the vast majority of γ-rays, emitted by (n,γ) product radioisotopes of biologically important trace elements, range from 50 to 2000 keV. So, very frequently, 4000 channel spectrometers are operated at a gain setting of 0.5 keV per channel, as in Fig. 3, which shows a HP-Ge spectrum of a multielement standard (human serum doped with 10 elements)[11] irradiated for approximately 75 hr at a neutron flux of 1.1×10^{12} n cm^{-2} sec^{-1} and measured for 15 hr, 18 days after the end of the irradiation. The exact location of the photopeaks is determined by calibration with radioisotope sources of two or more γ-rays of accurately known energies. The excellent resolution and the outstanding selectivity of the system are best illustrated by the individualization of the numerous photopeaks between 500 and 1000 keV (channel numbers 1000–2000) and 1000 and 1500 keV (channel numbers 2000–3000). As soon as their energies differ by 5 or 6 keV, photopeaks are adequately resolved to permit peak integration (calculation of the net counts of each peak) with reasonable accuracy. The 563.3 and 569.3 keV photopeaks of ^{134}Cs serve as an example. Only when their energies hardly differ by 2 or 3 keV, they largely overlap, as shown by the 602.7 keV photopeak of ^{124}Sb and the 604.7 keV photopeak of ^{134}Cs. For further information, the reader is referred to publications on the analysis of γ-spectra and processing of data.[12–14]

A spectral interference is defined as any γ-ray peak which appears among the channels used for an analysis and which is variable in intensity

[8] F. Adams and R. Dams, *J. Radioanal. Chem.* **3,** 99 (1969).

[9] G. Erdtman and W. Soyka, "The γ-Lines of the Radionuclides," Report Jül-1003-AC, three volumes. Kernforschungsanlage Jülich, Jülich, Federal Republic of Germany, 1973.

[10] G. Erdtman, "Neutron Activation Tables." Verlag Chemie, Weinheim, Federal Republic of Germany, 1976.

[11] R. Cornelis, A. Speecke, and J. Hoste, *Anal. Chim. Acta* **68,** 1 (1973).

[12] I. Perlman, *in* "Nondestructive Activation Analysis" (S. Amiel, ed.), p. 9. Elsevier, Amsterdam, 1981.

[13] J. P. Op de Beeck and J. Hoste, *in* "Physical Chemistry Series 2, Vol. 12: Analytical Chemistry. Part 1" (T. S. West, ed.), p. 151. Butterworths, London, 1976.

[14] H. P. Yule, *in* "Nondestructive Activation Analysis" (S. Amiel, ed.), p. 113. Elsevier, Amsterdam, 1981.

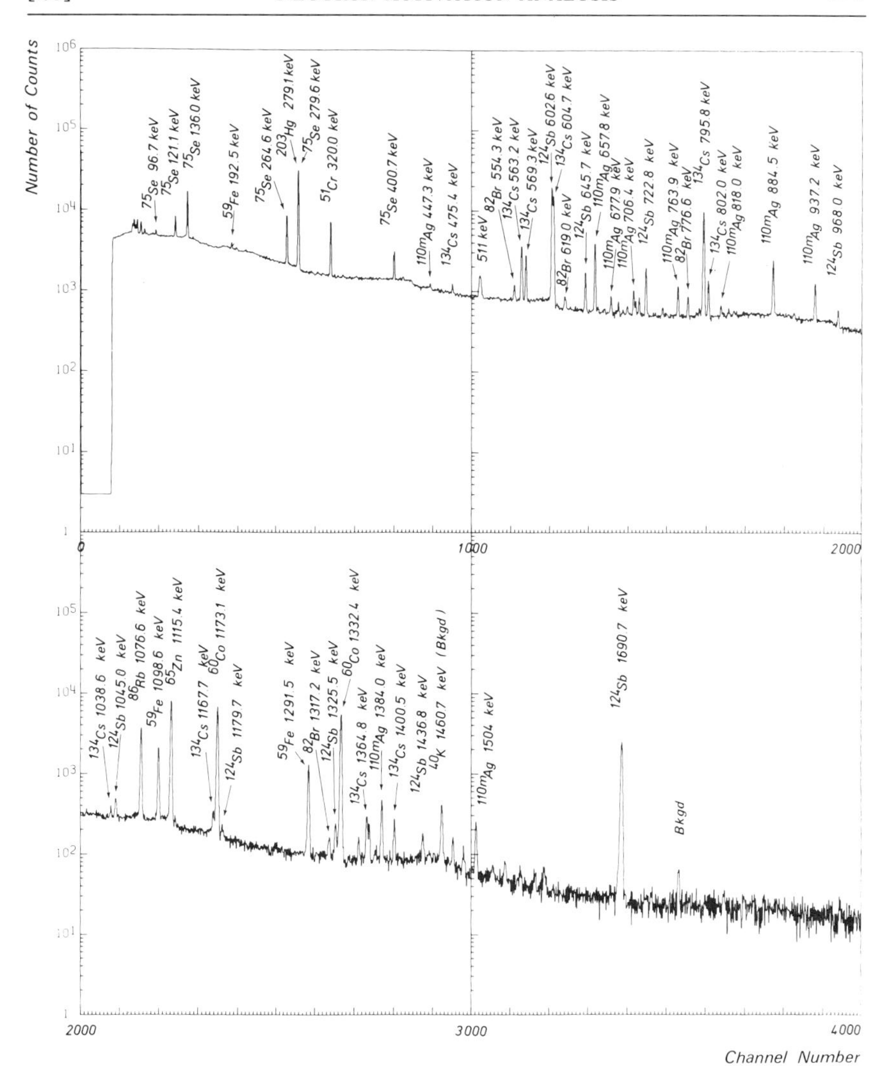

FIG. 3. HP-Ge spectrum of a multielement standard (human blood serum doped with 10 elements)[11] irradiated for approximately 75 hr (in periods of approximately 6 hr, two times a week) at a neutron flux of 1.1×10^{12} n cm^{-2} sec^{-1} and measured for 15 hr some 18 days after the end of the activation. The excellent resolution and the outstanding selectivity of the measurement system are illustrated by the individualization of the 563.2 and 569.3 keV and the 795.8 and 802.0 keV photopeaks of ^{134}Cs. Only when their energies differ only by 2 or 3 keV, photopeaks overlap as shown by the 602.6 keV photopeak of ^{124}Sb and the 604.7 keV photopeak of ^{134}Cs.

relative to the peak of interest.[12] In seeking out possible interferences, it is invaluable to have a catalog of the γ-lines of the radionuclides arranged according to their energy and a table of radioisotopes showing all γ-rays associated with their radioactive decay. Unfortunately, no catalog explicitly gives all the necessary information even though it is "complete" and even though the energies listed are accurate. For example, significant pile-up peaks (vide supra) and single (SE) or double escape (DE) peaks of a high energy γ-ray (e.g., 1732.1 keV—DE peak of the 2754.1 photopeak of ^{24}Na) are practically never listed.

Sensitivity

NAA is a technique that has shown great sensitivity in general trace element analysis. Absolute sensitivities depend on several factors: the flux of the nuclear reactor, the cross section and the fractional isotopic abundance of the target nuclide, the irradiation and decay periods, the counting efficiency of the detector, and so forth. Experimental sensitivities may be idealized because of matrix problems, difficulties in radiochemical separations, and other analytical complications resulting from the analysis of real, complex samples. Unless conditions are precisely defined, sensitivities obtained by different researchers cannot easily be compared.

"Best INAA detection sensitivities in the absence of interfering activities" were published by Guinn and Hoste.[3] They were determined for $\Phi = 10^{13}\ n\ \mathrm{cm}^{-2}\ \mathrm{sec}^{-1}$, $t_i = 5$ hr, $t_d = 0$, $t_c = 100$ min, a 40 cm^3 Ge(Li) detector, a sample-to-detector distance of 2 cm, and the γ-ray peak of the (n,γ) product of the element that provides the most sensitive detection of the element. In the text, the limit of detection is arbitrarily defined as the amount of the element (in μg) that gives 30 net photopeak counts. Values for biologically important trace elements are listed in Table I. In addition to the calculated best INAA detection sensitivities, the table also contains the calculated limits of detection (if any) for some other (n,γ) product that may be used for quantitative determination.

It should be noted that the detection limits may be considerably manipulated by changing the experimental conditions. They decrease if the neutron flux is reduced, the Ge(Li) detector has a lower detection efficiency, the counting geometry is poorer, the counting time is shorter, and—in the case of elements measured via activities with half-lives considerably longer than 5 hr—if the irradiation time is less than 5 hr. On the other hand, they improve if the neutron flux is pushed up, the Ge(Li) detector has a higher detection efficiency, the counting geometry is better, and—for the long-lived activities—if longer irradiation and counting

TABLE I
CALCULATED APPROXIMATE INAA DETECTION SENSITIVITIES[a] IN THE ABSENCE OF INTERFERING ACTIVITIES[b]

Element	(n, γ) Radioisotope	Half-life	γ-Ray energy (in keV)	Detection sensitivity (in μg)
Mn	^{56}Mn	2.582 hr	846.6	1.3×10^{-6}
Co	^{60m}Co	10.48 min	58.6	1.5×10^{-5}
	^{60}Co	5.272 years	1173.1	1.2×10^{-2}
Hg	^{197}Hg	64.1 hr	69–78	1.5×10^{-5}
	^{203}Hg	46.60 days	279.1	2.4×10^{-3}
Cs	^{134m}Cs	2.9 hr	127.4	1.6×10^{-5}
	^{134}Cs	2.06 years	604.7	2.4×10^{-3}
As	^{76}As	26.3 hr	559.1	2.8×10^{-5}
Cu	^{64}Cu	12.74 hr	511	3.5×10^{-5}
	^{66}Cu	5.10 min	1039	4.3×10^{-3}
V	^{52}V	3.755 min	1434.2	7.4×10^{-5}
Se	^{77m}Se	17.5 sec	161.9	5.5×10^{-4}
	^{75}Se	120 days	136.0	1.2×10^{-2}
Cd	^{111m}Cd	48.7 min	245.4	7.1×10^{-4}
	^{115}Cd	53.5 hr	527.9	6.3×10^{-3}
Ag	^{110}Ag	24.6 sec	657.8	1.0×10^{-3}
	^{108}Ag	2.41 min	632.9	1.1×10^{-3}
	^{110m}Ag	250.4 days	657.8	1.3×10^{-3}
Sn	^{123m}Sn	40.1 min	159.7	1.0×10^{-3}
	^{125m}Sn	9.2 min	332	5.3×10^{-3}
	^{117m}Sn	14 days	158.4	4.4×10^{-1}
Zn	^{69m}Zn	13.9 hr	438.9	1.2×10^{-3}
	^{65}Zn	243.7 days	1115.4	4.2×10^{-1}
Al	^{28}Al	2.246 min	1778.8	1.8×10^{-3}
Mo	^{101}Mo	14.6 min	191.9	3.1×10^{-3}
	^{99}Mo	66.02 hr	181.1	6.8×10^{-3}
Rb	^{86m}Rb	1.018 min	555.8	7.1×10^{-3}
	^{88}Rb	17.7 min	898	1.6×10^{-2}
	^{86}Rb	18.65 days	1076.6	8.4×10^{-2}
Ni	^{65}Ni	2.520 hr	366.5	7.3×10^{-3}
Cr	^{51}Cr	27.71 days	320.1	7.8×10^{-3}
Fe	^{59}Fe	44.6 days	1098.6	3.2

[a] For $\Phi_{th} = 10^{13}$ n cm^{-2} sec^{-1}, $t_i = 5$ hr, $t_d = 0$, $t_c = 100$ min, 40 cm^3 Ge(Li) detector, sample-to-detector distance = 2 cm, largest photopeak.
[b] From Ref. 3.

times are used. It should also be emphasized that the detection limits listed only apply when no other significant activities are present.

In practice, the (n,γ) reaction with the lowest detection limit does not invariably ensure the most convenient and most reliable analytical approach. So, in biological samples, cobalt is mostly determined via the $^{59}Co(n,\gamma)^{60}Co$ reaction,[15–17] selenium via the $^{74}Se(n,\gamma)^{75}Se$ reaction,[15,16,18] cesium via the $^{133}Cs(n,\gamma)^{134}Cs$ reaction[15,16,18] and cadmium via the $^{114}Cd(n,\gamma)^{115}Cd$ reaction[15,19] although, as shown in Table I, these reactions do not have the highest detection sensitivity under the conditions specified earlier in this section.

Other comments should be added. The first concerns aluminum: the element is included, although—as outlined earlier—its determination in biological matrices is very seriously hampered by the $^{31}P(n,\alpha)^{28}Al$ interference reaction. It can be calculated that, in the case of serum, to reduce the "apparent" aluminum concentration due to this interference reaction to ≤ 1 ng/ml, samples should be irradiated in a reactor position with a Φ_{th}/Φ_f flux ratio of at least 950. This is an exceptionally strict requirement. Indeed, the highest ratio in the Thetis reactor of the Laboratory for Analytical Chemistry in Ghent, Belgium, is only 110 (in channel 8—see Fig. 1); a look at the neutron activation facilities in other research reactors reveals that the highest values attained are 450–500 (DR 3 reactor in Risø National Laboratory in Roskilde, Denmark), 575 (DIDO reactor at the Atomic Energy Research Establishment in Harwell, U.K.), and 650 (NBSR reactor at the National Bureau of Standards in Gaithersburg, MD).[20] The second comment refers to the determination of molybdenum, cadmium, and tin: in actual trace element analyses of biological materials they are mostly measured via daughter isotopes (respectively ^{99m}Tc or ^{101}Tc, ^{115m}In, and ^{113m}In)[15,19,21,22] and not directly via the radioisotopes included in the table. Indeed, for a variety of reasons, these daughter isotopes turn out to be the most suitable nuclides for counting (vide supra).

In spite of its obvious limitations, the table gives a useful indication of the capability of neutron activation analysis to reach the levels of interest

[15] P. Lievens, J. Versieck, R. Cornelis, and J. Hoste, *J. Radioanal. Chem.* **37,** 483 (1977).

[16] C. Vanoeteren, R. Cornelis, J. Versieck, J. Hoste, and J. De Roose, *J. Radioanal. Chem.* **70,** 219 (1982).

[17] J. Versieck, J. Hoste, F. Barbier, H. Steyaert, J. De Rudder,and H. Michels, *Clin. Chem.* **24,** 303 (1978).

[18] J. Versieck, J. Hoste, F. Barbier, H. Michels, and J. De Rudder, *Clin. Chem.* **23,** 1301 (1977).

[19] J. Versieck, J. Hote, L. Vanballenberghe, and F. Barbier, *Atomkernenerg. Kerntech.* **44** (Suppl.), 717 (1984).

[20] K. Heydorn, "Neutron Activation Analysis for Clinical Trace Element Research," two volumes. CRC Press, Boca Raton, Florida, 1984.

[21] R. Cornelis, J. Versieck, A. Desmet, L. Mees, and L. Vanballenberghe, *Bull. Soc. Chim. Belg.* **90,** 289 (1981).

[22] J. Versieck, J. Hoste, F. Barbier, L. Vanballenberghe, J. De Rudder, and R. Cornelis, *Clin. Chim. Acta* **87,** 135 (1978).

in specified biological matrices. A paper has been published comparing the sensitivity limits of different analytical techniques and the concentration levels of both important essential and toxicological elements in human blood plasma or serum,[23] but its conclusions should be interpreted with extreme caution. Indeed, a comparison of the figures for the concentration levels in serum with reference values, selected from the literature on the basis of a critical evaluation of their reliability,[24,25] shows that some of the included data are grossly erroneous.

Quality of Analytical Data

In practice, significant experimental errors can occur in neutron activation analysis, as in other techniques. In general, they may be introduced at all stages of the analytical process: during the collection and preparation of the sample (additions by contamination with exogenous material, losses by volatilization), the irradiation (interfering reactions, neutron self-absorption, unequal flux at sample and standard), the postirradiation chemical processing (losses due to defective recovery of the sample, problems related to the addition of carrier), the radioactive counting (counting geometry, corrections for decay, counting statistics, background and peak area determination, different attenuation of γ-rays in sample and standard), and, finally, at the moment of the calculation and the interpretation of the results. In September 1982, the International Atomic Energy Agency convened an advisory group meeting around the subject: its members were requested to consider very carefully every conceivable cause of poor performance. For detailed information, the reader is referred to the published report.[26]

Neutron activation has a unique and invaluable advantage, namely its relative freedom from errors due to sample contamination—a formidable problem in biological trace element research.[27] Certainly, the collection and preparation of the samples for neutron activation studies must be done with the same amount of care (intelligently selected instruments, high-purity collection vessels cleaned with extraordinary care, clean room conditions, and so forth) as for other analytical techniques for trace element measurements.[27] There are even certain problems specifically linked with neutron activation analysis, namely irradiation and wet ashing

[23] G. H. Morrison, *CRC Crit. Rev. Anal. Chem.* **8,** 287 (1979).

[24] J. Versieck, *CRC Crit. Rev. Clin. Lab. Sci.* **22,** 97 (1985).

[25] J. Versieck and R. Cornelis, *Anal. Chim. Acta* **116,** 217 (1980).

[26] Advisory Group of the International Atomic Energy Agency, *Anal. Chim. Acta* **165,** 1 (1984).

[27] J. Versieck, F. Barbier, R. Cornelis, and J. Hoste, *Talanta* **29,** 973 (1982).

blanks.[28] On the other hand, preirradiation sample preparation can generally be kept to a minimum and—and this is of primary importance–blanks resulting from postirradiation handling and processing (mineralization of the sample, chemical separations) or from postirradiation airborne contamination do no longer affect the final result! In the author's opinion, this is the most decisive reason why the technique contributed to such an important extent to the establishment of reference values for trace element levels in a wide variety of biological matrices.

As a rule, the precision of a series of measurements is principally a question of the random error involved. Analytical determinations based on the measurement of radionuclides are subject to a special source of random variation, called counting statistics. The estimation rests on the assumption that radioactive decay follows a Poisson distribution. At low levels of the elements of interest, the contribution from counting statistics is a very significant part of the overall precision. In instrumental neutron activation analysis, other contributions come from the uncertainty about the quantity of sample and standard and their activation (flux homogeneity, flux gradients). In radiochemical neutron activation there are additional sources of variation, namely those related to the chemical separation process; if a yield determination is included, however, most are replaced by the uncertainty on the yield determination. Postirradiation carrier addition assures a constant, optimum level of the element to be determined; so, unlike other techniques, a radiochemical separation procedure is characterized by a precision which is independent of the concentration.

In neutron activation analysis it is possible to predict the standard deviation of a result before the analysis is carried out: it has been defined as the a priori precision.[20,29] With all sources of random variation identified and accounted for, it accurately refects the observed variability of replicate determinations, and the method is said to be in a state of statistical control. In the analysis of precision, described by Heydorn and coworkers,[20,29,30] the a priori precision is compared with the actually observed variation: it may be used as a method of continuous quality assessment.[31]

[28] R. Cornelis, J. Hoste, and J. Versieck, *Talanta* **29,** 1029 (1982).

[29] K. Heydorn, *in* "Metal Ions in Biological Systems: Vol. 16. Methods Involving Metal Ions and Complexes in Clinical Chemistry" (H. Sigel, ed.), p. 123. Dekker, New York, 1983.

[30] K. Heydorn and K. Nørgård, *Talanta* **20,** 835 (1973).

[31] K. Heydorn, *in* "Measurement, Detection and Control of Environmental Pollutants," p. 61. International Atomic Energy Agency, Vienna, Austria, 1976.

Limitations

Measurements by neutron activation are beset by certain problems. For example, there may be a long delay between the collection of a sample and the completion of the analysis, particularly if elements with long half-lives are to be determined. In those cases, samples must frequently be irradiated for several days, even up to 2 weeks, to attain the desired sensitivity; during this period, matrix elements or major components become highly radioactive so that long cooling periods (2–4 weeks) may be necessary to allow undesirable isotopes to decay. In addition, counting may also require a number of hours or even days. This long turnaround time means that the technique cannot compete with others in some fields, such as clinical chemistry. Fortunately, during irradiation some biologically important trace elements produce radionuclides with different half-lives so that, in a number of instances, there are alternative approaches. For example, selenium may be quantified not only via its ^{75}Se ($t_{1/2}$ = 120 days)[18] but also via its ^{81m}Se ($t_{1/2}$ = 57.3 min)[32] and ^{77m}Se ($t_{1/2}$ = 17.5 sec) isotopes.[33,34]

Moreover, neutron activation analysis is expensive in comparison with other techniques. Of course, the most serious problem is the need to have access to a research-type nuclear reactor. To reduce its running costs, very often it is shared by various departments in a university or various institutions and industries over a large area. Modern counting equipment is also high priced: a system with a high-resolution Ge(Li) or HP-Ge detector and a 4000 multichannel analyzer costs from $45,000 to $60,000, depending on the detection efficiency and resolution of the detector and the data reduction capabilities of the multichannel analyzer. Sample changers to count automatically and computers for data reduction increase the expenses further. However, the superiority of the technique with respect to the applicability for numerous trace and ultratrace determinations has largely offset its high cost.

Most biological samples contain considerable amounts of sodium which, after irradiation, emits γ-rays with great penetrating power. One hundred milligrams of dried (lyophilized) serum (roughly corresponding to 1 ml of fresh serum) contains approximately 3.25 mg of sodium which, activated for 5 hr in a neutron flux of 4×10^{12} n cm^{-2} sec^{-1} gives rise to approximately 1 mCi or 37 MBq of ^{24}Na. Under the same conditions, the

[32] K. Heydorn and E. Damsgaard, *Talanta* **20,** 1 (1973).

[33] R. C. Dickson and R. H. Tomlinson, *Clin. Chim. Acta* **16,** 311 (1967).

[34] H. K. J. Hahn, R. V. Williams, R.E . Burch, J. F. Sullivan, and E. A. Novak, *J. Lab. Clin. Med.* **80,** 718 (1972).

amount of chlorine present in serum (approximately 3.95 mg/ml) gives rise to approximately 0.75 mCi or 28 MBq of ^{38}Cl. The activity of the first (^{24}Na) decays with a half-life of 15.02 hr, that of the second (^{38}Cl) with a half-life of 37.2 min.

The foregoing makes it clear that the opening of a rabbit shortly after the end of an activation of biological samples requires careful monitoring to avoid excessive exposure to the investigator's eyes and hands as well as good ventilation to prevent the inhalation of radioactive dusts or gases. Furthermore, as long as matrix activities are not eliminated, radiochemical separation manipulations may have to be carried out behind a wall of 5 or 10 cm of lead but in an adequately equipped laboratory this does not entail a major problem. Besides, it is evident that several steps can readily be automated and this has been done by various groups.[15,35] Particularly when considerable numbers of samples of a similar nature are to be processed under a well-defined set of conditions, the devising and constructing of an automated separation system are worth serious thought.

In general, the hazards of handling reactor irradiated biological samples are seldom greater than those encountered in other work involving the use of radioactive isotopes.

Applications

A considerable body of literature has grown up about the application of neutron activation analysis to the trace element analysis of human body fluids and tissues and other samples of biomedical interest. At the University of Ghent, the technique has been used for the determination of many elements in serum,[6,17–19,22,25,36–38] packed blood cells,[6,18,38] urine,[21,39] liver,[15] and lung tissue.[16] There is no space here to oversee the numerous reports originating from other laboratories all over the world: they are scattered throughout the literature but they can relatively easily be located by consulting reviews[2,24,25] and books[20,40] where references from the literature are brought together.

The trace element concentrations in human serum, measured at the University of Ghent, are catalogued in Table II. A survey of the litera-

[35] P. S. Tjioe, J. J. M. de Goeij, and J. P. W. Houtman, *J. Radioanal. Chem.* **16,** 153 (1973).

[36] R. Cornelis, J. Versieck, L. Mees, J. Hoste, and F. Barbier, *Biol. Trace Elem. Res.* **3,** 257 (1981).

[37] J. Versieck and L. Vanballenberghe, *in* "Trace Elements in Man and Animals—TEMA 5" (C. F. Mills, I. Bremner, and J. K. Chesters, eds.), p. 650. Commonwealth Agricultural Bureaux, 1985.

[38] J. Versieck, F. Barbier, A. Speecke, and J. Hoste, *Clin. Chem.* **20,** 1141 (1974).

[39] R. Cornelis, A. Speecke, and J. Hoste, *Anal. Chim. Acta* **78,** 317 (1975).

TABLE II
TRACE ELEMENT CONCENTRATIONS IN HUMAN BLOOD SERUM MEASURED AT THE UNIVERSITY OF GHENT[a]

Element	Unit	Mean	SD	Range
V	ng/ml			0.024–0.939[b]
		0.031[c]	0.010[c]	0.016–0.139[c]
Cr	ng/ml	0.160	0.083	0.0382–0.351
Mn	ng/ml	0.57	0.13	0.38–1.04
Co	ng/ml	0.108	0.060	0.0394–0.271
Cu	μg/ml	1.07	0.24	0.73–1.99
Zn	μg/ml	0.94	0.13	0.69–1.21
As	ng/ml	0.958		0.088–5.488
Se	μg/ml	0.13	0.02	0.09–0.18
Br	μg/ml	4.87	2.02	1.28–7.48
Rb	μg/ml	0.17	0.04	0.09–0.27
Mo	ng/ml	0.58	0.21	0.28–1.17
Cd	ng/ml			<0.192
Cs	ng/ml	0.74	0.20	0.45–1.18

[a] Data are from Refs. 6, 17–19, 22, 25, and 36–38.
[b] Male.
[c] Female.

ture[25] shows that widely different figures have been published. Growing evidence suggests that the plasma or serum levels of certain elements (more particularly vanadium, chromium, manganese, cobalt, arsenic, molybdenum, and cadmium—but also nickel[41]) have been very seriously overestimated in a number of laboratories.

In clinical practice, most trace element determinations are made by techniques other than neutron activation analysis, more particularly by atomic absorption spectrometry (AAS)—a technique with sufficient sensitivity for the tasks required and better suited for routine applications. Neutron activation analysis has been chiefly used for a number of special purposes, e.g., for the determination of trace elements in blood platelets,[42,43] heart,[44] lung,[16,45] liver,[15,45] and other tissues.[45–47]

[40] G. V. Iyengar, W. E. Kollmer, and H. J. M. Bowen, "The Elemental Composition of Human Tissues and Body Fluids." Verlag Chemie, Weinheim, Federal Republic of Germany, 1978.

[41] F. W. Sunderman, M. C. Crisostomo, M. C. Reid, S. M. Hopfer, and S. Nomoto, *Ann. Clin. Lab. Sci.* **14,** 232 (1984).

[42] J. Kiem, H. Borberg, G. V. Iyengar, K. Kasperek, M. Siegers, L. E. Feinendegen, and R. Gross, *Clin. Chem.* **25,** 705 (1979).

Neutron activation analysis has been employed sporadically for the study of trace elements in serum and liver protein fractions, collagen, and nucleic acids as well as in biochemicals containing activable elements, such as cobalt in vitamin B_{12} and selenium in selenoamino acids but, as far as the author knows, only exceptionally for the study of metalloenzymes and metalloenzyme complexes.[48] Since measurements can be made with very small quantities of material, this type of application is expected to attract more researchers in the future.

Conclusions

Neutron activation analysis is a valuable technique for trace element determinations. In competent hands, it is able to give highly accurate and reliable results. It has an excellent sensitivity and remarkable specificity but, in the author's opinion, its most decisive advantage is its relative freedom from errors due to the contamination of samples with exogenous material which plague other techniques for trace element analyses. On the other hand, it is neither a panacea and this is best illustrated by the inconsistent results obtained in a number of laboratories.

Because of its relative inaccessibility, the use of neutron activation analysis has been restricted to a few (principally research) laboratories. Neutron activation analysis has contributed most to the advance of biomedical trace element research when it was employed to tackle selected problems, that could not be solved in other ways, or to check the validity of other, more easily available procedures. Probably, this will largely remain unchanged in the future.

Acknowledgments

The technical and/or secretarial assistance of Lidia Vanballenberghe, Antoine De Kesel, and Yvette Odent is gratefully acknowledged.

[43] K. Kasperek, G. V. Iyengar, J. Kiem, H. Borberg, and L. E. Feinendegen, *Clin. Chem.* **25,** 711 (1979).

[44] P. O. Wester, *Acta Med. Scand.* **178,** (Suppl.), 439 (1965).

[45] P. O. Wester, D. Brune, and G. Nordberg, *Br. J. Ind. Med.* **38,** 179 (1981).

[46] W. D. Ehmann, W. R. Markesbery, T. I. M. Hossain, N. Alauddin, and D. T. Goodin, *J. Radioanal. Chem.* **70,** 57 (1982).

[47] N. A. Larsen, H. Pakkenberg, E. Damsgaard, and K. Heydorn, *J. Neurol. Sci.* **42,** 407 (1979).

[48] F. Girardi, E. Marefante, R. Pietra, E. Sabbioni, and A. Marchesini, *J. Radioanal. Chem.* **37,** 427 (1977).

Section III

Analysis of Metals

[21] Aluminum

By JOHN SAVORY, SUE BROWN, ROGER L. BERTHOLF, NANCY MENDOZA, and MICHAEL R. WILLS

Introduction

Aluminum measurements in biological materials are essential to providing an understanding of the effects of this trace metal in health and disease. Accurate and precise analyses play a vital role in most investigations related to aluminum. The analyst is faced with the problems of detecting aluminum in the parts per billion (ppb) range, and must be able to accomplish collection, storage, processing, and final analysis without outside contamination from this ubiquitous element. The abundance of aluminum in the environment complicates efforts to provide a contamination-free system.

Many problems have existed with analytical methods for the measurement of aluminum, and have been detailed by Cornelis and Schutyser in a review of analytical problems related to these assays.[1] This review selected data published since 1974 on aluminum in serum (or plasma) and urine of normal healthy individuals. In one group, 19 mean values of serum aluminum were listed for their respective normal range studies. These mean values ranged from 2.1 to 42 μg/liter. Other reports have listed normal mean concentrations ranging from 72 to 1460 μg/liter, although interferences obviously were present in these methods. The literature survey of Cornelis and Schutyser[1] also provided normal whole blood mean aluminum concentrations that ranged from 1.6 to 22.1 μg/liter, and urine levels that varied between 4.7 and 1700 μg/liter. There were three studies for whole blood and seven reports for urine. In the absence of carefully controlled interlaboratory surveys with well-characterized quality control materials, the use of each laboratory's normal range provides a useful indicator of interlaboratory variability. The variability reported by Cornelis and Schutyser[1] is undoubtedly due to major analytical problems rather than to biological variation.

[1] R. Cornelis and P. Schutyser, *in* "Trace Elements in Renal Insufficiency" (E. A. Quellhorst, K. Finke, and C. Fuch, eds.), pp. 1–11. Karger, Basel, Switzerland, 1984.

Analytical Methods

Chemical and Physicochemical Methods

Aluminum can be measured using gravimetric, titrimetric, photometric, and fluorimetric methods. Interferences from other metals and contamination of reagents with aluminum are major problems with these techniques. All of these methods have as a basic requirement that the aluminum be isolated from interfering metals. The procedures used to do this isolation include either separation of the metals from aluminum or masking them in a manner that inhibits their interference. All of these methods were originally designed to measure the aluminum content in water, or in metal alloys. All determinations were performed in an aqueous environment. A spectrophotometric determination of aluminum in water utilizing catechol violet is still in general use by the regional water authorities in the United Kingdom. However, the complexity of the serum matrix, with its various constituent metals, potentially interferes with these methods. The proteins in serum would also precipitate with most of the reagents used in these methods, adding further to the difficulties of chemical or physicochemical analysis. For these reasons, together with their poor sensitivity, the various chemical and physicochemical methods are unsuitable for aluminum determinations in serum and other biological materials.

X-Ray Fluorescence

X-Ray fluorescence is associated with the K-shell electrons of metals. A thin film of the sample containing metals is placed on a small mylar sheet and dried. The sample is then bombarded with an electron beam and when an incident electron interacts with a K-shell electron in the metal, the K-shell electron is elevated to an unstable orbital state. As the K-shell electron returns to its stable orbital, it emits energy in the form of X rays that are characteristic of the metal involved. These X rays can be counted by an appropriate detector and the energies of the X rays correspond to different metals. Even though this method is very specific and capable of measuring aluminum, it does not appear to be sensitive enough in its present state to detect the trace levels of aluminum in serum.[2] However, one form of X-ray emission spectrography, that of electron probe X-ray microanalysis, has been used effectively to localize aluminum in both

[2] J. R. J. Sorenson, I. R. Campbell, L. B. Tepper, and R. D. Lingg, *Environ. Health Perspect.* **8,** 3 (1974).

bone and brain tissues.[3] Localization of aluminum in tissues will be discussed briefly in a later section.

Neutron Activation Analysis

Several investigators have used neutron activation analysis (NAA) to determine the aluminum content of biological specimens both with and without some chemical processing. Instrumental neutron activation analysis involves the bombardment of a sample with neutrons and the measurement of the radioactivity induced by nuclear reactions. No chemical processing is required. Upon activation, ^{27}Al (100% isotopic abundance) forms the radioactive ^{28}Al nuclide by a (n,γ) reaction. There are a number of attractive features in this technique, which include excellent sensitivity with relative independence from matrix effects and interferences. Also, there is relative freedom from contamination since the sample is analyzed directly with minimal handling. One major problem is the need to correct for fast neutron reactions on phosphorus that also produce ^{28}Al. A detailed description of the technique for the analysis of brain tissue has been reported recently.[4]

Other NAA methods involve the chemical separation of interfering ions[5]; in these the short half-life (2.24 min) of ^{28}Al makes postirradiation separations a problem. Preirradiation separation techniques have potential problems of contamination and losses during the separation phase.

The facilities required for NAA and the problems associated with either chemical separations or the spectral interferences in the instrumental NAA methods make this technique impractical for most investigators. Furthermore, NAA cannot be classified as a definitive or reference method since there are obvious interferences such as phosphorus which require some correction factor.

Atomic Emission Spectrometry

Aluminum can be measured by emission spectrometry using either a nitrous oxide–acetylene flame or an argon plasma.

Flame Emission. A method has been described[6] that was satisfactory for high aluminum levels, but was not sufficiently sensitive to detect the

[3] P. S. Smith and J. McClure, *J. Clin. Pathol.* **35,** 1283 (1982).

[4] W. D. Ehman, T. I. M. Hossain, D. T. Goodin, and W. B. Markesbery, *in* "Aluminum Analysis in Biological Materials" (M. R. Wills and J. Savory, eds.), pp. 23–33. University of Virginia Press, Charlottesville, Virginia.

[5] A. J. Blotcky, D. Hobson, J. A. Leffler, E. P. Rack, and R. R. Recker, *Anal. Chem.* **48,** 1084 (1976).

[6] M. Ihnat, *Anal. Biochem.* **73,** 120 (1976).

low concentrations that exist in serum. The difficulty with the nitrous oxide–acetylene flame method is that this type of flame is not hot enough to provide the energy needed to ionize all of the aluminum and, thus, is unsuitable for measurement of trace levels.

Inductively Coupled Plasma. Recently the lack of energy in flame emission has been overcome by the use of inductively coupled plasma (ICP) as an excitation source. In this technique a stream of argon is ionized by a high-energy radio-frequency field and the plasma of the ionized argon that is formed can reach temperatures of 10,000°. Inductively coupled plasma emission spectrometry is a multielement technique that is relatively free of chemical interferences. The matrix problems that exist in atomic absorption spectrometry are eliminated in ICP due to the very high excitation temperature of the sample. Thus, the technique is particularly useful for the determination of refractory elements such as aluminum and silicon. Inductively coupled plasma methods have been reported for the measurement of aluminum in biological materials[7–9] and could be an excellent alternative to electrothermal atomic absorption spectrometry for those laboratories possessing the appropriate instrumentation. A major problem with using the argon–plasma technique is the intense and broad emission spectrum of calcium, which increases the aluminum background and can raise the detection limits for this element. Correction for this problem is hampered because the effect of calcium varies with its concentration.

The sensitivity of ICP methods probably is comparable to electrothermal atomic absorption methods, and is reported to be about 2 μg/liter,[7] although relatively poor sensitivity (10 μg/liter) also has also been reported.[8] A recent report[9] describes an excellent study of the application of ICP to the analysis of blood, dialysis fluid, and water. The methods described in this report[9] are automated by use of an automatic sampler. There is computer-controlled displacement of the entrance slit allowing easy and precise measurement of the background for each sample. Variations in emission signal intensity in matrices of different compositions are cancelled by addition of cesium as a matrix modifier and by using gallium as an internal standard. The detection limit of the method is 0.3 μg/liter in pure solution with a large range of linearity and excellent precision. This technique offers good potential as a reference method for aluminum measurements. One drawback to ICP, however, is the relatively high cost of the instrumentation, which will limit its use in many routine laboratories.

[7] F. E. Lichte, S. Hopper, and T. W. Osborn, *Anal. Chem.* **52,** 120 (1980).

[8] P. Schramel, A. Wolf, and B. J. Klose, *J. Clin. Chem. Biochem.* **18,** 591 (1980).

[9] Y. Mauras and P. Allain, *Anal. Chem.* **57,** 1706 (1985).

Atomic Absorption Spectrometry

Atomic absorption spectrometry (AAS) has been widely used to analyze biological materials for aluminum content. Flame techniques, even with the hotter nitrous oxide–acetylene flame, do not perform as well as the flameless methods. However, flame atomic absorption methods have been used to analyze brain and cerebrospinal fluid,[10] rat tissues,[11] food, urine and feces,[12] heart muscle,[13] and plasma and tissues.[14–16]

The greatest degree of success of any technique for the determination of aluminum in biological specimens has been with electrothermal atomic absorption spectrometry (EAAS). In this technique the sample is placed in a graphite tube mounted in the light path of the spectrophotometer. The source of the light is usually a hollow cathode lamp that contains the metal being analyzed and emits characteristic wavelengths. First the graphite tube is heated with direct current to dry the sample at a low temperature, then the sample is ashed to destroy organic matter and burn off inorganic species that may interfere, and finally the temperature is quickly raised and the metal under analysis vaporizes and absorbs the light being passed through the graphite tube. Advantages of the graphite furnace include (1) sample pretreatment can usually be eliminated, (2) sample requirements are small (2–100 μl), (3) graphite furnaces are capable of attaining the high temperature needed to form ground state atoms, and (4) the atoms stay in the light path for a relatively long time, which results in increased sensitivity.

Several problems may be encountered in the electrothermal atomic absorption determination of aluminum, and these include difficulties with untreated graphite tubes,[17] matrix interferences,[17,18] and standardization procedures.[18]

[10] S. S. Krishnan, K. A. Gillespie, and D. R. Crapper, *Anal. Chem.* **44,** 1469 (1972).

[11] G. H. Mayor, S. M. Sprague, M. R. Hourani, and T. V. Sanchez, *Kidney Int.* **17,** 40 (1980).

[12] E. M. Clarkson, V. A. Luck, W. V. Hynson, R. R. Bailey, J. B. Eastwood, J. S. Woodhead, V. R. Clements, J. L. H. O'Riordan, and H. E. DeWardener, *Clin. Sci.* **43,** 519 (1972).

[13] B. Chipperfield, J. R. Chipperfield, and N. R. Bower, *Lancet* **1,** 755 (1977).

[14] G. M. Berlyne, D. Pest, J. Ben-Ari, J. Weinberger, M. Stern, G. R. Gilmore, and R. Levine, *Lancet* **2,** 494 (1970).

[15] G. M. Berlyne, R. Yagil, J. Ben-Ari, E. Knopf, G. Weinberger, and G. M. Danovitch, *Lancet* **1,** 564 (1972).

[16] G. Weinberger, R. Yagil, F. Popliker, and G. M. Berlyne, *in* "Uremia" (R. Fluthe, G. M. Berlyne, and B. Burton, eds.), pp. 128–137. Georg Thime Verlag, Stuttgart, Federal Republic of Germany, 1972.

[17] F. R. Alderman and H. J. Gitelman, *Clin. Chem.* **26,** 258 (1980).

[18] J. E. Gorsky and A. A. Dietz, *Clin. Chem.* **24,** 1485 (1978).

An investigation of two electrothermal AAS methods has been carried out in the authors' laboratory.[19] The first was a conventional approach using direct analysis of serum using a stabilized temperature platform in the graphite tube. The study demonstrated the necessity of using the method of standard additions as a means of standardization, since there were profound differences in the slopes of standard curves constructed from aqueous standards, sera from normal individuals, and sera from uremic patients. Thus, considerable analytical errors would result from using aqueous standards or standards made up in normal sera as a means of standardizing assays for uremic patients. The standard additions method allows each serum sample matrix to serve as its own standard and, therefore, provide more accurate analyses.

Leung and Henderson[20] have not been able to confirm these findings of grossly different curve slopes for different types of serum and standard samples. Recently, Bettinelli *et al.*[21] have made a thorough evaluation of the direct measurement of aluminum in serum using EAAS with the stabilized temperature furnace with the L'vov platform. Their recommendation was to use pyrolytically coated graphite tubes and to keep the method as simple as possible with minimal sample pretreatment. Slavin (personal communication) also has been unable to observe major slope change problems between aqueous and serum matrix samples and recommends a procedure similar to that of Bettinelli *et al.*,[21] Leung and Henderson,[20] and Brown *et al.*[19] Most EAAS procedures have been developed on Perkin-Elmer instruments with the use of autosampling to improve precision. Pyrolytically coated graphite tubes are recommended together with a pyrolytic graphite platform. Argon is preferred over nitrogen as the purge gas since argon produces a larger and less variable signal. Some type of background correction is recommended and the Zeeman correction system probably provides the most sensitive and reliable results. In the direct methods for serum analysis, standards and serum samples are diluted with an equal volume of an aqueous solution containing $Mg(NO_3)_2$ (2 g/liter). The autosampler is programmed to deliver a 20-μl aliquot of the sample onto the platform for final analysis.

Leung and Henderson[20] prepare their standards using a serum pool containing a minimal amount of endogenous aluminum, whereas Bettinelli *et al*[21] recommend aqueous standards. The use of simple aqueous standards is to be preferred, provided that there are no matrix effects from serum. A unique feature of the method proposed by Bettinelli *et al.*[21] and

[19] S. Brown, R. L. Bertholf, M. R. Wills, and J. Savory, *Clin. Chem.* **30**, 1216 (1984).
[20] F. Y. Leung and A. R. Henderson, *Clin. Chem.* **28**, 2139 (1982).
[21] M. Bettinelli, U. Baroni, F. Fontana, and P. Poisetti, *Analyst (London)* **110**, 19 (1985).

by Slavin (personal communication) is the use of oxygen to enhance charring of the sample. Substitution of oxygen for argon as the furnace purge gas during the ashing stage of the program results in more complete and efficient burning of organic matrix constituents. Not all EAAS instruments, however, are equipped to accommodate multiple purge gases.

Working Method for Serum Aluminum

Reagents. Matrix Modifier. Add 1.4 g $Mg(NO_3)_2$ to an acid-washed 1-liter flask. Add approximately 500 ml distilled water. Add 2 ml Triton X-100. Dilute to volume and let mix on stirring plate until Triton is dissolved (about 10 min). Transfer to an acid-washed polyethylene bottle. Stable at room temperature for 3 months.

Standards. 2000 μg/liter: Intermediate aluminum standard is prepared by diluting 2.0 ml of 1 mg/ml aluminum atomic absorption standard (Fisher Scientific Co., Fair Lawn, NJ 07410) to 1 liter of distilled water. This should be prepared in an acid-washed 1000-ml volumetric flask and transferred to an acid-washed polyethylene bottle after mixing well; it is stable at room temperature for 1 month. A series of working aluminum standards (0, 20, 40, 60, 80, 100, 200 μg/liter) are prepared daily by diluting 0, 1.0, 2.0, 3.0, 4.0, 5.0, 10.0 ml, respectively, of 2000 μg/liter intermediate standard to 100.00 ml with distilled water in acid-washed volumetric flasks.

Sample Preparation

1. Pipet 300 μl of aqueous aluminum standards, distilled water blank, quality control sample, or patient sample into a 2-ml polyethylene cup.
2. Dilute 1 : 1 by pipetting 300 μl of 1.4 g $Mg(NO_3)_2$/liter solution to the same cup, and mix thoroughly.
3. Set the spectrophotometer to 0 against the blank solution.

EAAS Analysis

The final analysis is carried out using a high quality atomic absorption spectrometer with electrothermal atomization capabilities. The authors use the Perkin-Elmer Model 5000 with HGA 500 electrothermal atomization unit (Perkin-Elmer Corporation, Norwalk, Conn. 06856). Pyrolytically coated graphite tubes with a platform are used. The volume of standards and treated sample pipetted into the graphite cuvette is 20 μl. Automatic sampling is recommended as a means of improving the precision of the assay. The instrument parameters are given in Table I.

This procedure for serum aluminum is rugged and precise. The linearity of the method is typically 0–125 μg/liter.

TABLE I
INSTRUMENT PARAMETERS FOR EAAS ALUMINUM ANALYSIS OF SERUM

Step	Description	Temperature (°C)	Ramp (sec)	Hold (sec)
1	Dry	130	1	45
2	Oxygen ash	600	30	55
3	Ash	600	1	25
4	Char	1700	15	25
5	Atomize	2400	0	6[a]
6	Clean	2600	1	6
7	Cool	20	1	20

[a] Gas flow = 0 ml/min.

Alternative EAAS for Serum Aluminum

The authors have developed a procedure that minimizes any potential matrix effects by protein precipitation.[19] This protein precipitation technique was originally developed for serum nickel[22] and markedly reduces matrix effects in the final atomic spectroscopic analysis. Briefly the procedure is as follows.

One milliliter of serum is pipetted into an acid-washed 4.5-ml polyethylene centrifuge tube (cat. no. 477, Walter Sarstedt, Inc., Princeton, NJ 08540). The tube is positioned on a Vortex mixer and mixing is begun at medium speed. Fifty microliters of ultrapure concentrated nitric acid is pipetted into the middle of the vortex. The tube is capped and mixing continued for 1 min followed by heating for 5 min in a 70° water bath, mixing for 10 sec by vortexing, and centrifuging for 10 min at 900 *g*. The supernatant is pipetted into a Teflon sampling cup with an acid-rinsed pipet tip and analyzed using electrothermal AAS with a stabilized platform. This protein precipitation technique is precise (5% CV) and is linear to 120 μg/liter.

Urine, Feces, and Tissue Analysis

For urine aluminum analysis, sample aliquots are diluted 1 : 1 with distilled water before application to the AAS stabilized temperature platform.[20] Fecal analysis requires considerably more complicated preparation steps than serum or urine. The procedure developed in the authors' laboratory[23] is summarized as follows. Frozen specimens are thawed and

[22] F. W. Sunderman, Jr., C. Crisostomo, M. C. Reid, S. M. Hopfer, and S. Nomoto, *Ann. Clin. Lab. Sci.* **14,** 232 (1984).

[23] S. Brown, N. Mendoza, R. L. Bertholf, R. Ross, M. R. Wills, and J. Savory, manuscript in preparation.

weighed in the original plastic container used for collection. Distilled water is added (1 ml/2 g feces) and the sample is homogenized in a sealed can on a paint shaker. A 10-ml aliquot is ashed at 500° in a muffle furnace, dissolved in dilute HNO_3, and analyzed by EAAS.

Solid tissues must be homogenized, dried, ashed, and/or dissolved to produce a liquid sample prior to EAAS analysis. Bone samples for analysis are washed free of marrow by a strong stream of distilled water. Any existing fat or muscle is scraped with an aluminum-free obsidian scalpel or blade. The bone sample can then be processed by several methods to obtain a solution for injection into the graphite furnace. In one method, the bone sample is placed into a quartz Kjeldahl flask. Five milliliters of ultrapure concentrated nitric acid is added and the flask is electrically heated to 200°. Destruction is complete when a clear solution is observed. The remaining liquid is evaporated, the flask allowed to cool, and 1 ml of concentrated nitric acid is added. The liquid is quantitatively transferred to a volumetric flask and diluted with distilled water and analyzed to EAAS.[24] A second method for processing bone samples is by extraction of aluminum in the sample with a saturated solution of disodium ethylenediaminetetracetate (EDTA).[25] The bone is washed as described above, allowed to air dry, ground in a Wiley mill, passed through a sieve, partially defatted with petroleum ether, reground, and passed through a sieve again. Preweighed samples are mixed with 5 ml of the EDTA solution and agitated for 2–4 hr. The supernatant is analyzed for aluminum and concentrations are calculated against standards prepared in the EDTA solution. Bone samples can also be ashed in the same manner as fecal samples, as described earlier. This method allows for the calculation of aluminum based on wet weight or dry weight.

Soft tissue samples, such as brain, liver, and muscle, must be homogenized before processing, and this can be accomplished easily by pummelling the tissue in a "Stomacher" blender (Fisher Scientific). Distilled water (5 ml) is added to the bag with the tissue, the bag is sealed, and then blended for 5–15 min, which completely homogenizes the sample.[26] The homogenate can then be processed as described for the fecal samples. Tissue samples can also be processed by the EDTA extraction method described for bone. Brain samples are dried and ashed, then extracted with EDTA. Muscle samples are dried, grounded, extracted in solvents, redried, and extracted in EDTA.[25] Tissue samples can also be wet digested (after homogenization) with hot nitric acid in a Kjeldahl flask as

[24] P. D'Haese, F. L. Van de Vyver, A. R. Bekaert, and M. E. De Broe, *Clin. Chem.* **31**, 24 (1985).

[25] G. R. LeGendre and A. C. Alfrey, *Clin. Chem.* **22**, 53 (1976).

[26] F. W. Sunderman, Jr., A. Marzouk, M. C. Crisostomo, and D. R. Weatherby, *Ann. Clin. Lab. Sci.* **15**, 299 (1985).

described above. Soft tissue can also be processed for aluminum analysis by using tetramethylammonium hydroxide (TMAH) as a dissolving solution.[27] Tissue samples are dried to constant weight, 2.0 ml of aqueous TMAH solution is added, and the mixture is heated in a hot-air oven (90°) for 1–2 hr, during which time the solution is mixed occasionally. This procedure will completely dissolve the tissue. The solution is then cooled, diluted with ethanol, and mixed well. Working aluminum standards are prepared by adding aliquots of an intermediate standard to 2.0 ml of aqueous TMAH and diluting to 10 ml with ethanol.

Localization of Aluminum in Tissues

Irwin[28] developed a histochemical stain for the demonstration of aluminum in tissues. The technique used ammonium aurintricarboxylate (aluminon), which forms a lake with a number of metallic ions under appropriate conditions. The lake formed using the Irwin technique has a cherry-red/purple color and is considered to be specific for the presence of aluminum. The technique developed by Irwin is extremely sensitive and specific as confirmed by electron probe X-ray microanalysis.[3] A modification of this aluminon staining procedure has been developed by Maloney *et al.*[29] for the histological quantitation of aluminum in iliac bone.

A few recent studies have used the scanning electron microscope to detect aluminum deposits by energy dispersive x-ray analysis. Smith *et al.*[30] located aluminum deposits within the glomerular basement membrane of humans. Perl *et al.*,[31] using the same technique, examined brain tissues from patients with amyotrophic lateral sclerosis and Parkinsonism dementia, and located aluminum in neurofibrillary tangle-bearing hippocampal neurons. Boyce *et al.*[32] and Cournot-Witmer *et al.*[33] similarly have localized aluminum in bone tissues.

An alternative approach to aluminum localization in tissues is to use the new technique of laser microprobe mass spectrometry (LAMMA). In this technique an intense laser beam is directed toward a specific organ-

[27] B. J. Stevens, *Clin. Chem.* **30,** 745 (1984).

[28] D. A. Irwin, *Arch. Ind. Health* **12,** 218 (1955).

[29] N. A. Maloney, S. M. Ott, A. C. Alfrey, N. L. Miller, J. W. Coburn, and D. J. Sherrard, *J. Lab. Clin. Med.* **99,** 206 (1982).

[30] D. M. Smith, Jr., J. A. Pitcock, and W. M. Murphy, *Am. J. Clin. Pathol.* **77,** 341 (1982).

[31] D. P. Perl, D. C. Gajdusek, R. M. Garruto, R. T. Yanagihara, and C. J. Gibbs, Jr., *Science* **217,** 1053 (1982).

[32] B. F. Boyce, H. Y. Elder, G. S. Fell, W. A. P. Nicholson, G. D. Smith, D. W. Dempster, C. C. Gray, and I. T. Boyle, *Scanning Electron Microsc.* **III,** 329 (1981).

[33] G. Cournot-Witmer, J. Zingraff, J. J. Plachot, F. Escaig, R. Lefevre, P. Boumati, A. Bourdeau, M. Garabedian, P. Galle, R. Bourdon, T. Drueke, and S. Balsan, *Kidney Int.* **20,** 375 (1981).

elle and metals thus vaporized from the region are analyzed in a mass spectrometer. This technique has been used to localize aluminum in the lysozomes of hepatocytes as well as Kupffer cells of liver tissue from patients on chronic hemodialysis.[34,35] In these studies aluminum was also localized in bone at the osteoid/calcified bone interface.

Both the LAMMA and energy dispersive X-ray analysis require very expensive sophisticated equipment, and both are extremely powerful techniques for studying metal localization in tissues. The sensitivity of the LAMMA technique appears to be better than energy dispersive X-ray analysis but the resolution of the latter technique is slightly better, as expected. Our understanding of the toxicity of aluminum will be aided considerably by the use of both of these localization techniques.

Specimen Collection and Control of Contamination

Blood Collection. Sample collection is, potentially, a major source of contamination. Needles, glass vacuum tubes, plastic syringes, anticoagulants, plastic tubes, and other equipment used in standard venipuncture methods are all potential sources of aluminum contamination and must be checked before introduction into routine use for specimen collection (see Table II).

The blood collection procedure that has been in use in the authors' laboratory involves drawing blood with a stainless-steel needle into a plain glass vacuum tube (No. 6430, Becton-Dickinson and Co., Rutherford, NJ 07070). The blood is allowed to clot for approximately 20 min before it is centrifuged; the serum is then transferred to a 17 $\times$ 100-mm polypropylene tube (Falcon, Oxnard, CA).

Transport and Storage. Blood drawn into glass vacuum tubes should be separated within 1 hr after collection. Variable results may occur if blood is allowed to remain in contact with glass.[36] The cap on the plastic tube should be tight fitting, or covered with Parafilm. If the tube is to be shipped, the tube's cap must be wrapped in such a way as to prevent any leakage during handling. There is no data available in the literature that storage of the specimen at room temperature results in any loss of aluminum when compared with storage at 4°. Specimens for aluminum analysis

[34] M. E. De Broe, F. L. Van de Vyver, A. B. Bekaert, P. D'Haese, G. J. Paulus, W. J. Visser, R. Van Grieken, F. A. de Wolff, and A. H. Verbueken, *in* "Trace Elements in Renal Insufficiency" (E. A. Quellhorst, K. Finke, and C. Fuchs, eds.), pp. 37–46. Karger, Basel, Switzerland, 1983.

[35] A. H. Verbueken, F. L. van de Vyver, R. E. Van Grieken, G. J. Paulus, W. F. Visser, P. D'Haese, and M. E. De Broe, *Clin. Chem.* **30,** 763 (1984).

[36] R. L. Bertholf, S. Brown, B. W. Renoe, M. R. Wills, and J. Savory, *Clin. Chem.* **29,** 1087 (1983).

TABLE II
SERUM ALUMINUM VALUES

Reference	Collection technique	Reference range (μg/liter)
a	Syringe/plastic tube	12–46
b	Syringe/plastic tube	2–15
c	Plastic tube	2.5–10.0
d	Plain glass vacuum tube	0–7.6
e	Plain glass vacuum tube	2–14
f	Plain glass vacuum tube	1–12
g	Heparinized vacuum tube	2–12

[a] J. E. Gorsky and A. A. Dietz (1978). *Clin. Chem.* **24,** 1485 (1978).
[b] I. S. Parkinson, M. K. Ward, and D. N. S. Kerr, *Clin. Chim. Acta* **125,** 125 (1982).
[c] O. Oster, *Clin. Chim. Acta* **114,** 53 (1981).
[d] F. R. Alderman and H. J. Gittleman, *Clin. Chem.* **26,** 258 (1980).
[e] F. Y. Leung and A. R. Henderson, *Clin. Chem.* **28,** 2139 (1982).
[f] R. L. Bertholf, S. Brown, B. W. Renoe, M. R. Wills, and J. Savory, *Clin. Chem.* **29,** 1087 (1983).
[g] W. D. Kaehny, A. C. Alfrey, R. E. Holmon, and W. J. Shorr, *Kidney Int.* **12,** 361 (1977).

have been received in the authors' laboratory both a room temperature and on ice packs. Several of these samples were from normal volunteers and the resulting aluminum values were all within our reference range. This finding suggests that there is not a contamination problem associated with shipping the specimen at room temperature.

Tissue samples should be placed into aluminum-free plastic containers. Brain tissue needs to be frozen until analysis.[37] Bone can be kept at room temperature.

Each step of the specimen collection procedure for aluminum analysis may potentially cause contamination and must be scrutinized to determine if it is adding any aluminum to the sample. Each laboratory should check all materials used in the collection process before collecting samples from patients and reporting any results. Once the collection procedure has been established, regular checks should be scheduled to verify that little or negligible contamination results from the technique. A quality control check for the blood collection technique might involve drawing blood from healthy individuals once a week. The individual chosen should be resident in the same area as the patients on whom aluminum levels

[37] D. R. Crapper, S. S. Krishnan, and S. Quittkat, *Brain* **99,** 67 (1976).

have been requested; this would enable monitoring of both the sample collection procedure and also the transport system for any spurious aluminum contamination. An aluminum value within the reference range would be acceptable for this specimen. Tissue samples from healthy, nondiseased persons at autopsy can serve as controls for bone, brain, and muscle aluminum determinations.

Sources of Contamination in Analysis

As in specimen collection, every item used during analysis should be regarded as a potential source of aluminum contamination. Glassware, pipet tips, plastic tubes, sample cups, the working environment, and the purified water supply must all be checked to ensure that they do not add measurable amounts of aluminum to the procedure. The room chosen for the analyses should have a limited access to ensure that dust in the working environment is being circulated as little as possible. Sample preparation should be carried out in an environmental laminar flow hood; this precaution helps minimize contamination by dust particles.

Water utilized for standard curve preparation, rinsing of glassware, and sample dilutions must be of high purity. The water should produce a resistivity of at least 18 MΩ; this purity should be continuously monitored.

All glassware should be aluminum free and this can be accomplished either by acid washing or by soaking glasware in saturated disodium EDTA. The acids used should be of ultrapure grade. The methods used in our laboratory include the acid cleaning technique described by Moody and Lindstrom[38] and the EDTA wash proposed by Leung and Henderson.[20]

Contamination is a major problem encountered in aluminum determinations. All items utilized during collection, transport, and assay should be checked for unwanted aluminum contribution to the procedure. Only by taking these stringent precautions will one be able to produce results with confidence.

[38] J. R. Moody and R. M. Lindstrom, *Anal. Chem.* **49,** 2264 (1977).

[22] Measurement of Total Calcium in Biological Fluids: Flame Atomic Absorption Spectrometry

By GEORGE N. BOWERS, JR. and THEODORE C. RAINS

Introduction

This chapter on the measurement of total calcium (Ca_T) in biological fluids by flame atomic absorption spectrometry (FAAS) describes what the authors believe to be one of the more accurate yet practical methods available today.[1-4] There are many alternative Ca_T methods that might be chosen for measuring the concentration of this analyte in biological fluid and a number of these are classified in Table I.[5-7] The most popular and readily available methods in the mid-1980s are those of the compleximetric/sectrophotometric types that are supplied as individual test kits from many suppliers or as an integrated part of the several large multichannel analytical systems used extensively in clinical laboratories for determining the Ca_T in human blood serum. In contrast to the well-documented analytical accuracy of FAAS (see below), the accuracy of most of the alternative methods for measuring Ca_T is less rigorously documented and, in general, is known primarily through national interlaboratory comparisons to FAAS as the comparative method.[8,9] For example, many compleximetric methods have been shown to be distorted by bilirubin and hemoglobin in human serum, yet these substances have very little or no effect on FAAS measurements.[10,11]

[1] J. P. Cali, G. N. Bowers, Jr., and D. S. Young, *Clin. Chem.* **19,** 1208 (1973).
[2] J. P. Cali, J. Mandel, L. Moore, and D. S. Young, *NBS Special Publ.* **260,** 1 (1972).
[3] S. S. Brown, M. J. R. Healy, and M. Kearns, *J. Clin. Chem. Clin. Biochem.* **19,** 395 (1981).
[4] J. Pybus, F. J. Feldman, and G. N. Bowers, Jr., *Clin. Chem.* **16,** 998 (1970).
[5] R. J. Henry, "Clinical Chemistry: Principles and Technics," pp. 356–378. Hoeber, New York, 1964.
[6] C. Baluja-Santos, A. Gonzalez-Portal, and F. Bermejo-Martinez, *Analyst (London)* **109,** 797 (1984).
[7] W. Slavin, *Anal. Chem.* **58,** 589A (1986).
[8] "Comprehensive Chemistry Surveys." College of American Pathologists, Skokie, Illinois, 1969–1986.
[9] G. N. Bowers, Jr., *in* "Medical Devices: Measurements, Quality Assurance, and Standards" (C. A. Caceres, H. T. Yolken, H. J. Jones and H. R. Piehler, eds.), ASTM Special Tech. Publ. No. 800, pp. 41–59. American Society for Testing Materials, Philadelphia, Pennsylvania, 1983.
[10] E. M. Brett, and J. M. Hicks, *Clin. Chem.* **27,** 1733 (1981).
[11] K. M. Chan *et al., Clin. Chem.* **29,** 1497 (1983).

TABLE I
METHODS FOR MEASURING TOTAL CALCIUM IN BIOLOGICAL FLUIDS

Classical manual methods
- Gravimetric analysis
  - Precipitation as CaC_2O_4 (oxalate) and conversion by dry-ashing to $CaCO_3$ or CaO
- Volumetric Analysis After Isolation of CaC_2O_4
  - Acidimetry after conversion of oxalate to $CaCO_3$
  - Titrimetry with hot permanganate (Clark-Callip)
- Manometric analysis
  - Gasimetry on released CO_2 after oxidation of oxalate
- Complexometric analysis
  - Titrations with EDTA and EGTA
  - Turbidimetric assays by formation of calcium oleate soap
  - Colorimetric/fluorimetric assays with murexide, Erichrome black T, Cal-Red, Calcein, calcofast blue 2G, cresolphthalein complexone, chlorophosphonazo III, methylthymol blue, etc.

Electrochemical methods
- Polarographic analysis (rarely used for calcium)
- Voltammetric analysis, anodic-stripping (rarely used for calcium)
- Potentiometric analysis with ion-selective electrodes (usually for ionized calcium but at least one analyzer also gives total calcium)

Radiometric methods
- Neutron activation analysis

Spectroscopic methods
- Colorimetric analysis
- Nephelometric analysis
- UV-visible absorption spectrometry
- Fluorimetry
- Mass spectrometry
- X-Ray fluorescence spectrometry
- Atomic emission spectrometry
  - Flame (FAES)
  - Flame (Dc arc, laser)
  - Inductively coupled plasmas (ICP-OES)
- Atomic absorption spectrometry
  - Flame (FAAS)
  - Electrothermal (ETAAS)
  - Inductively coupled plasmas (ICP-AAS)
- Fluorescence absorption spectrometry (FAS)

The FAAS method recommended in this chapter was specifically designed for Ca_T measurements in human serum.[4] It was studied intensively during the early method development stage to uncover and eliminate many sources of both random and systematic error. Subsequently, it was proposed as the candidate Reference Method (RM) for Ca_T[1] and intensively studied in comparison to a Definitive Method (DM) for Ca_T developed at the National Bureau of Standards (NBS).[12] Studies of the transferability of this RM within a network of cooperating clinical laboratories have established that the FAAS results on serum samples agree with the values assigned by the DM to better than ±2%.[2,3] This RM based on FAAS has now been credentialed by the Council of the National Reference System for the Clinical Laboratory (NRSCL) of the National Committee for Clinical Laboratory Standards (NCCLS) as an integral part of the National Reference System for Total Calcium (see Appendix).[13]

Biological Distribution and Concentrations of Calcium

Prior to discussing the analytical details of the FAAS method, we need to examine the wide range of Ca_T concentrations that exist in biological fluids and tissues. Calcium is the most abundant element in the whole body mass of many organisms. For example, NBS/SRM #915, calcium carbonate (99.9% pure) came from an ancient heap of oyster shells gathered by indians many centuries ago in the then unpolluted waters of the Chesapeake Bay.[14] Yet, for all of this calcium in the shell, desiccated oyster tissue itself contains only 0.15% calcium. Likewise, 99.7% (about 1 kg of 25,000 mmol) of calcium in man is found in the bone as the complex calcium phosphate salt, called apatite. While apatite is abundant, it is relatively inert in comparison to the remaining 0.3% of the calcium in the human body. This 0.3% translates into about 1.3 mg or 50 mmol of calcium circulating in the extracellular fluid spaces (25 mmol as ionized calcium, 20 mmol as protein bound calcium, and 5 mmol as several forms of complexed calcium). Only another 0.7 mg or 25 mmol of calcium is found within the cellular compartments in several different ionized and dynamically complexed forms. However, these multiple other forms, although minute in terms of the total mass of calcium in the body, are now known to be absolutely essential to life and are continuously involved in vital regulatory and metabolic changes in every cell of living organisms.[15,16] Thus, the amount of calcium available for measurement in different biological tissues and fluids can vary from moles in the skeletal struc-

[12] L. J. Moore and L. A. Machlan, *Anal. Chem.* **44,** 2291 (1972).

[13] "Status of Certified Reference Materials, Definitive Methods, and Reference Methods for Analytes," NRSCL7-CR. National Committee for Clinical Laboratory Standards, Villanova, Pennsylvania, 1985.

[14] "Certificate of Analysis: Standard Reference Material 915—Calcium Carbonate." Office of Standard Reference Materials, National Bureau of Standards, Gaithersburg, Maryland, 1973.

[15] H. Rasmussen, *N. Engl. J. Med.* **314,** 1094 and 1164 (1986).

[16] J. W. Cheung, J. V. Bonventre, C. D. Malis, and A. Leaf, *N. Engl. J. Med.* **314,** 1670 (1986).

ture to millimoles in extracellular fluids such as the blood serum to micromoles or even picomoles in discrete parts of the intracellular compartment. In general, FAAS is well suited to measurements in aqueous fluids with concentrations of Ca_T near 1 mmol/liter by simply diluting a noncellular fluid sample into the appropriate amount of lanthanum reagent. Table II gives an idea of the range of mmol/liter concentration levels that have been reported in various extracellular fluids from a number of different animal tissues.[17]

Although FAAS and other forms of AAS have been used to measure Ca_T concentrations in digested platelets[18] or in a subcellular localization within polymorphonuclear leukocytes,[19] the greater sensitivity required to measure intracellular calcium especially within the minienvironments of specialized organelles probably excludes the general use of FAAS in these investigations. Even the more sensitive forms of atomic spectrometry employing furnace and plasma techniques may also be too insensitive or require too much sample to be useful.[7] Methods to measure physiologically important subfractions of calcium, i.e., ionized calclium by ion-selective electrodes and intracellular calcium with fluorescent dyes, are included in subsequent chapters of this volume.[20,21]

The Element, Its Isotopes, and Mass Spectrometry

Calcium (Ca), atomic number 20 and nominal atomic weight 40, is the fifth most abundant element in the earth's crust and constitutes 3.6% of its mass. Six stable isotopes are found in all natural compounds of calcium and each contributes fractionally to the atomic weight of Ca. State-of-the-art mass spectrometry has been used to measure the relative abundances of each of these six isotopes with great accuracy. In 1983, the IUPAC Commission on Atomic Weights and Isotopic Abundances revised the atomic weight of calcium to 40.078 ± 0.004 based on the following geonormal distribution of isotopes: mass 40 ($96.941 \pm 0.013\%$), 42 ($0.647 \pm 0.003\%$), 43 ($0.135 \pm 0.003\%$), 44 ($2.086 \pm 0.005\%$), 46 ($0.004 \pm 0.003\%$), and 48 ($0.187 \pm 0.03\%$).[22,23]

[17] P. L. Atlman, *in* "Blood and Other Body Fluids" (D. S. Dittmer, ed.). Committee on Biological Handbooks, Federation of American Societies for Experimental Biology, Washington, D.C., 1961.

[18] T. Makino, *Clin. Chem.* **31,** 609 (1985).

[19] K. B. Raja, P. M. Leach, G. P. Smith, D. McCarthy, and T. J. Peters, *Clin. Chim. Acta* **123,** 19 (1982).

[20] S. F. Sena and G. N. Bowers, Jr., this volume [23].

[21] This series, in press.

[22] N. E. Holden and R. L. Martin, *Pure Appl. Chem.* **56,** 653 (1984).

[23] N. E. Holden, R. L. Martin, and I. L. Barnes, *Pure Appl. Chem.* **56,** 675 (1984).

TABLE II
TOTAL CALCIUM CONCENTRATIONS IN VARIOUS BIOLOGICAL FLUIDS[a]

Source	Species	Mean (mmol/liter)	Range
Amniotic fluid	Man	2.0	
Aqueous humor—eye	Man	1.8	1.4–2.6
	Rabbit	0.7	
	Ox	1.8	
	Horse	1.8	1.5–2.2
	Swine	2.0	
Blood, plasma/serum	Man	2.5	2.2–2.6
	Bovine	2.7	2.1–3.6
	Dog	2.6	2.1–3.3
	Dolphin	1.1	1.0–1.3
	Goat	2.6	
	Guinea pig	2.7	1.8–3.4
	Horse	3.1	
	Mouse	2.1	
	Rabbit	1.9	1.3–2.5
	Rat	3.1	2.7–3.6
	Swine	2.8	
	Other nonmammals		
	Chicken	4.5	3.3–5.9
	Pigeon	2.6	
	Crocodile	1.5	1.3–1.7
	Lizard	1.5	1.0–2.1
	Iguana	1.4	
	Snakes	1.7	1.1–2.5
	Turtle	2.1	1.2–3.0
	Frog	2.0	1.5–2.5
	Toad	1.9	1.8–2.0
	Carp	3.0	2.3–3.7
	Goosefish	0.3	
	Skate	3.0	2.4–3.6
Cerebrospinal fluid	Man	1.2	1.0–1.5
	Cat	1.5	
	Cattle	1.5	1.3–1.6
	Guinea pig	1.5	
	Horse	1.6	1.4–1.8
	Sheep	1.4	
Cystic fluid			
Echinoccal	Man	1.0	
Cysticerus	Rabbit	2.0	
Gastrointestinal fluids			
Saliva	Man	1.7	1.0–2.4
	Dog, mixed	2.4	1.5–3.3
	Dog, parotid	3.8	2.5–5.2
	Sheep	0.5	0.2–0.8

TABLE II (*continued*)

Source	Species	Mean (mmol/liter)	Range
Gastric juice	Man, fasting	1.6	1.0–2.4
	Dog	1.0	0.5–1.7
	Cat	1.6	0.8–2.6
Bile			
Gallbladder	Man	3.0	2.5–3.5
Liver	Man	1.6	1.0–2.2
Gallbladder	Dog	1.3	
Liver	Dog	2.5	1.8–3.6
Liver	Rabbit	2.4	2.0–4.8
Pancreatic fistula	Man	1.3	1.1–1.6
	Cat	2.0	1.1–2.5
	Dog	1.0	0.9–1.1
Duodenum, secretion stim	Man	3.2	
Jejunal, secretion stim	Man	2.4	1.6–3.2
	Dog	1.2	0.8–2.7
Ileal, secretion stim	Man	2.8	2.5–3.2
	Dog	2.6	2.5–2.8
Hemolymph	Crab	0.15	0.1–0.2
	Drayfish	0.15	0.1–0.2
	Lobster	0.20	0.15–0.25
Prostatic fluid	Man	3.0	
	Dog	0.3	
Semen	Man	10	4–15
	Dog	0.3	
	Cattle	12	5–10
	Chicken	2	
	Horse	5	
Sweat	Man	1.2	0.1–2.5
Synovial fluid	Man	1.8	1.0–2.7
	Cattle	1.8	1.3–2.2
Milk	Man	6	3–10
	Cattle	50	12–95
Urine	Man	5	0.7–7
	Cat	1.5	
	Cattle	1	0.1–2
	Dog	2	1–4
	Rabbit	2	1–6
	Rat	2	1–10
	Sheep	2	

[a] Modified from Ref. 17.

The inherent high accuracy of measuring isotopic ratios by mass spectrometry has been applied by scientists at the NBS to measure Ca_T in biological samples by determining the $^{40}Ca/^{44}Ca$ ratio by isotope dilution mass spectrometry (IDMS).[12] This technique depends on the spiking of the sample with ^{44}Ca, chemical dissolution and equilibration of the spike, separation of the analyte (Ca) from the matrix to eliminate potential mass spectrometric interferences, and then measurement of the altered isotope ratio by the thermal ionization technique. By bracketing unknowns with multiple standards, reducing the calcium blank of reagents, and rigidly controlling the thermal ionization steps, replicate IDMS measurements of isotope ratios were made to within 0.05%. The final Ca_T results on synthetic calcium solutions and serum samples were stated to agree with the true value to within $\pm$ 0.2% (95% limit of error for a single analysis). This calcium IDMS/DM as performed at NBS[12] and the high purity calcium carbonate of NBS/SRM #915[14] are the basic items that define the true value for Ca_T measurements within NRSCL/NCCLS. As described in the Appendix, the National Reference System for Total Calcium (NRS/CAL) includes the following: (1) the definitive method (DM) by IDMS,[12,24] (2) the certified reference material (CRM) as NBS/SRM #915,[14,25] (3) the reference method (RM) by FAAS,[1,26] and (4) other related reference materials carrying assigned values traceable to IDMS/DM as listed in Table III.[27,28] The individual analysts need no longer simply wonder about the accuracy of numerical results when concurrent analysis of these certified reference materials from NBS are so readily available.

True Value, Precision, and Accuracy

The concept of true value may seem relatively straightforward but is in reality quite complex when subject to critical review.[29] The true value

[24] "Development of Definitive Methods for Use in Clinical Chemistry for the National Reference System in the Clinical Laboratory: Tentative Guidelines," NRSCL1-T. National Committee for Clinical Laboratory Standards, Villanova, Pennsylvania, 1985.

[25] "Development of Certified Reference Materials for Use in Clinical Chemistry for the National Reference System in the Clinical Laboratory: Tentative Guidelines," NRSCL3-T. National Commitete for Clinical Laboratory Standards, Villanova, Pennsylvania, 1985.

[26] "Development of Reference Methods for Use in Clinical Chemistry for the National Reference System in the Clinical Laboratory: Tentative Guidelines," NRSCL2-T. National Committee for Clinical Laboratory Standards, Villanova, Pennsylvania, 1985.

[27] "Certificate of Analysis: Standard Reference Material 909—Human Serum, Lyophilized." Office of Standard Reference Materials, National Bureau of Standards, Gaithersburg, Maryland, 1973.

[28] "NBS Standard Reference Materials Catalog—1984–1985," NBS Special Publ. No. 260. Office of Standard Reference Materials, National Bureau of Standards, Gaithersburg, Maryland, 1973.

[29] C. Eisenhart, *J. Res. Natl. Bur. Stand., Sect. C* **67,** 2 (1963).

TABLE III
NBS/SRMs WITH ASSIGNED CALCIUM VALUES

SRM number	Material	Value
Primary reference material		
915	Calcium carbonate	99.9+% pure
Clinical reference material		
909	Human serum, lyophilized	3.02 mmol/liter
Biological reference material		
1549	Nonfat powdered milk	1.30 %
1566	Oyster tissue	0.15 %
1567	Wheat flour	0.019%
1586	Rice flour	0.014%
1572	Citrus leaves	3.15 %
1573	Tomato leaves	3.00 %
1575	Pine needles	0.41 %
1577a	Bovine liver	123 μg/g
Multielement reference materials		
1632a	Trace elements in coal (bituminous)	0.23 %
1633a	Trace elements in coal (fly ash)	1.11 %
1634a	Trace elements in fuel oil	(16) μg/g
1645	River sediment	(2.9)%
1646	Estuarine sediment	0.83%
Other reference materials carrying calcium values		
16-Ores		
13-Rocks minerals and refractories		
10-Glasses		
9-Cements		

(i.e., the actual value or *quaesitum*) of the property being measured is always uncertain. However, for daily operational purposes of a national measurement system that must turn concept into practice, the true value has been defined as the value determined by precise measurement methods that are free of systematic error, i.e., accurate methods, and these within the United States metrologic community are generally called Definitive Methods (DM).

In this chapter we will use the recently accepted NRSCL/NCCLS definition of *accuracy* which is "the difference between a result and the true value."[30] The true value for Ca_T is equated to the result obtained by measurements withthe NBS IDMS/DM.[12] The needed accuracy of a mea-

[30] "Nomenclature and Definitions for Use in the National Reference System for the Clinical Laboratory," NRSCL8-P. Vol. 5, No. 21. National Committee for Clinical Laboratory Standards, Villanova, Pennsylvania, 1985.

surement process is rationally determined by the degree of uncertainty that can be tolerated in the end use of the results. Thus, a much smaller degree of uncertainty is tolerable for the IDMS/DM of Ca_T than for the FAAS/RM and in turn the working methods used in clinical laboratories.[8] Therefore, a method's residual systematic error (or bias) must be small in relation to the desired end use of the results to be considered accurate for that particular analytical need.

The imprecision (or random error) of a method is determined by making a sufficient number of replicate analyses that can directly measure the following types of variability: (1) within-run, (2) between-runs or days (intralaboratory), and/or (3) between the same method in interlaboratory testing. Even when all types of random error are very small, this highly desirable analytical attribute of a method reveals very little about the accuracy or closeness to the true value of its results. Thus, the paradox that accurate methods must be precise but precise methods need not be accurate!

The FAAS method for Ca_T in this chapter has been carefully studied in comparison to the high accuracy of the IDMS/DM.[2,3] It rests on unique physical properties of calcium and the spectral characteristic inherent in high-resolution atomic absorption spectrometers to achieve isolation and quantitation. The simple chemical preparatory steps are designed to further reduce bias from the interfering substances that are often found in the sample or the reagents.

Theory of Atomic Absorption Spectrophotometry

The theory of atomic absorption spectrometry (AAS) is discussed at great length by many authors.[31-35] Since this chapter is intended as a practical source of information, only a brief discussion of the theoretical basis of AAS is presented.

The absorption of energy by ground-state atoms in the gaseous state is the basis of AAS. Understanding the process of energy absorption had its foundation in the study of the origin of the Fraunhofer dark lines in the sun's spectrum. When radiation of the proper wavelength passes through a vapor containing ground-state atoms, some of the radiation can be ab-

[31] A. Walsh, *Spectrochim. Acta* **7,** 108 (1955).

[32] C. T. J. Alkemade and J. M. W. Milatz, *J. Opt. Soc. Am.* **45,** 583 (1955).

[33] J. A. Dean and T. C. Rains (eds.), "Flame Emission and Atomic Absorption Spectrometry," p. 1. Dekker, New York, 1969.

[34] G. D. Christian and F. J. Feldman, "Atomic Absorption Spectroscopy: Applications in Agriculture, Biology, and Medicine." Wiley (Interscience), New York, 1970.

[35] C. T. J. Alkemade and R. Herman, "Fundamentals of Analytical Flame Spectroscopy." Adams Hilger, Bristol, England, 1979.

sorbed by excitation of the atoms, thus $M^0 + h\nu \rightarrow M^*$, and the intensity of the radiation at a wavelength corresponding to the energy of the proton $h\nu$ is decreased. If the concentration of M^0 in the vapor is increased, the decrease in radiant energy will be greater. Since each species of atoms can exist only in specific excited states, the photon energies required for each atomic species will be different, and thus will occur at different wavelengths. Only photons at the wavelength corresponding to specific excitation states will be absorbed in each case. Thus, photons of wavelength 422.7 nm can excite the calcium atom while those of wavelength 589.0 nm can excite the sodium atom. Therefore, each element requires photons of specific energy to produce excited atoms of that element.[31]

The magnitude of the AAS signal is directly related to the number of ground-state atoms in an absorption cell. The ground-state atoms are produced from the sample usually by a flame or electrothermal device. In each case, the sample must be dried or solvent evaporated, solids dissociated, and the neutral atoms produced.

The relative number of atoms in a particular energy state can be determined by the Boltzmann equation. Walsh calculated these ratios for the lowest excited states of several typical elements and several flame temperatures.[31]

The laws that govern the relationship between the amount of light absorbed by atoms in an absorption cell (flame or electrothermal) and thus concentrations are subject to the same rules as those for molecules in solution. This absorption–concentration relationship is summarized in the Beer–Lambert law[36]: absorbance $A = \log I_0/I = abc$, where I_0 is the incident radiation, I the radiation after passing through the samples gases, a the absorptivity, b the length of the absorption cell, and c the concentration of absorbing media.

Atomic absorption methods rely on the absorption of light by atoms. In principle all elements can be determined by AAS since the atoms of all elements can be excited and, therefore, capable of absorption. However, there are some practical limitations such as those elements emitting in the vacuum ultraviolet, strongly radioactive, and artificial elements.

Essential Features of Atomic Absorption Spectrometry

Sources of Radiation

The basic components of all atomic absorption spectrometers are depicted in Fig. 1. The function of the primary source of radiation in AAS is

[36] H. H. Willard, L. L. Merritt, Jr., J. A. Dean, and F. A. Settle, Jr., "Instrumental Methods of Analysis," 6th Ed., pp. 67–69. Van Nostrand, New York, 1981.

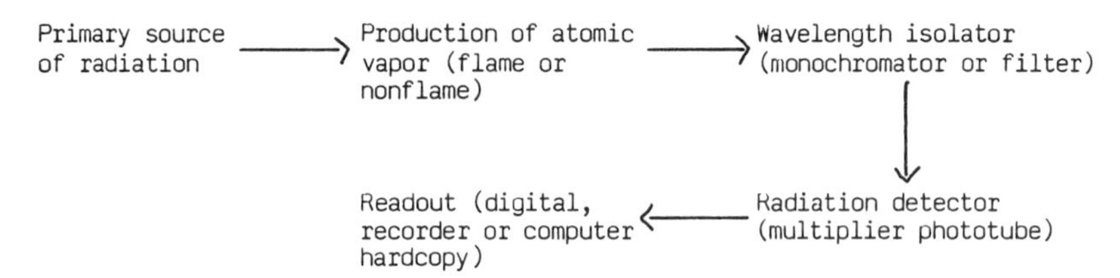

FIG. 1. Essential features of an atomic absorption spectrometer.

to generate radiation at the desired wavelength that is sharp, intense, and stable. There are many types of line sources used in AAS, however, the most commonly used source is a hollow cathode lamp (HCL).

Production of Atomic Vapor

There are two basic techniques (flames and electrothermal) for the production of atomic vapor in AAS. Walsh[31] was the first to introduce the nebulizer–burner–flame system and relatively few modifications have been made to his original design. The most commonly used type of burner for the production of atomic vapor in the premixed-gas or laminar-flow burner. In this system, the solution is aspirated into a pray chamber where the coarse and fine droplets are separated. The fine droplets are mixed with the flame gases and then pass into the flame where they are vaporized.

The major advantages of using a flame system are simplicity, speed of analysis, and a very stable atomization cell. A precision of better than 0.2% is obtainable if proper analytical technique is used. The major disadvantage of using a flame system is the loss of sensitivity.[37]

For those analytes below the detection limits of the flame system, electrothermal atomizers offer a very significant improvement in sensitivity. However, in biological tissue and fluids the calcium concentration is relatively high and flame systems would have adequate sensitivity. In a few rare cases where sample size is limited, electrothermal atomizers should be considered, however, the precision of the analytical measurement by this technique is 2 to 5%.

Optical Design

AAS instrumentation is of two basic designs—single beam and double beam. In a single-beam system the source (HCL) is focused by the means of two lenses through the flame and onto the entrance slit of the mono-

[37] J. E. Cantle (ed.), "Atomic Absorption Spectrometry." Elsevier, New York, 1982.

chromator. In a double-beam system, the light from the source is interrupted periodically and diverted around the flame and recombined before being focused on the entrance slit by a series of mirrors, lenses, and rotating mirrors. The chief advantage of a double-beam system is that the intensity of the resonance line is periodically monitored even while a sample is in the flame. Thus, any source noise, drift, or instability may be easily corrected electronically. However, HCLs have been improved to the extent that the flame is the limiting factor in precision in AAS.

In AAS, the monochromator is an essential part of the system. The monochromator must isolate the analytical line from the various other lines emitted by the source. Ideally, the monochromator should be capable of isolating the analyte resonance line while excluding all other wavelengths. For some analytes this is relatively easy. Lithium and sodium have relatively simple spectrums and the resonance lines of these elements can be easily isolated by the simplest of monochromators. For elements such as iron and nickel, which have a very complex spectrum, it becomes a very difficult task. As an example, nickel has strong nonresonance lines at 231.7 and 232.1 nm that are located on each side of the 232.0 nm resonance line. The ability to discriminate between the different wavelengths is thus a very important characteristic of the monochromator. In general, a typical monochromator in an AAS instrument has 0.1 nm spectral bandpass. Failure of the monochromator to isolate the resonance line of interest will lead to a nonlinear calibration curve.[37]

Photodetection and Readout

The multiplier phototube is by far the most widely used detector in AAS. A list and function of the most commonly used multiplier phototubes for AAS are described in the literature.[37,38]

For readout devices, digital displays are used in all modern instrumentation. Advantages are that operator bias in making readings is eliminated. A hard copy of the data produced by a printer or recorder eliminates the need for many tedious hours of sitting in front of an instrument taking data. The selecting of instrumental conditions, preparating of standards and diluting of samples, turning on of the instrument and adjusting flame conditions, preparing calibration curves, determining the concentration of analyte, and preparing reports are examples of technology that is now available for complete operation of the FAAS instrument by a computer. This new technology allows more efficient use of the analyte's time.

[38] J. A. Dean and T. C. Rains (eds.), "Flame Emission and Atomic Absorption Spectrometry," Vol. 2. Dekker, New York, 1971.

Procedure: Calcium in Biological Fluids by FAAS:

Principle

Calcium compounds in biological fluids are measured by diluting the sample in a solution of lanthanum. The diluted sample is then aspirated into an air–acetylene flame where the ground state atoms of calcium absorb 422.7 nm radiation and the absorbance is measured by a spectrometer.

Reagents

Only acid washed glassware is to be used (Note 1).

Ca stock standards (140 mmol/liter NaCl, 5.0 mmol/liter KCl, 2.00/2.50/3.00 mmol/liter $CaCO_3$, 1.0 mmol/liter Mg gluconate, 12 mmol/liter HCl).

1. Dry $CaCO_3$ for 4 hr at 200°, cool, and store in desiccator prior to use.
2. Add 8.18 g NaCl, 373 mg KCl, 450.6 mg gluconate (SRM #929), and 1.0 ml conc. HCl to 500 ml deionized water in three separate 1-liter flasks. Now add $CaCO_3$ (SRM #915), 200.2 mg to the first, 250.2 mg to the second, and 250.2 mg to the third. Mix until salts are dissolved, dilute to marks, and mix well.
3. Confirm accuracy of new standards using analysis by FAAS to compare with stock presently in use. Acceptable limits are ± 0.02 mmol/liter.
4. Label bottles for contents (i.e., concentration of Ca, and expiration date). Initial, date with the time of preparation, and store at room temperature. Stable at least 6 months.

Stock blank (140 mmol/liter NaCl, 5.0 mmol/liter KCl).

1. Add 8.18 g NaCl, 373 mg KCl to 500 ml deionized water in 1 liter.
2. Confirm by FAAS analysis, Ca should be 0.00 ± 0.02 mmol/liter.
3. Label and store at room temperature. Stable at least 6 months.

Working diluent (5 mmol/liter La_2O_3, 40 mmol/liter HCl).

1. Add 32.6 g La_2O_3 and 300 ml deionized water into a 600 ml acid-washed beaker.
2. While stirring slowly, add 133 ml conc. HCl. If La_2O_3 is not dissolved in 10–15 min, warm on a hot plate.
3. Transfer solution to an acid-washed 20-liter plastic carboy which has been calibrated and marked for 20 liters.
4. Dilute to volume (20 liters) with deionized water.
5. Label and store at room temperature. Stable at least 2 months.

Diluted set-up standard (for instrument set-up ONLY).

1. Add 20.0 ml Ca stock standard to a 1 liter acid-washed volumetric flask. Dilute to volume with deionized water (see Note 2).
2. Label bottle and store at room temperature.

Diluted set-up blank (for instrument set-up ONLY).

1. Add 20.0 ml stock blank to a 1-liter volumetric flask. Dilute to volume with deionized water.
2. Verify that readings for Ca is 0.00 ± 0.02 mmol/liter. Label bottle and store at room temperature.

Specimen Collection and Handling

Blood

1. Serum or heparinized plasma is acceptable. Collections using oxalate or EDTA (or patients receiving EDTA) are unacceptable (Note 3).
2. Fasting specimens are preferred.
3. Separate from red cells as soon as possible.

Urine

1. Twenty-four hour or random specimen may be analyzed.
2. Collection should be in acid (see procedure for processing nonacid collections of urine in Note 3).

Procedural Steps

Safety checks: Observe all isntructions in the instrument manual, and clear work area of all hazardous materials including corrosive liquids and flammable solvents.

Instrument set-up

1. *Warning:* Hollow cathode lamps and the burner flame emit dangerous UV radiation. The flame and lamps should never be viewed with the naked eye. If it is necessary to view the flame, safety glasses should be worn.
2. Follow the manufacturer's start-up procedure and verify that the hollow cathode lamp is on.
3. Allow lamp to warm-up a minimum of 10–15 min.
4. Set wavelength to 422.7 nm, and adjust lamp position and electrical controls until the signal is at maximum.

Flame ignition

1. Follow the manufacturers procedure. *Do not ignite flame unless both fuel and oxidant are available.*
2. Aspirate water at all times during warm-up time.
3. Allow flame and burner to warm-up for at least 20 min.

Sample preparation

1. Dilute blank, standards, controls, patient sera, and/or patient urine 1 to 50 with working diluent (preferably with an automated diluter). A minimum of 20 ml working blank and working standard is required for the daily run.

2. Mix well for 20 sec and cap diluted samples to prevent evaporation if time delay exists prior to analysis.

Calibration of isntrument

1. While aspirating blank, adjust to 0.00 ± 0.02 mmol/liter.

2. Aspirate set-up standard and adjust to find optimal aspiration rates, burner height, and flame adjustments (a fuel-rich, air–acetylene flame gives optimum sensitivity for calcium but a stoichiometric flame may give better precision).

3. Calibrate the instrument using the freshly diluted blank and each of the 2.00, 2.50, and 3.00 mmol/liter standards. Fine tune the instrument so that both the blank and the standards repeat to ± 0.02 mmol/liter.

Sample analysis

1. Aspirate diluted serum controls of known values. These must be within ± 1.0% of posted values before samples are analyzed.

2. Analyze unknown samples.

3. Check instrument calibration after every five samples by aspiration of blank and 2.50 standards. Adjust if necessary.

4. Dilute any specimen having a result greater than 3.0 mmol/liter by making a 1 : 100 dilution.

5. Aspirate water for 20–30 min at end of run to clean nebulizer.

Analysis of calculi for calcium

1. *Caution:* The dissolving solution for calculi is 5.3 mol/liter HNO_3. Handle with care!

2. Dilute the calculi dissolving solution according to the procedure outlined under sample preparation.

3. Analyze for Ca by the described procedure above.

Notes

1. Measurable amounts of Ca may be detected on glassware after routine washing. All glassware and plasticware used for reagents must first be washed in the routine manner and then immersed in 0.05 mol/liter HCl for at least 1 hr, followed by a thorough deionized water rinse. Pipets must be completely dry before use.

2. The set-up blank and standard should be used only during the initial instrument set-up for the purpose of adjustments of the stoichiometry of the flame and the correct burner and flame alignment.

3. Calcium salts may be precipitated if the pH is greater than four. Check pH of all urine specimens. If pH is less than four, specimen may be analyzed as received. When pH is greater than four, adjust pH to 4 with concentrated HCl. Mix well, and allow to stand for 1 hr before making dilution for analysis.

Interpretation

1. Precision: $s = \pm 0.03$ mmol/liter (at mean value of 2.50 mmol/liter).
2. Reference intervals: adults (serum), 2.20–2.58 mmol/liter; children (serum), 2.15–3.00 mmol/liter; urine, 2.3–7.5 mmol/24 hr.

Adaptation for Other Tissues and Fluids

Although this FAAS method was designed for measurements of Ca_T in serum, it can be used for any cell free fluid sample with varied amounts of protein (e.g., dialyzates to concentrates) and low to high solute concentrations (e.g., aqueous wash solutions to concentrated urine) if one makes adjustments in the calibrators. In any adaptations of the basic FAAS method, emphasis should be given to matrix matching of the calibrators to closely approximate the sample's composition. The use of diluents containing lanthanum to fully displace calcium from phosphate compounds is very important and must be reexamined carefully if the phosphate levels are markedly different than those of blood serum. Some biological tissues and fluids, e.g., the renal calculi mentioned above, will require wet-ashing in strong mineral acid to dissolve the stone or to digest protein before one can be sure that all calcium is in solution and available for complete atomization into the flame. Wet-ashing of renal calculi with nitric acid for the direct measurement of calcium and magnesium content by FAAS (see above) has been achieved routinely in the clinical laboratory at Hartford. Others have reported on the "no hazard" acid mixtures to destroy organic and inorganic constituents of calculi[39] or the "rapid and safe" digestion of dry plant materials[40] with nitric and perchloric acids in preparation for FAAS analysis. Tew *et al.* describe the rapid extraction of calcium from kidney tissue by sonification in a butanol–HCl–lanthanum media for FAAS analysis that avoids lengthy wet ashing.[41] FAAS measurements for Ca_T on homogenized and solubilized brain preparations,[42] after compara-

[39] M. A. E. Wandt, M. A. B. Pougnet, and A. L. Rodgers, *Analyst* (*London*) **109,** 1071 (1984).
[40] S. Allen, *Anal. Biochem.* **138,** 346 (1984).
[41] W. P. Tew, C. D. Malis, and G. Walker, *Anal. Biochem.* **112,** 346 (1981).
[42] W. J. Goldberg and N. Allen, *Clin. Chem.* **27,** 562 (1981).

tive washings of fingernails in nine solvents followed by dry and wet ashing,[43] and on specially ashed preparations of standard reference materials of human hair[44]; are only a few of the recent articles which describe detailed studies to bring tissue samples into solutions suitable for aspiration into the flame, yet hold down losses due to manipulations or contamination from reagents and equipment. The availability of the several different primary and secondary reference-type materials listed in Table III that carry a certified value for calcium content, especially those in complex materials that are similar to the sample under test, permits one to make concurrent runs to internally check on the analytical recovery and give one an idea of the accuracy of the measurement.

Summary

The measurement of total calcium (Ca_T) in biological fluids by flame atomic absorption spectroscopy (FAAS) is one of the more accurate yet practical methods available today. With attention to details at every analytical step, the imprecision of FAAS outlined above can be as low as ± 0.01 mmol/liter at the 2.65 mmol/liter upper reference level of Ca_T found in the serum of healthy young adults (a *CV* of 0.4%). The accuracy of FAAS as judged by measurements of NBS/SRM #909, a lyophilized serum material carrying Ca_T values assigned by isotope dilution mass spectrometry, can be within $\pm$ 1% (recovery of 99–101%) or less in skilled hands. As our brief section on application suggests, FAAS methods for Ca_T can be adopted to provide rapid and accurate results in many different biological fluids and tissues.

Appendix: National Reference System for Total Calcium (NRSCL/NCCLS)

Analyte: Calcium (in blood serum)
Updated: August 1, 1986
Status: Priority Analyte No. 5 by Council action 10/26/78 (see NRSCL4-CR)
Proposed Summary credentialed by Council action 7/14/86.

*Certified Reference Material (CRM)**

National Bureau of Standards SRM #915—Calcium Carbonate. Accepted by NRSCL Council on Jan. 18, 1979; for traceability see NBS/

[43] H. L. Bank *et al., Clin. Chim. Acta* **116,** 179 (1981).
[44] K. Okamoto *et al., Clin. Chem.* **31,** 1592 (1985).

SRM summary reports and primary notebooks in File No. R438 in OSRM at NBS. [For more detailed information see Certificate of Analyses on package insert and/or contact Mr. S. D. Raseberry, Chief of the Office of Standard Reference Materials, National Bureau of Standards, Gaithersburg, MD 20899 (Phone (301) 921-3479).] *See NRSCL3-T for Guideline to Development and Acceptance of CRM's into NRSCL.

*Definitive Method (DM)**

This isotope dilution mass spectrometry (IDMS) method for the definitive measurement of calcium was accepted by Council on Jan. 18, 1979. The supporting journal publication is by L. J. Moore and L. A. Machlan, *Anal. Chem.* **44,** 2291 (1972). The accuracy of this IDMS method by analysts at NBS is claimed to be 100 ± 0.02%. *See NRSCL1-T for Guideline to Development and Acceptance of DMs into NRSCL.

*Reference Method (RM)**

Flame atomic absorption spectroscopy (FAAS) is used in the reference method which meets the Guidelines described in NRSCL2-T*. It was accepted by Council on Jan. 30, 1980 (see NBS Spec Publ 1972; 260-36:1-121). The accuracy of this RM is claimed to be 100 ± 2% traceable to the DM values of 1.787, 2.147, 2.512, and 2.867 mmol/liter assigned by IDMS at NBS during the validation and transferability studies. *See NRSCL2-T for Guideline to Development and Acceptance of RMs into NRSCL.

Other Information

1. NBS/SRM #909, Human Serum, issued on 10/15/80 has a certified calcium value assigned by IDMS of 3.560 ± 0.013 mmol/liter/g (3.02 ± 0.011 mmol/liter) when the lypholized vial is reconstituted by the weight procedure "A" as directed.

2. Twelve collaborating European laboratories found a similar FAAS/RM to have a combined random and systematic bias of 100 ± 2% traceable to the IDMS/DM at NBS. Its major difference was in the use of a 5 ml sample rather than 10 ml (see *J. Clin. Chem. Clin. Biochem.* **19,** 395 (1981)).

3. The Comprehensive Chemistry Survey of the College of American Pathologist uses the mean value reported by FAAS methods as the target value because it is closest to the *true value* given by the IDMS/DM.

[23] Measurement of Ionized Calcium in Biological Fluids: Ion-Selective Electrode Method

By SALVADOR F. SENA and GEORGE N. BOWERS, JR.

Introduction

This chapter will describe the measurement of ionized calcium (Ca^{2+})[1] in blood, urine, and other biological fluids by direct (undiluted) potentiometry with ion-selective electrodes (ISEs). The major emphasis will be placed on serum, plasma, and whole blood samples, since the clinical utility of measuring Ca^{2+} in other biological fluids has not yet been established. The measurement of intracellular Ca^{2+} by ISEs will not be covered because of the much smaller and extremely varied concentrations of Ca^{2+} found in this environment[2,3] and the problems with unstable voltage, non-Nernstian slope, lack of Ca^{2+} selectivity, extreme miniaturization, and slow response time encountered with the microelectrodes used for these measurements.[4]

We will begin with a brief discussion of calcium homeostasis, the various calcium fractions in blood, the physiological significance of Ca^{2+}, the history and evolution of the Ca^{2+} ISE,[5] and finally, the clinical applications of the Ca^{2+} measurement. The majority of the chapter will be concerned with the actual measurement of Ca^{2+} by ISEs. This will include a description of sampling procedures (patient preparation, sample type, and use of anticoagulants), sample processing (storage and stability), sources of analytical bias (composition of calibrators, Ca^{2+} activity coefficient, Ca^{2+}–buffer association, and residual liquid junction potential), and the electrochemical cell itself (temperature, type of Ca^{2+} sensor, reference electrode, type and geometry of liquid junction, and bridge solution). Each of these factors represents a potential source of random or systematic error and must therefore be controlled in order to minimize the imprecision and inaccuracy of the measurement. Finally, the measurement of Ca^{2+} in urine and other biological fluids will be discussed briefly.

[1] Nonstandard abbreviations: Ca^{2+}, ionized calcium; ISE, ion-selective electrode; $1,25(OH)_2D_3$, 1,25-dihydroxycholecalciferol; PTH, parathyroid hormone; Ca_T, total calcium; E_j, liquid junction potential; ΔE_j, residual liquid junction potential.

[2] H. Rasmussen, *Clin. Endocrinol. Metab.* **1,** 3 (1972).

[3] H. Rasmussen, *N. Engl. J. Med.* **314,** 1094 and 1164 (1986).

[4] C. O. Lee and W. B. Im, *Kroc Found. Ser.* **17,** 157 (1984).

[5] G. N. Bowers, C. L. Brassard, and S. F. Sena, *Clin. Chem.* **32,** 1437 (1986).

Calcium Homeostasis

Calcium is the most abundant mineral element found in the human body. Of the total body calcium, 99% is localized in the skeleton and the remaining 1% is present in blood, extracellular fluids, and intracellular compartments. This small, nonskeletal fraction of calcium plays a vital role in many biochemical and physiological functions as both an intracellular and extracellular regulator and messenger.[2,3] These functions include bone formation and homeostasis,[6] maintenance of cell membrane integrity and permeability,[7] nerve excitability,[8] muscle excitation and contraction,[9] blood coagulation,[10] regulation of many enzyme and hormone reactions,[11] and the process of vision.[12]

In blood, calcium is present almost exclusively in the plasma phase and the concentration of plasma Ca^{2+} is maintained within narrow limits through hormonal control. The two most important hormones affecting calcium homeostasis are the vitamin D metabolite, 1,25-dihydroxycholecalciferol [1,25-$(OH)_2D_3$][13] and parathyroid hormone (PTH).[14] Calcium homeostasis is determined by the amount of calcium ingested in the diet, the amount of dietary calcium absorbed into the bloodstream by the gut (primarily under the regulation of 1,25-$(OH)_2D_3$), and the amount of bone resorption, renal reabsorption and urinary excretion of calcium (primarily under the regulation of PTH). Calcitonin, gonadal steroids, glucocorticoids, thyroid hormones, osteoclastic activating factor, growth factors, and prostaglandins are also involved in regulating plasma calcium levels,[15] although to a much smaller degree than 1,25-$(OH)_2D_3$ or PTH. In certain pathological conditions, other factors (e.g., interleukin-1, lymphotoxin, tumor necrosis factor) may cause significant disturbances in circulating levels of Ca^{2+}.[16]

[6] L. G. Raisz and B. E. Kream, *N. Engl. J. Med.* **309,** 29 and 83 (1983).
[7] J. W. Cheung, J. V. Bonventre, C. D. Malis, and A. Leaf, *N. Engl. J. Med.* **314,** 1670 (1986).
[8] A. M. Brown, N. Akaike, and K. S. Lee, *Ann. N.Y. Acad. Sci.* **307,** 330 (1978).
[9] C. van Breemen, P. Leijten, M. Yamamoto, P. Aaronson, and C. Cauvin, *Hypertension* **8** (Suppl. II), 89 (1986).
[10] E. K. Murer, *Semin. Hematol.* **22,** 313 (1985).
[11] J. R. Williamson, R. H. Cooper, and J. B. Hoek, *Biochim. Biophys. Acta* **639,** 243 (1981).
[12] D. F. O'Brien, *Science* **218,** 961 (1982).
[13] H. F. DeLuca, *Arch. Intern. Med.* **138,** 836 (1978).
[14] J. F. Habener and J. T. Potts, Jr., *N. Engl. J. Med.* **299,** 580 and 635 (1978).
[15] G. D. Aurbach, S. J. Marx, and A. M. Spiegel, *in* "Textbook of Endocrinology" (R. H. Williams, ed.), 7th Ed., pp. 1157–1161 and 1168–1169. Saunders, Philadelphia, Pennsylvania, 1985.
[16] M. Gowen and G. R. Mundy, *J. Immunol.* **136,** 2478 (1986).

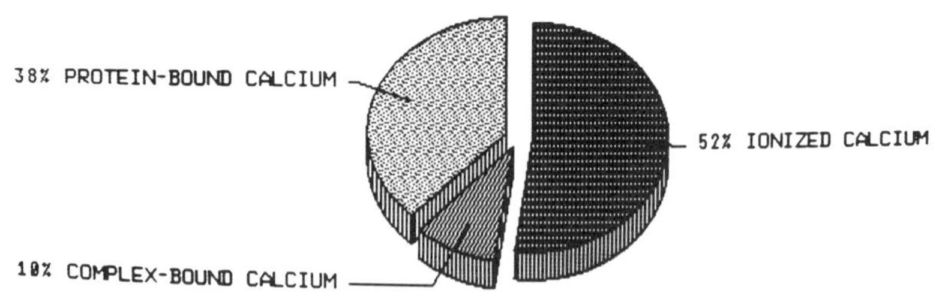

FIG. 1. Calcium fractions in blood.

Distribution of Calcium in Human Blood

As shown in Fig. 1, calcium is present in plasma and serum in three major fractions: protein-bound, complex-bound, and free or ionized calcium.[17,18] Together, the complex-bound and ionized calcium fractions are referred to as ultrafiltrable or dialyzable calcium. Protein-bound calcium is defined as calcium ions associated with proteins of MW 5000 or higher.[19] Assuming a mean total calcium concentration of 2.4 mmol/liter, approximately 0.9 mmol/liter (38%) is bound to plasma proteins (about 0.7 to albumin and 0.2 to globulins). The concentration of protein-bound calcium is directly proportional to the albumin concentration and is extremely pH dependent, since H^+ directly competes with Ca^{2+} for binding sites on albumin.

Complex-bound calcium refers to complexes of Ca^{2+} with anions in serum, such as bicarbonate, citrate, lactate, phosphate, and other inorganic and organic acids and peptides, having association constants higher than that between calcium and chloride ions.[19] This fraction is about 0.25 mmol/liter (10%) and consists mostly of Ca^{2+} bound to bicarbonate anion.

The remaining 1.25 mmol/liter (52%) of plasma or serum calcium consists of the free or ionized calcium fraction. This refers to calcium ions which are solvated with water molecules and also to electrostatically bound calcium ions, i.e., those which are deactivated or bound by electrostatic forces due to the presence of other plasma ions such as Na^+ and Cl^-. The fraction of calcium ions that are active is determined by the

[17] O. Siggaard-Andersen, J. Thode, and J. Wandrup, *in* "Blood pH, Carbon Dioxide, Oxygen and Calcium Ion: Proceedings of the 1980 Meeting of the IFCC Expert Panel on pH and Blood Gases" (O. Siggaard-Andersen, ed.), pp. 163–190. Private Press, Copenhagen, Denmark, 1981.

[18] O. Siggaard-Andersen, J. Thode, and N. Fogh-Andersen, *Scand. J. Clin. Lab. Invest.* **43** (Suppl. 165), 11 (1983).

[19] J. Toffaletti, *Proc. Meet. Eur. Working Group Ion Selective Electrodes* **1**, 17 (1984).

activity coefficient, which decreases with increasing ionic strength. By consensus, the ionic strength of normal plasma is 0.16 mol/kg of plasma water and the activity coefficient for Ca^{2+} in plasma is 0.30.[19]

Rigorously, an ISE responds to changes in the thermodynamic activity of ions, i.e., the Ca^{2+} ISE directly measures the molal activity of calcium ions. Furthermore, it is the activity and not the concentration of Ca^{2+} that is the physiological relevant quantity. However, units of molar concentration (and not molal activity) are commonly used for Ca^{2+} as well as for other plasma (serum) electrolytes in clinical practice. By convention, the measured Ca^{2+} activity is divided by the accepted activity coefficient of 0.3 and the resulting substance concentration (in mmol/liter) is reported. Since the purpose of the activity coefficient is merely to convert the measured ion activity to more conventional units of concentration, only one value is used, even though this value varies between samples.[19]

The ionized, complex-bound, and protein-bound calcium fractions are interrelated through several complex equilibria which are influenced by pH, temperature, ionic strength, magnesium ion concentration, total protein concentration, and albumin-to-globulin ratio.[17,20] Because of the marked dependence of Ca^{2+} on pH, these two parameters are usually measured concurrently and the measured Ca^{2+} is often corrected or normalized to the physiological pH of 7.4.[17,21]

Physiological Significance of Ionized Calcium

In 1934 and 1935, the landmark publications of McLean and Hastings and associates[22-25] on the measurement of Ca^{2+} by the contraction of frog heart muscle clearly demonstrated that this was the physiologically active fraction of calcium in blood. However, there was no reliable method for measuring Ca^{2+} at that time aside from this bioassay, which was not practical for the clinical laboratory. Instead, they suggested that Ca^{2+} be estimated from the total calcium (Ca_T) and total protein concentrations and published a nomogram for this calculation.[22] However, they noted that this estimation was only a first approximation, since "other factors not considered in the nomogram, such as temperature, pH, albumin-to-

[20] O. Siggaard-Andersen, J. Thode, and N. Fogh-Andersen, *Scand. J. Clin. Lab. Invest.* **43** (Suppl. 165), 57 (1983).

[21] J. Thode, N. Fogh-Andersen, P. D. Wimberly, A. M. Sorensen, and O. Siggaard-Andersen, *Scand. J. Clin. Lab. Invest.* **43** (Suppl. 165), 79 (1983).

[22] F. C. McLean and A. B. Hastings, *J. Biol. Chem.* **107,** 337 (1934).

[23] F. C. McLean and A. B. Hastings, *J. Biol. Chem.* **108,** 285 (1935).

[24] F. C. McLean and A. B. Hastings, *Am. J. Med. Sci.* **189,** 601 (1935).

[25] F. C. McLean, B. O. Barnes, and A. B. Hastings, *Am. J. Physiol.* **113,** 141 (1935).

globulin ratio, magnesium, and citrate, were known to affect the concentration of Ca^{2+} in the blood."[24] Subsequently, these nomograms were modified by replacing total protein with albumin and adding pH[20]; however, even the improved nomograms have been shown to be unreliable for calculating Ca^{2+} [26,27] and none has seen widespread use in the clinical laboratory. Instead, physicians simply relied on Ca_T as an indicator of calcium status in their patients, despite the several limitations of this approach.

The first ISE for the direct potentiometric measurement of Ca^{2+} was reported in 1967 by Ross[28] and was based on an organophosphate ion exchanger. The composition of this ISE was subsequently modified and its selectivity improved during the period from 1967 to 1975.[29,30] At about the same time, Simon and co-workers[31] described a different type of Ca^{2+} ISE based on a neutral carrier sensor. During the period immediately following Ross's discovery, Moore[32–34] investigated the clinical applications and utility of Ca^{2+} measurements and found that the ISE responded only to the free calcium ions in blood serum and not to the protein- and complex-bound calcium ions. He concluded, therefore, that the measurement of Ca^{2+} by ISE was a more clinically relevant indicator of calcium homeostasis than the Ca_T measurement. Since then, numerous reports have appeared in the literature supporting this conclusion. The clinical utility of Ca^{2+} measurements has been reviewed by Robertson and Marshall[35] and more recently by Toffaletti.[36] It is now widely accepted that Ca^{2+} is superior to Ca_T in monitoring the calcium status of patients, especially in cases of hyperparathyroidism,[37–40] hypercalcemia of

[26] J. H. Landenson, J. W. Lewis, and J. C. Boyd, *J. Clin. Endocrinol. Metab.* **46,** 986 (1978).
[27] T. F. White, J. R. Farndon, S. C. Conceicao, M. F. Laker, M. K. Ward, and D. N. S. Kerr, *Clin. Chim. Acta* **157,** 199 (1986).
[28] J. W. Ross, *Science* **156,** 1378 (1967).
[29] G. J. Moody, R. B. Oke, and J. D. R. Thomas, *Analyst (London)* **95,** 910 (1970).
[30] J. Ruzicka, E. H. Hansen, and J. C. Tjell, *Anal. Chim. Acta* **67,** 155 (1973).
[31] D. Ammann, M. Guggi, E. Pretsch, and W. Simon, *Anal. Lett.* **8,** 709 (1975).
[32] E. W. Moore, *J. Clin. Invest.* **49,** 318 (1970).
[33] E. W. Moore, *NBS Spec. Publ.* **314,** 215 (1969).
[34] E. W. Moore, *Gastroenterology* **49,** 43 (1971).
[35] W. C. Robertson and R. W. Marshall, *Crit. Rev. Clin. Lab. Sci.* **15,** 85 (1981).
[36] J. Toffaletti, "Ionized Calcium through Its First Decade: A Review of the Analytical Aspects and Clinical Importance of Ionized Calcium Measurements," NOVA Biomedical 1986. (Available from NOVA Biomedical Inc., 200 Prospect Street, Waltham, MA 02254-9141.)
[37] J. M. Monchik and H. F. Martin, *Surgery* **88,** 185 (1980).
[38] M. K. McLeod, J. M. Monchik, and H. F. Martin, *Surgery* **95,** 667 (1984).
[39] J. C. Boyd, J. W. Lewis, J. M. McDonald, E. Slatopolsky, and J. H. Ladenson, *Clin. Chem.* **27,** 574 (1981).

malignancy,[40,41] pancreatitis,[42,43] and critically ill patients.[44,45] Furthermore, Ca_T may be totally misleading in patients receiving blood transfusions,[46–48] on hemodialysis[49,50] or with end stage renal disease,[51] with extensive thermal burns,[52,53] and in infants for detecting and following neonatal hypocalcemia.[54–57] In these latter instances, Ca^{2+} is the only reliable measurement of calcium homeostasis.

Measurement of Ca^{2+} by ISEs

Patient Preparation

In vivo blood Ca^{2+} levels are affected by changes in diet, posture, and activity and the magnitude of these effects cannot be predicted from person to person.[58–60] Therefore, it is important to control these variables in order to obtain a meaningful and reproducible Ca^{2+} measurement. It is generally recommended that blood be collected after an overnight fast with the patient at rest in the supine position and with a minimum of venous stasis or forearm exercise. In addition, the sample should be

---

[40] J. H. Landenson, J. W. Lewis, J. M. McDonald, E. Slatopolsky, and J. C. Boyd, *J. Clin. Endocrinol. Metab.* **48,** 393 (1979).

[41] W. P. Shemerdiak, S. C. Kukreja, T. E. Lad, P. A. J. York, and W. J. Henderson, *Clin. Chem.* **27,** 1621 (1981).

[42] R. S. Croton, R. A. Warren, A. Stott, and N. B. Roberts, *Br. J. Surg.* **68,** 241 (1981).

[43] A. L. Warshaw, K. H. Lee, T. W. Napier, P. O. Fournier, D. Duchainey, and L. Axelrod, *Gastroenterology* **89,** 814 (1985).

[44] L. J. Drop and M. B. Laver, *Anesthesiology* **43,** 300 (1975).

[45] G. P. Zaloga, B. Chernow, D. Cook, R. Snyder, M. Clapper, and J. T. O'Brian, *Ann. Surg.* **202,** 587 (1985).

[46] T. R. Abbott, *Br. J. Anesthesiol.* **55,** 753 (1983).

[47] K. O. Pedersen and O. Juhl, *Scand. J. Clin. Lab. Invest.* **43** (Suppl. 165), 107 (1983).

[48] T. A. Gray, B. M. Buckley, M. M. Sealey, S. C. H. Smith, P. Tomlin, and P. McMaster, *Transplantation* **41,** 335 (1986).

[49] S. N. Asad, A. J. Olmer, and J. M. Letteri, *Miner. Elect. Metab.* **10,** 333 (1984).

[50] W. L. Henrich, J. M. Hunt, and J. V. Nixon, *N. Engl. J. Med.* **310,** 19 (1984).

[51] M. F. Burritt, A. M. Pierides, and K. P. Offord, *Mayo Clin. Proc.* **55,** 606 (1980).

[52] S. K. Szyfelbein, L. J. Drop, and A. J. Martyn, *Crit. Care Med.* **9,** 454 (1981).

[53] J. J. Fenton, M. Jones, and C. E. Hartford, *J. Trauma* **23,** 863 (1983).

[54] R. C. Tsang, I.-W. Chen, M. A. Friedman, and I. Chen, *J. Pediatr.* **83,** 728 (1973).

[55] D. R. Brown, B. H. Sterauka, and F. H. Taylor, *Am. J. Dis. Child* **135,** 24 (1981).

[56] D. R. Brown and D. J. Salsburey, *J. Pediatr.* **100,** 777 (1982).

[57] L. Larsson, O. Finnstrom, B. Nilsson, and S. Ohman, *Scand. J. Clin. Lab. Invest.* **43** (Suppl. 165), 21 (1983).

[58] J. H. Landenson and G. N. Bowers, Jr., *Clin. Chem.* **19,** 565 (1973).

[59] J. H. Landenson and G. N. Bowers, Jr., *Clin. Chem.* **19,** 575 (1973).

[60] R. S. Lazar, B. Deane, and G. N. Bowers, Jr., *Clin. Chem.* **29,** 1187 (Abstr. 308) (1983).

collected as anaerobically as possible, since the *in vitro* loss of carbon dioxide from the sample will cause an increase in pH and a concomitant decrease in measured Ca^{2+}. Therefore, vacuum tubes should be completely filled and air bubbles should be avoided when heparinized syringes are used for whole blood samples.

Sample Type

Most commercial Ca^{2+} analyzers accept serum, plasma, or whole blood[61] samples. The consensus in the literature is that anaerobic serum is the specimen of choice since it is more stable than whole blood and has less deleterious effects on the Ca^{2+} ISE. The use of whole blood is advantageous in instances where the need for rapid results precludes the time required for clotting and centrifugation, e.g., monitoring Ca^{2+} during cardiopulmonary bypass surgery or liver transplantation. In addition, whole blood samples may be necessary in screening and monitoring Ca^{2+} in neonates, where the amount of blood removed from the patient should be minimized.

However, if excessive amounts of heparin are used, plasma and whole blood samples may give substantially lower Ca^{2+} results than serum due to binding of Ca^{2+}, as explained in the next section. Furthermore, when measuring Ca^{2+} in whole blood, a positive bias of 4–5% relative to plasma has been reported when a saturated or concentrated KCl solution is used as the salt bridge. This was attributed to the effect of erythrocytes on the liquid junction potential (E_j) and was found to depend on the hematocrit, i.e., the effect was more pronounced as the volume of packed red cells increased. This bias can be reduced or eliminated altogether by increasing the cross-sectional area of contact between blood and KCl, monitoring the change in potential immediately after the liquid junction has been established,[17] or by using a mixed-salt bridge solution[62] or a special bridge solution composed of concentrated sodium formate.[63,64]

Anticoagulants

Heparin is the only anticoagulant that can be used for whole blood (or plasma) Ca^{2+} samples. Citrate, oxalate, and EDTA must be avoided, since they strongly bind calcium ions and will substantially lower the Ca^{2+}

[61] The ISE measures only the Ca^{2+} in the *plasma* of whole blood.

[62] O. Siggaard-Andersen, N. Fogh-Andersen, and J. Thode, *Scand. J. Clin. Lab. Invest.* **43** (Suppl. 165), 43 (1983).

[63] O. Siggaard-Andersen, N. Fogh-Andersen, J. Thode, and T. F. Christiansen, *Proc. Meet. Eur. Working Group Ion Selective Electrodes* **1**, 149 (1984).

[64] J. Wandrup and J. Kvetny, *Clin. Chem.* **31**, 856 (1985).

measured by the ISE. Although heparin itself also binds Ca^{2+}, it does so to a much lesser degree and will have a minor and clinically insignificant effect on Ca^{2+} when used at low concentrations. The exact amount of heparin to be used is a matter of some debate. Most workers agree that a heparin concentration of 50 IU/ml of whole blood is too high and yields Ca^{2+} results for on whole blood which are 6% lower than those obtained for corresponding serum samples.[17] However, some state that a heparin concentration of 20 IU/ml of whole blood does not have an appreciable effect on Ca^{2+},[35] while others recommend even a lower concentration (5–10 IU/ml).[17,65,66] Also, the amount of Ca^{2+} binding has been reported to vary from one heparin preparation to another, with some brands of heparin lowering Ca^{2+} in plasma or whole blood samples by as much as 15–25% at a heparin concentration of 100 IU/ml.[36] An increasingly popular practice has been the use of a calcium-containing heparin solution.[66] This preparation is an aqueous solution of sodium heparin (875 IU/ml) and NaCl (155 mmol/liter) which has been titrated with $CaCl_2$ to a final measured (by ISE) Ca^{2+} of 1.25 mmol/liter and is commercially available from Radiometer.[67]

Sample Stability and Storage

There have been numerous studies in the literature on the stability and storage of serum, plasma, and whole blood samples for Ca^{2+} measurements.[65,66,68,69] The most important consideration in the specimen processing is keeping the sample anaerobic, i.e., guarding against the loss of carbon dioxide before analysis. As mentioned earlier, care should be exercised at the time of sample collection to fill the vacuum tube as much as possible. Although a small amount of dead air space is usually unavoidable even in a full tube (about 8 mm between the bottom of the rubber stopper and the top of the sample in a 7 cm^3 tube), the tube should be filled to at least 75% of its total volume (no more than 35 mm of air space in a 7 cm^3 tube).[65]

Anaerobically stored blood samples are still subject to metabolic changes, i.e., production of lactic acid. The accompanying decrease in pH

[65] G. Graham, Paper presented at the international symposium "Ionized Calcium: Its Determination and Clinical Usefulness," February 27–28, 1986.

[66] P. Urban, B. Buchmann, and D. Scheidegger, *Clin. Chem.* **31,** 264 (1985).

[67] "Radiometer ICA1 Ionized Calcium Analyzer User's Handbook." Radiometer A/S, Emrupvej 72, DK-200 Copenhagen NV, Denmark, or Radiometer America, Inc., 811 Sharon Drive, Westlake, OH 44145.

[68] J. Thode, N. Fogh-Andersen, F. Aas, and O. Siggaard-Andersen, *Scand. J. Clin. Lab. Invest.* **45,** 131 (1985).

[69] J. Toffaletti, N. Blosser, and K. Kirvan, *Clin. Chem.* **30,** 553 (1984).

causes a displacement of calcium ions from albumin (and bicarbonate) and an increase in Ca^{2+}. Fortuitously, this effect is almost exactly offset by the binding of Ca^{2+} by lactate ion. Therefore, the Ca^{2+} concentration in serum or plasma that is allowed to sit anaerobically on the red blood cells at 25° has been reported to be stable for at least 2 hr by some workers[69] and for 4 hr by others.[65] Similarly, blood samples may be stored anaerobically at 4° for at least 6 hr[69] (and probably longer[65]) without a significant change in Ca^{2+}. Serum may be removed anaerobically from the red blood cells and stored in a sealed syringe at 4° for at least 24 hr.[69] Recently, a study on the effect of storage on the serum Ca^{2+} of uremic patients recommended that these samples should be analyzed for Ca^{2+} within 2 hr of collection if stored at 25° and within 6 hr if stored at 4°.[70] In any case, samples should never be stored at temperatures much greater than 25°, since the resulting increase in metabolic rate will cause much more rapid changes in Ca^{2+}.[65]

Samples that have not been processed anaerobically may be handled in two ways, depending upon how much CO_2 has been lost and how elevated the pH has become. First, many Ca^{2+} instruments concurrently measure both Ca^{2+} and pH and then correct or normalize the Ca^{2+} to a standard pH of 7.40 ($Ca^{2+}_{7.4}$). This value can usually be used in place of the actual measured Ca^{2+} in the pH range 7.20–7.60, where the relationship between Ca^{2+} and pH is linear and well defined.[17] Other instruments (NOVA Models 7 and 8, Baker Ana-Lyte+2) calculate the $Ca^{2+}_{7.4}$ over a wider pH range (6.9–8.0); however, the corrections at the low (less than 7.2) and high (greater than 7.6) extremes of this range can often be misleading. Alternatively, it is possible to adjust the pH to a value close to 7.4 by equilibrating the sample with CO_2 to a normal P_{CO_2} of about 5.3 kPa.[70] The Ca^{2+} and pH of the reequilibrated sample can then be measured and the calculated $Ca^{2+}_{7.4}$ can be used as a reasonable approximation of the *in vivo* Ca^{2+}. In any case, a sample with a pH of greater than 8.0 is not suitable for analysis (even if the pH is readjusted) since precipitation of calcium phosphate salts will occur and cause falsely low results.

Instrumentation

The Ca^{2+} ISE has been incorporated into several commercial instruments for Ca^{2+} measurements in the clinical laboratory. A list of available Ca^{2+} analyzers that simultaneously measure Ca^{2+} and pH in undiluted serum, plasma, or whole blood samples is given in Table I. Because of the close relationship between Ca^{2+} and pH,[17,21] most state-of-the-art Ca^{2+}

[70] J. Thode, N. Fogh-Andersen, and O. Siggaard-Andersen, *Proc. Meet. Eur. Working Group Ion Selective Electrodes* **1,** 71 (1984).

TABLE I
Ca^{2+} ANALYZERS WITH SIMULTANEOUS pH MEASUREMENT

Year	Manufacturer	Model	Sample size (μl)	Ca^{2+} ISE	Comment
1980	Radiometer A/S (Denmark)	ICA1	125	PVC membrane organophosphate ion exchanger	Calculates $Ca^{2+}_{7.4}$; intro. US 1982
1982	NOVA Biomedical (US)	Model 8	350	PVC membrane neutral carrier	Calculates $Ca^{2+}_{7.4}$
1984	Nova Biomedical (US)	Model 7	400	PVC membrane neutral carrier	$Ca^{2+}_{7.4}$; also does Ca_T by ISE
1984	Baker Insts. (US)	Ana Lyte+2	250	PVC membrane neutral carrier	$Ca^{2+}_{7.4}$ Horiba/Japan
1985	Sentech (US)	ChemPro-1000	150	Disposable solid state ISE card	$Ca^{2+}_{7.4}$; also Na^+, K^+ doctor's office
1987	Ciba-Corning (UK)	Model 634	35	PVC membrane neutral carrier	Calculates $Ca^{2+}_{7.4}$

analyzers concurrently measure Ca^{2+} and pH and also calculate $Ca^{2+}_{7.4}$. Most of these second-generation[5] instruments are microprocessor controlled with an anaerobic flow system, semiautomated measurement cycle, miniature flow-through (or flow-by) electrodes, and a thermostatted measurement chamber. Some of the newer instruments feature semiautomatic or fully automatic calibration. All analyzers use the neutral carrier type of Ca^{2+} sensor with the exception of the Radiometer ICA1, which uses the organophosphate ion-exchanger. Both types of ISEs demonstrate near-Nernstian slopes and good selectivities for Ca^{2+} over H^+, Na^+, K^+, and Mg^{2+}.[30,31] As mentioned earlier, it is desirable to make simultaneous Ca^{2+} and pH measurements on all samples.

Electrochemical Cell

A schematic representation of the basic electrochemical cell used to measure Ca^{2+} is shown in Fig. 2. The Ca^{2+} ISE consists of an internal reference or indicator electrode, e.g., Ag/AgCl, in contact with an internal solution of fixed Ca^{2+} activity and separated from the analyte solution by the calcium ion-selective liquid membrane. This liquid membrane consists of either the organophosphate ion-exchanger or neutral carrier material dissolved in a solid support matrix, usually poly(vinyl chloride) (PVC). The cell is completed by the external reference electrode, usually a saturated calomel or Ag/AgCl electrode, which also serves as the reference

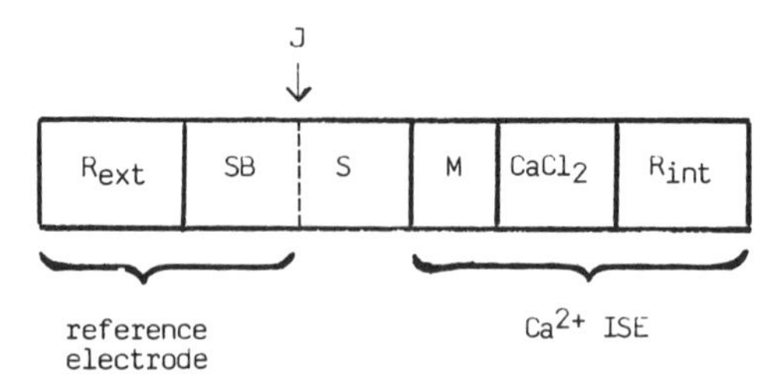

FIG. 2. Electrochemical cell with Ca^{2+} ISE. R_{int}, internal reference (indicator) electrode for Ca^{2+} ISE; $CaCl_2$, aqueous calcium chloride internal filling solution; M, Ca^{2+}-selective liquid membrane; S, sample solution; SB, salt bridge solution; J, liquid junction between SB and S; R_{ext}, external reference electrode.

for the pH electrode. Electrical contact between the sample solution and the reference electrode occurs at the liquid–liquid junction. This is accomplished by allowing a very small, controlled leakage of the salt bridge solution (concentrated KCl or sodium formate) into the sample stream.

The Ca^{2+} ISE will develop an electromotive force, or potential which is logarithmically related to the *activity* of calcium ions in the sample solution by the Nernst equation.[18] If the electrode is calibrated with standard solutions having an ionic strength equal to that of the sample, i.e., if the Ca^{2+} activity coefficients in the sample and standard solutions are equal, then the cell can be calibrated in terms of concentration.[17] As mentioned earlier, by consensus this ionic strength is 0.16 mol/kg and the value for the Ca^{2+} activity coefficient at this ionic strength is 0.30.

Calibration

There is no universally accepted reference system for the measurement of Ca^{2+} by ISEs.[71,72] Each manufacturer has taken a slightly different approach to the calibration of the Ca^{2+} ISE. Consequently, the Ca^{2+} results obtained on different types of analyzers do not agree exactly since each instrument uses different calibration solutions.[72]

The Ca^{2+} ISE is commonly calibrated with two aqueous solutions supplied by the manufacturer of the Ca^{2+} analyzer. These secondary or working calibrators usually consists of $CaCl_2$ in an aqueous matrix of

[71] A. B. T. J. Boink, B. M. Buckley, T. F. Christiansen, A. K. Covington, A. H. J. Maas, O. Muller-Plathe, C. Sachs, and O. Siggaard-Andersen, Proposed reference method submitted by members of the European Working Group on Ion Selective Electrodes to the International Federation of Clinical Chemistry, Scientific Committee, Expert Panel on pH and Blood Gases (A. H. J. Maas, personal communication).

[72] A. O. Ohorodudu, S. F. Sena, R. W. Burnett, and G. N. Bowers, in preparation.

NaCl and a pH buffer. The buffer serves to fix the pH at a constant, known value so that the pH electrode may be calibrated concurrently with the Ca^{2+} ISE. (Even on analyzers that do not measure pH, the working calibrators contain a small amount of buffer so that the pH of the solution is maintained within the optimal working range of the Ca^{2+} ISE.) The pH buffer used is commonly one of the zwitterionic sulfonate buffers,[73] e.g., HEPES, MOPS, TES, or BES.[74] The total ionic strength of the calibrator is adjusted to 0.16 mol/kg with NaCl.

These secondary calibration solutions have been referenced to primary solutions by the manufacturer. For pH, the two national Bureau of Standards (NBS) phosphate buffers (pH 7.386 and 6.841 at 37°)[75] are usually used as the primary calibrators. The primary Ca^{2+} standards are simply aqueous solutions of pure $CaCl_2$ and NaCl at an ionic strength of 0.16 mol/kg. A very small amount (1 mmol/liter) of pH buffer is added to these primary solutions in order to maintain a pH near 7. The calcium content of a secondary calibrator is then empirically adjusted to give the same measured Ca^{2+} activity as the primary calibrator.

Typically, $CaCl_2$ *in excess of the theoretical amount* must be added to the secondary standard to compensate for Ca^{2+} that is bound by the pH buffer (and therefore not measured by the Ca^{2+} ISE) and also to offset the effect of residual liquid junction potential (ΔE_j) between the primary and secondary standard solutions. A ΔE_j of about 0.7–0.9 mV (depending on the salt bridge) arises between a serum matrix and the simple aqueous matrix described above for the primary calibration solution because of the difference in ionic mobilities between the two matrices.[72] This translates into a bias of 5–7% in Ca^{2+} concentration and could, in theory, be eliminated either by matching the calibrator matrix to serum with respect to E_j[72] or by using an electronic adjustment. However, this bias caused by ΔE_j is usually accepted as an inherent bias of the calibration system and ignored. Ideally, the best calibrator matrix for the measurement of Ca^{2+} in serum, plasma, or whole blood by ISEs would be a plasma or serum protein-based matrix, since this would eliminate the bias caused by ΔE_j. Unfortunately, as mentioned above, a reference method does not exist by which to assign a true value of Ca^{2+} to such a calibrator.

[73] N. E. Good, G. D. Winget, W. Winter, T. N. Connolly, S. Izawa, and R. M. M. Singh, *Biochemistry* **5,** 467 (1966).

[74] Abbreviations: HEPES, 4-(2-hydroxyethyl)-1-piperazineethanesulfonic acid; TES, 2-[[2-hydroxy-1,1-bis(hydroxymethyl)ethyl]amino]ethanesulfonic acid; BES, 2-[bis(2-hydroxyethyl)amino]ethanesulfonic acid; MOPS, 4-morpholinepropanesulfonic acid.

[75] Certificate of analysis accompanying NBS SRM 186-I-c and SRM 186-II-c. See also R. A. Durst, *NBS Spec. Publ.* **260-53,** 27 (1975).

Precision, Accuracy, and Quality Control

When properly maintained and calibrated, modern commercial Ca^{2+} analyzers are capable of performing with a day-to-day, intralaboratory coefficient of variation (*CV*) of around 1% for serum samples and 2% for whole blood samples.[5] The corresponding interlaboratory *CV*s were found to be approximately twice the intralaboratory *CV*s in a recent large survey.[76]

Strictly speaking, the *absolute* accuracy of Ca^{2+} measurements is difficult to assess due to the lack of an accepted reference method. Only the *relative* accuracy of a particular Ca^{2+} ISE can be evaluated by comparing it to another ISE. Assuming that the various precautions regarding sample collection, processing, and storage are taken, the calibration of the ISE is by far the most important factor determining the absolute accuracy of the Ca^{2+} measurement.

Quality control solutions can be used to monitor the short- and long-term precision and the relative accuracy of Ca^{2+} measurements. Most manufacturers of Ca^{2+} analyzers provide their own aqueous-based quality control solutions for monitoring the performance of the Ca^{2+} (and pH) electrodes at the normal and low and high abnormal levels of the physiological range. Recently, one manufacturer has introduced a set of three bovine albumin-based controls which are claimed to have stable values of Ca^{2+} and pH.[77]

Ca^{2+} in Urine and Other Biological Fluids

As mentioned in the introduction, Ca^{2+} has been measured mostly in blood, since its clinical significance in other biological fluids is still largely unknown. However, Ca^{2+} measurements have been performed in a wide variety of nonblood samples, including urine, cerebrospinal fluid, saliva, bile, and miscellaneous extracellular fluids. This work has been reviewed by Robertson and Marshall,[35] and will not be discussed here in detail except for a brief description of the use of urine Ca^{2+} measurements. However, some general comments on the measurement of Ca^{2+} in nonblood samples are in order, since most Ca^{2+} analyzers are intended for use only with serum, plasma, and whole blood samples.

In principal, the concentration of Ca^{2+} in any biological fluid can be measured by direct potentiometry with an ISE. However, careful consideration must be given to the calibration of the Ca^{2+} ISE. It is extremely important that the matrices of the calibration and sample solutions be

[76] A. B. T. J. Boink and R. Sprokholt, personal communication (1985).

[77] NOVA Stat-Controls, NOVA Biomedical, Inc., 200 Prospect Street, Waltham, MA 02254-9141.

closely matched. As mentioned earlier, it is the calcium ion activity which is actually measured by the ISE and is the physiologically relevant quantity. However, ISE analyzers are calibrated in terms of substance concentration, since this is the quantity that is used clinically. The measured activity is converted to concentration by the instrument microprocessor by dividing by the activity coefficient, which is assumed to be constant. Therefore, the ionic strength of the calibrator(s) and the sample should be the same in order for the Ca^{2+} activity coefficients in both solutions to be equal. If the ionic strength of the calibrator and the sample differ appreciably from each other, a substantial error in the results will occur. For example, if the ionic strength of the sample is much greater than that of the calibrator, then the Ca^{2+} activity coefficient will be considerably smaller in the sample and the calculated concentration result will be falsely low.

Thode *et al.*[78,79] have investigated the measurement of Ca^{2+} in urine in both healthy subjects and in patients with kidney stones. Using a stepwise discriminant analysis in which urine Ca^{2+} excretion, urine pH, plasma albumin, plasma phosphate, creatinine clearance, and nephrogeneous cyclic adenosine monophosphate were among the parameters considered, they found a significantly higher Ca^{2+} *activity* and a higher pH in the patients with kidney stones compared to the normal controls and proposed that Ca^{2+} and pH be measured in a 4 hr fasting urine specimen for the evaluation of idioopathic hypercalciuria. Due to the wide range of ionic strengths encountered in urine, the activity coefficient for Ca^{2+} varies markedly from one urine sample to another, making calibration of the ISE in terms of Ca^{2+} concentration impractical. Therefore, the authors elected to simply measure the Ca^{2+} *activity* in the urine.

It is also possible to use ISEs to measure Ca^{2+} in buffered solutions used for tissue culture media. Frequently, the calcium content of these solutions appears adequate for culture growth when checked by measurement of total calcium, yet the Ca^{2+} concentration measured by ISE is markedly reduced due to binding of Ca^{2+} by albumin, anions, and pH buffers.[80]

Summary

There is now a consensus that Ca^{2+} measurements are more physiologically and clinically meaningful than Ca_T measurements. Ca^{2+} in se-

[78] J. Thode and O. Siggaard-Andersen, *Proc. Meet. Eur. Working Group Ion Selective Electrodes* **1,** 75 (1984).

[79] J. Thode, manuscript in preparation.

[80] G. N. Bowers, Jr., unpublished observations.

rum, plasma, whole blood, and other biological fluids can be measured by direct potentiometry with ion-selective electrodes and a number of reliable instruments are commercially available for this measurement. Several factors affect the Ca^{2+} concentration and must be carefully controlled for the results to be meaningful. The most important of these considerations are the anaerobicity of the sample, the need to concurrently measure pH, and the concentration of heparin, if whole blood or plasma samples are used. The calibration of the Ca^{2+} ISE is critical to the accuracy of the measurement. The matrix of the calibrator should match that of the sample as closely as possible, particularly in regard to ionic strength and liquid junction potential. The measurement of Ca^{2+} in urine is complicated by the wide variation in ionic strength encountered in this type of sample; thus, it is more meaningful to standardize the ISE in terms of Ca^{2+} activity instead of concentration.

[24] Chromium

By CLAUDE VEILLON

The determination of chromium in biological materials represents one of the most difficult analyses one can choose to perform.[1] The levels of chromium in many samples, particularly biological fluids like serum, plasma, and urine, are extremely low, down in the 0.1 ng/g range, tenths of a part per billion, one part in 10 billion. To put this in perspective, it is equivalent to measuring a single second in the past 317 years! Yet, with modern instrumentation, adequate contamination control, and sufficient attention to detail, it is possible to measure chromium directly in some samples at the 0.1 ppb level.

At these very low levels, there are perhaps only three analytical methods with sufficient sensitivity for the determination. These are neutron activation analysis, mass spectrometry, and graphite furnace atomic absorption spectrometry. The first two are not widely available, and the third, as we shall see, has only recently been able to achieve accurate results. It is also the technique most susceptable to matrix interference effects.

[1] Specific manufacturers' products are mentioned throughout this chapter solely to reflect the personal experience of the author and do not constitute their endorsement nor that of the Department of Agriculture.

So, if chromium levels are so low, and the determination so difficult, why bother at all? Well, chromium is an essential trace nutrient for humans, and is involved in glucose metabolism and/or the mechanism of action of the pancreatic hormone insulin. This has been dramatically demonstrated in a supplementation experiment involving one type of impaired glucose tolerance.[2]

By 1978, it had become obvious that serious problems existed in the determination of chromium in biological materials. For the reasons stated above, almost all of the literature values were obtained by graphite furnace atomic absorption spectrometry (GFAAS). Reported values for the same or very similar materials differed by well over an order of magnitude. This was particularly true in the case of urinary chromium, the excretion of which had been proposed as a means of assessing the chromium nutritional status of individuals.[3] Most of the values appeared to be in the 2–20 ng/ml range, which presented a serious dilemma from a nutritional standpoint. The amount of chromium absorbed had been shown to be on the order of 0.5% by radiotracer experiments, so daily urinary excretions on the order of 10 μg/day meant that dietary chromium intake had to be greater than 1 mg/day. No reasonable diet in the United States is capable of supplying much more than 100 μg/day.

During the past few years, several advances in atomic absorption instrumentation made it possible to resolve this problem. In 1978, Guthrie *et al.*[4] demonstrated conclusively that GFAAS with then state-of-the-art deuterium lamp background correctors were inadequate for urinary chromium determinations, because of low intensity at the chromium wavelength. A direct correlation between background absorption and apparent chromium concentration was found (Fig. 1).

Figure 1 implies that GFAAS measurements of urinary chromium (with the best instrumentation available at the time) are not that at all, but merely a measure of the background produced in the furnace by the urine matrix. This casts serious doubt on the validity of all previously reported urinary chromium values obtained by GFAAS.

Fortunately, we gained access to a novel continuum source, echelle monochromator, wavelength-modulated atomic absorption spectrometer[5] which had extraordinary background correction capabilities. With this instrument, urinary chromium concentrations of about 0.5 ng/ml were observed, which make far more sense when intakes and fractional absorb-

[2] K. N. Jeejeebhoy, R. C. Chu, E. B. Marliss, G. R. Greenberg, and A. Bruce-Robertson, *Am. J. Clin Nutr.* **30,** 531 (1977).

[3] W. Mertz, *Phys. Rev.* **49,** 163 (1969).

[4] B. E. Guthrie, W. R. Wolf, and C. Veillon, *Anal. Chem.* **50,** 1900 (1978).

[5] J. M. Harnly and T. C. O'Haver, *Anal. Chem.* **41,** 2187 (1977).

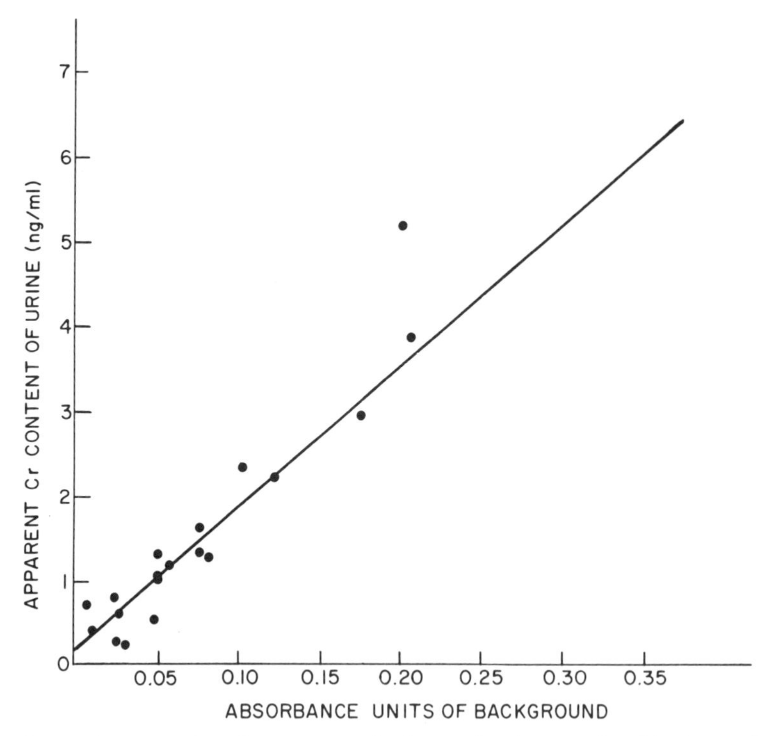

FIG. 1. Apparent chromium content of urine vs absorbance units of background.

ance are considered. However, one cannot take a value lower than ever seen before and obtained with a unique instrument and make a strong case for saying that previous determinations were incorrect. So, the values were confirmed by an independent method, namely, stable isotope dilution, isotope ratio mass spectrometry.[6] Further evidence came independently from other laboratories. At about the same time, Kayne *et al.*[7] modified an instrument similar to the one used by Guthrie and co-workers[4] and greatly enhanced its background correction capabilities; they too observed urinary chromium concentrations below 1 ng/ml. By carefully controlling furnace conditions and by adding hydrogen to the furnace gas, Routh[8] was able to reduce background absorption from the urine matrix sufficiently and obtained values below 1 ng/ml.

While this sudden lowering of the urinary chromium levels about 10-fold resolved the intake–absorption–excretion discrepancies of the earlier

[6] C. Veillon, W. R. Wolf, and B. E. Guthrie, *Anal. Chem.* **51,** 1022 (1979).
[7] F. J. Kayne, G. Kormar, H. Laboda, and R. E. Vanderlinde, *Clin. Chem.* **24,** 2151 (1978).
[8] M. W. Routh, *Anal. Chem.* **52,** 182 (1980).

literature, it also increased sensitivity requirements and blank and contamination requirements by the same amount. But as mentioned earlier, given modern instrumentation and taking care to minimize sample and blank contamination, chromium can be determined in biological samples at the very low levels present. Since 1978, we have gained considerable experience in measuring chromium levels in urine, serum, foods, and tissues, using commercially available instrumentation. These procedures will be described, along with some of the results we have learned from experimentation.

Urine

Urinary chromium can be determined directly by GFAAS using the method of additions. The accuracy of the method has been established[9] and it has been proposed as a selected method for clinical laboratories.[10] Few reagents are used and very little sample preparation and manipulation are required, greatly reducing the incidence of sample contamination. The method of additions compensates for sensitivity changes as the graphite furnace tube ages and for the widely different matrices encountered in urine samples. Furnace operating parameters must be carefully controlled, and detection limits on the order of 0.03 ng/ml can be obtained.

Sample Collection and Handling

Urine samples are collected in appropriately sized plastic containers checked for chromium contamination and cleaned if necessary. No acids or preservatives are added. Homogeneous aliquots for analysis are stored at room temperature in sterile disposable polypropylene tubes. Samples for analysis by the method of additions are prepared in disposable polystyrene sample cups that are rinsed with deionized water before use.

Reagents

The only reagents used are water, hydrochloric acid, and a standard solution of known chromium content. The water is prepared by demineralization and should contain no detectable chromium. Several suppliers produce high-purity hydrochloric acid with suitably low chromium blanks when diluted in 1 mol/liter. Dilutions of the standard solution are made with this diluted HCl.

[9] C. Veillon, K. Y. Patterson, and N. A. Bryden, *Anal. Chim. Acta* **136,** 233 (1982).
[10] C. Veillon, K. Y. Patterson, and N. A. Bryden, *Clin. Chem.* **28,** 2309 (1982).

Apparatus

The primary tool required is a modern GFAAS instrument with sufficient background correction capabilities. Our experience is limited to the Perkin-Elmer 5000/500 system, but other instruments have been demonstrated to be suitable for this determination.

All pipetting is done with Eppendorf "Standard" fixed-volume pipets. The exposed internal parts of these are all plastic. Most other brands, and the Eppendorf tip-ejector models, have exposed stainless steel parts that corrode and lead to serious contamination problems. Disposable pipet tips are rinsed several times before use by pipetting deionized water or 1 mol/liter HCl if necessary.

All sample preparation is done in a Class-100 laminar flow work station (free of exposed stainless-steel parts), to prevent airborne contamination.

Parameters

Standard instrument operating conditions are used. The furnace program includes a 1200° (max) charring step after drying and atomization from the tube wall at 2700° with internal flow reduced to 50 ml/min. Pyrolytically coated furnace tubes are used, and peak height recorded.

Procedure

The general procedure is as follows.

1. Place four 1-ml aliquots of well-mixed sample in new, disposable polystyrene sample cups previously rinsed with deionized water, or with 1 mol/liter HCl if necessary.
2. To each aliquot, add 10 μl of 1 mol/liter HCl containing 0, 20, 50 or 100 μg of chromium per liter, respectively.
3. Mix each by several uptakes and discharges with a 1000 μl disposable-tip pipet immediately prior to analysis.
4. Introduce 25-μl aliquotes of each into the furnace for analysis.
5. Plot results and determine initial concentration, neglecting the 1% volume change due to the standard addition.

Discussion

Standards must be checked periodically for accuracy, and it is recommended that one or more urine pools be maintained and run with each batch of samples for quality control (see this volume [7]).

In addition to strict contamination control at every step in the procedure, the most important variables are the background correction capability of the instrument, the charring or "ashing" temperature, and the use of the method of additions.

Inadequate background correction will manifest itself in abnormally high and erratic values, in negative peaks upon atomization, and in odd-shaped or multiple atomization peaks.

Related to this problem is the charring temperature. At temperatures much below 1200°, too much of the sample matrix can remain and overwhelm the background corrector during atomization. At temperatures above 1300°, chromium is rapidly lost.

The method of additions is important for two reasons. First, urine samples vary widely in matrix composition, so the slope of the analytical curve usually differs from that of aqueous standards, and even from that of other urine samples. Second, the slope of the additions curve for a given sample decreases as the furnace tube ages, and this is compensated for by the method of additions. Full details of the method are given in Refs. 9 and 10. The detection limit is on the order of 0.03 μg/liter, below that of most normal urines, which typically are in the 0.1–0.5 μg/liter range. Accuracy of the method has been established by stable isotope dilution mass spectrometry. The method should prove suitable for urinary chromium determinations in routine clinical use.

Serum

Much of what was said earlier for urinary chromium determinations is applicable for serum. In developing a method for serum chromium,[11] based on the methodology for urinary chromium, I predicted that it would be easier, because serum is of a more uniform matrix composition. This prediction proved to be grossly in error.

True, it is very uniform in matrix composition, eliminating the need to use the method of additions on each sample. Collection of the samples without contamination is a more formidable task. The chromium level is at best as high as urinary levels, but usually is slightly lower. There is considerable inorganic material in the matrix, but more importantly, the graphite tube furnace does not deal well with the considerable organic matter in the matrix. This pretty well rules out direct serum chromium determinations and requires that the organic matter be destroyed somehow, and without contaminating a sample with an analyte content on the order of 0.1–0.2 μg/liter.

A workable procedure has been arrived at and will be described. Because of the strict contamination control precautions taken, the procedure should be very useful for other elements in serum as well, and these will

[11] C. Veillon, K. Y. Patterson, and N. A. Bryden, *Anal. Chim. Acta* **164,** 67 (1984).

generally be of higher concentration. If the serum has not been contaminated for chromium, it is probably not contaminated for other elements. Plasma can also be analyzed, provided a sufficiently "clean" anticoagulant can be obtained.

A small amount of magnesium nitrate is added as an ashing aid and matrix modifier. Samples are lyophilized and dry-ashed in silanized quartz tubes. Residue is dissolved in 0.1 mol/liter HCl and chromium determined by GFAAS. The uniform nature of sera permits the use of an additions calibration curve prepared in a single serum sample, or preferably in a reference serum pool sample employed also as a quality control material. Because of the very low analyte levels, strict contamination control measures must be used throughout the procedures for collection, storage, preparation, and quantitation. The detection limit of the method is about 0.03 μg/liter and the accuracy has been established by stable isotope dilution mass spectrometry.

Reagents

Generally, the same requirements are necessary as before: deionized water free of detectable chromium; high-purity HCl diluted to 0.1 mol/liter and likewise free of detectable chromium; magnesium nitrate (high-purity, Johnson Matthey Chemicals, Royston, Herts, UK); chromium standard solutions diluted from a stock solution prepared by dissolving elemental chromium in HCl; 5% dichlorodimethylsilane (DCDMS, Pierce Chemical, Rockford, IL) in toluene; and regent grade methanol.

Instrumentation

The instrumentation is the same as for the urinary chromium procedure described earlier.

Apparatus

Sera are collected in all-plastic syringes (Monovette, Sarstedt, Princeton, NJ) which were free of measurable chromium when soaked with water or serum. Needles for venipuncture are critical because most are made of stainless steel. Satisfactory ones are the "butterfly" type, a short, *siliconized* needle attached to a length of small-bore poly(vinyl chloride) tubing (Minicath, Deseret Medical, Sandy, UT). Apparently the siliconizing imparts sufficient hydrophobicity to the needle surface to prevent serum contact. A single low-chromium serum sample can be repeatedly drawn through several of these needles in succession with an insignificant increase in chromium content.

The syringes have a removable plunger handle so the samples can be centrifuged directly in the collection device. Following centrifugation serum is removed with a sterile polyethylene transfer pipet and aliquots stored in sterile polypropylene tubes at −20°.

Samples are ashed in 10 × 100-mm quartz test tubes, fabricated from Vitreosil tubing (Thermal American Fused Quartz, Montville, NJ). Tubing is first cleaned with 3% HF, rinsed, air-dried, and fabricated into test tubes by standard glassblowing techniques, taking care not to contaminate the surface with bare hands, placing on unclean surfaces, etc. After fabrication, tubes are again cleaned with 3% HF, rinsed, heated for 24 hr at 100° in 10% HNO_3, and stored in this solution at room temperature until ready for use. They are then thoroughly rinsed with water, drained, and dried by lyophilization (see below).

Note: As many of these and the following operations should be carried out in a Clean Room or a class-100 laminar flow hood (no stainless-steel parts) as possible, to minimize airborne contamination. Surfaces touched by containers should be clean, preferably plastic, and powder-free vinyl gloves should be worn at all times.

Dry tubes are silanized by contact with 5% DCDMS in toluene, rinsed with methanol, followed by six water rinses. Serum samples (1–2 ml) are weighed into the tubes, and magnesium nitrate added [10 μl of a 0.186 g/ml solution of $Mg(NO_3)_2 \cdot 6H_2O$ in water per ml of serum]. An equal volume of water is used as a blank and treated like the samples.

Frozen sample and blanks are lyophilized in a polycarbonate vacuum chamber attached to a stainless-steel-free freeze dryer (Model FD-6-84-VP, FTS Systems, Stone Ridge, NY). Tubes with the dried samples are placed under quartz beakers in a muffle furnace. This furnace (51894, Lindberg, Watertown, WI) has a molded ceramic chamber which also completely embeds the heating elements, preventing the serious contamination observed from designs with exposed heating elements. To further reduce airborne contamination, the furnace is placed within a Class-100 clean area. Samples are heated at 100° for 1 hr, then increased 50° each hour to 250°. After 1 hr at 250°, the furnace controller is set to 480° for an overnight ash.

Procedure

Ash in the quartz tubes is dissolved in 0.1 mol/liter HCl (volume equal to serum volume) and 25-μl aliquots pipetted into the graphite furnace for analysis, as before.

Discussion

It was mentioned earlier that chromium losses begin to occur above about 1200–1300° during the char cycle of the graphite furnace. The addition of magnesium nitrate to the sample serves two purposes. It acts as a matrix modifier in that it permits much higher charring temperatures, up to about 1400–1500°, without chromium loss. This has the benefit of further reducing the amount of inorganic sample matrix in the furnace tube, making things easier for the background correction system during atomization. The matrix modifier also serves as an ashing aid.

The silanizing of the quartz tubes serves more of a physical benefit during sample ashing. It allows the dry serum plug to pull away from the tube wall and shrink into a spherical pellet having minimal contact with the quartz tube.

The ashed serum matrix cause an enhancement effect of the signals compared to aqueous standards. However, since the matrix is so uniform, a standard additions curve made up in a single serum sample will serve for all of the other samples in a given run. We utilize a large bovine serum pool for this purpose, which also serves as a quality control material for both short- and long-term analyses. It behaves identically to human serum, and can be obtained in large quantities without contamination.

With suitable equipment, apparatus, precautions, and sufficient attention to details, serum chromium determinations can be performed accurately on a routine basis. About 30–40 determinations per day can be made. Because of the measures taken to prevent contamination, the sample preparation procedure should be suitable for a number of other trace elements in serum.

Foods and Tissues

The chromium levels in diets, foods, and various tissues are usually very much higher than that of serum, perhaps 1–3 order of magnitude. This renders their analysis very much easier, from a contamination control standpoint alone. Sample collection and homogenization techniques offer some opportunities for pretty spectacular contamination. The secret is to stay away from stainless-steel devices or those of other chromium-containing alloys. Drop a tissue sample in a blender with high-speed blades made of stainless steel, turn it on, and pull a sample every few minutes. Then analyze these for chromium. Or maybe analyze an autopsy tissue sample chopped up with a stainless-steel scalpel and convince yourself that the chromium found is all endogenous.

Devices which have been used to overcome this problem include blenders (not to the ones with stainless-steel containers) with blades replaced by low-chromium tool steel alloy, food processors (plastic container) with custom blades made of titanium, Teflon ball mills operated at cryogenic temperatures, grinding up lyophilized material with an agate mortar and pestle, sealing lyophilized material in a sturdy plastic bag and pounding it.

It would not be practical to discuss the numerous options for various situations, but the initial sample collection and handling must be considered in terms of not contaminating it. This can usually be handled with some thought given the matter, and with perhaps some preliminary testing.

Assuming that we have a meaningful sample to be analyzed for chromium at relatively high levels, the sample preparation procedures are less stringent than for urine and serum. The procedure described for serum could be used, but would perhaps be "overkill." Acid digestions are a definite possibility, given the commercial availability of high-purity reagents. One method we have used is a combination of wet and dry ashing.[12] It is not particularly fast, but requires very little operator time.

Samples are weighed into acid-cleaned borosilicate test tubes and ashed for 24–48 hr in a muffle furnace at 350°. Then a small quantity of HNO_3 and H_2O_2 is added to the residue and samples are heated in a heating block at 100°. This is usually sufficient to completely destroy the organic matter. Residue is then dissolved in dilute HCl and diluted to volume for analysis. For chromium determinations, GFAAS would be used. For other elements present at higher concentrations, methods like flame AAS could be used.

Very small quantities of reagents are used, contributing to low blank values. If the method were to be used for elements like aluminum, one might wish to use quartz tubes rather than borosilicate glass. In any event, the method is not labor intensive and works well for chromium analysis of materials of this type.

Conclusions

Chromium analyses of biological materials is difficult because of the very low levels present in many samples. Contamination control is essential. However, instrumentation and procedures now exist to make these determinations on a routine basis, for this trace element so essential to our well-being. New biological reference materials are available and more are on the way with established chromium levels near those in real samples. These will all aid us in unraveling the role of this essential trace element in man and allow meaningful comparisons of data from different laboratories, perhaps for the first time.

[12] A. D. Hill, K. Y. Patterson, C. Veillon, and E. R. Morris, *Anal. Chem.* **58,** 2340 (1986).

[25] Determination of Cobalt by Atomic Absorption Spectrometry

By ROBERT SHAPIRO and MARK T. MARTIN

Introduction

Cobalt is present in plant and animal tissues at relatively low concentrations, typically below 0.1 $\mu g/g$ on a dry weight basis. It serves its major established biological role as a component of vitamin B_{12}. To date only two metalloproteins not carrying vitamin B_{12} have been reported to contain intrinsic cobalt in their natural states.[1,2] Nonetheless, cobalt has proved to be extremely important in metallobiochemistry. Functional studies of numerous zinc, iron, and copper proteins have been facilitated by replacing these metals with cobalt. In particular, substitution of the chromophoric, paramagnetic cobalt(II) ion for zinc, which is spectroscopically silent, has provided a valuable probe in the investigation of the catalytic mechanisms of a number of enzymes including carboxypeptidase A, carbonate dehydratase, alkaline phosphatase, and alcohol dehydrogenase.[3]

Cobalt concentrations have been measured by a variety of techniques, including atomic absorption spectrometry (AAS), atomic emission spectrometry,[4] neutron activation analysis,[5] chemiluminescence,[6] and colorimetry.[7-9] Of those techniques widely available to the biochemist, AAS is the most rapid, versatile, and reliable. For some applications where atomic absorption instrumentation is not available, the colorimetric procedures cited may be adequate. It should be recognized, however, that all of these are considerably less sensitive than electrothermal AAS. Further, only the method of Evans[9] is relatively simple and rapid, and its use is restricted to samples lacking significant amounts of zinc, iron, and several other metals.

[1] D. B. Northrop and H. G. Wood, *J. Biol. Chem.* **224,** 5801 (1969).

[2] J. J. G. Moura, I. Moura, M. Bruschi, J. Le Gall, and A. V. Xavier, *Biochem. Biophys. Res. Commun.* **92,** 962 (1980).

[3] B. L. Vallee, *Adv. Exp. Med. Biol.* **40,** 1 (1973).

[4] K. W. Olsen, W. J. Haas, and V. A. Fassel, *Anal. Chem.* **49,** 632 (1977).

[5] J. Versieck, J. Hoste, F. Barbier, H. Steyaert, J. De Rudder, and H. Michels, *Clin. Chem.* **24,** 303 (1978).

[6] D. F. Marino, F. Wolff, and J. D. Ingle, Jr., *Anal. Chem.* **51,** 2051 (1979).

[7] D. W. Dewey and H. R. Marston, *Anal. Chim. Acta* **57,** 45 (1971).

[8] R. M. Pearson and H. J. Seim, *Anal. Chem.* **49,** 580 (1977).

[9] C. H. Evans, *Anal. Biochem.* **135,** 335 (1983).

METHODS IN ENZYMOLOGY, VOL. 158

General Considerations

Selecting a Method

In deciding how to best quantitate cobalt in a particular sample, the major factors to consider are the likely concentration, the amount of sample available, and the nature of the matrix. These will dictate the extent of sample processing necessary prior to analysis and whether flame or electrothermal (graphite furnace) atomization should be employed. Electrothermal AAS can accurately quantitate as little as ~20 pg of cobalt (e.g., 20 μl of a 1 ng/ml solution), whereas concentrations ~100-fold higher and volumes considerably greater are required for the flame technique. (Sensitivity will vary with both the instrument used and its condition.) Nonetheless if material is not limiting and the cobalt concentration is above 100 ng/ml, the flame method may be preferable because of its greater speed and lower susceptibility to interference effects.

For many biological samples, e.g., serum, the extremely low cobalt concentrations present as well as the complexity of the matrix will make extensive pretreatment necessary before analyses can be performed. In general this involves ashing and either ion-exchange chromatography or extraction (see below). It is, however, preferable to avoid such pretreatments if at all possible since they may increase sample loss and introduce contamination. As long as the cobalt concentration in the unprocessed sample is significantly above the detection limit, it is advisable to evaluate matrix interference effects by standard additions (see below) before electing to subject the sample to elaborate purification procedures.

Contamination Control

Measurements performed by the authors on 50 m*M* solutions of four common laboratory chemicals (reagent grade)—sodium phosphate, sodium chloride, HEPES, and Tris—showed only the first to contain more than 1 ng/ml cobalt. Indeed neither tap water (in Boston, MA) nor sea water[10] contains amounts of cobalt detectable by electrothermal AAS. These observations might appear to suggest that contamination control is not a major consideration when working with this metal. However, the levels of cobalt present in many samples are themselves near or below the detection limit, necessitating 10- to 100-fold concentration prior to analysis. In this case adventitious cobalt can become a serious problem. For example, contamination is likely responsible for much of the 3 order of

[10] J. M. Lo, J. C. Yu, F. I. Hutchison, and C. M. Wal, *Anal. Chem.* **54,** 2536 (1982).

magnitude variation in serum cobalt values reported (reviewed in Ref. 11). Thus for such applications the standard precautions for reducing trace metal contamination described previously in this volume should be taken. These include the use of high-purity water, e.g., that obtained using a Milli-Q system (Millipore Corp, Bedford, MA), ultrapure acids (J. T. Baker Chemical Co., Phillipsburg, NJ), metal-free buffers, noncolored pipette tips (Bio-Rad Laboratories, Richmond, CA), and acid-washed laboratory ware. For blood collection, ordinary stainless-steel needles and vacutainer tubes should be avoided. Suitable alternatives are siliconized stainless-steel needles (21 gauge × 1.9 cm, Travenol Laboratories Inc., Deerfield, CN) or Teflon-polyethylene intravenous cannulae (22 gauge × 2.5 cm, Abbott Labs. Inc., Chicago, IL) and polypropylene syringes (Peel-A-Way Sci. Co., South El Monte, CA). The first several milliliters of blood passing through the infusion set should be collected in a separate syringe and discarded.

Interferences

For any sample being analyzed, it is necessary to determine whether interferences are present which may artificially enhance or suppress the absorption signal. This can be achieved by standard additions. Briefly, standard curves are constructed in the presence and absence of sample. If the lines obtained are not parallel, this indicates that there are interferences. Such effects can frequently be eliminated by sample dilution, chemical additions, or alteration of flame or furnace conditions. If this is not possible, it is necessary to either separate the cobalt from interfering agents or use the standard addition curve for quantitation.

Interferences are much more prevalent in general with electrothermal than with flame AAS. The authors have examined the extent of interference produced in both systems by four common buffers and salts and by protein. Up to at least 50 m*M* concentrations, sodium phosphate, HEPES, and Tris, all at pH 7.5, and sodium chloride have no significant effect on the peak produced by 0.3 μg/ml cobalt using the flame technique. A bovine serum albumin (BSA) concentration of 10 mg/ml was also without effect. Substances other than those tested here may cause interference, and in such cases the addition of 75% saturated 8-hydroxyquinoline[12] or 60 m*M* potassium cyanide[13] has been reported to obivate these problems.

[11] J. Versieck and R. Cornelis, *Anal. Chim. Acta* **116,** 217 (1980).
[12] M. Suzuki, K. Hayashi, and W. E. C. Wacker, *Anal. Chim. Acta* **104,** 389 (1979).
[13] M. M. M. El-Defrawy, J. Posta, and M. T. Beck, *Anal. Chim. Acta* **115,** 155 (1980).

Using electrothermal AAS (20 μl sample aliquots), BSA concentrations as high as 10 mg/ml again did not significantly influence the measurements. However, HEPES, Tris, sodium phosphate, and sodium chloride all suppressed the signal observed for 30 ng/ml cobalt. At 1 and 10 m*M*, both HEPES and Tris caused 30–40% suppression, although neither interfered appreciably at 100 m*M* concentration. Sodium phosphate at 10 and 100 m*M* gave 20–30% suppression. Sodium chloride concentrations of 10 and 100 m*M* were essentially without effect, but a 74% decrease in signal was observed with 300 m*M*. Addition of 1% ascorbic acid[14] eliminated the effects of Tris and HEPES, and vastly reduced the suppression due to 300 m*M* NaCl and 100 m*M* sodium phosphate.

The effects of several other metals—Ca, Cu, Fe, Mg, Mn, and Ni—on electrothermal AAS were also examined (using the 240.7 nm cobalt absorption line). At the 100 μg/ml level, none of these significantly influenced the signal produced by 30 ng/ml cobalt.

Working Methods

Instrumentation

For flame AAS, the authors employ a Perkin-Elmer Model 5000 or 2280 spectrophotometer (Perkin-Elmer Corp., Norwalk, CT) equipped with a deuterium background correction lamp. An air–acetylene flame is used with oxidant and fuel flow rates of 17.5 and 5.0 liters/min, respectively. The hotter nitrous oxide–acetylene flame has been recommended for some applications involving particularly refractory matrices or certain Co(III) compounds.[15] It should be noted, however, that the sensitivity is decreased considerably in this system. For electrothermal AAS, we employ a Perkin-Elmer Model 5000 spectrophotometer equipped with a Model AS-40 automatic sampling system, Model HGA-500 atomization unit, Model 56 strip-chart recorder, Model PRS-10 digital printer, Intensitron cobalt hollow cathode lamp, and deuterium background correction lamp. If available, the Zeeman background correction system is somewhat preferable. Pyrolytically coated graphite tubes (Perkin-Elmer) are strongly recommended. The simultaneous use of both strip-chart recorder and digital printer allows the measurement of both peak area and peak height, as well as the observation of baseline noise levels and peak shape abnormalities. The latter may relect either matrix effects or a need to replace the graphite tube.

[14] D. J. Hydes, *Anal. Chem.* **52,** 959 (1980).

[15] K. Fujiwara, H. Haraguchi, and K. Fuwa, *Anal. Chem.* **44,** 1895 (1972).

TABLE I
GRAPHITE FURNACE HEATING PROGRAM FOR DETERMINATION OF COBALT

Step	Temperature (°C)	Ramp time (sec)	Hold time (sec)	Argon flow (ml/min)
1	110	5	10	300
2	150	10	5	300
3	1000	5	10	300
4	1000	1	5	40
5	2700	0	3	40
6	2700	0	5	300

The wavelength generally employed for cobalt determination is 240.7 nm. However, the analytical sensitivity using the 242.5 nm absorption line is only slightly lower and the standard curve shows somewhat greater linearity. A narrow slit width of 0.2 nm should be used. A scale expansion of 5× will be suitable for measurements in the range of 5–50 ng/ml with electrothermal AAS (20-μl sample aliquots). A reading time of 3 sec is sufficient to record the entire peak.

The optimal graphite furnace temperature program will depend on the particular sample being analyzed. The program shown in Table I has worked well for 20-μl aliquots of a variety of samples containing common buffers and salts and amounts of protein up to 10 mg/ml. If larger sample aliquots are used, the hold times at 110 and 150° in the drying portion of the program should be increased. If higher protein concentrations or particularly complex matrices are employed, it may be advisable to increase the ashing time at 1000°.

Standards

Certified cobalt atomic absorption standards (1 mg/ml) are available from Fisher Scientific Co. (Fairlawn, NJ) and J. T. Baker Chemical Co., among others. For flame AAS, convenient working standards, generally at concentrations of 1–5 μg/ml, are made by diluting the 1 mg/ml standard solution with water. For electrothermal AAS, an intermediate standard solution of 2 μg/ml is prepared. This solution and the flame AAS standard solutions are stable for at least 2 months at room temperature. Working standards for electrothermal AAS, generally 10–50 ng/ml, are prepared daily from the intermediate standard.

During the course of an extended series of measurements by electrothermal AAS, it is advisable to generate standard curves periodically, since the measured absorbance can decrease markedly as the graphite tube ages.

Useful Concentration Ranges

With flame AAS, concentrations of ~0.1–7 μg/ml can be quantitated accurately. Above this level, the standard curve ceases to be linear, and for maximum accuracy samples containing higher cobalt concentrations should be diluted prior to analysis.

With electrothermal AAS (20 μl sample aliquots), the standard curve is linear to about 50 ng/ml and quantitation of as little as 1 ng/ml is achievable. This range will of course vary with aliquot size.

Specific Applications

Buffers, Purified Proteins, and Other Simple Solutions

Analysis can usually be performed without any special preparation using the general methods detailed above.

Serum and Blood

There is presently no standard method for determination of cobalt in blood or serum, nor is there agreement concerning what are normal values. Average serum values ranging from 0.04 to 146 ng/ml have been reported, with the majority falling between 0.1 and 2.0 ng/ml.[11,16] These discrepancies undoubtedly stem for the most part from analytical and contamination problems rather than real differences among the various populations studied.

Due to the extremely low levels of cobalt in serum and marked matrix effects, samples must be subjected to purification and concentration prior to analysis. We feel that the following procedure, adapted from Lidums[16] and Thiers *et al.*,[17] should yield reliable results.

Ashing. Place 10 ml of sample in a 100 ml Pyrex beaker and add 2/3 volume of ultrapure nitric acid and 1% (v/v) sulfuric acid. Boil gently on a hot plate until the sample is dispersed, then evaporate to dryness and increase the heat to ~250° until the residue turns black. Heat on a hot plate at 350° until vapor evolution ceases. Remove the beaker from the hot plate, cover, and allow to cool. Next add 0.8 ml nitric acid and heat to 350°. Continue heating for 10–15 min after the residue becomes dry. Re-

[16] V. V. Lidums, *At. Absorpt. Newsl.* **18,** 71 (1979).
[17] R. E. Thiers, J. F. Williams, and J. H. Yoe, *Anal. Chem.* **27,** 1725 (1955).

peat as necessary with lower volumes of nitric acid until the residue is white.

Ion Exchange. Add 5 ml 4 *M* HCl to the sample and heat to dissolve the ash. Decrease the volume to ~1 ml by evaporation, add 2 ml 9 *M* HCl and allow the solution to cool. At this point it may be necessary to pass the solution through a sintered glass filter in order to remove insoluble material (which does not contain cobalt). Next apply the sample to a 12 × 150 mm column of Dowex 1-X8 resin, 50–100 mesh, which has previously been washed with 90 ml of 0.005 *M* HCl followed by 25 ml of 9 *M* HCl (flow rate 1 ml/min). The beaker and filter should be rinsed with at least three 1–2 ml portions of 9 *M* HCl and the washings also applied to the column. Pass 25 ml of 6 *M* HCl through the column to remove alkali metals, nickel, manganese, and numerous other ions. Finally elute cobalt with 30 ml of 4 *M* HCl, collecting into a weighed 50 ml beaker. Evaporate this fraction just to dryness and redissolve in 2 ml of 0.01 *M* HCl. Evaporate to ~1 ml, weighing the beaker to determine the amount of solution remaining. The sample can then be analyzed by electrothermal AAS. In some cases it may be necessary to inject up to 100-μl aliquots. If so, the hold times at 110 and 150° listed in Table I should be increased to 30 and 15 sec, respectively. Since there are numerous steps during which contamination might be introduced it is critical to also analyze a blank control sample that has been through identical procedures.

Plant and Animal Tissues

Cobalt levels of 0.01–1.5 μg/g dry weight have been reported for a variety of plant and animal tissues.[18,19] Thus digests of these materials will in general contain amounts of cobalt readily measurable by electrothermal AAS. Such a digest can be prepared, for example,[20] by adding 10 ml of 50% (v/v) nitric acid to 1 g of sample, heating at 80° until the volume is ~3 ml, filtering, and diluting to 25 ml in a volumetric flask. This type of direct determination method will be adequate for many samples. However, interference effects can be considerable and standard additions will in general have to be employed in order to obtain accurate concentrations. If these effects are severe, purification of cobalt by either ion-exchange chromatography or extraction should be performed. For example, the ion-exchange procedure described above for blood and serum can be applied with minor modification. Alternatively, numerous methods using

[18] W. J. Simmons, *Anal. Chem.* **47,** 2015 (1975).

[19] F. J. Jackson, J. I. Read, and B. E. Lucas, *Analyst (London)* **105,** 359 (1980).

[20] S. Slavin, G. E. Peterson, and P. C. Lindahl, *At. Absorpt. Newsl.* **14,** 57 (1975).

extraction have been developed which appear to give excellent recoveries. The following is adapted from Simmons.[18]

Digestion. Add 15 ml nitric acid and 2.5 ml of 72% w/v perchloric acid to 0.5 g dry sample in a 50-ml Erlenmeyer flask and heat on a hot plate. After white perchloric acid fumes appear, swirl the flask occasionally, and continue heating at 180° for 1 hr. (If the flask becomes dry, allow it to cool somewhat, add 0.5 ml $HClO_4$, and continue heating.) Allow the flask to cool and then rinse the walls with 4–5 ml water. Warm the solution to dissolve any precipitate.

Extraction. Add 1 ml 40% w/v sodium citrate and adjust the pH to 5.3–5.7 with concentrated ammonium hydroxide, correcting any overshoot with HCl. Add 1 ml 30% hydrogen peroxide, followed by 0.3 ml of 1% w/v 2-nitroso-1-naphthol in ethanol. Transfer to a 15 ml stoppered tube. Rinse the flask twice with 1.5 ml water and add to the tube contents. Bring the volume to 11–12 ml with water. Add 1 ml isoamyl acetate and vigorously shake for 30 sec. The top layer, containing the complex of cobalt with 2-nitroso-1-naphthol, can now be analyzed by electrothermal AAS.

[26] Measurement of Copper in Biological Samples by Flame or Electrothermal Atomic Absorption Spectrometry

By MERLE A. EVENSON

Introduction

In biological materials copper is one of the easiest elements to measure with adequate accuracy and precision if one uses atomic absorption spectrometry (AAS). The instrumentation required is less expensive, the methods are more sensitive, and the sample preparation less complicated than that for emission methods. Either a flame (FAAS) or the more sensitive electrothermal atomic absorption spectrometric (ETAAS) volatilization procedure can be used for the measurement of copper in a variety of biological samples.

In general terms, the principles upon which AAS are based are not complicated or difficult to understand. A copper hollow cathode lamp is used to provide the resonance line at 324.7 nm. This line is focused through the sample plasma produced by the flame or the electrothermal volatilization device. The neutral copper atoms in the plasma will absorb the 324.7 nm copper resonance line and the amount of absorbance of this line is proportional to the number of neutral copper atoms in the plasma.

More details about the instrumental and chemical principles of AAS can be found in instrumental analysis textbooks.[1]

Advantages of flame and/or electrothermal atomization AAS for copper analysis include high accuracy, high precision, high sensitivity, minimal chemical and/or matrix effects, (interferences) relatively inexpensive instrumentation, and operating costs that are less than most other instrumental methods. Relatively inexperienced laboratory workers can be quickly taught the complete analytical method and can rapidly obtain accurate copper results.

As in all trace element analyses careful attention to details to prevent or reduce trace element and metal contamination requires constant awareness by all involved with the analysis. In the case for copper, the sample treatment, preparation of ultrapure volatile acids, acid-cleaned laboratory ware, highly purified water, accurate calibration standards, and access to adequate National Bureau of Standard–Standard Reference Materials (NBS-SRM) materials are all essential to be able to obtain accurate and precise copper analytical results.

Prior to analysis some thought should be given to the method of reporting the results. The analysis report will be determined by the type of biological sample analyzed, the sample pretreatment, and finally the expected use of the analytical result. Occasionally, copper reported as wet weight, as in the case of human liver analysis, may be convenient to obtain and can be compared directly to older literature values. Dry weight, ashed weight, and lipid-free weight are other choices used for the reporting of copper in tissues. Ratios of copper to another trace metal, of copper to measured total protein, of copper to an amino acid, i.e., proline, or of copper to the amount of DNA in the tissue analyzed have all been used as methods of reporting copper measurements. In solutions, concentrations in moles per liter or micrograms per liter or milligrams per liter are often the units of choice. Older spectrometry and industrial literature often used parts per million as the unit for reporting results (ppm = 1 μg of copper per gram of solid or liquid; 1 ml of water is often taken to weigh 1 g.

Reagents, Laboratory Ware, and Instrumentation

Although high-quality clean room facilities are not necessary for copper analysis in biological materials, care should be taken to prevent laboratory dust from contacting the sample, the containers, and the reagents used in the analysis of copper.

[1] H. H. Willard, H. H. Hobart, L. L. Merritt, J. A. Dean, and F. A. Settle, "Instrumental Methods of Analysis," 6th Ed., p. 146. Wadsworth, Belmont, California, 1981.

Commercially available water purification systems, i.e., Milli-Que Systems, Millipore Corporation, Bedford, MA 01730 connected to the house-prepared distilled, deionized, or reverse osmosis water will prepare high-quality water where the copper present will be below the detection limit of the instrumental measurement.

The priority order of materials used for the preparation of the laboratory ware are Teflon, new linear noncolored polyethylene or polypropylene, followed by quartz and finally Pyrex glass. But, even Pyrex glass if properly cleaned, can be used with minimum contamination for copper. Most laboratories find Teflon to be the best but may be too expensive for the whole laboratory. Hence, polyethylene and polypropylene are often selected as the materials of choice. Knives, blades, and needles made of stainless steel can usually be used if they are washed with acetone, water, and 2 *M* nitric acid followed by exhaustive rinsing with high-quality water. The above type of cleaning will enable these tools to touch the samples without causing measurable contamination.

It is recommended that all laboratory ware be washed with acetone to remove oils, lipids, and nonpolar materials, then rinsed with water, soaked for 24 hr in 2 *M* nitric acid, followed by exhaustive rinsing with purified water. The glassware should then be air-dried in a environment protected from the laboratory dust.

Extreme high-purity volatile acids can be easily prepared by subboiling distillation methods.[2] One approach is the "cold thumb" method described using all Teflon materials.[3] Another approach is an isothermal distillation method achieved by placing 2 plastic beakers side by side in a desiccator. One beaker will contain the concentrated acid and the other one an equal volume of ultrapure water. The concentrated acid will evaporate and dissolve into the purified water in the other beaker. If fresh concentrated acid is placed in the desiccator each day, then concentrated acid will result in a few day in the beaker that previously contained only the highly purified water. As before all glassware including the desiccator must be acid washed as described previously.

A very pure concentrated acid can be prepared in this manner.[4] These ultrapure acids should then be used for all preparations of standards, controls, and all sample preparation steps.

Calibration standards for copper can be conveniently purchased from a company such as J. T. Baker Chemical Co., Phillipsburg, NJ 08865. These standards are usually available in concentrations of 1,000–10,000 μg/ml of copper metal in dilute acid. The acid concentration usually

[2] E. C. Kuehner, R. Alvarez, P. J. Paulsen, and T. J. Murphy, *Anal. Chem.* **44,** 2050 (1972).
[3] J. M. Mattinson, *Anal. Chem.* **44,** 1715 (1972).
[4] C. Veillon and D. C. Reamer, *Anal. Chem.* **53,** 549 (1981).

ranges from 1 to 6 *M* in the purchased standards. Another source of standard materials is to purchase the spectrographically pure metal from a company such as Johnson Matthey Inc., Seabrook, NH 03874. The pure copper metal can be cleaned, weighted and dissolved in 6 *M* high-purity nitric acid. Working standards are prepared by making serial dilutions using 10 m*M* nitric acid. A comfortable working concentration for AAS is a standard that contains 0.1–1.0 ng/μg of copper. A 5 point standard working curve should be prepared. Each point should be measured in triplicate, the means used to examine the linearity using least-squares regression (linear regression analysis).

Most scientific hand calculators have least-square programs that work adequately for this purpose. The working curve concentrations should bracket the samples to be analzyed. Each time an analysis is to be conducted at least 2 NBS-SRM materials should be analyzed as controls. For copper an alloy from NBS (SRM number C1100) and bovine liver (SRM number 1577) both work well.

For the copper alloy an electric drill can be used to produce some shavings. These shavings are then washed in acetone, water, rinsed in 1.0 *M* nitric acid for 30 sec, rinsed exhaustively with purified water, and brought to constant weight in a 105° oven. After being equilibrated to room temperature in a desiccator the shavings are accurately weighed on an analytical balance. The clean shavings are then dissolved in highly purified 0.5 *M* nitric acid. The stock alloy standard is then diluted with 10 m*M* nitric acid. For the bovine liver sample, a sample is lyophilized to obtain the dry weight. Next 5.0 mg of the powder is dissolved in 1.0 *M* nitric acid then further diluted with 10 m*M* nitric acid to the concentration desired.

Sample Preparation

To measure copper in tissue by the ETAAS method a sample of about 15 mg (wet weight) is placed in an acid-washed, oven dried, preweighed 3 ml Pyrex centrifuge tube.[5] The wet tissue is oven dried at 105° for 24–48 hr until a constant weight is obtained. The dry weight of the sample is then calculated. The dry weight will usually be about 30% of the wet weight. Next 1.0 ml of 1 *M* ultrapure nitric acid is added, loosely covered, and heated at 80° just to dryness. The sample is then reconstituted with 2 ml of 10 m*M* purified nitric acid, mixed vigorously, covered with parafilm, and allowed to stand for 4–6 hr for the solids to dissolve. Usually the yellowish solution will be clear and homogeneous. Occasionally small amounts of insoluble materials will remain. However, analysis of the slightly turbid

[5] M. A. Evenson and C. T. Anderson, Jr., *Clin. Chem.* **21,** 537 (1975).

supernate will yield results on NBS materials that are accurate. With tissue samples standard addition will provide essentially the same results as the direct analysis. For plant materials or for samples that contain significant amounts of silica, a wet digestion scheme using hydrofluoric, nitric, perchloric, and sulfuric acids is recommended.[6]

Sample preparation for solutions containing copper can usually be prepared for acetylene–air FAAS by dilution with 10 m*M* nitric acid.[7] Use of other gases such as hydrogen–air and different burners will produce different flame temperatures and conditions and different sample preparation schemes will be necessary to minimize the matrix interferences. Most often the instruments used for FAAS are the Perkin-Elmer models 303, 403, 503, and 603 or the Instrumental Laboratories Inc. double beam models using a laminar flow burner head that uses acetylene–air as gases.

If a total consumption burner using hydrogen–air as gases is used then sample preparation for copper analysis consists of a 1 : 10 dilution of the sample with a 75% saturated solution of 8-hydroxyquinoline.[8] A solvent extraction procedure could also be necessary if the matrix interferences are great, or if the viscosity of the diluted samples is too much greater than the calibration standards.[8]

Quality Assurance

Accuracy

Perhaps the most important performance criterion for any analysis is accuracy. Accuracy includes the linearity of the working curve, freedom from matrix effects that are potential interferences, minimum blank effects, high-yield standard addition studies, and analysis of at least 2 NBS-SRM samples. Tris(hydroxymethyl)aminomethane (Tris) buffers in the pH range of 7–8 will enhance the copper standard signal when the Tris concentration is below 1 m*M*, will quench the signal at 50 m*M*, and will again enhance the signal when the concentration is 100 m*M*.[7] Therefore, care should be taken when selecting the buffer system to make certain accuracy is maintained in that buffer.

Precision

Two types of precision are meaningful in evaluation of analytical methods: between-run and within-run precision. On a single day the same

[6] T. J. Murphy, "NBS Special Publ. No. 492, P. 6. U.S. Government Printing Office, Washington, D.C., 1977.

[7] M. A. Evenson and B. L. Warren, *Clin. Chem.* **21,** 617 (1975).

[8] K. H. Falchuk, M. Evenson, and B. L. Vallee, *Anal. Biochem.* **62,** 255 (1974).

sample can be analyzed 10–20 times, the mean, the standard deviation, and the coefficient of variation are then calculated. The analysis of one aliquot of a single large or pooled sample once a day for 10–20 days will provide the data for between-run precision. The same calculated values are produced as described above. The between run values may be a much as double the within-run precision values. The between-run precision is a measure of the stability of the sample and the analysis while the within-run value is the best precision possible and is due to mostly instrumental variation.

Sensitivity

The limit of detection (LOD), as a measure of the sensitivity, should be defined as 2 or 3 times the standard deviation of the blank signal. The limit of quantitation (LOQ) is often defined as 10 times the standard deviation of the blank signal.[9] These standard performance criteria should always be provided when an analytical method is being described.

Summary

Guidelines presented here allow for copper analysis of biological materials by methods that are very sensitive, that require little sample preparation, that have few chemical or spectral interferences, that are inexpensive, and that require only usual care in contamination control. The commercial instruments for FAAS and ETAAS from Perkin-Elmer, from Varian, and from Instrumentation Laboratories Inc. (Allied Analytical Systems) all work well in either the flame or the flameless mode. Background correction techniques are not essential for copper analysis if care is taken with the sample preparation to minimize the background signals. Different types of burners will work adequately if one makes certain that the viscosity of the sample and the control products are similar to the calibration standards. Further, dilution of samples is preferred over increasing the viscosity of the calibration standards by the addition of a protein containing solution or a substance such as glycerol. A 1 : 10 dilution of blood plasma or serum with dilute nitric acid or water is all that is necessary for copper analysis by the FFAS methods.

Cation and anion effects should be tested by bracketing the concentrations of the ions found in the sample with known amounts of ions in the sample solutions. Increasing the concentrations of the ions thought to interfere while keeping the copper concentration constant is another way to test for ion interferences.[5,7]

[9] D. MacDougall and W. B. Crummett *et al.*, *Anal. Chem.* **52,** 2242 (1980).

Instrumentally, the measure of copper in biological samples can be achieved with high accuracy, high precision, without background correction, and with minimum sample pretreatment if care is taken to carefully plan and implement all the critical steps in the analysis procedure.

Standard addition studies of copper will yield results from 90 to 105% in liver and 97 to 110% in human serum. Measured values of the NBS-SRM samples will provide answers that are 99% of the certified value for copper in bovine liver (SRM 1577) by ETAAS and 97% of the certified value for copper in the alloy (SRM 157) by FAAS. The between-run coefficient of variation for ETAAS will range from 2 to 6% and the within run precision will range from 2 to 4%. The FAAS methods will have similar precision values. This type of method performance for copper analysis in biological materials is adequately accurate and precise, requires a minimum sample preparation, and at a reasonable cost for most cases when copper analysis is required.

[27] Rapid Colorimetric Micromethod for the Quantitation of Complexed Iron in Biological Samples

By WAYNE W. FISH

The array of methods available for quantitating iron in an experimental sample often bewilders the investigator occasionally faced with the need for such quantitation. This volume and an earlier volume[1] in this series describe various methods which are particularly applicable for the quantitation of iron in biological materials. However, because of the development of more sensitive chromophoric chelators and simpler techniques, new methodologies continually evolve. This chapter details one of these more recent methodologies for the quantitative determination of iron with ease, sensitivity, and simplicity.

Methods commonly employed for the quantitation of complexed iron in biological samples rely on an initial treatment which releases the complexed iron for its subsequent quantitative determination.[1] Dry ashing in a furnace or wet ashing with hot concentrated acid releases the iron, but both procedures take time and present hazards, particularly for the researcher with only an occasional need to determine the iron content of a biological sample. Thus, the novelty of the procedure described in this

[1] H. Beinert, this series, Vol. 54, p. 435.

chapter lies in the digestion procedure which uses acid–permanganate to release the complexed iron.[2] Furthermore, the use of a water-soluble Fe(II) chelator whose molar absorptivity with ferrous iron[3] equals 27,900 provides a highly sensitive iron-specific reagent and eliminates the need for an organic solvent extraction step in the procedure. The reagents described in the following procedure show some modification from those in the original report.[4]

Reagents

Reagent A: Iron-Releasing Reagent

0.6 *N* HCl
2.25% (w/v) (0.142 *M*) $KMnO_4$

Prepare reagent A immediately before use as it cannot be stored. This reagent releases chlorine gas and is best prepared and pipetted in a fume hood. It is most easily prepared by mixing equal volumes of 1.2 *M* HCl and 4.5% (w/v) (0.285 *M*) $KMnO_4$. The potassium permanganate solution should be stored in an amber bottle.

Reagent B: Reducing, Iron-Chelating Reagent

Twenty-five milliliters of solution contains
80 mg Ferrozine[5] (final concentration in reagent = 6.5 m*M*)
80 mg neocuproine[6] (final concentration in reagent = 13.1 m*M*)
8.8 g ascorbic acid (final concentration in reagent = 2 *M*)
9.7 g ammonium acetate (final concentration in reagent = 5 *M*)

To prepare reagent B, first dissolve the ammonium acetate and ascorbate in water, add the Ferrozine and neocuprine, and add water to the solution so that the final volume equals 25 ml. When necessary, use the slightly different preparative procedure, described later, for the pretreatment of ammonium acetate and ascorbate to remove contaminating iron. The light amber-colored solution should be stored in an amber bottle. A limited shelf-life of only 3–6 weeks necessitates keeping track of the age of this reagent. The solution becomes darker with age, which increases the value of the blank reading. The reagent retains its efficacy in the iron assay for up to 6 weeks. However, we routinely discard unused reagent

[2] G. B. Tennant and D. A. Greenman, *J. Clin. Pathol.* **22,** 301 (1969).
[3] L. L. Stookey, *Anal. Chem.* **42,** 779 (1970).
[4] M. E. May and W. W. Fish, *Arch. Biochem. Biophys.* **190,** 720 (1978).
[5] Ferrozine, Disodium 3-(2-pyridyl)-5,6-bis(4-phenyl sulfonate)-1,2,4-triazine.
[6] Neocuproine, 2,9-Dimethyl-1,10-phenanthroline.

after 3 weeks as the absorbances of the blanks become only marginally acceptable after this time.

Iron Standard. Six micrograms Fe/ml. Ferrous ethylenediammonium sulfate, primary standard grade, in 0.01 *N* HCl. Forty-one milligrams/liter of this primary standard gives the requisite iron concentration.

Procedure

The procedure described in the following paragraph illustrates a simpler format for iron quantitation. It utilizes acceptable working volumes of samples and reagents while still providing high sensitivity (0.5–6 μg Fe; see Fig. 1). A slightly more cumbersome procedure of greater sensitivity appears later which provides for the quantitation of iron down to 0.1 μg.

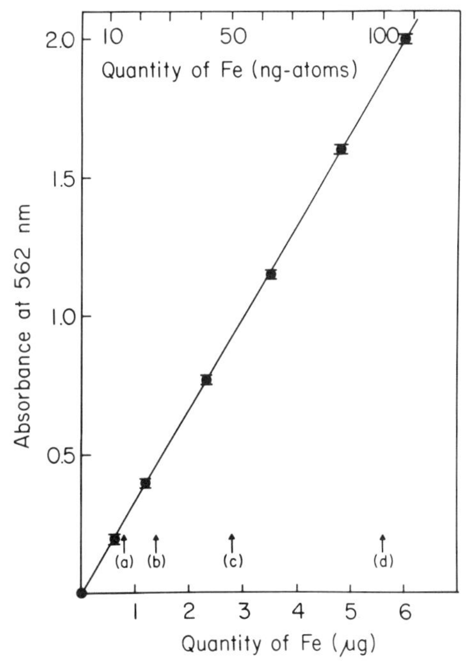

FIG. 1. A typical standard curve for the regular iron assay procedure detailed in the text. Each experimental point represents duplicate samples whose values fall within the error bars above and below the point. The coefficient of determination for these data equals 0.9996. Each arrow represents the anticipated color yield for 1 mg of a hypothetical iron-containing protein which possesses 1 g-atom of Fe per (a) 70,000 MW, (b) 40,000 MW, (c) 20,000 MW, and (d) 10,000 MW.

Add 0.5 ml of freshly prepared reagent A to 1.0 ml of solution to be assayed and allow this digestion mixture to incubate in a capped tube for 2 hr at 60°. Following digestion, add 0.1 ml of reagent B and mix. Allow the resulting solution to stand at room temperature for at least 30 min. Complete formation of the magenta-colored Fe(II) $[\text{Ferrozine}]_3^{2+}$ species takes about 15 min at room temperature, and the absorbance of the solution remains stable for at least 20 hr under these same conditions. After this second period of incubation, measure the absorbances of standards and experimental samples at 562 nm.

Convert the absorbance values for the samples to μg Fe or ng-atoms of Fe using a standard curve generated by the various quantities of the iron standard (Fig. 1). Correct the absorbances of standards and experimental samples for a reagent blank. Use water as the blank correction to the standards and the solubilizing buffer for the compound (protein) of interest as the blank correction to the experimental samples.

In those instances where quantities of sample are sufficiently small that they preclude the use of the procedure just detailed, the sensitivity of the assay may be increased 3- to 5-fold by decreasing the final volume of the assay (Fig. 2). Lyophilization of the standards and unknowns in their respective assay tubes before addition of the reagents will adjust any initial differences in volumes. Subsequent addition of the usual volumes of reagents A and B provides a precise quantitation of iron down to 0.1 μg (1.8 ng-atoms). Following the heating step, allow the samples to cool to room temperature. Then centrifuge each tube in order to clear from the sides of the tube water which evaporated from the digestion mixture and subsequently condensed on the sides during the 60° incubation. For a hypothetical iron-containing protein which binds 1 Fe per 50,000 MW, 100 μg of protein provides sufficient material for a precise measurement (see other hypothetical cases illustrated in Fig. 2).

Remarks Regarding Reagents

Accurate quantitation of functionally relevant iron requires the minimization or elimination of the confounding effects of nonfunctional, contaminating iron. This necessitates the utilization of virtually iron-free reagents and glassware. Contamination by iron originates from three sources: the solvent system, the glassware, and trace iron contamination of the reagents themselves. Use metal-free water at all times to minimize contamination from all three sources.

To further ensure the removal of any possible soluble or loosely bound iron from the sample and its attending solvent, treat the sample with chelating agents and follow with dialysis. However, monitoring a sizable

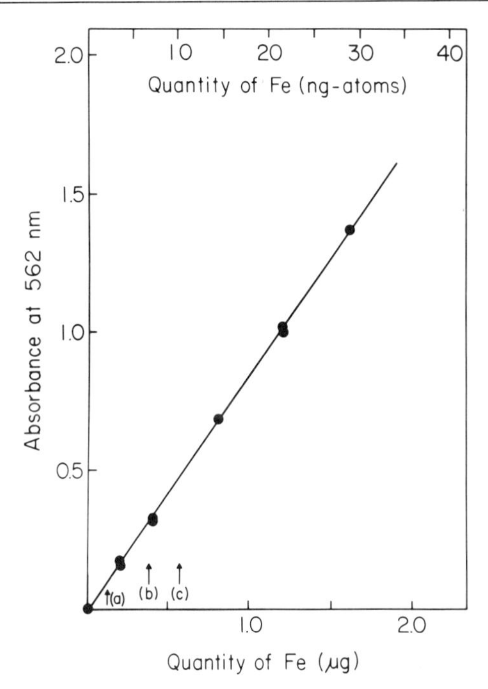

FIG. 2. A typical standard curve for the iron assay at the higher level of sensitivity. The coefficient of determination for these data equals 0.9991. Each arrow represents the anticipated color yield for 100 μg of a hypothetical iron-containing protein which possesses 1 g-atom of Fe per (a) 50,000 MW, (b) 20,000 MW, and (c) 10,000 MW.

number of fractions for iron from a column chromatography procedure renders the dialysis procedure impractical. Instead, use buffer systems with low intrinsic iron contamination to insure reliable column monitoring. A more detailed discussion of buffers appears later.

Acid treat all glassware utilized in the procedure, from reagent bottles to test tubes, to remove any possible iron contamination. Directions for acid treatment of glassware may be found in most quantitative analysis texts as well as in an article on iron determination[1] which appeared in an earlier volume of this series. Several brands of metal-free disposable polyethylene test tubes and pipet tips may be used directly from the package.[7] This provides a possible alternative to using acid-washed glassware and thus further simplifies the procedure.

[7] Falcon Plastics and Sarstedt manufacture iron-free disposable polyethylene test tubes. Gilson and Sarstedt manufacture iron-free pipet tips.

High, but reproducible absorbance values for the blanks and standards provide evidence that traces of iron contaminate one or more of the reagents. Of the reagents utilized in this procedure, ascorbic acid and ammonium acetate serve as the most probable sources of contamination; however, only stepwise elimination determines the true culprit. By utilizing reagent grade chemicals of stated low iron content and, if necessary, by employing the chelation treatment described below, the reagent blank for the regular assay procedure should give an absorbance versus H_2O at 562 nm of no more than 0.05 (1 cm light path). Use commercial ultrapure HCl of low metal content and reagent grade $KMnO_4$ without further purification. Likewise, use the iron and copper chelators from most commercial sources without further purification. Recrystallization of the chelators from water provides a simple and effective method for purification.[3] On the other hand, both ascorbic acid and ammonium acetate, which generally contain insignificant quantities of contaminating iron when obtained in reagent grade, serve as the most likely potential contributors of contaminating iron to the reagent blank. The relatively large quantities of these two chemicals used in the assay provide the principal reason for this potential contribution. Treatment of a solution of 2 *M* ascorbic acid and 5 *M* ammonium acetate with Chelex[8] before the addition of the two metal chelators significantly reduces any contaminating iron and thus reduces the absorbance value of the blank. To treat with Chelex, prepare 35 ml of the 5 *M* ammonium acetate, 2 *M* ascorbic acid and use 7 ml of this solution to wash and equilibrate a column of about 2 ml volume (1 g Chelex resin from the bottle). Save the remainder of the solution after it has passed through the chelating column, and to 25 ml of this treated solution add the appropriate quantities of Ferrozine and neocuproine.

Remarks Regarding Procedure

The digestion process utilized in this procedure to release iron proves successful with a number of proteins and other organic sources of iron.[2,4] The quantity of protein utilized per sample assay, up to at least 4 mg of protein, exhibits no adverse effect on the accuracy of the iron estimation, and this capacity of the digestion reagent should prove sufficient for nearly all cases encountered during experimentation (see hypothetical cases in Figs. 1 and 2). This, however, by no means precludes the possibility of encountering a protein or other type of biological molecule resistant to the acid–permanganate process of iron release. This weakness of the digestion procedure reveals the potential drawbacks of not completely ashing the sample.

[8] Bio-Rad Laboratories of Richmond, California, manufacture Chelex 100, a polystyrene resin which possesses iminodiacetic acid exchange groups.

A second reagent, reagent B, which contains ascorbic acid, Ferrozine, and ammonium acetate integrates with the acid–permanganate digestion. The ascorbate reduces excess permanganate and released metals. The Ferrozine forms a water-soluble, highly stable, intensely colored ferrous complex. The ammonium acetate buffers the assay mixture to a final pH of 4.5–5. This pH falls within the pH range 4–9 in which the Fe(II) $[\text{Ferrozine}]_3^{2+}$ species completely forms when in aqueous solution.[3]

As illustrated by Figs. 1 and 2, the regular assay procedure best suits quantitation of iron in the range of 0.5–6 μg while the more sensitive assay procedure best suits quantitation in the range of 0.1–2 μg of Fe. In general, the amount of iron in a sample can be estimated to within $\pm 5\%$ with a precision of $\pm 3\%$ by these procedures.[4] The proportionality between absorbance and iron content up to at least an absorbance of 2 represents another advantage of this methodology. At this point, there still exists a 3-fold excess of Ferrozine after 6 μg of iron has been chelated into the tris complex. This ability to cover a 20- to 30-fold range in quantity of iron reduces the possibility of misjudging the amount of material necessary to yield data in the functional range of the assay.

Apart from the concerns mentioned earlier regarding iron contamination of the sample (or sample buffers), exercise care to ensure that the procedure eliminates other substances which might interfere with the assay. Introduction of these substances may occur with the sample or as a part of the buffer system. Metal ions enter into the assay as contaminants of buffers and/or salts or bound to the protein under investigation. Only copper(I) and cobalt(II) form a colored complex with Ferrozine.[3,9,10] Because of its more likely presence in materials of biological origin and as a result of the spectral properties of the Cu(I) $[\text{Ferrozine}]_2^{+}$ complex,[9,10] copper most often interferes as a contaminant. Introduction of neocuproine in reagent B effectively reduces the severity of the error encountered in the estimation of Fe(II) in the presence of Cu(I)[9] (Fig. 3). Utilization of neocuproine permits a quantitation of iron with less than $+5\%$ error in the presence of up to a 2 to 1 weight ratio of Cu to Fe[9] (Fig. 3). Co(II) interferes less than $+5\%$ in the iron assay for quantities of Co(II) up to a 5 to 1 weight ratio over Fe.

The limited information available regarding possible interference by various buffer systems with the iron assay supplies few absolutes in determining possible problems with such systems. Contrary to the warnings against the use of phosphate buffers in other assay procedures[1] and against the use of azide reported for an earlier modification of this procedure,[4] no interference by 0.05 *M* phosphate buffers, by 0.02% sodium

[9] P. Carter, *Anal. Biochem.* **40,** 450 (1971).

[10] S. K. Kundra, M. Katyal, and R. P. Singh, *Anal. Chem.* **46,** 1605 (1974).

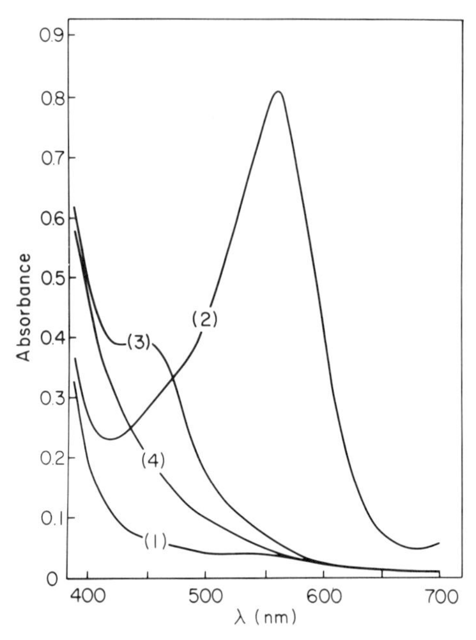

FIG. 3. Visible absorption spectra of reagent B and three of its metal complexes. (1) Reagent blank from the regular assay procedure, (2) 43.5 ng-atoms (2.43 μg) of Fe in the regular assay, (3) 86.7 ng-atoms (5.51 μg) of Cu in the regular assay, (4) 174 ng-atoms (10.25 μg) of Co in the regular assay.

azide, or by 10 m*M* EDTA has been observed for the iron assay procedure outlined above. The presence of nonionic detergents or sodium laural sulfate also offers no interference. Sodium deoxycholate represents the only detergent thus far observed to interfere with the assay. Of course, the simplicity of the assay procedure easily accommodates a preliminary investigation to examine for possible interference or contamination by any buffer system under consideration.

In conclusion, the procedure detailed above offers the investigator without access to more sophisticated instrumentation a simple and sensitive colorimetric method for the quantitation of iron in biological material.

Acknowledgments

I thank James Collawn and Michael May for their helpful scientific discussions and Sandra Carnie for her skillful editorial assistance in the preparation of the manuscript. A USPHS grant, Hl19491, and Medical University of South Carolina funds for research provided support for this work.

[28] Atomic Absorption Spectrometry of Magnesium

By MARK T. MARTIN and ROBERT SHAPIRO

Introduction

As well as being essential to photosynthetic organisms as a part of the pigment chlorophyll, magnesium is also an abundant and critical element in man. A 70-kg human body contains approximately 20 g of magnesium, a quantity that ranks second only to calcium among the divalent metal ions.[1] In light of this abundance, it is not surprising that magnesium participates in a diverse array of nonenzymatic and enzymatic functions. Nonenzymatically, magnesium stabilizes DNA, RNA, and ribosomes, influences membrane fluidity through phospholipid binding, and is a component of bone structure.[2,3] Numerous enzymes are activated by magnesium either as a result of direct binding to the enzyme as with enolase (EC 4.2.1.11)[3,4] and alkaline phosphatase (EC 3.1.3.1),[5] or by forming a magnesium–substrate complex that is acted on more efficiently than the free substrate. Examples of the latter type of activation are the kinases which hydrolyze ATP only as the MgATP complex.[3]

Fortunately for biochemists whose studies require the quantitation of magnesium, atomic absorption spectrometry (AAS) is exceptionally amenable to determination of this metal. The sensitivity—defined as the concentration required for 1% absorbance—is about 0.01 μg/ml for flame AAS at the 285.2 nm resonance line, while electrothermal (graphite furnace) AAS has a sensitivity of 0.17 pg.[6] Although there are a fairly large number of potential interferences, most can be avoided simply by choosing the proper flame and diluent.[6–9]

[1] H. A. Schroeder, A. P. Nason, and I. H. Tipton, *J. Chronic Dis.* **21,** 815 (1969).

[2] W. E. C. Wacker and A. F. Parisi, *N. Engl. J. Med.* **278,** 658 (1968).

[3] H. Ebel and T. Gunther, *J. Clin. Chem. Clin. Biochem.* **18,** 257 (1980).

[4] L. D. Faller, B. M. Baroudy, A. M. Johnson, and R. A. Ewall, *Biochemistry* **16,** 3864 (1977).

[5] W. F. Bosron, R. A. Anderson, M. C. Falk, F. S. Kennedy, and B. L. Vallee, *Biochemistry* **16,** 610 (1977).

[6] R. E. Sturgeon and S. S. Berman, *Anal. Chem.* **55,** 190 (1983).

[7] B. Welz, *in* "Atomic Absorption Spectroscopy," pp. 155–158. Verlag Chemie, New York, 1976.

[8] G. D. Christian and F. J. Feldman, *in* "Atomic Absorption Spectroscopy: Applications in Agriculture, Biology, and Medicine," p. 239. Wiley (Interscience), New York, 1970.

[9] T. V. Ramakrishna, J. W. Robinson, and P. W. West, *Anal. Chim. Acta* **36,** 57 (1966).

The convenience of magnesium AAS benefits not only those in basic research, but also clinicians who are able to correlate magnesium levels in biological fluids or tissues with various disease states. Magnesium concentrations in serum are altered in alcoholism, cardiovascular disease, acute pancreatitis, diabetic ketoacidosis, and various renal and gastrointestinal diseases.[10,11] Urine magnesium levels correlate with a number of diseases as well.

The objectives of this chapter are to recommend methods for the determination of magnesium in some commonly assayed biological materials and to provide guidelines for the general determination of magnesium in biological samples.

Instrument Parameters: Standard Conditions

The sensitivity of magnesium measurement by flame atomic absorption is sufficiently high that the graphite furnace instrument is infrequently required for biological determinations. Indeed, electrothermal AAS has the disadvantage that the large dilution of the sample often required to enter the linear working range may contaminate the sample and compromise accuracy and precision.[12,13] Further, interference is in general more severe with electrothermal AAS.

The flame most often used is air–acetylene since it is sensitive, prohibits many interferences, and is readily available. A cooler flame, such as air–propane, is susceptible to interferences and is not recommended.[14] The hotter nitrous oxide–acetylene flame has been favored by some since it eliminates certain interferences observed with the air–acetylene flame.[13,15] However, sensitivity with the nitrous oxide–acetylene flame is about 2-fold lower than with air–acetylene[16] and, as described below, simple alternative methods of eliminating interferences are available.

While a fuel-rich air–acetylene flame has a greater sensitivity toward magnesium, an oxidizing (fuel-lean) flame is less susceptible to some interferences.[17] Thus, as a general rule, an oxidizing flame is recommended except when extreme sensitivity is required.

The two most commonly used wavelengths for Mg AAS are 285.2 and 202.6 nm. The sensitivity of the 285.2 nm resonance line is about 0.01 μg/

[10] W. E. C. Wacker and A. F. Parisi, *N. Engl. J. Med.* **278,** 712 (1968).
[11] W. E. C. Wacker and A. F. Parisi, *N. Engl. J. Med.* **278,** 772 (1968).
[12] J. P. Matoušek and B. J. Stevens, *Clin. Chem.* **17,** 363 (1971).
[13] S. K. Bhattacharya and J. C. Williams, *Anal. Lett.* **12,** 397 (1979).
[14] D. J. Halls and A. Townshend, *Anal. Chim. Acta* **36,** 278 (1966).
[15] S. K. Bhattacharya, J. C. Williams, and G. M. A. Palmieri, *Anal. Lett.* **12,** 1451 (1979).
[16] M. D. Amos and J. W. Willis, *Spectrochim. Acta* **22,** 1325 (1966).
[17] W. W. Harrison and W. H. Wadlin, *Anal. Chem.* **41,** 374 (1969).

TABLE I
GRAPHITE FURNACE HEATING PROGRAM FOR DETERMINATION OF MAGNESIUM

Step	Temperature (°C)	Ramp time (sec)	Hold time (sec)	Argon flow (ml/min)
1	110	5	10	300
2	150	10	5	300
3	800	10	15	300
4	800	1	5	40
5	2500	0	3	40
6	2500	0	5	300

ml which corresponds to an optimum detection range of 0.05–2.0 μg/ml. The 202.6 nm absorbing line is not as sensitive; approximately 2.0 μg/ml gives 1% absorbance. With an optimum detection range of 50–500 μg/ml, this wavelength is attractive for measuring higher magnesium concentrations.[8]

A spectral bandwidth of 0.7 nm is most frequently employed.

For some samples, it may be necessary to use electrothermal AAS due to either extremely low magnesium concentrations or paucity of material. In such cases the furnace temperature program in Table I should be suitable for 20-μl aliquots of samples in simple matrices. With larger aliquots, the drying times at 110 and 150° should be increased.

Interferences

The worst interferences in magnesium AAS are observed with metals that form acid oxides that are stable at high temperature.[18] These elements, which include Al, Si, Ti, and Zr, cause severe interferences only when present in relatively high concentrations.[18] Many other metals have been reported to interfere with magnesium AAS under various conditions including Li, Na, K, Rb, Cr, Sn, Be, Fe, V, Mo, Cs, Sr, Ca, and Ba.[9,14] Chloride, oxalate, ethylenediaminetetraacetic acid (EDTA), and 8-hydroxyquinoline have been reported as anion interferences.[9,14]

Although a large number of interferences exist, most are easily overcome. An air–acetylene flame prevents interferences due to Na, K, Ca, phosphate, and Fe.[19] The presence of 0.1–1% (w/v) lanthanum chloride or strontium chloride eliminates the remaining interferences except for those

[18] W. T. Elwell and J. A. F. Gidley, *in* "Atomic Absorption Spectrophotometry," pp. 67–75. Macmillan, New York, 1961.

[19] J. B. Willis, *Spectrochim. Acta* **16,** 273 (1960).

caused by Cr and Ti.[9] EDTA (0.4%) overcomes the interference of Cr.[9] Many of these interferences (Al, Ti, Zr, for example) are absent in a nitrous oxide–acetylene flame.[16,17]

Some commonly used biological buffers cause little or no interference of magnesium AAS. High (nonphysiological) concentrations of phosphate have been reported to interfere.[9,19] However, this interference is not seen with either an air–acetylene flame or a strontium or lanthanum diluent.[9,19] Other buffers also do not appear to interfere in an air–acetylene flame. Metal-free solutions of 50 m*M* tris-(hydroxymethyl)aminomethane (Tris), 2-(*N*-morpholino)ethanesulfonic acid (Mes), or *N*-(2-hydroxyethyl)piperazine-*N*′-2-ethanesulfonic acid (HEPES) do not interfere with the determination of 0.25 μg/ml magnesium.[20]

Protein has occasionally been reported to interfere, especially in serum samples.[21,22] In these cases, this effect is probably due to high sample viscosity which may alter the sample aspiration rate. Dilution of the sample is generally a convenient remedy to this problem.

Factors that exacerbate interferences are a cool flame (air–propane for instance) and the presence of nitrate or sulfate.[14,18] Elwell and Gidley found that at even less than 0.2 g/100 ml, nitrate or sulfate drastically worsened interference by aluminum.[18] However, as stated above, interference by aluminum can be obviated by a lanthanum or strontium diluent.

Reference Standards

A 1 mg/ml magnesium reference solution can be purchased from numerous chemical companies (e.g., J. T. Baker Chem. Co.) or made up in the following manner. A ribbon of magnesium (obtainable from J. T. Baker and other sources) is cleaned by dipping in 50 g/liter HNO_3, rinsed in H_2O, and dried between two pieces of filter paper. One gram of magnesium ribbon is then dissolved in metal-free 6 *M* HCl. Dilution to 1 liter with 1% (v/v) HCl gives a 1 mg/ml solution.[22]

Sample Preparation

Pure or Partially Pure Biochemical Samples

The magnesium content in an aqueous solution of protein, nucleic acid, or other pure or partially pure biological fraction can be determined

[20] M. T. Martin and R. Shapiro, unpublished results.

[21] W. J. Price, *in* "Analytical Atomic Absorption Spectrometry," pp. 163–166 and 201. Heyden, New York, 1972.

[22] G. Stendig-Lindberg, J. Penciner, N. Rudy, and W. E. C. Wacker, *Magnesium* **3,** 50 (1984).

without difficulty. The sample should be diluted to a final magnesium concentration of 0.05–2.0 μg/ml (if the 285.2 nm resonance line is used) with a solution of strontium chloride such that the final concentration of Sr^{2+} is 0.25% (w/v). This dilution serves two purposes; it lowers the viscosity and introduces Sr^{2+} which eliminates a number of potential interferences. If a sample contains a high concentration of protein or nucleic acid but an extremely low concentration of magnesium, dilution may not be feasible and the sample can be prepared as described below for milk, with the exception that the final dilution factor may differ depending on the amount of magnesium present.

Biological Fluids

Serum. Normal serum contains approximately 1.8 meq/liter (43.7 μg/ml) of magnesium. Sample preparation consists of a 50-fold dilution of serum with a diluent such as 0.1% La^{3+}, as $LaCl_3$ or 0.25% Sr^{2+}, as $SrCl_2$.[19,22] Although the apparent serum magnesium concentration is only somewhat dependent on the diluent used, Stendig-Lindberg *et al.*[22] suggest the convention of using 0.25% Sr^{2+}.

Urine. A 1 : 200 or 1 : 500 dilution with metal-free water is sufficient for preparation of urine for magnesium determination.[21,23]

Cerebrospinal Fluid. A 1 : 100 dilution with 0.25% Sr^{2+} is recommended for magnesium analysis of this fluid.[24]

Milk. Since milk contains a large amount of protein, it is necessary to either dry-ash or deproteinate the sample. Dry-ashing is preferred.[21] Typically, 1.0 ml of milk is dried at 90° for 16 hr in a platinum crucible, then ashed for 4 hr at 480° in a muffle furnace. The cooled ashes are taken up in a minimal amount of concentrated HCl and diluted to 200 ml with either 1% La^{3+} or 0.25% Sr^{2+}. To deproteinate milk, 1.0 ml is mixed with 1.0 ml 25% trichloroacetic acid and centrifuged. The supernatant is diluted 100-fold with 1% La^{3+} or 0.25% Sr^{2+}.

Solid Tissues

Magnesium determinations have been carried out on many solid tissues including bone, hair, skeletal muscle, and brain. All of these solids require ashing or acid extraction.

Soft Tissue. Soft tissue may be prepared for AAS either by dry-ashing[25] or by extraction in 0.5 *M* HNO_3.[24] For dry-ashing, 0.7–0.9 g of tissue

[23] J. B. Willis, *Anal. Chem.* **33,** 556 (1961).

[24] M. W. B. Bradbury, C. R. Kleeman, H. Bagdoyan, and A. Berberian, *J. Lab. Clin. Med.* **71,** 884 (1968).

[25] B. J. Hunt, *Clin. Chem.* **15,** 979 (1969).

is dried in a platinum crucible for 16 hr at 105° and atmospheric pressure, then for an additional 12 hr at the same temperature *in vacuo*. The sample is reweighed, then heated in a muffle furnace for 16 hr at 600°. After cooling, the ashes are taken up in 2 ml 5 *N* metal-free HCl. The clear solution is diluted to reach a suitable magnesium concentration with 0.25% Sr^{2+} (typically 500-fold for skeletal muscle or brain). For HNO_3 extraction, 500 mg of tissue are dried for 48 hr at 105°, reweighed, and ground to a powder with a flat-ended glass rod. One milliliter of 0.5 *M* HNO_3 is added per 100 mg wet tissue, and the sample is stirred for 25 hr at room temperature, then allowed to stand for an additional 24 hr at 5°. The mixture is then centrifuged and the supernatant is diluted to the proper magnesium concentration, typically 100-fold, with 0.25% Sr^{2+}.[24]

Bone. Bone can be dry-ashed as follows.[25] Approximately 100–120 mg of bone is scraped clean of any soft tissue including marrow, swabbed with gauze dampened with 0.9% NaCl, and immersed in ether for 10 min. After air-drying, the bone is dried to a constant weight in a platinum crucible, then ashed for 16 hr at 600°, and at 800° for an additional 4 hr. After cooling, the ashes are taken up in 2.0 ml 6 *N* metal-free HCl, mixed, and diluted 1000-fold with 0.5% La^{3+}.

Hair. Preparation of hair for magnesium atomic absorption is described in detail by Harrison *et al.*[26] Approximately 0.5 g of hair is trimmed from the nape of the neck and cut into pieces of less than 1 cm in length. The specimen is washed by agitation in 150 ml of a 1% solution of 7X-O-Matic detergent (Linbro Chem. Co., New Haven, Conn.) for 30 min, and rinsed with 1 liter metal-free distilled water. After overnight drying at 110°, 6 ml concentrated HNO_3 is slowly added to a 0.3–0.8 g sample followed by 1 ml perchloric acid, and the mixture is heated at 200° until dense white fumes evolve.[26] The digested sample is diluted to a convenient magnesium concentration with 0.25% Sr^{2+}.

Acknowledgment

M.T.M. was supported by National Research Service Award 1-F32-HL06965-01A1 (BI-2) from the National Institutes of Health.

[26] W. W. Harrison, J. P. Yurachek, and C. A. Benson, *Clin. Chim. Acta* **23,** 83 (1969).

[29] Molybdenum

By JEAN L. JOHNSON

Molybdenum[1] is present in trace amounts in nearly all materials of biological origin. It is a redox active metal that functions in a select group of enzymes either in association with molybdopterin as the molybdenum cofactor or, in nitrogenase, as a part of the iron–molybdenum cofactor. The most common inorganic form of the metal is the oxyanion molybdate. The analysis and quantitation of molybdenum can be effected by a number of procedures. Colorimetric analysis is simple and reliable and subject to few interferences. It can readily be applied to samples of purified enzymes or tissue samples where adequate material is available. More sensitive analyses require sophisticated and elaborate instrumentation, i.e., atomic absorption with heated graphite furnace or neutron activation capabilities. Each of these methods will be discussed in some detail below. In addition, other less common methods which have been employed for molybdenum analysis will be cited.

The element tungsten shares many chemical and physical properties with molybdenum and serves as a biological antagonist of molybdenum uptake and utilization. *In vivo,* tungsten can be incorporated into enzyme active centers in place of molybdenum with the tungsten analogs being catalytically inactive.[2] However, in at least one case, a naturally tungsten-containing enzyme has been shown to exhibit catalytic activity while the corresponding molybdenum derivative is nonfunctional.[3] Because of the biological interactions which exist between molybdenum and tungsten and the need which often arises to assay one in the presence of the other, methodology for tungsten analysis will be included in the discussion below.

Colorimetric Analysis

Procedures for colorimetric analysis of molybdenum generally depend on complexation of the metal with 4-methyl-1,2-dimercaptobenzene, also referred to as toluene-3,4-dithiol, followed by extraction of the complex into an organic solvent. This reagent was first used for molybdenum anal-

[1] This work was supported by United States Public Health Service Grant AM-35029 from the National Institutes of Health and by United States Department of Agriculture Competitive Research Grant 85-CRCR-1-1599.

[2] J. L. Johnson, H. J. Cohen, and K. V. Rajagopalan, *J. Biol. Chem.* **249,** 5046 (1974).

[3] I. Yamamoto, T. Saiki, S. M. Liu, and L. G. Ljungdahl, *J. Biol. Chem.* **258,** 1826 (1983).

ysis by Hammence[4] and for tungsten determination by Miller.[5,6] The mercaptide complexes are suggested[7] to have the following structures:

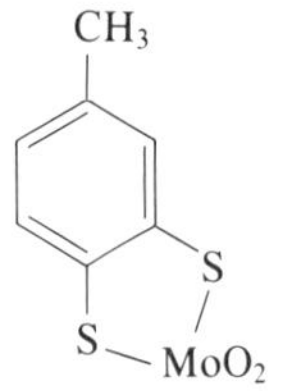

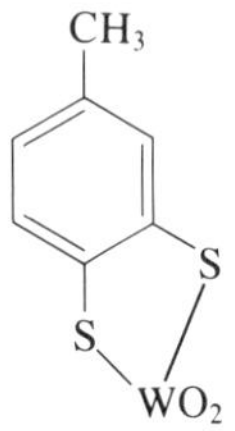

Molybdenum Analysis

For determination of molybdenum in biological materials, the following procedure is most convenient.

Sample Preparation. Biological materials may be wet-ashed by acid digestion or dry-ashed in a muffle furnace. For wet ashing, the sample is heated in the presence of concentrated sulfuric acid in a glass test tube (18 × 150 mm) on an electric burner for several hours. When the organic material has been completely digested, a few drops of H_2O_2 (30%) are added to decolorize the sample and heating is continued to drive off any residual H_2O_2. Use of HNO_3 and $HClO_4$ is to be avoided as they have been reported to interfere with the dithiol reaction.[6,8] For dry ashing, samples are heated in acid-washed porcelain crucibles for several hours in a muffle furnace at 80–100°. The temperature is then raised to 450°, and the samples are ashed overnight. The residue is dissolved in 4 *N* HCl and transferred quantitatively to a glass test tube for color development.

Reagents for Color Development and Extraction. Dithiol reagent. Toluene-3,4-dithiol is obtained from Eastman Organic Chemicals. Once opened, the liquid is dispensed in 100-μl aliquots and stored at −100° to prevent decomposition. An aliquot of dithiol is thawed and combined with 40 ml of 0.25 *M* NaOH. The mixture is stirred well on a magnetic stirrer (the dithiol may persist as small dispersed droplets) and 200 μl thioglycolic acid is added. The resulting opalescent solution is used for 1 day and then discarded.

[4] J. H. Hammence, *Analyst (London)* **65,** 152 (1940).
[5] C. C. Miller, *J. Chem. Soc.* **151,** 792 (1941).
[6] C. C. Miller, *Analyst (London)* **69,** 109 (1944).
[7] C. F. Bickford, W. S. Jones, and J. S. Keene, *J. Am. Pharm. Assoc.* **37,** 255 (1948).
[8] L. J. Clark, and J. H. Axley, *Anal. Chem.* **27,** 2000 (1955).

HCl, 4 *N*
KI, 50% solution, freshly prepared
$Na_2S_2O_3$, 10% solution
Potassium tartrate, 40% solution
Isoamyl acetate

Standards. A 1 *M* solution of $Na_2MoO_4 \cdot 2H_2O$ is prepared and diluted appropriately to provide standards from 5 to 100 nmol.

Procedure. Wet-ashed samples are made up to 14 ml total volume with 4 *N* HCl. Residues from dry ashing are transferred to glass test tubes with 14 ml 4 *N* HCl. KI solution (0.5 ml) is added to reduce any ferric iron in the sample. After 5 min, the iodine formed is decolorized by addition of one drop $Na_2S_2O_3$ solution. Tartrate solution (0.5 ml) is added to ensure that any tungsten present in the sample will not complex with the dithiol reagent. Dithiol reagent (2.0 ml) is added and mixed. After 10 min the molybdenum dithiol complex is extracted from the aqueous phase by vigorous vortexing with 1.5 ml amyl acetate. The organic layer is carefully transferred to a cuvette and the $A_{680\,nm}$ determined.

Tungsten Analysis

The conditions required for formation of the tungsten–dithiol complex are somewhat more rigorous than those necessary for molybdenum complexation. The ionic strength, pH, and temperature must all be more stringently controlled to assure complete complexation of tungsten by the dithiol reagent. Thus, for quantitation of tungsten in samples which contain little or no molybdenum the following procedure is applicable.

Sample Preparation. Samples are ashed as described earlier. Dry-ashed samples are transferred from the crucibles to glass test tubes with salt-solution (see below).

Reagents for Color Development and Extraction. Dithiol reagent, same as described above.

Salt solution. Dissolve 300 g ultrapure ammonium sulfate in 750 ml H_2O. Add 40 ml 85% phosphoric acid and adjust pH to 1.8 with ammonium hydroxide. Make volume up to 1 liter with H_2O. This reagent is stable for several months at room temperature.

Butyl acetate.

Standards. A 1 *M* solution of $Na_2WO_4 \cdot 2H_2O$ is prepared and diluted appropriately to provide standards from 5 to 100 nmol.

Procedure. Wet-ashed samples are made up to 6.5 ml total volume with salt solution. Residues from dry ashing are transferred to glass test tubes with 6.5 ml salt solution. Additions of 2.5 ml H_2O and 1.0 ml dithiol

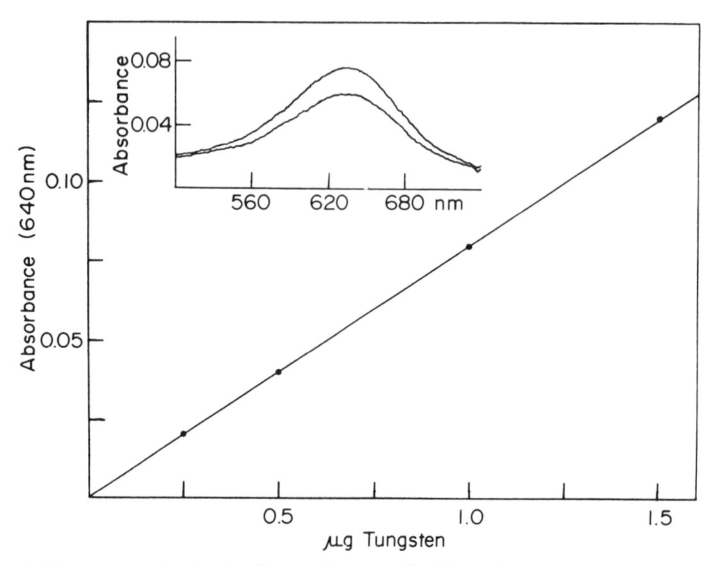

FIG. 1. Tungsten analysis of sulfite oxidase purified from livers of tungsten-fed rats. The main figure shows the absorbance at 640 nm of the tungsten–dithiol complex obtained from tungsten standard solutions. The inset shows absorption spectra of the tungsten–dithiol complex obtained from 1 μg of a tungsten standard (upper curve) and from 1.5 mg of purified sulfite oxidase (lower curve). From Johnson *et al.*[20]

reagent are made to each sample and the tubes are placed in a boiling water bath for 30 min. After cooling, the dithiol complexes are extracted by vigorous vortexing with 1.5 ml butyl acetate. The organic layer is carefully transferred to a cuvette and the A_{640} determined.

Figure 1 illustrates an application of this method. In this case, sulfite oxidase from livers of tungsten-fed rats was shown to yield exclusively the tungsten–dithiol complex absorbing at 640 nm.

The procedure described for tungsten analysis may also be applied to molybdenum although the dithiol complex of the latter still exhibits the charcteristic 680 nm absorption (see Fig. 2). The advantages of method 1 over method 2 for molybdenum quantitation are its simplicity (4 *N* HCl rather than the salt solution and room temperature color development) and the inclusion of KI to reduce ferric iron eliminating possible interference from this metal.

Quantitation of Tungsten in the Presence of Molybdenum

Quantitation of tungsten in those samples which may contain high levels of molybdenum is somewhat more difficult since the absorption

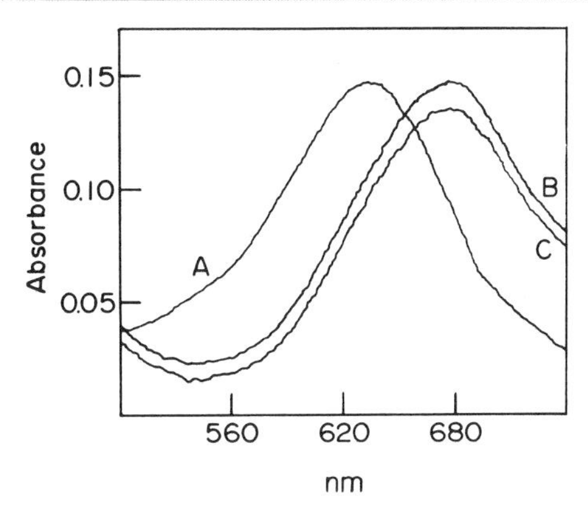

FIG. 2. Absorption spectra of toluene–3,4-dithiol complexes of tungsten and molybdenum. The spectra were obtained from 1.92 μg tungsten (A) and 1.00 μg molybdenum (B) (equimolar quantities of the two metals). C is the spectrum of the molybdenum-dithiol complex obtained from 3 g of ashed rat liver. From Johnson *et al.*[20]

profiles of the complexes of the two metals overlap. Tungsten may be analyzed for in the presence of molybdenum by method 2 if duplicate samples are prepared and analyzed with and without the addition of 0.25 g citric acid. The sample without citrate measures molybdenum + tungsten, and the sample with citrate measures molybdenum alone; tungsten is quantitated by the difference.

A method for quantitation of both molybdenum and tungsten in a single sample have been outlined by Cardenas and Mortenson.[9] The sample is first treated in such a way as to complex molybdenum but not tungsten. After the molybdenum–dithiol complex has been extracted into organic solvent and physically removed, the sample conditions are altered to accommodate tungsten complexation. The method recommended by Cardenas and Mortenson[9] is as follows.

Sample Preparation. Samples are digested by wet ashing as described earlier.

Reagents for Color Development and Extraction. Dithiol reagent. One gram toluene-3,4-dithiol is dissolved in 500 ml of 1.0% NaOH. The mixture is stirred for 1 hr. Thioglycolic acid is added drop by drop (about 8.0 ml) until a faint opalescent turbidity is formed. The reagent is stored in a polyethylene bottle at 4° for up to 3 months. If a precipitate forms, the reagent is filtered before use.

Isoamyl acetate.

Sulfuric acid.

[9] J. Cardenas and L. E. Mortenson, *Anal. Biochem.* **60,** 372 (1974).

Standards. Standards are prepared as described earlier.

Procedure. Samples are adjusted with sulfuric acid to a final concentration of 8 N (including the volume of the dithiol reagent), and 3.0 ml dithiol reagent is added and mixed. After 15 min the molybdenum–dithiol complex is extracted by vigorously vortexing with 3 ml amyl acetate. The organic layer is carefully transferred to a cuvette and the $A_{680\,nm}$ determined. The remaining solution is washed three times with 3 ml amyl acetate to remove residual molybdenum. The acidity of the sample is adjusted to 1 N with ammonium hydroxide, 3.0 ml dithiol reagent is added to each sample, and the tubes are placed in a boiling water bath for 30 min. After cooling, the tungsten–dithiol complexes are extracted by vigorous vortexing with 3.0 ml amyl acetate. The organic layer is carefully transferred to a cuvette and the $A_{630\,nm}$ determined.

Atomic Absorption Spectroscopy

While colorimetric procedures can conveniently assay 5–100 nmol of molybdenum, the need often arises to quantitate much lower amounts of the metal. A technique which is widely used for analyses of molybdenum in amounts less than a nanomole is atomic absorption with heated graphite atomization. Flame atomization is of virtually no value for molybdenum analysis due to the high temperatures (>2500°) required to volatilize the metal.

In the heated graphite furnace, the sample is introduced into a carbon tube which is electrothermally heated. Volatilization of the metal proceeds from molybdate to MoO_3 to Mo_2C to MoC to Mo and then to the vapor phase.[10] The molybdenum carbides are somewhat refractory to atomization with the result that even with high temperatures for atomization some metal tends to remain on the tube causing "memory effects." These effects can be minimized by charring at temperatures below 2100° minimizing MoC formation,[11] by atomizing at temperatures of 2650° or higher (although high temperatures severely limit the lifetime of the carbon tubes), by applying the sample to the tube wall rather than onto a temperature stabilized L'vov platform, and by including a high-temperature burnoff step in the furnace program.

When molybdenum has been converted to the atomic vapor, it absorbs light at a wavelength of 313.3 nm. Light at this wavelength is provided by a hollow cathode lamp where the cathode is a hollow cylinder of molybde-

[10] J. Sneddon, J. M. Ottaway, and W. B. Rowston, *Analyst (London)* **103,** 776 (1978).
[11] W. Wendl and C. Muller-Vogt, *Spectrochim. Acta* **39B,** 237 (1984).

num. When a sufficient voltage is applied, inert gas ions produced at the anode are accelerated toward the cathode where they sputter out excited atoms of the cathode material. As these atoms decay to ground state, they emit the monochromatic light appropriate for absorption by the analyte.[12]

The atomic absorption instrument used in this laboratory for quantitation of molybdenum is a Perkin-Elmer Zeeman/3030, and the methodology described has been developed specifically for this instrument. However, atomic absorption spectrometers with nearly identical capabilities are available from other suppliers and very similar protocols should be applicable.

Sample Preparation

Samples of enzymes, tissue homogenates, or biological fluids may be diluted with water to a molybdenum concentration of 5–20 ppb. With a volume of 20 μl injected, this corresponds to 1–4 pmol of the metal. Samples with an estimated molybdenum content lower than 5 ppb or those containing particulate matter can be dry ashed in a muffle furnace as described for colorimetric analysis. The residue in the porcelain crucible after ashing is dissolved in 0.2% nitric acid for introduction into the graphite furnace. Occasionally, dilution of a sample to the 5–20 ppb molybdenum range results in a solution with an extremely low concentration of protein which can be problematic. We have found with sulfite oxidase, for example, that dilution to the appropriate molybdenum concentration leads to a significant loss by adsorption of the dilute enzyme to polystyrene autosampler cups, plastic Eppendorf tubes, and glass test tubes. There are two alternative sample preparation procedures which overcome this problem and yield accurate molybdenum values for sulfite oxidase. In the first, the enzyme (which must be free of potassium) is diluted into a solution of 0.05% sodium dodecyl sulfate (electrophoresis grade). The presence of sulfate ions at this concentration depresses the molybdenum absorption by about 50% (see below), but this presents no problem if standards and sample are prepared in the same solution. In the second procedure, the enzyme solution is diluted with an equal volume of concentrated nitric acid and heated in a sealed tube in a boiling water bath for 6 hr. After wet-ashing in this manner, the sample may be diluted with water to the appropriate molybdenum concentration and read against a standard curve prepared with molybdenum in 0.2% nitric acid. Wet ash-

[12] S. S. M. Hassan, "Organic Analysis Using Atomic Absorption Spectrometry." Wiley, New York, 1984.

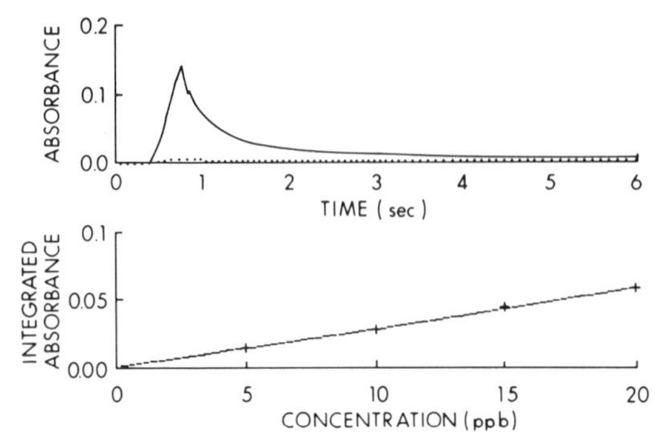

FIG. 3. Upper: Absorption of light at 313.3 nm by atomic molybdenum. The dotted line is the background signal obtained with Zeeman background correction applied. The solid line is the analyte signal obtained as the total uncorrected signal *minus* the background signal. The time on the X axis corresponds to step 3 of the graphite furnace program (Table II). Lower: Standard curve for molybdenum diluted in 0.2% nitric acid.

ing in nitric acid is also recommended for preparation of plant samples for atomic absorption analysis since the presence of silica in the plant material can lead to incomplete dissolution of dry ashed elements.[13]

Standards

Molybdenum certified atomic absorption standard (1000 ppm $\pm$ 1%) from Fisher Scientific is diluted serially into 0.2% nitric acid to yield a working standard solution of 0.02 ppm. The dilutions are made by transfer of 100 (or 200) μl with an Eppendorf pipettor with plastic tip into 900 (or 800) μl nitric acid in a plastic Eppendorf centrifuge tube. A fresh solution of the working standard is prepared daily. A standard curve is most conveniently prepared by autosampler dilution of the working standard with 0.2% nitric acid to give standards of 5, 10, 15 and 20 ppb (see Fig. 3). A linear standard curve can be prepared up to 100 ppb molybdenum, but because of memory effects it is technically more difficult to work in this range.

Instrument Setup and Operating Conditions

The Zeeman/3030 is set up for molybdenum analysis with the appropriate hollow cathode lamp monitoring 313.3 nm absorption with a 0.7 nm band pass. Argon is used as the inert purge gas; flow is interrupted during atomization. A new pyrolytically coated carbon tube is inserted and con-

[13] M. Hoenig, Y. Van Elsen, and R. Van Cauter, *Anal. Chem.* **58,** 777 (1986).

TABLE I
GRAPHITE FURNACE PROGRAM FOR CONDITIONING CARBON TUBE

Step number	Furnace temperature	Time (sec)		Internal gas flow (ml/min)
		Ramp	Hold	
1	2650	60	2	300
2	20	1	20	300
3	2650	10	10	300
4	20	1	20	300
5	2650	10	10	300
6	20	1	20	300
7	2650	10	10	300

ditioned as recommended by the manufacturer (Table I). The instrument is programmed to calculate peak area during the entire atomization step with Zeeman background correction. The autosampler is programmed to prepare a standard curve by appropriate dilution of the 20 ppb working standard and to inject samples. A blank solution of 0.2% nitric acid is run before the standard curve, before the first sample, and between each subsequent sample to prevent accumulation of molybdenum and to monitor the condition of the carbon tube. A 5% carryover of molybdenum from standard or sample to the blank is acceptable; when carryover approaches 10%, a new carbon tube is used. As a further check of the condition of the carbon tube, a 20 ppb standard is run as a sample after every tenth injection. This has been found to be an essential step, since the sensitivity of the carbon tube does not remain constant. Typically the tube response for a given standard solution will remain at the initial 100% level for 20–30 injections, may increase abruptly to 150% or more and then gradually decay to about 80%. It will remain constant at this level for some time and then decay further, concomitant with an increase in carryover to the next sample or blank. The tube is usable during all but the last stage described, but the standard curve must be reevaluated whenever the response for the 20 ppb standard changes by more than 5%. A solution of 0.1% Triton X-100 with 1% nitric acid is used to flush the autosampler lines between injections. The furnace program employed is outlined in Table II. A typical standard curve and absorption profile are shown in Fig. 3.

Interferences

A number of elements are known to interfere with analysis of molybdenum by atomic absorption. Zirconium, niobium, and tungsten interfere

TABLE II
GRAPHITE FURNACE PROGRAM FOR MOLYBDENUM ANALYSIS

Step number	Furnace temperature	Time (sec)		Internal gas flow (ml/min)
		Ramp	Hold	
1	110	10	20	300
2	1800	10	20	300
3	2650	0	6	0
4	20	5	30	300
5	2650	5	5	300

if present in 10^3-fold excess over molybdenum because the molybdenum carbide is incorporated within the lattice of the Zr, Nb, or W carbides.[11] It should be noted that tungsten cannot be analyzed by atomic absorption due to its low volatility. Zr and Nb are not likely to be present in biological materials. Both phosphate and sulfate cause strong suppression of the molybdenum signal. As noted above enzyme samples can be analyzed after dilution in 0.05% SDS (1.7 m*M* SO_4^{2-}) if the standards are also prepared in SDS. The absorption profile will be reduced by about 50% in the presence of this concentration of sulfate. Phosphate presents a more serious problem, however, and is to be avoided wherever possible. At a concentration of 1.7 m*M*, phosphate suppresses the molybdenum signal greater than 75%. It has been reported that phosphate also interferes with analysis of copper by atomic absorption spectroscopy, but that the interference can be eliminated by acidifying samples to pH 2 or below, since phosphoric acid is volatilized at 213° and driven off during the charring step.[14] We find that acidification of samples for molybdenum analysis does not alleviate phosphate suppression and infer that a phosphate–molybdenum complex is formed with atomization properties which differ from that of the uncomplexed metal. Generally, it is not difficult to avoid the use of phosphate buffers during sample preparation. In those samples where endogenous phosphate concentration may be high, such as in physiological fluids, the method of standard additions should be employed for molybdenum analysis.

Neutron Activation Analysis

Both molybdenum and tungsten are amenable to analysis in the ng range by neutron activation. In this technique, samples and standards are

[14] J. H. Freedman and J. Peisach, *Anal. Biochem.* **141,** 301 (1984).

irradiated with thermal neutrons produced in a nuclear reactor. Characteristic γ and X rays are emitted by the radioactive isotopes which are produced and these are quantitated by number and energy to determine the concentration of the element of interest.

Neutron activation analysis is particularly advantageous in that it can be applied to almost any sample matrix with no pretreatment of the sample. In addition it is possible to quantitate a large number of trace elements using a single sample. Its main disadvantage is in the sophisticated, expensive equipment required for irradiation and analysis. As a result of this constraint, samples must be transported to a neutron activation facility and analyzed on a fee per sample basis. While the cost for analysis of a single samples is reasonable, the cost is too high to allow routine analysis of column fractions or other samples generated in large numbers by experimental manipulation. Neutron activation analysis has been of great value in surveying the trace metal content of newly purified proteins or enzymes and can serve as an excellent reference technique to verify the accuracy of metal analyses carried out by other methods.

Other Methods

A few alternative methods for molybdenum quantitation will be mentioned in this section but no attempt will be made to present a comprehensive survey of the literature in this area. They are cited to indicate that there are alternatives to conventional methodology that may be particularly applicable in certain specific situations.

Other colorimetric procedures have been reported for molybdenum quantitation. One that may be particularly useful for biological applications involves reduction of molybdenum(VI) to molybdenum(V) with ascorbic acid, reaction with thiocyanate in HCl to yield the Mo(V)–SCN complex and extraction into benzene with *N,N'*-diarylbenzamidines.[15] The method is reported to be sensitive and highly selective with almost no interference from other elements.

Molybdenum has been quantitated in serum plasma and urine by cathode-ray polarography.[16] The metal was extracted from dry-ashed samples and quantitated in 2 *M* potassium nitrate at pH 1.6–2.2. The method is quite sensitive and specific.

Both molybdenum and tungsten can be quantitated by inductively coupled plasma analysis (ICP).[17] This technique is particularly advanta-

[15] K. S. Patel, R. M. Verma, and R. K. Mishra, *Anal. Chem.* **54,** 52 (1982).

[16] G. D. Christian and G. J. Patriarche, *Analyst (London)* **104,** 680 (1979).

[17] W. Slavin, *Anal. Chem.* **58,** 589 (1986).

geous for the more refractory trace elements even though the detection limits for molybdenum are somewhat higher with ICP than with graphite furnace atomic absorption.

Finally, the molybdenum content of rat liver and rat liver mitochondria has been examined by EPR spectroscopy.[18,19] While this method is obviously limited in its applications, it is of interest nonetheless. In 1971 Peisach *et al.*[18] observed an EPR signal of molybdenum(V) near $g = 1.96$ in preparations of rat liver and rat liver mitochondria and attributed it to one of the soluble molybdoproteins. Several years later Kessler *et al.*[19] showed conclusively that the signal was due to sulfite oxidase and could be used to monitor levels of sulfite oxidase molybdenum in various tissues and different animal species. In addition, it was found by Johnson *et al.*[20] that administration of tungsten to rats led to a decrease in signal amplitude that exactly paralleled the decrease in hepatic sulfite oxidase activity. Thus, by this approach, it is possible to monitor specifically the molybdenum content of sulfite oxidase in crude preparations (even in intact pieces of liver tissue).

[18] J. Peisach, R. Oltzik, and W. E. Blumberg, *Biochim. Biophys. Acta* **253,** 58 (1971).
[19] D. L. Kessler, J. L. Johnson, H. J. Cohen, and K. V. Rajagopalan, *Biochim. Biophys. Acta* **334,** 86 (1974).
[20] J. L. Johnson, K. V. Rajagopalan, and H. J. Cohen, *J. Biol. Chem.* **249,** 859 (1974).

[30] Nickel Analysis by Electrothermal Atomic Absorption Spectrometry

By F. WILLIAM SUNDERMAN, JR., SIDNEY M. HOPFER, and M. CRISTINA CRISOSTOMO

Introduction

The advantages and limitations of various techniques for analysis of nickel in biological materials have been evaluated in several reviews.[1-6]

[1] M. Stoeppler, *in* "Nickel in the Environment" (J. O. Nriagu, ed.), p. 661. Wiley (Interscience), New York, 1980.
[2] M. Stoeppler, *in* "Analytical Techniques for Heavy Metals in Biological Fluids" (S. Facchetti, ed.), p. 133. Elsevier, Amsterdam, 1981.
[3] M. Stoeppler, *in* "Nickel in the Human Environment" (F. W. Sunderman, Jr., ed.), p. 459. International Agency for Cancer Research, Lyon, France, 1984.
[4] F. W. Sunderman, Jr., *Pure Appl. Chem.* **52,** 527 (1980).

The electrothermal atomic absorption spectrometric (EAAS) procedure of Brown *et al.*,[7] which was developed through collaborative interlaboratory trials by biochemists in 13 nations,[8-10] has been accepted as a reference method by the Internationl Union of Pure and Applied Chemistry (IUPAC) and the International Agency for Research on Cancer (IARC).[6,7] This procedure involves (1) digestion with mixed nitric, sulfuric, and perchloric acids, (2) chelation of nickel with ammonium tetramethylenedithiocarbamate, (3) extraction of bis(1-pyrrolidinecarbodithioato)nickel(II) into 4-methyl-2-pentanone, and (4) measurement of nickel in the solvent extract by EAAS. The simpler EAAS procedures that are described in this chapter[11-13] yield analytical results that agree closely with measurements by the reference procedure of Brown *et al.*[7] The present EAAS procedures have the advantages of greater analytical sensitivity and precision, and they are considerably more rapid and convenient than the reference procedure. Three procedures are presented.

Acid digestion technique,[11] in which organic constituents of the sample are oxidized with mixed nitric, sulfuric, and perchloric acids prior to EAAS analysis. This technique is used to determine nickel in tissues, diets, and feces, as well as any biological samples that contain nickel that might be incompletely released from macromolecules after acidification with hot nitric acid (e.g., nickel in the active site of urease[14]) or nickel in a volatile form that would be lost by vaporization during the preliminary

[5] F. W. Sunderman, Jr., *in* "Techniques and Instrumentation in Analytical Chemistry" (A. Vercruysse, ed.), Vol. 49, p. 279. Elsevier, Amsterdam, 1984.

[6] F. W. Sunderman, Jr., *in* "IARC Monographs on Environmental Carcinogens—Selected Methods of Analysis" (I. O'Neill, ed.), Vol. 7, p. 319. International Agency for Cancer Research, Lyon, France, 1986.

[7] S. S. Brown, S. Nomoto, M. Stoeppler, and F. W. Sunderman, Jr., *Clin. Biochem.* **14,** 295 (1981).

[8] D. B. Adams, S. S. Brown, F. W. Sunderman, Jr., and H. Zachariasen, *Clin. Chem.* **24,** 862 (1978).

[9] S. S. Brown, A. D. diMichiel, and F. W. Sunderman, Jr., *in* "Nickel Toxicology" (S. S. Brown and F. W. Sunderman, eds.), p. 167. Academic Press, New York, 1980.

[10] F. W. Sunderman, Jr., S. S. Brown, M. Stoeppler, and D. Tonks, *in* "Collaborative Interlaboratory Studies in Chemical Analyses" (T. S. West and H. Egan, eds.), p. 25. Pergamon, London, 1982.

[11] F. W. Sunderman, Jr., A. Marzouk, M. C. Crisostomo, and D. P. Weatherby, *Ann. Clin. Lab. Sci.* **15,** 299 (1985).

[12] F. W. Sunderman, Jr., M. C. Crisostomo, M. C. Reid, S. M. Hopfer, and S. Nomoto, *Ann. Clin. Lab. Sci.* **14,** 232 (1984).

[13] F. W. Sunderman, Jr., S. M. Hopfer, M. C. Crisostomo, and M. Stoeppler, *Ann. Clin. Lab. Sci.* **16,** 219 (1986).

[14] N. F. Dixon, C. Gazzola, R. L. Blakeley, and B. Zerner, *J. Am. Chem. Soc.* **97,** 4143 (1975).

drying or ashing steps of EAAS analysis [e.g., nickel present as a Tris-Ni(II) complex[15]].

Protein precipitation technique,[12] in which proteins are precipitated with nitric acid and heat prior to EAAS analysis. This technique is used to determine nickel in serum, plasma, whole blood, saliva, bile, and cerebrospinal fluid.

Direct technique,[13] in which the sample is acidified with nitric acid prior to EAAS analysis. This technique is used to determine nickel in urine, water, and protein-free aqueous samples.

The detection limits for nickel determinations by these techniques, expressed as three times the standard deviation (SD) of the reagent blanks, are 0.45 μg/liter for urine, 0.1 μg/liter for whole blood, 0.05 μg/liter for serum or plasma, and 10 μg/kg (dry weight) for tissues.[11-13] If necessary, greater analytical sensitivity can be achieved by differential pulse absorption voltametry (DPAV) with a dimethylglyoxime-sensitized mercury electrode, which provides detection limits of 1 ng/liter for nickel determinations in whole blood, urine, saliva, and tissue homogenates.[16] However, the DPAV method is cumbersome and time-consuming, compared to the EAAS procedures.

Materials and Methods

Apparatus for Specimen Collection

Metal-free scalpels (for tissue dissection). Scalpels with obsidian blades, prepared by fracture of volcanic glass, are obtained from Fracture Mechanics, Ltd. (520 Marine St., Boulder, CO).

Polyethylene specimen bags (76 μm thickness, 10 cm width, 15 cm length) with round bottom and self-sealing closure device. These bags are impermeable to water and withstand pummelling during sample blending (catalog No. 10-0303, Tekmar Co., Box 371856, Cincinnati, OH).

Blood collection equipment. Teflon-polyethylene intravenous cannulae (22 gauge, 2.5 cm length, "Clear Cath," Abbott Labs., Inc., No. Chicago, IL), polypropylene syringes (10 ml capacity, "Monovette," Walter Sarstedt, Inc., Princeton, NJ), and polyethylene test tubes are used for blood collection and specimen storage.

[15] T. H. Grove and F. W. Sunderman, Jr., *Ann. Clin. Lab. Sci.* **8,** 495 (1978).

[16] P. Ostapczuk, P. Valenta, M. Stoeppler, and H. W. Nurnberg, *in* "Chemical Toxicology and Clinical Chemistry of Metals" (S. S. Brown and J. Savory, eds.), p. 61. Academic Press, London, 1983.

Apparatus for Sample Processing

Laboratory blender, "Stomacher-80" blender, motor-driven, with two steel paddles that pummel the sample bag (120 blows/min) (Tekmar Co., Box 371856, Cincinnati, OH).

Digestion appartus, electrically heated, including an aluminum block with 42 holes (18 mm diameter, 80 mm depth) to accommodate digestion tubes, and a continuously variable temperature regulator (to 300°) (Scientific Products, Inc., Evanston, IL). Aluminum foil is packed in the bottom of each hole to facilitate heat transfer to the hemispherical base of the digestion tube, and thereby to minimize "bumping." The digestion apparatus is operated in a fume hood. The temperature during digestion is monitored by a thermometer suspended in a digestion tube that contains 4 ml of concentrated sulfuric acid.

Digestion tubes, borosilicate glass (16 mm outside diameter, 125 mm length, 13.5 ml capacity) with Teflon-lined screw caps, calibrated at 4 ml volume (Corning Glass Works, Corning, NY).

Centrifuge tubes, conical, clear polystyrene (4.5 ml capacity, catalog No. 477 with caps No. 809, Walter Sarstedt, Inc., Princeton, NJ).

Vortex-mixing apparatus, variable speed ("Vortex-Genie," Fisher Scientific Co., Pittsburgh, PA), totally contained in a polyethylene bag to prevent nickel contamination from wear particles generated by the motor.

Water bath, 70°, with polyethylene basin, which is acid washed and filled with distilled water each week. The water bath is kept covered to minimize nickel contamination.

Centrifuge (model RT-6000, E.I. DuPont Co., Wilmington, DE) with plastic trunion cups that are fitted with air-tight caps. The trunion cups and caps protect the samples against nickel contamination from wear particles generated by the motor.

Sample cups, clear polystyrene (1.5 ml capacity, catalog No. 641 with caps No. 649, Walter Sarstedt, Inc., Princeton, NJ).

Piston-displacement pipettors, constructed so that only plastic surfaces are exposed ("Finn-Pipets," Vanguard International Corp., Neptune, NJ), with polyethylene pipettor tips.

Electrothermal atomic absorption spectrophotometer, model 5000-Z, Perkin-Elmer Corp., Norwalk, CT) with the following accessories: (1) automatic sampling system (model AS-40), (2) Zeeman background correction system, (3) heated graphite atomizer (model HGA-500), (4) strip-chart recorder (model 56), (5) digital printer (model PRS-10), (5) pyrolytic graphite tubes, and (6) cylinder of argon, ultrapure (99.99%, Linde Division, Union Carbide Corp., New York, NY).

Reagents for Specimen Collection

Heparin solution (10,000 USP units/ml). The contents of a vial that contains 25,000 USP units of sodium heparin from bovine lung (catalog No. H-9133, Sigma Chemical Co., St. Louis, MO) are dissolved in 2.5 ml of ultrapure water. The heparin solution is stored at 4° for up to 1 month. One drop of heparin solution (50 μl) is sufficient anticoagulant for 5 ml of whole blood. (*Note:* each batch of heparin solution is analyzed to ensure that the nickel concentration is <1 μg/liter, resulting in negligible contamination of blood samples. Heparin prepared from bovine lung generally contains much less nickel than that prepared from porcine intestine.)

Reagents for Sample Processing

Acid digestion mixture. Into a glass-stoppered, borosilicate glass bottle (250 ml capacity) are placed successively 120 ml of nitric acid (650 g/kg, relative density 1.40), 40 ml of sulfuric acid (690 g/kg, relative density 1.84), and 40 ml of perchloric acid (760 g/kg, relative density 1.67). Ultrapure acids (J. T. Baker Chemical Co., Phillipsburg, NJ, or E. Merck Co., Darmstadt, FRG) are essential.

Nitric acid, ultrapure (650 g/kg, relative density 1.40).

Dilute nitric acid. Into a polypropylene volumetric flask (500 ml capacity) is placed 3.2 ml of ultrapure, concentrated nitric acid; the contents are diluted to volume with ultrapure water.

Dilute hydrochloric acid. Into a polypropylene volumetric flask (100 ml capacity) is palced 10 ml of ultrapure, concentrated hydrochloric acid; the contents are diluted to volume with ultrapure water.

Reagents for Nickel Analysis

Ultrapure water is prepared by successive use of deionization and double distillation from a quartz still. The target nickel concentration in the ultrapure water is <0.05 μg/liter, as determined by successive evaporation of four 5-ml volumes to dryness in a digestion tube for analysis by the direct EAAS technique.

Nickel stock standard solution (100 mg Ni/liter). Into a tared borosilicate glass beaker (25 ml capacity) is weighed 50 mg of nickel powder (99.99% pure, Ventron Corp., Beverly, MA). Ultrapure water (5 ml) and ultrapure concentrated nitric acid (5 ml) are added and the nickel powder is dissolved by cautiously warming the beaker. The solution is cooled, transferred to a polypropylene volumetric flask (500 ml capacity), and diluted to volume with ultrapure water. This solution, stored in a screw-capped polyethylene bottle, is stable for at least 1 year.

Nickel intermediate standard solution (400 μg Ni/liter). Into a polypropylene volumetric flask (500 ml capacity) is pipetted 2 ml of nickel stock standard solution and 2 ml of concentrated ultrapure nitric acid. The contents are diluted to volume with ultrapure water. When stored in a screw-capped polyethylene bottle, this solution is stable for at least 3 months.

Nickel working standard solutions (ranging from 1 to 80 μg Ni/liter, depending upon the samples for analysis) are prepared by pipetting the requisite volume (0.25–20 ml) of nickel intermediate standard solution and 20 ml of dilute nitric acid solution into polypropylene volumetric flasks (100 ml capacity). The contents are diluted to volume with ultrapure water. These standard solutions are prepared every 2 weeks.

Procedures

Sampling of serum and whole blood. A polyethylene cannula is inserted in a suitable vein. The stylus of the cannula is removed and the cannula is flushed with 2 ml of blood, which is discarded. A polypropylene "Monovette" syringe is then used to collect 5 ml of blood. To prepare serum, the syringe barrel (a polypropylene test tube) is capped with a polyethylene stopper. After the blood has clotted, the test tube is centrifuged at 900 *g* for 15 min. Serum is transferred to a polyethylene test tube and stored at 4° until the time of analysis. To prepare heparinized blood, one drop (50 μl) of heparin solution is added to the "Monovette" syringe prior to blood collection. Heparinized whole blood can be stored at 5° for up to 48 hr prior to analysis. (*Note:* steel needles and rubber-stoppered tubes should not be used for blood collection.)

Urine collection and storage. Urine is collected directly into an acid-washed polyethylene bottle, with care to avoid contamination from feces, hair, or dust. Concentrated, ultrapure nitric acid is added (1 ml of acid per 100 ml of urine); the acidified urine specimen is refrigerated at 4° for up to 1 week, or frozen at −20° for longer storage.

Acid digestion technique (for tissues, foods, feces, etc.). Tissue slices (e.g., liver, kidney, spleen, brain, lung) are minced with plastic forceps and obsidian scalpels; samples (1 g, wet weight) are placed in homogenizer bags with 5 ml of distilled water. Each bag is sealed, placed in the Stomacher blender, and pummelled (5–15 min) until the contents are completely blended. Samples of dry foods, animal diets, or dried, pulverized feces (0.5 g, dry weight) are treated similarly. Bone samples cannot be homogenized in the Stomacher blender; pieces of bone are excised, scraped, dried at 105° for 3 days, pulverized in an agate mortar, weighed, and transferred to duplicate digestion tubes. Samples (1 ml) of blended tissue, food, or feces are transferred to duplicate digestion tubes. Into seven pairs of digestion tubes are transferred 1 ml of standard solutions containing 0, 4, 8, 16, 24, 32, and 40 μg Ni/liter. The acid digestion mixture (1 ml) is dispensed into all of the digestion tubes. The tubes are placed into holes in the aluminum blocker heater at ambient temperature. The tubes are heated initially at 110° for 30 min. This is the most critical stage of digestion, because of the possibility of sample loss by foaming. The digestion then proceeds stepwise as follows: 1 hr at 140°, 30 min at 190°, and finally 1 hr at 300°. At the conclusion of the 3-hr digestion period, the contents of the tubes should be clear and colorless; the sample volumes should be approximately 0.05 ml. After the tubes have cooled to ambient temperature, the contents are diluted to the 4-ml calibration mark by addition of dilute hydrochloric acid. Concentrated nitric acid (0.2 ml) is added to each tube. The tubes are capped; after the contents have been mixed by swirling, the samples are ready for EAAS analysis.

Protein precipitation technique (for serum, plasma, whole blood, etc.). Duplicate 1 ml samples of serum, plasma, or whole blood are placed in centrifuge tubes. Whole blood samples are diluted by addition of 0.9 ml of ultrapure water. While the contents of each tube are being mixed with a vortex mixer, concentrated nitric acid (50 μl for serum or plasma; 200 μl for whole blood) is added directly into the vortex with a piston-displacement pipettor. Mixing is continued for 1 min after acidification. Each tube is stoppered and placed in a water bath at 70° for 5 min. The tubes are mixed for 10 sec with a vortex mixer and centrifuged at 900 g for 10 min. The protein-free supernatants are transferred into sample cups and capped until EAAS analysis. Protein-free extracts of whole blood are recentrifuged, if necessary, to remove any flecks of precipitated protein. Into a set of six sample cups are pipetted 1 ml of nickel working standard solutions (0, 1, 2, 4, 8, and 12 μg Ni/liter) and 50 μl of concentrated nitric acid. The sample cups are capped until EAAS analysis.

Direct technique (for urine). Into the requisite number of centrifuge tubes are pipetted, respectively, 1 ml of each urine specimen and 1 ml of seven nickel working standard solutions (0, 4, 8, 20, 40, 60, and 80 μg Ni/liter). Dilute nitric acid solution (1 ml) is added to each sample; the tubes are capped and mixed with a vortex mixer. If an acidified urine sample is turbid or contains a precipitate, the tube is centrifuged at 900 g for 10 min. The samples are decanted into sample cups and capped until EAAS analysis.

Procedure for EAAS analysis. The sample cups are placed in the automatic sampler of the electrothermal atomic absorption spectrophotometer. The spectrophotometer is adjusted to the following settings:

(1) sample volume: 50 μl for digested samples and protein-free extracts of serum, plasma, or whole blood; 20 μl for acidified urine; (2) wavelength: 232.0 nm; (3) spectral band-width, 0.2 nm; (4) Ni lamp current, 25 mA; (5) time constant: 0.3 sec; (6) chart speed: 8 cm/min; (7) recorder range: 0.2 or 0.5 A (full-scale). The following temperature program is used: (1) 1 sec ramp from 25 to 100° (2) hold temperature at 100° for 1 sec; (3) 60 sec ramp from 100 to 140°; (4) hold temperature at 140° for 10 sec; (5) 30 sec ramp from 140 to 190°; (6) hold temperature at 190° for 5 sec; (7) 80 sec ramp from 190 to 1200°; (8) hold temperature at 1200° for 50 sec; (9) atomize at 2600° for 5 sec; (10) 1 sec ramp from 2600 to 2700°; and (11) hold temperature at 2700° for 3 sec. The atomization temperature of 2600° is calibrated daily by optical pyrometry. The argon flow is set at 300 ml/min throughout the temperature program, excepting the last 5 sec of step (8) and 5 sec of step (9). During this 10 sec interval, the argon flow is reduced to 30 ml/min. The automatic baseline adjustment is programmed to occur 6 sec prior to atomization; the spectrophotometer is programmed to integrate the Zeeman-corrected atomic absorption signal during the 5 sec of atomization. When nickel concentrations in samples exceed the highest standard, the samples are diluted and the analyses are repeated.

Computations

The calibration chart of absorbance peak areas versus nickel concentrations is linear in the range from 0 to 0.6 ng of nickel injected into the graphite tube. Nickel concentrations in the samples are computed by proportionality to the standards and expressed as μg/liter, μg/day (for timed collections of urine and feces), or μg/kg (dry weight) of tissues, diets, or feces. The dry weights are determined by transferring duplicate aliquots (1 or 2 ml) of each blended sample into tared weighing boats and drying at 105° for 48 hr. The results of nickel concentrations in protein-free extracts are multiplied by correction factors of 1.85 for whole blood and 0.94 for serum.[12] These correction factors compensate for the dilutions of whole blood, serum, and standard samples and adjust for volume displacement from precipitated proteins. Since whole blood water averages 0.84 kg/liter, nickel in the 1 ml blood sample is transferred into 1.94 ml of protein-free extract (0.84 ml of blood water + 0.9 ml of added water + 0.2 ml of HNO_3). Since serum water averages 0.94 kg/liter, nickel in the 1 ml serum sample is transferred into 0.99 ml of protein-free extract (0.94 ml of serum water + 0.05 ml of HNO_3). The 1 ml nickel standard samples are diluted to 1.05 ml by addition of 0.05 ml of HNO_3. Therefore, the correction factor for whole blood analysis is $1.94 \div 1.05 = 1.85$; the correction factor for serum analysis is $0.99 \div 1.05 = 0.94$.[12]

Discussion

Procedural Notes

To minimize nickel contamination, the analyses should be performed in a scrupulously clean laboratory room that is solely devoted to determinations of trace elements. Cigarette smoking should be prohibited. Pipetting should be performed in a laminar-flow hood to prevent nickel contamination from dust. To achieve the requisite analytical sensitivity, the atomic absorption spectrophotometer must be in perfect optical, mechanical, and electronic adjustment; attentive preventive maintenance is essential. Each pyrocoated graphite tube should be conditioned by six cycles through the temperature program before the first assay. The graphite tube should be routinely replaced after 100 assays. Although the digital output of the absorbance integrator is used to compute nickel concentrations, the strip-chart recorder tracing is required to monitor the baseline noise level and to ensure that the atomic absorption peaks have the expected conformation.

Before each use, polypropylene syringes, polyethylene tubes, bottles, flasks, pipettor tips, centrifuge tubes, and sample cups are scrubbed in hot detergent solution (e.g., "7-X Cleaning Solution," Limbro Chemical Co., New Haven, CT), diluted 30-fold with hot tap water. The plasticware are placed in polythylene canisters (2 liter capacity) and rinsed in batch fashion by filling and decanting three times with tap water and six times with deionized water, without contact of the contents with the analyst's hands. After the last rinse with deionized water, concentrated ultrapure HCl (50 ml) is poured into the canister. The canister is capped tightly, shaken, and allowed to stand at room temperature for 1 hr. The canister is then filled with deionized water, shaken, and allowed to stand for 20 min. The contents are rinsed five times with deionized water and twice with ultrapure water. The canister is placed with lid ajar in an oven at 100° until the contents are dry. After acid washing, plasticware should be handled with plastic gloves that have been rinsed with ultrapure water, in order to avoid nickel contamination from perspiration on the analyst's hands.

Precision, Recovery, and Interference Tests

Run-to-run analytical precision, expressed as coefficients of variation (*CV*), averaged 5.8% for urine (based upon analyses of a single urine specimen containing 15 μg Ni/liter in 10 consecutive runs) and 8.1% for serum (based upon analyses of a single serum specimen containing 3.5 μg Ni/liter in 33 consecutive runs).[12,13] Recovery of nickel added in concentration of 20 μg/liter to 14 urine specimens averaged 99 ± 5%; recovery of nickel added in concentration of 8 μg/liter to 16 serum specimens averaged 97 ± 3%; recovery of nickel added in concentration of 200 μg/kg (dry weight) to homogenates of 7 rat kidneys averaged 101 ± 8%.[11-13] No interferences in the described procedures were observed by additions of Fe, Cu, or Zn (0.5 m*M*) or As, Au, Ba, Bi, Cd, Co, Cr, Hg, Mn, Pb, or V (50 μ*M*) as soluble salts to serum or urine specimens.[12,13] Standard additions curves, prepared by additions of nickel to specimens of serum, whole blood, urine, and tissue homogenates, were consistently parallel to the slopes of the calibration lines, so that computations by the method of standard additions are unnecessary.

[31] Determination of Selenium in Biological Matrices

By S. A. LEWIS

Introduction

There has been an increasing need for methods to determine selenium in biological matrices because of its toxicological and physiological importance and its potential carcinopreventive and anticarcinogenic roles. There has also been a demand for the determination of selenium species, both organic and inorganic. Many reviews of selenium methodologies[1-15]

[1] J. H. Watkinson, *in* "Selenium in Biomedicine" (O. H. Muth, J. E. Oldfield, and P. H. Weswig, eds.), p. 97. AVI Publ., Westport, Connecticut, 1967.

[2] O. E. Olson, I. S. Palmer, and E. I. Whitehead, *Methods Biochem. Anal.* **21,** 39 (1973).

[3] J. E. Alcino and J. A. Kowald, *in* "Organic Selenium Compounds: Their Chemistry and Biology" (D. L. Kalyman and W. H. H. Gunther, eds.), p. 1049. Wiley, New York, 1973.

[4] W. C. Cooper, *in* "Selenium" (R. A. Zingaro and W. C. Cooper, eds.), p. 615. Van Nostrand-Reinhold, New York, 1974.

[5] A. D. Shrendrikar, *Sci. Total Environ.* **3,** 155 (1974).

[6] O. E. Olson, *Proc. Symp. Selenium–Tellurium Environ., Univ. Notre Dame, Indiana* **May 11–13** (1976).

[7] N. T. Crosby, *Analyst (London)* **102,** 225 (1977).

[8] T. M. Florence, *Talanta* **29,** 345 (1982).

[9] H. Robberecht and R. Van Grieken, *Talanta* **29,** 823 (1982).

[10] S. E. Raptis, G. Kaiser, and G. Tölg, *Anal. Chem.* **316,** 105 (1983).

[11] H. J. Robberecht and H. A. Deelstra, *Talanta* **31,** 497 (1984).

[12] L. Fishbein, *Int. J. Environ. Anal. Chem.* **17,** 113 (1984).

[13] G. Tölg, "Selenium Analysis in Biological Materials: Trace Element Analytical Chemistry in Medicine and Biology," Vol. 3. de Gruyter, Berlin, Federal Republic of Germany, 1984.

[14] M. Verlinden, H. Deelstra, and E. Adriaenssens, *Talanta* **28,** 637 (1981).

[15] S. A. Lewis and C. Veillon, *in* "Selenium" (M. Ihnat, eds.). CRC Press, Boca Raton, Florida, in press.

have been published and the reader is encouraged to scrutinize these. However, it is not the purpose of this chapter to provide a comprehensive review, but rather to cover a few methods that are used to determine selenium in specific biological matrices, emphasizing applicability but at the same time pointing out the potential pitfalls and limitations of these methods.

Isotope dilution gas chromatography/mass spectrometry (GC/MS) and fluorometry will be discussed in detail. Using the classification scheme established by the International Federation of Clinical Chemistry,[16] the GC/MS method can be considered a definitive technique representing the ultimate in quality. Isotope dilution GC/MS employs an ideal internal standard—another isotope of the same element—negating the need for quantitative sample preparation and external standardization.[17] Definitive methods are generally considered to be so sophisticated that they are outside the realm of most laboratories. However, with the advent of small, simple-to-use, benchtop mass spectrometers coupled to capillary gas chromatographs, this technique is now both feasible and affordable for routine selenium analyses.

Reference methods are those that generally can be considered as representing the best method available for routine use under carefully controlled conditions. The Association of Official Analytical Chemists' (AOAC) fluorometric method for the determination of selenium in plants and foods[18] can be classified as a reference method. This method has been adapted for many other biological matrices; however, these adaptations do not necessarily confer reference method status to these other sample types.

Some routine methods may or may not have analytical biases or systematic errors. The following methods fall into this category and will be discussed in less detail: atomic absorption spectrometry using either hydride generation or electrothermal atomization (preferably with Zeeman background correction) and gas chromatography coupled to detection systems other than mass spectrometry. However, emphasis will be placed on methods that have been developed for the analysis of specific biological matrices for selenium using these techniques.

In the discussion of all of these methods, the limit of detection, imprecision, and suitability for the determination of selenium in specific matrices will be examined.

[16] J. Bittner, R. Bolton, J. A. Boutwell, and P. M. G. Broughton, *Clin. Chem.* **22,** 532 (1976).

[17] C. Veillon and R. Alvarez, *in* "Metal Ion in Biological Systems" (H. Sigel, ed.). Dekker, New York, 1983.

[18] "Official Manual of Analysis." Association Official Analytical Chemists, Washington, D.C., 1984.

Quality Assurance

All aspects of any analytical method, from sampling through method validation, must be incorporated if a quality assurance program is to be successful. Sample collection and subsequent sample manipulation must be documented. Standards and reagents must be chosen carefully and the judicious use of sample and reagent blanks is mandatory.

Contamination is not generally a concern with selenium determinations; however, analyte loss because of the volatility of selenium can present a problem especially for biological matrices. Sample containers of borosilicate glass are reported to cause less problems through adsorption and desorption than plastics.[19,20]

The ideal way to establish the validity of a method is by the analysis of samples by two or more methods that utilize different physical characteristics. However, this procedure is generally beyond the scope of most laboratories. In the author's opinion, the single most effective way of assuring the accuracy and measuring the imprecision of a method is by consistent use of well characterized, matrix-suitable quality control materials with similar analyte concentrations that are treated exactly like samples.

A list of available and potentially available quality assurance materials of varying biological matrices is shown in Table I. However, in spite of the availability of these reference materials, many authors do not report the use of suitable reference materials or the establishment of imprecision.

Isotope Dilution Gas Chromatography/Mass Spectrometry (GC/MS)

Introduction

Combined GC/MS has the advantages of shorter analysis time and less expensive equipment over other definitive methods such as other mass spectrometry techniques, and neutron activation analysis. This is especially true since the advent of benchtop capillary GC/MS systems. Another limitation is the lack of suitable chelates for many elements of biological interest; however, a suitable chelate and GC/MS method for selenium has been developed.[21] This method has been logically extended to a double-isotope technique for use with nutritional tracer studies—the

[19] J. R. Moody and R. M. Lindstrom, *Anal. Chem.* **49,** 2264 (1977).

[20] M. Ihnat, Y. Thomassen, M. S. Wolynetz, and C. Veillon, *Acta Pharmacol. Toxicol.* submitted for publication (1985).

[21] D. C. Reamer and C. Veillon, *Anal. Chem.* **53,** 2166 (1981).

TABLE I
BIOLOGICAL REFERENCE MATERIALS FOR Se

Source[a]	Reference material	Matrix	Se concentration
Bowen	Bowen's kale	Kale	0.134 μg/g
NBS	SRM 1566	Oyster tissue	2.1 μg/g
NBS	SRM 1577A	Bovine liver	0.71 μg/g
NBS	SRM 1567	Wheat flour	1.1 μg/g
NBS	SRM 1568	Rice flour	0.4 μg/g
NBS	SRM 1549	Nonfat milk	0.11 μg/g
NBS	SRM 1572	Citrus leaves	0.025 μg/g
NBS	SRM 2670	Urine	
		Normal	30 μg/liter
		Elevated	460 μg/liter
NBS	SRM 909	Human serum[b]	105 μg/liter
NBS	RM 8419	Bovine serum	16 μg/liter
NBS	RM 1589	Bovine serum[c]	
NBS	RM 8431	Mixed diet	0.242 μg/g
NBS	RM 50	Albacore tuna	3.6 μg/g
NYE	105	Serum	90.7 μg/liter
NYE	108	Urine	49.4 μg/liter
IAEA	MA-A-1	Copepoda	3.0 μg/g
IAEA	MA-A-2	Fish flesh	1.7 μg/g
IAEA	H-5	Bone	0.0537 μg/g
IAEA	HH-1	Human hair	1.7 μg/g
IAEA	H-4	Animal muscle	0.28 μg/g
IAEA	A-11	Milk powder	0.034 μg/g
IAEA	MA-M2	Mussel tissue	2.27 μg/g
IAEA	A-13	Animal blood	0.24 μg/g
IAEA	V-10	Hay powder	0.022 μg/g
IAEA	H-8	Horse kidney	4.67 μg/g
IAEA	V-9	Cotton cellulose	0.015 μg/g
NPCC	TORT-1	Lobster	6.88 μg/g
NIES	CRM-6	Mussel	1.5 μg/g
NIES	CRM-5	Human hair	1.4 μg/g
IUPAC[d]	Seronorm[e] (Batch 103)	Lyophilized serum	91 μg/liter
BCR	CRM-150	Milk powder (spiked)	0.127 μg/g
BCR	CRM-151	Milk powder (spiked)	0.125 μg/g
BCR	CRM-063	Milk powder	0.088 μg/g

[a] Bowen, Reading, England; NBS, National Bureau of Standards, Gaithersburg, MD; NY, Nygaard & Company AS, Oslo, Norway; IAEA, International Atomic Energy Agency, Vienna, Austria; NRCC, National Research Council of Canada, Ottawa, Canada; NIES, National Institute for Environmental Studies, P.O. Yatabe, Tsukaba Ibaraki, Japan; IUPAC, International Union of Pure and Applied Chemistry; BCR, Community Bureau of Reference, Brussels, Belgium.

[b] R. Alvarez (National Bureau of Standards, Gaithersburg, Maryland), personal communication.

[c] Under development.

[d] See Ref. 20.

[e] US Distributors: Accurate Chemical & Scientific Corporation, Westbury, NY.

primary reason for its development.[22] The samples are "spiked" with a known amount of an enriched stable isotope of selenium (^{82}Se) and are digested to destroy the organic matter and acidified with hydrochloric acid to convert the selenium to the tetravalent oxidation state. Undigested lipids are extracted with chloroform. The selenite is reacted with 4-nitro-*o*-phenylenediamine (NPD) to form the nitropiazselenol (Se-NPD) and this chelate is then extracted into chloroform. Aliquots of the chloroform extract are injected into the GC/MS and the individual ions are measured. The total selenium and specific isotopes, if required, are quantified from the isotope ratios, and the enriched stable isotope spike. The method presented herein contains the latest modifications to the original work.

Instrumentation

In the development work, a Finnigan 4000 quadrupole GC/MS capable of monitoring multiple ions was used. However, this method is not limited to this equipment.[23]

Chromatographic Conditions

Injector, 175°; column, 160°; transfer line, 175°; carrier gas, He (20 ml/min); column, 1.2 m × 2 mm i.d. silanized glass; packing, 3% OV-101 on 100/120 mesh Supelcoport (or equivalent). Sample volume, 5 μl in chloroform.

Reagents

Enriched ^{82}Se was obtained in elemental form from Oak Ridge National Laboratory, Oak Ridge, TN. A standard containing approximately 100 μg/g was prepared in hydrochloric acid. The exact concentration was obtained by a reverse-isotope dilution method.[21] The chelating agent, NPD, was obtained from Aldrich Chemical Company, Milwaukee, WI and was prepared by dissolving 1 g in 100 ml of 10% hydrochloric acid and extracting the excess NPD five times with chloroform. The reagent was then stored in the refrigerator. (*Note:* crystals may form.) Reagent grade nitric acid, phosphoric acid, formic acid, and 30–50% hydrogen peroxide were used. It was found necessary to use Ultrex grade hydrochloric acid in tracer studies because of unidentifiable background interference for the ^{74}Se-NPD peak. Alternatively, extracting reagent grade HCl several times with chloroform also removes the interference. However, monitoring of

[22] D. C. Reamer and C. Veillon, *J. Nutr.* **113,** 786 (1983).

[23] M. J. Christensen (Brigham Young University) and W. R. Wolf (U. S. Department of Agriculture), personal communications.

the ^{74}Se-NPD peak was only necessary in those studies utilizing ^{74}Se as a tracer.

Digestion

The appropriate amount of sample (0.5–10 g or 0.5–20 ml) was weighed or pipetted into 100-ml Kjeldahl flasks containing two glass boiling beads and spiked with a known quantity of ^{82}Se. Five milliliters of concentrated nitric acid and 1 ml of phosphoric acid were added to the sample and allowed to stand at least 1 hr (or overnight) to partially digest the organic material and reduce foaming. The digests were then boiled. When the dark fumes of nitrogen dioxide subsided, hydrogen peroxide (30–50%) was added dropwise from a disposable plastic pipet. Hydrogen peroxide was continually added until the volume was reduced to about 1 ml and no visible nitrogen dioxide fumes were given off. The samples were removed from heat, cooled, 1 ml formic acid added, and then heated until the brown fumes of nitrogen dioxide were given off. The formic acid removed the residual nitric acid by reduction to nitrogen dioxide. Two milliliters of concentrated hydrochloric acid was then added and the samples were boiled gently for 10 min to convert the selenium to the tetravalent state. Ten milliliters of water was added and if lipids were present, the digests were extracted with one or two volumes of 3–5 ml of chloroform (10 min on a wrist action shaker). Derivatization with NPD was carried out by adding 0.2 ml of the NPD reagent and shaking for 10 min. The Se-NPD was extracted with chloroform, dried in a vacuum oven (60°, 1/3 atm), and reconstituted with chloroform immediately prior to injection into the GC/MS.

Tuning

The individual channels of the multiple ion monitor on the GC/MS are adjusted to the specific Se isotopes using a standard preparation from Se-NPD in the natural (unenriched) ratio. This is only needed to assess the tuning of the individual channels.

Discussion

In the hands of the personnel at the Vitamin and Mineral Laboratory, USDA, Beltsville, Maryland, this method proved to be precise and extremely versatile for the analysis of total selenium in a wide variety of matrices. These included many foods, human breast milk, infant formulas, plasma and serum, red blood cells, feces, and urine. The sample preparation and digestion does not use perchloric acid,[24] and hence negates the need for a perchloric acid hood and other safety devices; however it is not as suitable for samples with a high fat content. The use of a stable selenium isotope as an internal standard added prior to digestion eliminates the need for quantitative sample recoveries—always a problem because of the volatility of selenium. There has been some criticism that this digestion procedure, as opposed to the perchloric acid procedure, does not completely digest some organic forms of selenium, but in the analysis of some 5000 samples of varying matrices this has not been found to be so.

The availability and use of the many quality control materials shown in Table I permitted us to use GC/MS in a routine fashion. There is no good quality control material available for feces. Although we were involved in the development of several reference materials (bovine serum and mixed diet),[25,26] we felt there was limited need for a fecal reference material. We did attempt to use BCR CRM 144 sewage sludge. However, because of a high silicon content the matrix proved to be dissimilar to fecal materials. We, therefore, used RM 8431 (mixed diet) on the premise that the matrices are similar. However, the analyte concentrations (especially selenium) for the diet are generally lower than the expected fecal content. We routinely used NBS RM 8419 bovine serum and NBS SRM 1577A bovine liver as a control for red blood cells. However, the IAEA A-13 animal blood could be used.

This method had the accuracy one expects from a definitive method and over a 1.5-year period had an imprecision of about 2% at 100 ng/ml. The absolute detection limit is about 50 pg. As the required sample size ranges from 0.5 g (0.5 ml) to 10 g (10 ml), this requirement effectively removes this method as an effective technique for small samples such as encountered in the analysis of pediatric specimens.

Fluorometry

Introduction

The reader is urged to consult the AOAC Official Methods of Analysis for the fluorometric determination of selenium in foods and plant material. The fluorometric method is well suited for many applications. The only

[24] D. C. Reamer and C. Veillon, *Anal. Chem.* **56,** 605 (1983).

[25] C. Veillon, S. A. Lewis, K. Y. Patterson, W. R. Wolf, J. M. Harnly, J. Versieck, L. Vanballenberghe, R. Cornelis, and T. C. O'Haver, *Anal. Chem.* **57,** 2106 (1985).

[26] N. J. Miller-Ihli and W. R. Wolf, *Anal. Chem.* **58,** 3225 (1986).

equipment and resources needed are a fluorometer and a perchloric acid hood. The organic material is digested in a perchloric acid/nitric acid mixture. The selenium must be converted to the tetravalent state and then reacted with 2,3-diaminonaphthalene (DAN) to form the fluorescent piazselenol. This is extracted under subdued light into a suitable organic solvent. Selenium is then quantified by comparing the fluorescence of the unknowns to that of suitably prepared standards (excitation wavelength = 369 nm, emission wavelength = 525 nm). This method has been automated,[27] however the automated procedure does not automate the digestion which is the labor intensive part of the method.

Instrumentation and Facilities

1. A fluorometer capable of excitation of 369 nm and emission at 525 nm.
2. A perchloric acid hood.

Method

The reader can do no better than to consult the 1984 AOAC, Official Methods of Analysis for details of reagents, digestion, and precise instructions for apparatus, analytical conditions, and procedure. The methods are specifically designed for the analysis of selenium in foods and plants. However, the method has been adapted to determine selenium in a wide variety of matrices, namely, serum, urine, red blood cells, and/or whole blood, milk and infant formulas, feces, and hair.[28-32] The method has also been modified for the analysis of some inorganic matrices such as soils and sediments.[33,34]

Discussion

The fluorometric method has the advantage that it is inexpensive to set up; however it is very labor intensive. It is mandatory that the digestion

[27] M. W. Brown and J. H. Watkinson, *Anal. Chem. Acta* **89,** 29 (1977).
[28] R. C. Ewan, C. A. Bauman, and A. L. Pope, *J. Agric. Food Chem.* **16,** 212 (1968).
[29] N. L. Zabel, J. Harland, A. T. Gormican, and H. E. Ganther, *Am. J. Clin. Nutr.* **31,** 850 (1978).
[30] A. Geahchan and P. Chambon, *Clin. Chem.* **26,** 1272 (1980).
[31] J. A. Watkinson, *Anal. Chem.* **38,** 92 (1966).
[32] Analytical Methods Committee, *Analyst (London)* **104,** 778 (1979).
[33] "The Fluorometric Determination of Selenium—A Literature Review." Turner Associates, 1972. Turner Associates, Palo Alto, CA.
[34] L. Lalonde, Y. Jean, K. D. Roberts, A. Chapdelaine, and G. Bleau, *Clin. Chem.* **28,** 172 (1982).

procedure be quantitative—sometimes a difficult task due to the volatility of selenium. It requires the use of perchloric acid in the digestion mixture necessitating the use of a perchloric acid hood. Again, it is imperative to ensure that all of the selenium is in the tetravalent state prior to reaction with DAN. Reamer and co-workers developed a digestion procedure using a nitric acid/phosphoric acid mixture and this procedure can be used instead of the perchloric acid digestion for the fluorometric determination of selenium. However, this digestion procedure is not good for samples with high lipid content and there can be a positive fluorometric interference from the lipids.[35] There are several disadvantages to the fluorometric method. The glassware, etc., used must be rigorously cleaned to avoid contamination with other fluorescing compounds. This method is not very good for very low concentrations of selenium. Detection limits of 10–100 ng have been reported[10,28,32] and imprecisions range from 2 to 100%.[28,32] However, the consensus of opinion is that above 100 ng/g the percentage RSD should be better than 8%. Recoveries range from 75 to 110%,[28,32] and are very dependent on the concentration and the complexity of the matrix. Again, many authors do not report detection limits, imprecisions, and recoveries, or the consistent use of reference materials.

In the proper hands, with close attention to detail, and the enforcement of a good quality assurance program, the fluorometric method for selenium determinations is probably the simplest, least expensive, and most versatile method available. However the limitations of imprecision and detection limits must be realized.

Atomic Absorption Spectrometry (AAS)

The development of AAS in 1955 by Walsh[36] brought about a revolution in the field of trace element analysis. Flame AAS can only be used for samples containing high concentrations of selenium as the detection limit varies from 0.05 to 3 mg/liter. Generation of volatile selenium hydride coupled to an atomic absorption spectrometer, and electrothermal atomic absorption spectrometry (especially with graphite furnace atomization) can both be used to determine selenium at a much lower concentration.

Hydride generation atomic absorption spectrometry (HG-AAS) offers the advantage of good sensitivity and relatively simple instrumentation for the determination of selenium in a wide variety of matrices. In HG-AAS, sodium tetraborohydride is added to an acidic solution of the sample containing selenium in the tetravalent state forming the gaseous selenium hydride, which is then stripped from solution with a gas and

[35] D. C. Reamer and C. Veillon, *Anal. Chem.* **54,** 1605 (1983).
[36] A. Walsh, *Spectrochim. Acta.* **7,** 108 (1955).

atomized—generally in a heated quartz cell. There are reports that the accuracy of this method is very dependent on the decomposition technique used. Welz and co-workers[37–39] have investigated the problems of sample digestion and have shown that provided adequate care is taken to ensure complete sample digestion and conversion of the selenium to selenite, this technique can be accurate. Siemer and Koteel[40] have examined different HG-AAS techniques and mention methods of optimizing systems to obtain maximum sensitivity. Reamer and co-workers[41] used radiotracers to evaluate losses in different hydride generation systems.

The HG-AAS approach has been used to determine selenium in waters, blood and blood products, urine, environmental samples, foods, and feeds. It is reported not to be a good technique for the analysis of all biological samples for selenium.[10]

The HG-AAS technique has been shown to have absolute detection limits ranging from 0.1 to 60 ng.[14] Imprecision and recovery vary according to care and expertise of the user but have been reported to be 1–5% RSD and 70–110%, respectively.

Graphite Atomic Absorption Spectrometry (GFAAS)

Graphite furnace atomic absorption spectrometry would seem to be the most appropriate technique to analyze samples for small amounts of selenium. As many semiautomated systems are now available, this technique could be used for the rapid direct determination of selenium in large batches of routine samples. However, GFAAS is not simple nor free from interferences—both spectral and matrix—and/or losses due to the volatility of selenium. The use of pyrolytically coated or total pyrolytic graphite tubes, a L'vov platform, peak area integration, matrix modification, and Zeeman background techniques (ZAAS) have enabled some advances to be made. Spectral interferences from iron and/or phosphate are compensated for by Zeeman background correction. The use of matrix modification—generally nitrates of Ni, Cu, Pd, Pt, Ag, Mo, and/or Mg—help to thermally stabilize selenium and allow for ashing temperatures of up to 1200°.[10,14,42–44] However, only a few sample types are suitable for

[37] B. Welz, M. Melcher, and G. Schlemmer, *in* "Trace Element Analytical Chemistry in Medicine and Biology," Vol. 3. de Gruyter, Berlin, Federal Republic of Germany, 1984.

[38] B. Welz, M. Melcher, and J. Neve, *Anal. Chem. Acta* **165,** 131 (1984).

[39] B. Welz and M. Verlinden, submitted for publication.

[40] D. Siemer and P. Koteel, *Anal. Chem.* **49,** 1096 (1977).

[41] D. C. Reamer, C. Veillon, and P. T. Tokousbalides, *Anal. Chem.* **53,** 245 (1981).

[42] W. Slavin, G. R. Carnrick, D. C. Manning, and E. Pruszkowska, *At. Spectrosc.* **4,** 69 (1983).

[43] R. D. Ediger, *At. Absorpt. Newsl.* **14,** 127 (1975).

[44] F. J. Fernandez, S. A. Myers, and W. Slavin, *Anal. Chem.* **52,** 721 (1980).

direct analyses even with the use of matrix modification. Blood, serum, semen, yeast, and food supplements and waters have been analyzed directly.[10,14,42] A ZAAS method was shown to correlate well with the definitive IDMS method ($r = 0.987$) for the analysis of 30 plasma samples for selenium, but was less precise.[45] Many commonly used digestion procedures and acids have been reported to cause matrix interferences for GFAAS.[46] Pretreatments involving the formation of selenium chelates and subsequent analysis by GFAAS again raise the question of quantitative sample recoveries. Contamination, generally a problem with many GFAAS techniques, is of less concern for selenium. GFAAS has good detection limits (20–50 pg) and reported imprecisions ranging from 2 to 15% RSD.

Gas Liquid Chromatography (GLC)

The determination of selenium by GLC is usually based on the detection of a volatile piazselenol formed by the reaction of selenite, after destruction of the organic matrix, with an aromatic diamine, like 4-nitro-*o*-phenylenediamine. This again presents the problem of quantitative sample recoveries and conversion to the tetravalent state. Various detectors[10] have been used with this system, including flame ionization, thermal conductivity, atomic absorption spectrometry, mass spectrometry (see mass spectrometry method), and electron capture detectors (ECD).[47,48] When using GC/ECD, the absolute detection limit can be as low as 1 pg. Imprecisions are reported to be from 2 to 4% RSD.[48]

This method has been adapted for the determination of selenium in the following matrices: water, blood, and blood products, grain, fish, milk, hair, urine, and soft tissues.[10,47,48] The method has also been used to detect selenium species in biological materials and in waters.[9]

Summary

The definitive IDMS method has the advantage of good detection limits and precision. The advent of the new, smaller, easy-to-use GC/MS system should make this method more popular. The fluorometric method remains the method of choice for the determination of selenium in many matrices if the limitations of the method are understood.

The newer methods include gas chromatography, HG-AAS, and GFAAS. The gas chromatography methods show promise because of the

[45] W. Slavin and D. C. Manning, *Prog. Anal. At. Spectrosc.* **5,** 243 (1982).
[46] S. A. Lewis, N. S. Hardison, and C. Veillon, *Anal. Chem.* **58,** 1272 (1986).
[47] T. P. McCarthy, B. Brodie, J. A. Milner, and R. F. Bevill, *J. Chromatogr.* **225,** 9 (1981).
[48] C. J. Cappon and J. C. Smith, *J. Anal. Toxicol.* **2,** 114 (1978).

realtively simple instrumentation needed and the fact that the analyte is separated from the matrix. HG-AAS offers good sensitivity provided care is taken to ensure complete sample digestion and conversion of selenium to selenite. The advent of Zeeman background correction systems for GFAAS has greatly facilitated selenium determinations, particularly in biological matrices where iron and phosphorus are also present.

The reference materials now available, used as part of a quality assurance program, should help to ensure accurate determinations, permit method validation, and allow performance evaluation in interlaboratory trials.

[32] Vanadium

By DONNA M. MARTIN and N. DENNIS CHASTEEN

Introduction

Studies of the metabolism, biochemistry, clinical pathology, and environmental toxicology of vanadium require reliable analytical techniques for the determination of this element in biological materials. The ideal technique for vanadium analysis would permit trace amounts to be determined rapidly and economically but with sufficient sensitivity to detect normal and subnormal levels. Because vanadium concentrations in biological materials can vary widely, all these requirements cannot be met with a single analytical method.

In the present chapter, we survey the most common techniques for determining vanadium and discuss some of the advantages and disadvantages of each. Often the instrumentation at hand dictates the analytical method to be employed. While most of this chapter is devoted to analysis of tissues and fluids, the methods are readily adapted to other types of samples. To our knowledge, methods for vanadium analysis of biomaterials have not been reviewed previously. Since much of the earlier work on vanadium analysis, particularly at the sub parts per million level, is of questionable reliability, this review focuses on the more recent literature.

Atomic Absorption Spectrometry (AAS)

Graphite furnace atomic absorption is the most conventional and economical technique for analyzing vanadium in biological samples in the pbb range. Analysis requires only a small amount of sample and matrix matching due to spectral interferences is usually not a problem. The instrumentation is commonplace. Accordingly, we discuss in some detail the protocols for vanadium analysis by AAS.

Pyrolytically coated graphite tubes are used in AAS of vanadium since the absorption signal is improved nearly 5-fold compared to uncoated tubes.[1] This enhancement is probably due to reduced vanadium penetration of the graphite tube, avoiding the formation of undissociable vanadium carbides. Wendl and Müller-Vogt studied the reactions of vanadium in the graphite furnace using absorption measurements, electron microscopy, and X-ray diffraction techniques.[1] They concluded that an ashing temperature of 1300 K is optimum for decomposition of complex matrices without loss in sensitivity. In this temperature region, vanadium forms carbides which decompose into the element during atomization.

The presence of nitrate ions in acidic solutions strongly affects the precision and sensitivity of AAS.[2] Buchet *et al.*[3] employ cupferron to chelate vanadium in wet ashed urine, extracting the resultant complex into 4-methylpentan-2-one (MIBK). In this way, interferences from the nitric acid digest are avoided and the lifetime of the pyrolytic coating is extended. The following procedure allows precise determination of vanadium in urine in the 0.1–500 ppb range. Urine (100 ml) is evaporated to dryness on a hot plate at 80–90°. Ten milliliters of 65% nitric acid is added and, after the reaction has subsided, the acid solution is refluxed for 24 hr at 120–130° and again brought to dryness. The dried residue is dissolved in demineralized water and the pH adjusted to 2–2.5 by adding 0.5 *M* HCl. The solution is then diluted to 20 ml and a 10-ml aliquot is mixed with 1 ml of aqueous 1% w/v of cupferron. The complex is extracted by shaking for 10 min with 2 ml of MIBK saturated with water at pH 2–2.5. A standard stock solution (19.63 m*M*) is prepared by dissolving 1 g vanadium(V) oxide in 1 liter of 0.5 *M* NaOH. Working standards (100 μg V/liter) are prepared by dilution of the stock solution with demineralized water. Absorption measurements are made on 50 μl of sample which is injected into the furnace within 1 hr of the extraction. Buchet *et al.* obtained a linear response up to a concentration of 60 ppb with a coefficient of variation of 7.5% at the 10 ppb level.[3] The detection limit was 0.1 ppb. For the 11 subjects studied, a concentration range of 0.3–0.7 ppb vanadium in urine was determined.

Another solvent extraction procedure has been developed by Ishizaki and Ueno.[4] After wet digestion of the sample, vanadium is chelated with

[1] W. Wendl and G. Müller-Vogt, *Spectrochim. Acta* **40B,** 527 (1985).
[2] E. M. Sutter and M. J. F. Leroy, *Anal. Chim. Acta* **96,** 243 (1978).
[3] J. P. Buchet, E. Knepper, and R. Lauwerys, *Anal. Chim. Acta* **136,** 243 (1982).
[4] M. Ishizaki and S. Ueno, *Talanta* **26,** 523 (1979).

N-cinnamoyl-*N*-2,3-xylylhydroxylamine (CXA) and extracted into carbon tetrachloride from 6 *N* HCl. The samples are digested with nitric acid and perchloric acid, diluted to 30 ml with water, and transferred to a separatory funnel. To maintain vanadium in the +5 oxidation state, 0.02 *M* $KMnO_4$ is added dropwise until a pink color persists for 5 min. One milliliter of 1% CXA in CCl_4 and 30 ml of concentrated HCl are added. The funnel is agitated for 3 min and 20 μl of the extract is introduced into the furnace. The sample is dried by raising the current from 0 to 25 A at a 1 A/sec rate (final temperature 100°), ashed at 150 A (1850°) for 20 sec, and atomized at 300 A (2800°) for 10 sec. Argon is used as the purge gas at a flow rate of 2.6 liters/min. After the addition of HCl, the extraction must be done quickly since partial reduction of vanadium(V) to vanadium(IV) occurs within 2 min. Vanadium(IV) does not complex with the CXA. Vanadium(V) standards used for recovery experiments were freshly prepared by dissolving 0.148 g of ammonium metavanadate in a minimum volume of ammonia followed by dilution to 500 ml with water.

Table I illustrates the results obtained by Ishizaki and Ueno for various types of biological samples.[4] The average recovery was 94% with a coefficient of variation ranging from 3.1 to 13.7% depending upon the sample. The sensitivity at 1% absorption was estimated to be 0.04 ng with a detection limit of 0.1 ng/g. In a study of possible interfering substances, it was found that 50 ng of vanadium could be determined in the presence of a large excess of foreign ions, namely 40 mg each of Ca(II), Mg(II), and Zn(II), 20 mg of phosphate, 100 μg of Al(III), Cu(II), Ni(II), and Sn(IV), and 10 μg of As(V), Ba(II), Cd(II), Co(II), Cr(IV), Hg(II), Mo(VI), Pb(II), Sb(V), Se(IV), and Ti(IV). This method, if substantiated by a referee method, offers a straightforward procedure for measuring trace amounts of vanadium in a wide range of biomaterials.

Mousty *et al.* evaluated a procedure for the routine determination of vanadium in human tissues and biological fluids.[5] Radiotracer experiments using ^{48}V were performed to evaluate the retention and vaporization in the graphite furnace. Neutron activation was used as a reference technique for the determination of vanadium in urine samples of oil-fired power plant workers.

Urine was mineralized in a Teflon bomb, preconcentrated on a Chelex 100 column, and eluted from the resin. Recovery experiments during different steps of the procedure were carried out using urine from rats injected with 1 μg of ^{48}V label. The eluates were divided into two aliquots and were analyzed by AAS and neutron activation analysis (NAA). The standard addition method was used in AAS to avoid possible interfer-

[5] F. Mousty, N. Omenetto, R. Pietra, and E. Sabbioni, *Analyst* (*London*) **109,** 1451 (1984).

TABLE I
VANADIUM IN VARIOUS BIOLOGICAL SAMPLES USING AAS[a]

Sample (weight)	V added (ng)	Mean (ng)	CV^b (%)	Recovery (%)
Apple (5 g)	0	37.4	4.5	—
	30	67.9	3.2	101.7
Parsley (1 g)	0	111	9.2	—
	100	204	4.0	93.0
Potato (1 g)	0	25.1	13.7	—
	20	44.5	4.0	97.0
Undaria pinnatifida (1 g)	0	40.3	3.6	—
	50	87.7	3.1	94.8
Jack mackerel (2 g)	0	30.1	5.8	—
	100	131	4.4	100.9
Pacific saury (1 g)	0	26.1	12.7	—
	100	122	4.6	95.9
Sardine (2 g)	0	246	5.4	—
	100	337	1.6	91.0
Pig liver (10 g)	0	60	4.4	—
	50	107	5.5	94.8
Pig kidney (10 g)	0	48	4.7	—
	50	97	7.9	97.8
Human blood (10 ml)	0	4.7	4.3	—
	10	15.4	9.6	107.0
Human urine (50 ml)	0	12.9	11.7	—
	10	22.9	7.6	100.0
Human hair (5 g)	0	770	6.3	—
	0.500	1250	5.4	96.0

[a] Reproduced with permission from Ishizaki and Ueno.[4]
[b] Coefficient of variation.

ences from Fe, Ca, and phosphate. The reader is referred to the article for details on the procedures and instrumentation used.[5] Good agreement between the two methods were obtained (Table II). An absolute sensitivity of 0.03 ng (for 1% absorption) was found using AAS with an internal flow rate of 50 ml/min and 50 μl of sample in HNO_3. The relative standard deviation of this analysis was about 10%. It should be noted that the mean value of the control group (3 ng/ml) was higher than the level typical of unexposed subjects (0.5–1.5 ng/ml). The authors propose this discrepancy could be due to the fact the control group samples were also taken at the power plant on subjects that were perhaps also exposed to vanadium to some extent. In addition, a few samples showed vanadium contents significantly higher by AAS than by the NAA method, probably due to contamination (Table II).

TABLE II
DETERMINATION OF VANADIUM IN HUMAN URINE BY NEUTRON ACTIVATION ANALYSIS AND ATOMIC ABSORPTION SPECTROSCOPY[a]

Group	Urine sample number	Vanadium content (ng/ml)	
		NAA	AAS
Control group	1	0.76	0.7
	2	3.15	3.1
	3	2.00	3.3
	4	3.30	3.0
	5	4.32	4.3
Exposed group	1	3.68	4.0
	2	4.76	12.0
	3	6.75	7.0
	4	16.25	29.0
	5	21.70	21.5

[a] Reproduced with permission from Mousty *et al.*[5]

Neutron Activation Analysis (NAA)

NAA is the most sensitive and reliable method for the determination of vanadium below the ppb level. Irradiation with thermal neutrons produces the radioactive isotope ^{52}V by the nuclear reaction $^{51}V(n,\gamma)^{52}V$. The resultant principal γ ray is detected at 1.434 MeV. The neutron capture cross section (σ) is relatively large (4.8 barns).

When vanadium is present in trace amounts, chemical separation is necessary in order to eliminate the interferences from the presence of ^{24}Na, ^{38}Cl, and ^{80}Br. The concentration range of vanadium dictates whether a preirradiation or postirradiation chemical separation is used. Separation prior to neutron activation gives rise to higher sensitivity since counting can begin almost immediately after irradiation. However, in this case, sample handling before analysis introduces the risk of contamination, necessitating the use of blanks.

A postirradiation chemical separation scheme minimizes sample contamination, thus eliminating the need for blanks; however, due to the short half-life of ^{52}V ($t_{1/2}$ 3.76 min), rapid separation is required. To minimize blank values, polyethylene irradiation vials and Teflon beakers are leached with hot HNO_3 before use and the same minimum volumes for digestion are used. Blank values range typically from 2 to 8 ng vanadium, depending upon the sample size and nature of the matrix.[6] When working at 0.1 ppm vanadium and higher, such blank levels usually can be tolerated and a preirradiation separation can be used.

Marine biological samples such as shrimp, crab, and oyster taken from industrialized and nonindustrialized areas have been analyzed using a preirradiation technique.[7] The values ranged from 0.07 to 1.64 μg vanadium/g dry weight, with the higher values found in samples taken near industrialized areas. It should be noted, however, that simpler and more economical techniques such as AAS could be used for such samples at these levels of vanadium.

When vanadium is present at ppb concentrations, a postirradiation separation must be used to obtain reliable results. The most success has been obtained with an irradiation–ashing–extraction scheme.[8-10] In this scheme, the matrix is rapidly ashed to a form which can be quantitatively extracted without loss of the analyte. Wet-ashing with concentrated acids is preferred over dry-ashing since the latter may result in loss of vanadium due to its low volatility in certain organic complexes such as porphyrins. Accordingly, dry-ashing procedures must be checked for loss of vanadium. The extraction efficiency is assessed using a ^{48}V tracer which is an emitter at 0.986 MeV. The addition of a tracer is not necessary for every sample once a quantitative and reproducible recovery scheme is established. This is advantageous since ^{48}V is an expensive isotope and has a half-life of only 16 days. The collection and handling of samples demand caution and the reader should consult the original articles for the techniques used.[8-11]

Byrne and Kosta[10] developed a rapid, postirradiation chemical separation scheme using *N*-benzoyl-*N*-phenylhydroxylamine (BPHA) as a chelating agent for extraction of vanadium(V) from a 5 M HCl–1.8 M H_2SO_4 medium. The separation requires 12–14 min for completion with an average extraction yield of 96.3 $\pm$1.2% based on 10 determinations of seven different samples. The new method was checked by using alternative techniques and the role of Mn and Cr as possible nuclear interferents from the reactions $^{52}Cr(n,p)^{52}V$ and $^{55}Mn(n,\alpha)^{52}V$ was investigated.

The sample is placed in a polyethylene tube and doubly wrapped in polythene foil to avoid possible contamination from the pneumatic trans-

[6] R. O. Allen and E. Steinnes, *Anal. Chem.* **50,** 1553 (1978).

[7] A. J. Blotcky, C. Falcone, V. A. Medina, E. P. Rack, and D. W. Hobson, *Anal. Chem.* **51,** 178 (1979).

[8] M. Simonoff, Y. Liabador, A. M. Peers,and G. N. Simonoff, *Clin. Chem.* **30,** 1700 (1984).

[9] R. Cornelius, J. Versieck, L. Mees, J. Hoste, and F. Barbier, *J. Radioanal. Chem.* **55,** 35 (1980).

[10] A. R. Byrne and L. Kosta, *J. Radioanal. Chem.* **44,** 247 (1978).

[11] K. Heydorn, E. Damsgaard, and B. Rietz, *Anal. Chem.* **52,** 1045 (1980).

fer system. The sample is irradiated for 5 min at a thermal neutron flux of 4.5×10^{12} n/cm^2-sec. After irradiation the sample is quantitatively transferred to a 100-ml long-necked quartz Kjeldahl flask containing 2 ml concentrated sulfuric acid and 200 μg V(V) carrier. The sample container is then washed with 5 ml of concentrated nitric acid which is also transferred to the flask. The flask is then heated under two gas flames while being agitated. More HNO_3 is added when the solution begins to char; this procedure is repeated until a yellow solution is obtained. The beaker is cooled, 30% H_2O_2 is added dropwise, and the solution reheated. If necessary, more H_2O_2 is added followed by heating until the solution is colorless. The flask is cooled by immersion in a waterbath, and a few ml of H_2O is added followed by boiling to eliminate excess peroxide. At this point the volume of solution should be about 4–7 ml. The contents are poured into a separatory funnel and the flask washed successively with 10 ml of 10 *M* HCl followed by a few ml H_2O. The washings are likewise transferred to the separatory funnel. Water is added to obtain a total volume of 19 ml followed by 1 ml 0.1 *N* $KMnO_4$ oxidizing agent. After shaking for 30 sec the solution is extracted with 10 ml BPHA (0.1% in toluene) for 30 sec. The organic phase is scrubbed with 10 ml 5 *M* HCl for 15 sec.

Table III lists the results of Byrne and Kosta[10] along with literature values for the standard samples. The values obtained by Byrne and Kosta have acceptable standard deviations. Generally there is good agreement among workers for the vanadium content of Bowen's kale and NBS bovine liver. However, a wide range of values has been reported for NBS orchard leaves (Table III). Heydorn *et al.*[11] investigated different decomposition procedures using separation before and after irradiation. Their findings show that part of the vanadium in orchard leaves is present in an acid-insoluble form associated with siliceous material, accounting for the discrepancies found in Table III. Moreover, in recovery experiments using ^{48}V as a tracer, only some of the insoluble vanadium is exchanged, depending on the duration of contact with the tracer solution and perhaps upon surface area of the solid. Therefore the suitability of using NBS-1571 orchard leaves as a standard reference material is questionable.

Cornelius *et al.*[9] determined vanadium in human serum in 22 male and 17 female subjects. One gram of lyophilized serum contained in a 30 mm high, 14-mm-diameter ultraclean Spectrosil quartz vial was ashed in a Simon-Muller oven. The temperature was held at 100° for 1 hr and then gradually increased to 450° where it was maintained for 24 hr. Upon cooling, concentrated HNO_3 was added and the ash heated to 450° for 24 hr. Another aliquot of acid was added followed by heating at 150° for 2 hr. The quartz vials were sealed with a polythene lid. Ashing temperatures

TABLE III
VANADIUM CONTENT OF BIOLOGICAL STANDARD OR PROPOSED STANDARD REFERENCE MATERIALS

Sample	Vanadium ppm dry weight for (*n*) determinations
Bowen's kale	0.376 ± 0.013 (5)[a]
	0.414 ± 0.013 (16)[a]
	0.337 ± 0.039 (5)[a]
	0.332 ± 0.011 (9)[a]
	0.375 ± 0.071 (3)[a]
	0.37[a]
NBS bovine liver (SRM 1577)	0.0586 ± 0.016 (5)[a]
	0.04[a]
	0.0615 ± 0.002[b]
	0.056 ± 0.007 (5)[a]
	0.02[a]
NBS orchard leaves (SRM 1571)	0.471 ± 0.014 (9)[a]
	0.622 ± 0.023 (7)[a]
	0.614 ± 0.077 (3)[a]
	0.4 ± 0.1[a]
	0.64 ± 0.31 (6)[a]
	0.64 ± 0.10[c]
	0.54[c]
	0.68 ± 0.14 (3)[d]
	0.409 ± 0.041 (7)[a]
	0.58[a]
	0.7[a]
	0.9 ± 0.2[a]
	0.361 ± 0.09[a]
	0.6 ± 0.15[c]
	0.61[c]
NBS bovine liver (SRM 8419)	1.4 ± 0.7 μg/liter[e,f]
	4.0 μg/liter[f]
	1.2 μg/liter[f]
NBS oyster tissue (SRM 1566)	2.316 ± 0.006 (6)[g]
NBS citrus leaves (SRM 1572)	0.245 ± 0.005 (6)[g]
NBS bovine liver (SRM 1577a)	0.0987 ± 0.002 (5)[g]
NBS human serum (SRM 909)	0.00263 ± 0.00031 (7)[g]
NBS spinach (SRM 1570)	1.13 ± 0.01 (4)[a]
	1.5 ± 0.2[c]
	1.2 ± 0.1[c]
	1.20 ± 0.06 (5)[a]
	1.3 ± 0.3[c]

(*continued*)

TABLE III (*continued*)

Sample	Vanadium ppm dry weight for (*n*) determinations	
NBS tomato leaves (SRM 1573)	1.27	± 0.035 (5)[a]
	1.3[c]	
	1.5	± 0.2[c]
	1.27	± 0.03[c]
IAEA animal muscle H-4 (1975)	0.0038	± 0.0007 (5)[a]
	0.00276	± 0.0021 (7)[b]
NBS pine needles (SRM 1575)	0.346	± 0.018 (5)[a]
BCR (water pest) RM No. 60	5.2	± 0.6[c]
BCR (water moss) RM No. 61	12.7	± 1.0[c]
BCR (olive leaves) RM No. 62	0.93	± 0.11[c]
IAEA potato V-4 (1972)	0.0094	± 0.0006 (5)[a]
IAEA wheat flour V-2/1 (1972)	0.0438	± 0.0035 (5)[a]
IAEA animal blood A-2 (1972)	0.414	± 0.012 (5)[a]
IAEA milk powder A-8 (1972)	0.0195	± 0.002 (6)[a]

[a] A. R. Byrne and L. Kosta, *J. Radioanal. Chem.* **44,** 247 (1978).
[b] E. Sabbioni, E. Marafante, and R. Pietra, *in* "Nuclear Activation Techniques in the Life Sciences 1978," p. 167. IAEA, Vienna, Austria, 1979.
[c] P. Schramel and X. Li-qiang, *Fresenius Z. Anal. Chem.* **314,** 671 (1983).
[d] L. Vos and R. Van Grieken, *Int. J. Mass. Spectrom. Ion Phys.* **47,** 303 (1983).
[e] S. A. Lewis, T. C. O'Haver, and J. M. Harnly, *Anal. Chem.* **57,** 2 (1985).
[f] C. Veillon, S. A. Lewis, K. Y. Patterson, W. P. R. Wolf, J. D. M. Harnly, J. Versieck, L. Vanballenberghe, R. Cornelis, and T. C. O'Haver, *Anal. Chem.* **57,** 2106 (1985).
[g] J. D. Fassett and H. M. Kingston, *Anal. Chem.* **57,** 2474 (1985).

higher than 450° were not used since etching of the quartz vial was observed. Samples were irradiated for 8 min at neutron flux of 8.0×10^{13} n/cm^2-sec in the presence of a Ni flux monitor. A 1 g vanadium standard was also irradiated and the conversion factor between the ^{52}V activity and the Ni flux monitor was determined. The postirradiation separation was then achieved by dissolving the ash in 2.5 ml of hot 3 *M* HCl and rinsed twice with the same solution and transferred into a separatory funnel containing

15 ml H_2O, 1 mg vanadium carrier, and the radio tracer ^{48}V. Fresh 6% cupferron (2.5 ml) was then added followed by 10 ml $CHCl_3$. After agitating the mixture for 30 sec, the organic phase was placed in a counting vial. The $^{48,52}V$ activities were counted for 5 min with a Ge(Li) detector coupled to a 1024 multichannel analyzer. The analysis of two packed blood cell samples indicated that approximately 68% of the total vanadium in blood is present in serum. The mean concentration was 0.33 ±0.01 ppb. These values are the lowest V concentrations for serum reported in the literature.

Another recent investigation of vanadium in human serum by NAA was done by Simonoff *et al.*[8] Nonlyophilized serum was dried for 24 hr at 60° and then ashed for 8 hr at 300° followed by 4 hr at 500°. The product was transferred into a quartz tube (70 mm high × 20 mm in diameter) using three 1-ml portions of 3 *M* HCl. The tube was heated at 100° for 24 hr then at 200° for 3 hr. The quartz tubes were placed in a sealed aluminum container and irradiated for 1–2 min with a neutron flux of 2×10^{14} n/cm^2-sec. A second quartz tube containing 50 ng of vanadium was simultaneously irradiated for normalization purposes. The chemical separation procedure was the same as that of Cornelius *et al.*[9] The vanadium carrier was NH_4VO_3; $^{48}VO_2Cl$ was used to calculate the extraction yield. The activities were also counted on a Ge(Li) detector. The results of this laboratory, 0.26–1.3 ppb, overlap those of Cornelius and co-workers.[9] Simonoff *et al.* state that their method appears five times more sensitive due to improvements in detector characteristics, increased neutron flux and serum weight.[8]

Multielement NAA techniques are not discussed here. Vanadium determination by multielement analysis has included samples such as bone,[12,13] rat brain tissue,[14] biological and botanical materials,[15] amniotic fluid,[16] biological standard material,[17,18] and botanical and zoological standard reference materials.[19]

Some multielement work on biological samples has been done using

[12] G. J. Willems, R. A. Palmans, J. Colard, and P. Ducheyne, *Analysis* **12,** 443 (1984).

[13] H. Bem and D. E. Ryan, *Anal. Chim. Acta* **135,** 129 (1982).

[14] A. W. K. Chan, M. J. Minski, and J. C. K. Lai, *J. Neurosci. Methods* **7,** 317 (1983).

[15] S. H. Harrison, *NBS Spec. Publ. (U.S.)* **492,** 15 (1977).

[16] N. Ward, D. Bryce-Smith, M. Minski, J. Zaayman, and B. Pim, *Trace Elem. Anal. Chem. Med. Biol., Proc. Int. Workshop* **2,** 483 (1983).

[17] A. R. Byrne, *Radiochem. Radioanal. Lett.* **52,** 99 (1982).

[18] J. R. W. Woittiez and H. A. Das, *Trace Elem. Anal. Chem. Med. Biol., Proc. Int. Workshop* **1,** 701 (1980).

[19] K. N. Desilva and A. Chatt, *J. Trace Microprobe Tech.* **1,** 307, (1982–1983).

particle induced X-ray emission (PIXE).[20–24a] The sample is bombarded by a proton beam; characteristic X rays corresponding to the elements are then detected. To date, PIXE offers no practical advantages over neutron activation in analyzing for vanadium in biomaterials.

Inductively Coupled Plasma (ICP)

All metals can be determined at the ultratrace level on microgram samples using a plasma excitation source.[25] Because of the high dynamic concentration range of the ICP method, determinations can be made from the percentage to the μg/g level and below. An advantage to using inductively coupled plasma, in comparison to other methods such as spark source mass spectrometry (SSMS), dc atomic emission spectroscopy (DC-AES), and X-ray fluorescence (XRF), is that the requirement of matrix matching need not be stringent. Calibration standards are prepared from high purity chemicals.

Schramel and Liqiang describe a method in which NBS-1571 orchard leaves was used as a multielement standard to detect 14 elements.[26] Two NBS (National Bureau of Standards) and three BCR (Community Bureau of Reference of the European Communities) samples were analyzed: NBS-1570 spinach, NBS-1573 tomato leaves, BCR 60 water pest, BCR 61 water moss, and BCR 62 olive leaves.[26] Following a wet digestion procedure,[26] elements present in low concentrations, such as Cr, Ni, Cd, and V, were spiked (1.2 μg V) into the digested orchard leaves to provide a wide concentration range for the calibration curves. The measured intensities obtained for the working curve were fitted to a linear or a second-order polynomial regression equation. The recalibration standards were used to validate the calibration curve in later measurements. In this way the intensities of the samples were normalized before the calibration coefficients were applied. Unless an interelement interference correction from Mn was performed, vanadium could not be determined. Aside from stray light effects, the interference was due to overlapping of the nearby the Mn line with the 311.071 nm V line.

[20] W. Maenhaut, L. De Reu, and V. Tomza, *Anal. Chim. Acta* **136,** 301 (1982).
[21] A. J. J. Bos, R. D. Vis, F. Van Langevelde, F. Ullings, and H. Verheul, *Nucl. Instrum. Methods* **197,** 139 (1982).
[22] W. Maenhaut and L. De Reu, *IEEE Trans. Nucl. Sci.* **NS-28,** 1386 (1981).
[23] D. D. Cohen, E. Clayton, and T. Ainsworth, *Nucl. Instrum. Methods* **188,** 203 (1981).
[24] E. Rokita, W. Maenhaut, and J. Cafmeyer, *IEEE Trans. Nucl. Sci.* **N3-30,** 1601 (1983).
[24a] R. D. Vis, C. C. A. H. Van der Stap, and A. J. J. Bos, *Turun Yliopiston Julk., Sar. D,* **17** 108 (1984); CA 104:2982m.
[25] V. A. Fasseland and R. N. Kniseky, *Anal. Chem.* **46,** 1110A (1974).
[26] P. Schramel and X. Liqiang, *Fresenius Z. Anal. Chem.* **314,** 671 (1983).

The values for the NBS tomato leaves and spinach are in good agreement with values from different laboratories as shown in Table III. In both cases, however, the values reported using neutron activation are slightly lower. More independent studies will have to be performed on the BCR and IAEA (International Atomic Energy Agency) reference materials since insufficient data are available to make a comparison. Because a significant portion of the vanadium in SRM-1571 orchard leaves is present in an acid-insoluble siliceous form,[11] which is discarded before measurement, the values obtained using the above preparation scheme largely reflect the vanadium content of the soluble portion of botanical materials.

Jones and O'Haver studied the effects of pH and digestion conditions on the separation of trace elements from tissue digests prior to ICP-AES analysis.[26a] The results show that vanadium recovery appears to be digestion dependent with nitric–perchloric–sulfuric acid digestion giving the best recovery. Additional studies are needed to investigate the complexation and release of vanadium on Chelex 100 resin.

Barnes *et al.* have studied trace element analysis of digested bone,[27] urine,[28] and serum[29] using a polydithiocarbamate resin,[28] which strongly complexes Cd, Cu, Mo, Ni, Ti, and V at pH 5. Because the resin does not complex alkali and alkaline earths, the stray light and background from these elements common to biological matrices are eliminated. In a recent paper by Mianzhi and Barnes,[29] resin complexation and element recoveries were measured as a function of pH for acid digested serum. In this study, commercially prepared bovine liver (0.2–0.5 g) or human serum (2–5 ml) were wet-ashed in a Teflon bomb and the residue was then diluted to 50 ml with distilled water.

Changes in pH which occur as the sample solution passes through the column can seriously affect the resin recovery. To ensure a constant pH, 2 ml of pH 4.5 sodium acetate–acetic acid buffer solution was added beforehand. These solutions were passed through the column containing 50 mg of resin. The resin was then digested upon addition of 2 ml of H_2O_2 and 5 ml of HNO_3 and the concentration of each element was determined by ICP-AES. The limit of detection for vanadium (concentration giving a signal equal to three times the standard deviation of the backround) for the 309.3 nm line was reported to be 9.6 ng/ml at 0.9 kW power and a 1.0 liters/min gas flow rate. The values obtained for the serum were high compared to literature values. This discrepancy is probably due to contamination which can occur during commercial preparation of the samples

[26a] J. W. Jones and T. C. O'Haver, *Spectrochim. Acta* **40B,** 263 (1985).
[27] H. S. Mahanti and R. M. Barnes, *Anal. Chim. Acta* **151,** 409 (1983).
[28] R. M. Barnes, P. Fodor, K. Inagaki, and M. Fodor, *Spectrochim. Acta* **38B,** 245 (1983).
[29] Z. Mianzhi and R. M. Barnes, *Appl. Spectrosc.* **38,** 635 (1984).

which were chosen for the analysis. However, V in native serum samples could not be detected by this method since the concentration of vanadium in serum is on the order of 0.3 ppb, which is lower than the detection limit of 9.6 ppb reported.

ICP-AES offers several advantages over AAS; it is a rapid, reliable, multielement technique which can be employed over a wider concentration range. The plasma has a higher temperature and therefore atomization is more complete. There are fewer chemical and spectral interference problems. ICP-AES is not as sensitive as AAS and the instrument is expensive and the analysis cost is high due to the large quantities of Argon gas required. Sample preparation requires the same amount of time as AAS.

dc Arc Emission

The primary advantage of using a spectrographic method such as dc arc emission lies in the rapidity and ease of surveying the trace element content of a sample. Spectra of the unknown and standard samples are usually photographed on the same plate under identical conditions. The concentration of the analyte is estimated by comparing line intensities of unknowns and standards. The accuracy depends upon maintenance of constant excitation and exposure conditions. The relative standard deviation of the method is rather large, usually 10–15%,[30] and it is necessary to ascertain the major elemental composition of the sample matrix prior to analysis. The synthetic standard electrodes which match the unknown closely can then be prepared. This is achieved by prior quantitative analysis of the material for it's major components.

dc arc emission was used by Bhale *et al.*[30] to analyze for vanadium and nine other trace elements in plant materials. Samples were ashed in a muffler furnace for 6 hr to remove organic matter. After semiquantitative analysis for the major components (i.e., Na, K, Al, Mg, Si, and Ca), a synthetic matrix was prepared using the oxide salts of the major elements. The unknown sample was prepared by mixing the graphite, plant ash, and the internal standard. Volatilization studies suggested two exposure times were needed, 10 A/30 sec for volatile and medium volatile elements and 10 A/60 sec for nonvolatile elements. The linear portion of standard curves indicated the useful concentration range for each element; for vanadium, this was 5–50 ppm (on an ash basis) using the 318.54 nm line. The data obtained were in good agreement with 52 participating laboratories which

[30] G. L. Bhale, V. P. Bellary, A. Thomas, and M. N. Dixit, *Indian J. Pure Appl. Phys.* **17**, 830 (1979).

used different trace analysis techniques on the same sample. The results show dc arc emission gives sufficient accuracy to be useful in routine environmental and agricultural studies where high precision is not required.

Marinov[31] developed a statistical method for optimizing excitation conditions in a matrix model using dried blood serum. Certain concentrations of elements were introduced (e.g., 0.1 mg/liter V) into the matrix. The parameter chosen to optimize the system was the differences in intensities of the analytical lines and their mean backround. By varying excitation conditions, sample size, arc current, gas flow, and electrode geometry, and other parameters, an optimum region for analysis was determined. The reader is referred to the paper for details of the statistics employed. Marinov's work shows that the signal increases with increasing arc current in the range 6–30 A, and is independent of the sample size and gas flow rates in the ranges of 50–200 μl and 2–6 liters/min, respectively. Operating conditions are given for analyzing trace elements in a dry residue of serum using the standard addition method. Relative precision of 4–14%, depending upon the element, is obtained.

dc Plasma Atomic Emission Spectroscopy (DCP-AES)

DCP-AES is useful when a reliable analytical procedure is required to routinely survey many elements in biological substances. A technique for the simultaneous determination of 14 elements, including vanadium, in organ tissues has been developed by Frank and Petersson.[32] Five grams of tissue sample was wrapped in a filter paper and placed in an ashing tube of borosilicate glass. Following wet digestion, the samples were prepared for measurement.[32] Spectral interferences from other elements present in relatively high concentrations in the matrix, i.e., Ca, Cu, Fe, K, Mg, Mn, Na, P, and Zn were investigated. The elements contributing to the emission backround were measured and their contributions at each analytical line determined. Linear relationships for the false signal contribution for all the elements (including vanadium) except for the effect of Ca on Al, Pb, and W were established. These signals were expressed as spectral interference correction coefficients (SICC) on a concentration equivalent basis. The backround correction was then calculated from the SICC. The working concentration range for vanadium was 0.003–1.0 μg/ml using the 437.92 nm analytical line. SICC corrections for vanadium included Al, Ca, and Fe as interfering elements.

[31] M. I. Marinov, *Fresenius Z. Anal. Chem.* **319,** 307 (1984).

[32] A. Frank and L. R. Petersson, *Spectrochim. Acta* **38B,** 207 (1983).

A comparative study was done by Pyy and co-workers[33] to test the suitability of using DCP-AES instead of atomic absorption spectrometry for the analysis of concentrated solutions of vanadium in serum and urine. Samples were wet-ashed and extracted with ammonium 1-pyrrolidinecarbodithioate (APDC) into methyl isobutyl ketone. The detection limits using the 437.924 nm vanadium line were 0.5 μg/liter by atomic absorption versus 2.5 μg/liter by DCP-AES when the samples were concentrated 4-fold following acid digestion. It was not possible to analyze for vanadium at the 309.311 nm line because of spectral interference from Mg at 309.299 nm. The liquid–liquid extraction procedure offers the advantage of eliminating interfering inorganic ions such as Na^+, PO_4^{3-}, and K^+ from the matrix, thus avoiding the need for interelement corrections. Urine can be analyzed directly for vanadium levels above 10 μg/liter provided the standard addition method is used. This is adequate as a screening method for subjects industrially exposed to vanadium. The proposed technique is not sensitive enough to detect vanadium in unexposed subjects by DCP-AES or AAS.

Even though DCP-AES is well suited for determinations in biological materials above the 10 ppb range, the method has disadvantages. To obtain maximum sensitivity in a multielement analysis, the plasma position must be optimized to obtain a maximum signal at each of the analytical wavelengths. Also quantitative studies of spectral interferences, background correction, and data manipulation make the procedure time consuming when initially being developed for routine analysis. By using an extraction procedure spectral interferences are minimized, but the multielement advantage of DCP-AES is sacrificed and AAS is the more sensitive technique.

Mass Spectrometry (SSMS)

Fassett and Kingston[33a] of the National Bureau of Standards determined vanadium in four biological standard reference materials using isotope dilution thermal ionization mass spectrometry where ^{50}V was spike enriched to 64 atom% and ion counting detection was employed. The results are listed in Table III. The concentration of vanadium in human serum (SRM 909) is the lowest certified concentration of any reference material with a value of 2.63 ppb V.

SSMS does not lend itself easily to trace quantitative analysis of biological materials. Due to the overwhelming matrix interferences, matrix

[33] L. Pyy, E. Hakala, and L. H. J. Lajunen, *Anal. Chim. Acta* **158,** 297 (1984).
[33a] J. D. Fassett and H. M. Kingston, *Anal. Chem.* **57,** 2474 (1985).

matching of the biomaterial becomes a necessity. Since the matrix of the synthetic standard must closely match that of the unknown, SSMS analysis is a very time consuming as well as expensive technique.

Photographic plate detection is used to achieve the resolution necessary to distinguish between mass lines. To obtain electrodes of sufficient conductivity, a graphite to ash weight ratio of 60/40 cannot be exceeded.[34] Dry-ashing, wet digestion, wet digestion in a Teflon bomb, or low temperature ashings are used in preparing biological samples for incorporation into conducting electrodes in SSMS and emission spectroscopy. Vos and Van Grieken[34] recently evaluated each procedure. Both dry-ashing procedures are prone to losses by volatilization and simple dry-ashing suffers from contamination during the preparation of electrodes. Wet digestion with a concentrated 2:1 mixture of HNO_3 and H_2SO_4 acids in an open flask is the method of choice, allows for homogeneous mixing, and gives better precision and lower blanks than the teflon bomb.

Matrix effects and interferences can seriously affect sensitivity and accuracy. The use of internal standards, i.e., Ir, In, In/Zr, requires that the sensitivity of each unknown element remain in a fixed ratio to that of the internal standard in different matrices. If there are changes in sensitivity for elements from one matrix to another, considerable errors in the analysis can result. Therefore the RSC (relative sensitivity coefficient) defined as the ratio of the measured element content to the known element content is usually calculated for each element in each matrix type to compensate for such errors.

Verbueken *et al.*[35] evaluated the total analysis of plant and biological tissue by SSMS using NBS SRM-1571 orchard leaves (0.6 ppmw V) and NBS SRM-1577 bovine liver (not certified for vanadium) reference materials. The graphite standards used were doped with 20 ppmw of 20 elements. In this work vanadium was not detected in either NBS standard reference material. However, in subsequent work vanadium was detected in the orchard leaves, 0.68 ± 0.14 ppmw.[36] The method was not sensitive enough to detect trace vanadium in the bovine liver. As the major composition of samples differs from that of the synthetic standard, the accuracy of the analysis decreases.

Bacon and Ure[37] studied the effects of simple matrices (NaCl, Na_2CO_3, Al_2O_3, $CaCO_3$, SiO_2) on the RSC of the elements and found that the sensitivity depends markedly on the matrix. Their work suggests that

[34] L. Vos and R. Van Grieken, *Anal. Chim. Acta* **164,** 83 (1984).

[35] A. Verbueken, E. Michiels, and R. Van Grieken, *Fresenius Z. Anal. Chem.* **309,** 300 (1981).

[36] L. Vos and R. Van Grieken, *Int. J. Mass. Spectrom. Ion Phys.* **47,** 303 (1983).

[37] J. R. Bacon and A. M. Ure, *Adv. Mass. Spectrom.* **94,** 362 (1980).

research on natural complex matrices is needed to further optimize SSMS as a viable means to obtain accurate quantitative results in a multielement trace analysis of biological materials.

Spectrophotometry

Spectrophotometry has been used to determine vanadium in environmental samples, plants, and animal tissues.[38–42] The spectrophotometric method is convenient, rapid, selective, and sensitive for materials containing vanadium in the ppm range. The method typically involves dry- or wet-ashing of the sample followed by preconcentration via an extraction procedure using $CHCl_3$ or *o*-dichlorobenzene and a chelating agent specific for vanadium. Common interferences from Ti(IV), Zr(IV), Mo(VI), and W(VI) can be eliminated by adding oxalate and citrate prior to the extraction step.[42] Successful chelating agents used by various laboratories are summarized in Table IV along with pertinent analytical data. Sample size depends on the concentration of vanadium present. Sample sizes as large as 50 g have been used in some analyses[42,43] which furthers the risk of introducing significant amounts of adventitious vanadium.

A reverse-phase HPLC method using spectrophotometric detection for vanadium has been reported.[43a] With appropriate sample pretreatment the methods may be applicable to biomaterials. Metal ions, e.g., Al(III), Co(II), Cu(II), Ga(III), Cr(III), Fe(III), Mn(II), Mo(VI), Ni(II), V(V), and Zn(II), are first derivatized using a Schiff base ligand 4,4′-d-*N,N*-diethyl-PBS which is not commonly available but is easily synthesized from PBS [PBS ≡ *N,N*′-*o*-phenylenebis(salicylaldimine)]. The Schiff base is dissolved in dimethylformamide (DMF) and 1 ml added to a 1 ml solution of the metal ions in 1 *M* sodium acetate buffer, pH 5. The resultant solution is heated in a water bath for 30 min at 60°. Upon cooling, a 100-μl aliquot is injected onto a HPLC reverse phase Shim-Pack FLC-ODS column and the chromatogram developed as detailed in the original article.[43a]

Of the 11 metal ions, only V(V), Co(II), Cu(II), and Ni(II) gave well-defined elution peaks, the other metal ion complexes possibly decompos-

[38] S. A. Abbasi, *Int. J. Environ. Stud.* **18,** 51 (1981); *C.A.* **96,** 27808h.

[39] A. K. Shrivastava and S. G. Tandon, *Indian J. Environ. Health* **24,** 89 (1982); *C.A.* **98,** 49884d.

[40] R. D. Roshania and Y. K. Agrawal, *Chem. Anal.* (*Warsaw*) **26,** 191 (1981); *C.A.* **96,** 79008x.

[41] R. R. Nanewar and U. Tandon, *Talanta* **25,** 352 (1978).

[42] Y. K. Agrawal and K. P. S. Raj, *Microchem. J.* **25,** 219 (1980).

[43] T. Kiriyama and R. Kuroda, *Analyst* (*London*) **107,** 505 (1982).

[43a] M. Kanbayashi, H. Hoshino, and T. Yotsuyanagi, *J. Chromatogr.* **386,** 191 (1987).

TABLE IV
SPECTROPHOTOMETRIC CHELATING AGENTS FOR VANADIUM ANALYSIS

Chelating agent	λ_{max} (nm)	ε $(M)^{-1}$	Beer's law range	Analyte
N-benzylbenzohydroxamic acid[a]	510	4300 ± 50	0.7–12 ppm	Blood, liver, urine
N-(*p*-*N*,*N*-Dimethylanilino-3-methoxy-2-naphtho)hydroxamic acid[b]	—	—	—	—
N-Benzoyl-*N*-phenylhydroxylamine[c]	530	4950 ± 25	10.3–30.0 μg V	Plant
N-(*p*-Chlorophenyl-2-naphthyl)-hydroxamic acid[d]	535	5600	—	Rocks, biological material
N-(*m*-Tolyl-*o*-methoxybenzo)-hydroxamic acid[e]	550	6500	0.05–15 μg V	Plant and animal tissue
4-(2-Pyridylazo)resorcinol[f]	585	110000	0.05–0.5 ppm	Plant materials

[a] R. R. Nanewar and U. Tandon, *Talanta* **25,** 352 (1978).
[b] S. A. Abbasi, *Int. J. Environ. Stud.* **18,** 51 (1981); *C.A.* **96,** 27808h.
[c] A. K. Shrivastava and S. G. Tandon, *Indian J. Environ. Health* **24,** 89 (1982); *C.A.* **98,** 49884d.
[d] R. D. Roshania and Y. K. Agrawal, *Chem. Anal.* (*Warsaw*) **26,** 191 (1981); *C.A.* **96,** 79008x.
[e] Y. K. Agrawal and K. P. S. Raj, *Microchem. J.* **25,** 219 (1980).
[f] J. Minczewsi, J. Chwastowska, and T. H. M. Pham, *Analyst* (*London*) **100,** 708 (1975).

ing on the column. The reagent is highly selective for vanadium, forming a chromophoric complex with an absorbance maximum at 448 nm. Vanadium analyses can be performed in the 5.5 ppb to the 1.1 ppm range with a detection limit of 0.3 ppb. Neither a 100-fold molar excess of Al(III), Cd(II), Cr(III), and Pb(II) nor a 50-fold molar excess of Cu(II), Fe(III), Mn(II), and Ni(II) interferes with the determination of vanadium at the 1 ppm level.

Simultaneous spectrophotometric determination of more than one element requires that they be separated prior to analysis. For example Co and V have been measured in orchard leaves, bovine liver, and standard reference materials.[43] After sample digestion, the metals were adsorbed on an Amberlite CG-400 (SCN^-) anion exchange column from a dilute NH_4SCN–HCl solution. V and Co were separated by elution with 12 *M* HCl and 2 *M* $HClO_4$, respectively. The complex of 4-(2-pyridylazo)resorcinol with each element was formed followed by spectrophotometric determination. The precisions reported were 3–7% for V and 3–10% for Co. dc arc emission or dc plasma emission spectroscopies are better suited for multielement analysis.

Energy Dispersive X-Ray Fluorescence (EDXRF)

The detection limit for quantitative analysis in multielement trace analyses of biological samples using EDXRF is 10 ppm.[44] Combined with the problem of low sensitivity, is the need to minimize absorption and enhancement effects through matrix matching or use of an internal standard. Therefore, this method is rarely chosen for analysis of vanadium.

Lieser and co-workers studied the trace element concentrations of plants and foodstuffs using wheat flour as a standard.[44] Vanadium concentrations were detected in plant ash (200 ppm) and *Platanus acerifolia* (20 ppm). Vanoeteren *et al.*[45] used SEM coupled to an energy dispersive spectrophotometer for the qualitative analysis of particles deposited in living tissue. Vanadium among other elements could be detected.

Electron Paramagnetic Resonance Spectroscopy (EPR)

Electron paramagnetic resonance (EPR) spectroscopy is normally used to measure spin concentrations of paramagnetic ions or radicals. It can, however, be used to determine total vanadium, provided the metal is

[44] K. H. Lieser, R. Schmidt, and R. Bowitz, *Fresenius Z. Anal. Chem.* **314,** 41 (1983).

[45] C. Vanoeteren, R. Cornelius, R. Dams, and R. Ryckaert, *Trace Elem. Anal. Chem. Med. Biol., Proc. Int. Workshop* **3,** 407 (1984).

entirely converted to EPR active vanadyl(IV) ion, VO^{2+}, and interfering signals from other paramagnetic species are absent. In the course of EPR studies of proteins it is often more convenient to measure total vanadium by EPR than by another instrumental method. Aspects of vanadium analysis by EPR have been discussed in more detail elsewhere.[46–48]

The following is an EPR method for measuring vandium(V) in ultrafiltrates and dialyzates.[49] The method is based on the quantitative reduction of vanadium(V) to vanadium(IV) by ascorbate in acid solution followed by measurement of the eight-line EPR spectrum of the uncomplexed VO^{2+} ion. Linear standard curves are obtained with correlation coefficients typically greater than 0.999 in the 10–200 μM range of vanadium concentration. The relative precision of the vanadium determination is ±6% (95% confidence level) at a metal concentration of 0.1 mM. The detection limit for measurements performed at room temperature (see below) is approximately 5 μM vanadium (0.3 ppm), corresponding to a signal-to-noise ratio of 2 : 1 on a Varian E-4 spectrometer. A lower detection limit (50 ppb) is obtained for samples measured at 77 K in a 3-mm-i.d. by 4-mm-o.d. tube using a liquid nitrogen dewar insert.

Ten microliters of 1 M HCl is added to 50 μl of unknown or standard solution followed by the addition of 10 μl of freshly prepared 0.1 M ascorbic acid. The peak-to-peak amplitude of the fourth EPR line from low field is measured at room temperature in a 1-mm-i.d. by 2-mm-o.d. quartz capillary fitted inside a 2-mm-i.d. by 4-mm-o.d. quartz tube permanently positioned in the TE_{102} rectangular EPR cavity. This arrangement ensures reproducible positioning of the capillary each time. Standards are prepared from ammonium metavanadate in the same buffer as for the unknown. The E-4 instrument parameters used in the room temperature analysis are field set = 330 mT, scan range = 100 mT, scan time = 30 min, gain = 10^4, 100 KHz modulation amplitude = 1.0 mT, time constant = 3 sec, power = 100 mW, and frequency = 9.47 GHz. Only the EPR line of interest is scanned. Weak background cavity signals, if present, are computer subtracted from the signals of the standards and unknowns.

Acknowledgment

This work was supported by Grant GM 20194 from the National Institute of General Medical Sciences.

[46] J. J. Fitzgerald and N. D. Chasteen, *Anal. Biochem.* **60,** 170 (1974).
[47] B. A. Burgess, N. D. Chasteen, and H. E. Gaudette, *Environ. Geol.* **1,** 171 (1975–1976).
[48] N. D. Chasteen, *Biol. Magn. Reson.* **3,** 53 (1981).
[49] N. D. Chasteen, J. K. Grady, and C. E. Holloway, *Inorg. Chem.* **25,** 2754 (1986).

[33] Determination of Zinc in Biological Samples by Atomic Absorption Spectrometry

By K. H. Falchuk, K. L. Hilt, and B. L. Vallee

Introduction

The ability to measure zinc precisely in biological samples has resulted in the recognition of its essentiality to human nutrition, its role in disease, and its critical role in catalytic, structural, and regulatory functions in proteins and other macromolecules.[1-3] Numerous analytical techniques are now available for zinc determination (Table I). The majority of these methods are spectroscopic. For example, chemical methods rely upon the absorption of zinc chelates in the visible or ultraviolet ranges of the electromagnetic spectrum. X-Ray fluorescence measures the intensity of characteristic X rays emitted by zinc atoms exposed to a source of X rays or accelerated charged particles. Neutron activation changes stable zinc isotopes in a sample to ones that decay by characteristic γ-emission. Nonspectroscopic methods are also available, e.g., voltammetric analysis which relies on the reductive potential of Zn^{2+} ions and the current generated relative to an applied voltage to differentiate it from other ions.

The variety of methods available provides flexibility in terms of the size and number of samples to be measured, the range of zinc concentration over which measurements are linear, whether or not the sample is destroyed by the analysis, and whether other elements are to be simultaneously measured. At present, the most widely employed methods are based on atomic spectrometry—the interaction of analyte atoms with electromagnetic radiation. These procedures fall into three categories: atomic emission, atomic absorption, and atomic fluorescent spectrometry. The features of the most readily available of these techniques, i.e., flame and electrothermic atomic absorption spectrometries, their advantages and their application to the analysis of biological samples will be described in the following sections. In addition, methods for preparing zinc-free buffers and reagents, standards, and specific biological fluids and tissues for analysis will be detailed.

[1] B. L. Vallee, *Physiol. Rev.* **39,** 443 (1959).

[2] T.-K. Li and B. L. Vallee, *in* "Modern Nutrition in Health and Disease" (R. S. Goodhart and M. E. Shils, eds.), 6th Ed., p. 408. Lea & Febiger, Philadelphia, Pennsylvania, 1980.

[3] B. L. Vallee and A. Galdes, *Adv. Enzymol. Relat. Areas Mol. Biol.* **56,** 283 (1984).

TABLE I
DETECTION LIMITS[a] FOR ZINC OF VARIOUS ANALYTIC TECHNIQUES[b]

Technique	Absolute detection limit (g)
Flame atomic absorption spectrometry	
Air–acetylene flame	6×10^{-10}
Boat technique	3×10^{-11}
Delves technique	5×10^{-12}
Electrothermal atomic absorption spectrometry	5×10^{-14}
Chemical analyses	1×10^{-6}
Polarography	1×10^{-7}
Neutron activation analysis	1×10^{-8}
Atomic fluorescence	2×10^{-10}
X-ray fluorescence	4×10^{-11}
Proton induced X-ray fluorescence	1×10^{-11}
Anodic stripping voltammetry	4×10^{-12}
Inductively coupled plasma emission spectrometry	2×10^{-12}

[a] The concentration which produces an analytical signal equal to twice the standard deviation of the background signal.
[b] Adapted from Ref. 4.

Preparation of Buffers and Solutions

Zinc is ubiquitous in the environment. Avoidance of contamination of samples from water, containers, reagents, etc. is an important factor in performing zinc analyses. The following procedures will reduce the amount of contamination by adventitious zinc ions.

Water

Ultrapure water is essential. The best method for preparing large volumes of high purity water (resistivity of ~18 MΩ and zinc concentration of ~0.3 ppb) is by using a combination of reverse osmosis, organic adsorbtion, and ion-exchange chromatography.[4]

Containers and Other Laboratory Ware

Plastic laboratory ware usually contain less zinc than glassware and are recommended for use whenever possible for storing or transferring liquids and solids. Many commercially available disposable plastic pipets, tubes, and containers can be used either directly or after rinsing with zinc-

[4] C. Veillon and B. L. Vallee, this series, Vol. 49, p. 446.

free water. Nondisposable plastic items that need to be cleaned should be soaked overnight in metal-chelating detergent solutions such as "Count Off" or its equivalent, followed by thorough rinsing in metal-free water. Cleaning with acid (as described below for glassware) may actually accentuate rather than resolve the problem of zinc contamination, since many plastics can become ion exchangers and, therefore, a source of zinc contamination.[4] If glass must be employed, Pyrex should be used. Soft glass is unsuitable since it contains a surfeit of metals. Glass requires prolonged soaking in 6 *N* HNO_3 acid for a period of 12 hr to as long as 2–3 weeks. Prior to use, the glassware should be rinsed thoroughly with metal-free H_2O. Platinum is the only metal that is suitable for spatulas or similar devices in manipulation of samples. Stainless-steel spatulas and scissors should only be used if Teflon coated. Plastic instruments are preferable. Rubber stoppers or tubing must be avoided throughout as they are serious sources of contamination.

Prior to using any container, however, it is essential that the amount of zinc contamination contributed by it be ascertained. An aliquot of zinc-free buffer, water, or 0.01 *M* HCl is added to the cleaned container. The liquid is allowed to remain in the device for one or more hours and its zinc content is analyzed.

Buffers and Solutions

Chemicals of the highest purity are used in preparing buffers and other reagents. Inorganic salts are available from companies such as Johnson Matthey Chemicals Ltd. (Royston, England) which contain negligible amounts of zinc. They can be used directly to prepare solutions. Alternatively, adventitious zinc ions are removed from buffers and other solutions prepared from reagent grade chemicals. This is accomplished using dithizone/carbon tetrachloride (CCl_4) extraction[5] or cation-exchange chromatography. With these approaches, used either alone or in combination, it is possible to obtain solutions containing from 1×10^{-7} to 3×10^{-8} *M* Zn^{2+}.

Dithizone extraction is carried out in a hood to avoid exposure to CCl_4. The pH of the solution to be extracted is adjusted to 7.0–7.5. One milliliter of 0.01% (w/v) dithizone in pure grade CCl_4 is added to each 10 ml of solution in a separatory funnel. The solutions are mixed by shaking until the green color of the organic phase turns pink or red. The latter is removed after the organic and inorganic solutions have separated in the funnel. The procedure is repeated with new aliquots of dithizone/CCl_4

[5] R. E. Thiers, *Methods Biochem. Anal.* **5,** 273 (1957).

solution until there is no perceptible color change. Traces of dithizone in the aqueous phase are removed by adding 1 ml of pure CCl_4 to each 10 ml of solution, and back-extracting as above. Any CCl_4 remaining in the aqueous solution is removed by carefully applying a vacuum to the top of the separatory funnel for as long as 1–5 hr. During this step, care is taken to prevent implosion of the separatory funnel, intake of dust into the aqueous solutions, and evaporative losses by connecting a water trap to the bottom of the funnel.

Ion-exchange chromatography is carried out with a chelating resin, such as Chelex 100 (Bio-Rad Laboratories, Richmond, CA). A suitable column and its inlet and outlet attachments are acid cleaned and rinsed with ultrapure water prior to filling with cation exchanger. The resin is washed in a sintered glass funnel with five bed volumes of H_2O, followed by two bed volumes of 1 *N* HCl, five volumes of H_2O, two volumes of 1 *N* NaOH, and five volumes of H_2O. The resin then is packed in the column and equilibrated with five volumes of metal-free water (the pH of the effluent is ~10). The solution to be extracted is chromatographed at a flow rate of 4–10 ml/min/cm^2. The first two column volumes are discarded and the subsequent eluate is collected in a zinc-free container. The pH of the eluate should be measured and adjusted with metal-free acid, if necessary.

Sample Preparation

Serious analytical errors are often encountered when zinc is determined in samples containing large amounts of organic material. The problems originate from interactions between the organic material and the analyte zinc atoms, on the one hand, and the relative absence of such interactions with the inorganic zinc standard solutions employed for calibration of the measurement system, on the other. Hence, it is necessary to dilute or remove the organic matrix for zinc and most other trace element determinations. This is usually accomplished by diluting the sample with H_2O, precipitating out interfering substances, such as proteins, with trichloroacetic acid, or dry or wet ashing.

For the preparation of samples with any of the techniques described below it is necessary to use polyethylene gloves (without talcum) to prevent zinc contamination from the skin. Whenever possible it is advsiable to perform sample manipulations in a clean air space, preferably a laminar flow hood. Ordinary chemical fume hoods are not suitable work areas as they draw dust with metal contaminants into the samples and should only be used when necessary to remove noxious chemicals. Parafilm should be used to seal containers to further obviate air-borne contamination.

Biological Fluids

Plasma, serum, amniotic fluid, urine, and cerebrospinal fluid are among several biological fluids which may only need to be diluted with water (usually 1 : 4 to 1 : 10) and/or undergo treatment with trichloroacetic acid prior to analysis by flame or electrothermal atomic absorption spectrometry.[6-9] To precipitate serum, for example, equal volumes of sample and 25% (v/v) trichloroacetic acid are mixed and allowed to stand 10 min at room temperature. The mixture then is centrifuged at 3000 *g* for 5 min and the clear supernatant analyzed.[9]

Enzymes

Removal of adventitious zinc from enzymes prior to zinc analysis is critical to the identification of such proteins as zinc metalloenzymes. Most of the preparative work in the isolation of enzymes is not carried out under metal-free conditions. Hence, adventitious contamination by nanogram to microgram amounts of zinc is expected. This contamination may exceed the stoichiometric quantities of zinc found in most pure metalloenzymes and proteins, particularly when very small (microgram or less) total amounts of enzymes are analyzed, as is now the case routinely. Dialysis or gel exclusion chromatography can be used for removal of such adventitious zinc. The selection of one over the other method is dependent on the amount of enzyme available.

Dialysis can be used when milligram or greater amounts of enzyme are available. Buffers are rendered zinc free as previously described. The buffer pH must be selected such that zinc is not lost from either active or structural sites of the enzyme. In many cases, pH 7–8 is suitable. In addition, the dialysis buffer should not include components which may remove zinc from the enzyme, e.g., metal chelators (ethylenediamine tetraacetic acid) or sulfhydryl reducing agents (dithiotreitol or mercaptoethanol). The dialysis tubing itself is rendered zinc free by heating at 70° in several changes of metal-free water. The treated tubing then may be stored in metal-free water at 4° until used.

Gel-exclusion chromatography may be used to remove zinc and other metal contaminants from microgram quantities of enzymes in microliter

[6] K. H. Falchuk, M. Evenson, and V. L. Vallee, *Anal. Biochem.* **62,** 255 (1974).

[7] "Analytical Methods for Atomic Absorption Spectrophotometry." Perkin-Elmer, Norwalk, Connecticut, 1976.

[8] R. Laitinen, A. S. I. Siimes, E. Vuori, and S. S. Salmela, *Biol. Trace Elem. Res.* **6,** 415 (1984).

[9] A. S. Attiyat and G. D. Christian, *Clin. Chim. Acta* **137,** 151 (1984).

volumes.[10] Commercially available sizing gels, however, contain zinc,[11] which must be removed before chromatography. The gel is swollen in zinc-free water and loaded into an acid-cleaned chromatography column. The column then is washed with several column volumes of 10 m*M*, 1,10-phenanthroline in the buffer to be used. Subsequently, the column is equilibrated with buffer without chelating agent. The sample is loaded and collected in zinc-free tubes.

Specific activity should always be checked before and after using either of these methods since these procedures may remove zinc from the enzyme itself.

Biological Solids

Cells and tissues require either dry or wet ashing. Before either can be performed, the sample must be freed of any external contaminating zinc. Hair and nail samples, for example, can be washed with a nonionic detergent known to be zinc free, followed by several rinses with metal-free water. Plant cells, whose cell wall is sturdy, as well as other tissues, such as teeth, can be rinsed extensively with metal-free water alone. Mammalian tissues, such as liver, red blood, and other cells, should be rinsed with a zinc-free isotonic salt solution.

Dry Ashing. Platinum crucibles are cleaned by placing several ml of 6 *N* HNO_3 in each one. The acid in the crucible is heated for 10–15 min in a hood. The crucibles, then, are rinsed with metal-free H_2O. An aliquot (usually ~1 g) of the material to be ashed is placed in the acid-cleaned platinum crucible and slowly dried to a constant weight either in an electric oven at ~114° or beneath a heat lamp. Some samples, e.g., bone, may need to be ground into a powder with a mortar and pestle prior to this first step. The crucibles then are transferred to a muffle furnace or plasma reactor. The muffle furnace should be lined with a quartz liner to preclude zinc contamination from flakes off the furnace walls. Such liners may have to be custom designed and manufactured. The temperature is raised slowly to 400° and then to 500°, a temperature which must not be exceeded. This slow temperature increase is critical to avoid spattering of the sample out of the crucible as the water in the sample is evaporated. After 12–24 hr, most samples are reduced to a white ash. For some samples, however, repeated ashing still leaves small black carbonaceous particles in the crucibles. These particles can be dissolved by adding small quantities (2–3 ml) of concentrated metal-free HCl (Aesar, Seabrook,

[10] H. Kawaguchi and D. S. Auld, *Clin. Chem.* **21,** 591 (1975).

[11] R. S. Morgan, N. H. Morgan, and R. A. Guinavan, *Anal. Biochem.* **45,** 668 (1972).

NH) and heating slowly on a hot plate (under a hood) until the solutions are evaporated. An acid-cleaned glass rod may be used to break up these particles. The ash in each crucible then is solubilized with several drops of concentrated metal-free HCl and transferred to a suitable sized volumetric (usually 5 or 10 ml). The crucibles then are rinsed several times with a 1% (v/v) metal-free HCl solution, the rinses being quantitatively transferred to the volumetrics. These samples then are diluted with H_2O or dilute metal-free acid. In general, the dry ashing procedure is simple and has a low probability of zinc contamination.

Wet Ashing. Wet digestion constitutes another means of decomposing biological samples. The sample is heated or refluxed with a solution of a strong oxidizing agent such as concentrated H_2SO_4, HNO_3, $HClO_4$, or some combination thereof. HNO_3 and H_2SO_4 are preferable whenever possible, since $HClO_4$ acid may become explosive in the presence of organic matter. However, when using HNO_3 and $HClO_4$, HNO_3 should always be added first.

Depending on the specific tissue used and its acid solubility, wet is usually faster than dry ashing. Moreover, volatilization losses of zinc and other metals are minimized compared to dry ashing because of the relatively low temperatures (100–200°) employed. However, it is not easy to obtain highly purified metal-free acids commercially. Nitric acid, in particular, often contains trace amounts of metals owing to the high solubility of their nitrates. Even when such acids are certified to be metal free this may not constitute adequate validation, owing to the methods of analysis which are usually unknown. It is recommended, therefore, that the acids used be analyzed to ensure that they are free of zinc contamination. Further purification of acids may be necessary. This may present serious problems and may limit this approach in some instances.

The following procedures describe methods for wet ashing a number of different biological materials:

Animal Tissues and Cell Pellets. The tissue (~5 g) is placed in a 125-ml Erlenmeyer flask containing an equal weight of acid washed glass beads and 25 ml of water. Ten milliliters of concentrated HNO_3 : concentrated $HClO_4$ (1 : 1) is added and the solution boiled until it is clear. The solution then is transferred to a 100-ml volumetric flask and brought to volume with water.[7]

Blood. Add 1 ml 70% (w/w) HNO_3, 0.5 ml 96% (w/w) H_2O_4, and 1 ml 70% (w/w) $HClO_4$ to 1 ml of blood in a reflux condenser. The mixture is boiled. When the solution becomes colorless it is evaporated to near dryness and the residue brought up in 5 ml of 0.25% HNO_3.[12]

[12] R. N. Khandekar and U. C. Mishra, *Z. Anal. Chem.* **319,** 577 (1984).

Plant Tissues. The sample (~1 g) is placed into a beaker. Five milliliters of 70% (w/w) HNO_3 is added first, followed by 2 ml 70% (w/w) $HClO_4$. The mixture is covered with a watch glass and digested by heating to a final volume of 3–5 ml. The resultant solution is diluted with 10–15 ml of water and filtered, if necessary.[7]

Teeth. Five milliliters of 70% (w/w) HNO_3 and 2 ml of 70% (w/w) $HClO_4$ are added to ~250 mg sample. After slow heating, a clear, colorless solution is obtained and evaporated to near dryness. The residue then is taken up in about 25 ml of 0.25% HNO_3.[12]

Hair. Samples are cut into pieces ~1 cm in length. They are washed in a 500-ml polyethylene bottle containing 150 ml of a 1% solution of nonionic detergent for 30 min on a mechanical shaker. The hair is transferred to a polyethylene filter crucible and rinsed with 1 liter water. It then is dried overnight at 110°. An aliquot (~0.5 g) is placed into a 50-ml Erlenmeyer flask to which 6 ml of concentrated HNO_3 is added. The mixture is allowed to react at room temperature. The digest is warmed slowly and 1 ml concentrated $HClO_4$ is added. The partial digest is heated at 200° until dense white fumes are evolved. The resultant solution, which should be clear, is transferred to a 5 ml volumetric and brought to volume with water.[7]

Fingernails. Surface dirt is removed by scraping with Teflon-coated forceps. The sample is placed in a polyethylene bottle containing 25 ml of 1% nonionic detergent and shaken for 30 min. The nails are rinsed with 1 liter water and dried overnight at 105°. The sample then is digested in a 10-ml Erlenmeyer flask using 1 ml concentrated HNO_3 and 0.5 ml concentrated $HClO_4$. The digest is placed into a 5-ml volumetric flask and brought to volume with water.[7]

Preparation of Standard Solutions of Zinc

The response of any instrument used for analysis must be calibrated using samples of known zinc concentration. Primary standards for zinc should be prepared from spectrographically pure zinc such as that supplied by Johnson-Matthey Chemicals, Ltd. (Royston, England), and others, and spectrographically pure hydrochloric acid. Ideally, HCl should be prepared by bubbling hydrogen chloride gas into metal-free water. An accurately weighed amount of zinc is dissolved in the acid and then diluted with metal-free water to yield a primary standard solution. Alternatively, primary standards are available commercially (Fisher Scientific Company, Medford, MA).

Aliquots of the primary standard then serve to prepare a set of secondary standards with which the instrument is calibrated. Secondary standards are prepared prior to each analysis. They can be stored for short periods of time. However, storage of dilute solutions (<1 μg/ml) may result in gain or loss of zinc due to desorption or absorption from the storage container.

Method Verification

Considering the number of unknown variables introduced when complex biological samples are analyzed for their zinc content, it is recommended that verification of the sample preparation and measurement be carried out either by analysis of a known material with the same method an/or of the unknown sample by an alternate, independent approach. The former can be achieved by the preparation and analysis of National Bureau of Standards standard reference materials (SRM) which have a known, validated amount of zinc. A number of these reference materials are available including oyster tissue (SRM 1566), wheat flour (SRM 1567), rice flour (SRM 1568), citrus leaves (SRM 1527), tomato leaves (SRM 1573), bovine liver (SRM 1577a), and others.[13] Since the type and nature of the standards available change, it is advisable to contact the NBS for an updated list at any given time.

Flame Atomic Absorption Spectrometry (FAAS)

This technique is presently one of the most widely used for routine analyses of zinc. It is precise, simple, sensitive, and relatively free from matrix interferences compared to other methods, e.g., electrothermal atomic absorption spectrometry (EAAS) or inductively coupled plasma emission spectrometry (ICPES). The design and description of flame atomic absorption instruments are detailed elsewhere in this volume. Briefly, a hollow cathode tube emits a narrow line spectrum specific for the element to be measured. A monochromator isolates the particular wavelength to be employed (usually 213.9 nm for zinc). A detection system converts the radiant energy received into an electrical signal. In the absence of sample, the readout is displayed as 100% transmittance or 0% absorbance. A liquid sample (1–2 ml) is aspirated into a flame-nebulizer system where the zinc atoms are atomized. The zinc atoms absorb the radiation emitted by the hollow cathode tube in proportion to their concentration. The absorption decreases the incident radiation received by the detection system. For most instruments, the proportionality between

[13] "NBS Standard Reference Materials Catalog 1984–85" (C. H. Hudson, ed.), Special Publ. N. 260. U.S. Government Printing Office, Washington, D.C., 1984.

zinc and absorbance follows Beer's law and is linear, in the range of 0.1–1.0 μg Zn/ml. The sensitivity of the system (the amount of zinc required to give 1% absorption) can be extended to lower than 0.1 μg Zn/ml by use of various modifications of the sample delivery system, e.g., use of a slotted burner, or a small sampling device such as the Delves Cup or boat device.[4]

Interferences

The analysis of zinc in biological fluids is more complex than that in inorganic solutions. This is due to the fact that the matrices of the biological samples to be analyzed and of the standard zinc solutions differ. These differences present serious analytical problems since the physical and chemical compositions of the solutions analyzed affect the rate of uptake, extent of nebulization, as well as the degree of chemical interferences. This is exemplified by the analytical difficulties observed with serum samples which are viscous owing to their high protein content. Appropriate dilution of sera (~1 : 10) with metal-free water can be used to obviate the effects of viscosity on zinc analysis.[6] Particular care must be taken in each case to match the matrices of the sample and standards as closely as possible. It may be necessary, for example, to prepare the secondary standards with the appropriate matched constituents, i.e., buffers, glycerol, nonionic detergents, organic solvents, etc. found in the sample to be analzyed.[14] Several commercially available solutions have been found to be suitable for specific cases, e.g., Auto Cal 1 (Instrumentations Lab, Inc.)

Biological samples, including serum, also differ from the standards in that they contain a variety of cations and anions, usually in unknown concentrations. Some ions, e.g., phosphates, interfere with zinc analyses by forming insoluble metal complexes of high boiling point and low volatility. The fraction of zinc atomized by the flame is thereby reduced. Two approaches can be used to obviate these interferences. The first and most frequently used method is simple dilution of the sample (e.g., dilution of serum 1 : 10 or more with water obviates any effects of Na^+, K^+, Ca^{2+}, Fe^{3+}, Mg^{2+}, Cl^-, PO_4^{2-}, NO_3^-, and SO_4^{2-}). The second utilizes protective agents such as the chelators 8-hydroxyquinoline, EDTA, or APDC (the ammonium salt of 1-pyrrolidinecarbodithioic acid). These agents form stable, but volatile species with zinc.[6,15]

[14] T. Uchida and B. L. Vallee, *Anal. Sci.* **2,** 71 (1986).

[15] D. A. Skoog and D. M. West, *in* "Fundamentals of Analytical Chemistry" (J. Vondeling, P. L. Smith, and L. Kesselring, eds.), 4th Ed., p. 580. CBS College Publ., New York, 1982.

Advantages and Limitations

The method is relatively free of interferences compared to other techniques, it is easy to analyze multiple samples rapidly (time of analysis is approximately 15 sec), and the cost of the instrument is inexpensive compared to other approaches. However, the sample volume of 1–2 ml may be a problem when only small quantities of material are available. This problem has recently been addressed by design of a discrete nebulization device allowing for analysis of as little as 0.1 ml sample.[16]

Electrothermal Atomic Absorption Spectrometry (EASS)

Electrothermal atomic absorption spectrophotometry is an alternative to the use of the flame for atomization of the sample. The latter is placed in the graphite tube furnace and dried by briefly raising its temperature to approximately 120°. It then is both ashed and atomized by increasing the temperature to ~3000°. Absorption of the hollow cathode radiation by atomized zinc atoms is measured at 213.9 nm. Total analysis time is less than 2 min per sample. The detection limit for zinc is about 1.5 pg.[16] The increased sensitivity of this method over that of FAAS is due to complete atomization of the sample, the oxygen-free atmosphere used, the reducing conditions afforded by the presence of hot carbon, the small furnace volume, and the relatively long residence time (1 sec or more) of zinc atoms in the optical path.

Advantages and Limitations

The major advantage of this technique is its increased sensitivity with small sample volumes (5–50 μl). This allows it to be used for measuring zinc content of purified enzymes where quantities of material for analysis may be extremely limited.[16] However, while the method is more sensitive (see Table I), precision is in the range of 5–10% compared to the 2–5% with FAAS.[15]

The known limitations of this technique are due to problems arising when samples are vaporized and atomized under rapidly changing conditions while the analytical measurement is in progress. Evaporating and dissociating a sample in a complex matrix, while in contact with the graphite furnace wall and in an atmosphere undergoing rapid temperature change, can lead to serious matrix effects. Any component of the matrix which affects the evaporation or dissociation processes, or their kinetics, may adversely affect the accuracy of the method unless duplicated ex-

[16] T. Kumamaru, J. F. Riordan, and B. L. Vallee, *Anal. Biochem.* **126,** 214 (1982).

TABLE II

THE EFFECT OF $NH_4H_2PO_4$ ON INTERFERENCE BY BUFFERS IN THE DETERMINATION OF ZINC[a,b,c]

Added compound	$-NH_4H_2PO_4$			+50 m*M* $NH_4H_2PO_4$		
	1 m*M*	10 m*M*	35 m*M*	1 m*M*	10 m*M*	35 m*M*
NaCl	−15	−55	−40	0	0	−15
K_2SO_4	0	+10	+15	0	0	0
$(NH_4)_2SO_4$	+20	+30	+35	0	0	0
$NaNO_3$	+20	+10	+15	0	0	0
Sodium borate	0	−65	−55	0	−20	−35
Sodium acetate	0	0	−10	0	0	−10
Sodium citrate	0	0	−15	0	0	−20
HEPES	0	+10	+30	0	0	0
MES	0	+25	+15	0	0	−15
Tris	−10	−15	−70	0	−35	−40
Veronal	+35	+30	−25	0	0	−10

[a] Zn^{2+}, 10 ng/ml (0.01 *M* in HNO_3).
[b] Expressed as percentage increase or decrease of zinc absorption signal measured by the peak height method.
[c] From Ref. 16.

TABLE III

EFFECT OF $NH_4H_2PO_4$ ON INTERFERENCE BY INORGANIC IONS IN THE DETERMINATION OF ZINC[a,b,c]

Ion	Added as	$-NH_4H_2PO_4$		+50 m*M* $NH_4H_2PO_4$	
		Peak height	Peak area	Peak height	Peak area
Mg^{2+}	$Mg(NO_3)_2$	−30	0	0	0
Ca^{2+}	$CaCl_2$	0	−10	+10	0
Mn^{2+}	$Mn(NO_3)_2$	−20	−15	0	0
Fe^{3+}	$FeCl_3$	+40	+35	+10	0
Co^{2+}	$Co(NO_3)_2$	+25	−10	+15	0
Ni^{2+}	$Ni(NO_3)_2$	+10	0	0	0
Cu^{2+}	$Cu(NO_3)_2$	+30	0	0	0
Cd^{2+}	$Cd(NO_3)_2$	+20	0	0	0
Hg^{2+}	$HgCl_2$	0	0	0	0
As(III)	H_3AsO_3	+10	0	0	0
Se(IV)	H_2SeO_3	+15	+10	0	0
Mo(VI)	H_2MoO_4	+15	0	0	0

[a] Zn^{2+}, 10 ng/ml; inorganic ions, 10 μg/ml; HNO_3, 0.01 *M*.
[b] Expressed as percentage increase or decrease of zinc absorption signal measured by both peak height and peak area integration methods.
[c] From Ref. 16.

actly by the standards. The effects are concentration dependent and the direction (i.e., whether the zinc absorption signal increases or decreases) varies for each compound. This is exemplified by the interference with the zinc absorption signal induced by commonly used buffers (Table II) and inorganic and metal ions (Table III). Interferences by a number of compounds and ions at 35 m*M* or less concentration can be overcome or reduced by the addition of 50 m*M* ammonium dihydrogen phosphate to the samples.[16] Moreover, the signal for zinc absorption from this instrument can be read out either as peak height or by the peak area integration method. The interferences are less and the ability of ammonium dihydrogen phosphate to overcome them greater when the peak area integration method is used (Table III).

Conclusion

The analysis of zinc in most biological samples can be carried out effectively by using either flame or electrothermal atomic absorption spectrometry. The availability of the instruments, their cost, ease of use, and range of sensitivities for zinc determination make them the methods of choice for routine use in biological work on this element. Special situations or problems may require the use of other techniques (Table I). The complexities of these other techniques, however, limit their usefulness to specialists skilled in the respective fields.

Author Index

Numbers in parentheses are footnote reference numbers and indicate that an author's work is referred to although the name is not cited in the text.

A

B

H

I

J

K

L

M

N

O

P

Q

R

S

T

U

V

W

X

Y

Z

Subject Index

A

D

E

K

L

M

N

O

P

Q

R

S

T

U

V

W

X

Z